COMPUTATIONAL MATERIALS SCIENCE

NATO Science Series

A series presenting the results of scientific meetings supported under the NATO Science Programme.

The series is published by IOS Press and Kluwer Academic Publishers in conjunction with the NATO Scientific Affairs Division.

Sub-Series

I.	Life and Behavioural Sciences	IOS Press
II.	Mathematics, Physics and Chemistry	Kluwer Academic Publishers
III.	Computer and Systems Sciences	IOS Press
IV.	Earth and Environmental Sciences	Kluwer Academic Publishers
V.	Science and Technology Policy	IOS Press

The NATO Science Series continues the series of books published formerly as the NATO ASI Series.

The NATO Science Programme offers support for collaboration in civil science between scientists of countries of the Euro-Atlantic Partnership Council. The types of scientific meeting generally supported are “Advanced Study Institutes” and “Advanced Research Workshops”, although other types of meeting are supported from time to time. The NATO Science Series collects together the results of these meetings. The meetings are co-organized by scientists from NATO countries and scientists from NATO’s Partner countries – countries of the CIS and Central and Eastern Europe.

Advanced Study Institutes are high-level tutorial courses offering in-depth study of latest advances in a field.
Advanced Research Workshops are expert meetings aimed at critical assessment of a field, and identification of directions for future action.

As a consequence of the restructuring of the NATO Science Programme in 1999, the NATO Science Series has been re-organized and there are currently five sub-series as noted above. Please consult the following web sites for information on previous volumes published in the series, as well as details of earlier sub-series:

http://www.nato.int/science
http://www.wkap.nl
http://www.iospress.nl
http://www.wtv-books.de/nato_pco.htm

Series III: Computer and Systems Sciences - Vol. 187 ISSN 1387-6694

Computational Materials Science

Edited by

Richard Catlow

Davy Faraday Laboratory of the Royal Institution of Great Britain, London, United Kingdom

and

Eugene Kotomin

Institute of Solid State Physics, University of Latvia, Riga, Latvia
and
Max-Planck Institut für Festkörperforschung, Stuttgart, Germany

Amsterdam • Berlin • Oxford • Tokyo • Washington, DC

Published in cooperation with NATO Scientific Affairs Division

Proceedings of the NATO Advanced Study Institute on
Computational Materials Science
9–22 September 2001
Il Ciocco, Italy

ISBN 1 58603 335 2 (IOS Press)
ISBN 4 274 90603 5 C3055 (Ohmsha)
Library of Congress Control Number: 2003106040

Publisher
IOS Press
Nieuwe Hemweg 6B
1013 BG Amsterdam
Netherlands
fax: +31 20 620 3419
e-mail: order@iospress.nl

Distributor in the UK and Ireland
IOS Press/Lavis Marketing
73 Lime Walk
Headington
Oxford OX3 7AD
England
fax: +44 1865 75 0079

Distributor in the USA and Canada
IOS Press, Inc.
5795-G Burke Centre Parkway
Burke, VA 22015
USA
fax: +1 703 323 3668
e-mail: iosbooks@iospress.com

Distributor in Germany, Austria and Switzerland
IOS Press/LSL.de
Gerichtsweg 28
D-04103 Leipzig
Germany
fax: +49 341 995 4255

Distributor in Japan
Ohmsha, Ltd.
3-1 Kanda Nishiki-cho
Chiyoda-ku, Tokyo 101-8460
Japan
fax: +81 3 3233 2426

PRINTED IN THE NETHERLANDS

Preface

Computer modelling techniques are now widely used in all scientific disciplines. A particularly fruitful area of application is the field of materials science, where computational methods are now indispensable tools in studying the structure, properties and reactivities of materials. The field has expanded very rapidly in recent years owing both to the continuing exponential growth in hardware performance and the ongoing developments in techniques and algorithms.

The present volume is based on a NATO Advanced Study Institute held in Il Ciocco, Lucca, Italy in September 2001. The volume aims to give a survey of the fundamental methodologies and an illustrative range of contemporary applications. The field is now so large that the latter is necessarily selective, and our focus is on inorganic (especially oxide) materials and semiconductors, where there have been many important and exciting recent developments.

The editors are grateful to the NATO Scientific Affairs Division for their sponsorship of the ASI.

Richard Catlow
Eugene Kotomin

Contents

Computational Materials Science
C.R.A. Catlow and E.A. Kotomin (Eds.)
IOS Press, 2003

Computer Modelling of Materials: an Introduction

Richard CATLOW

Davy Faraday Research Laboratory, Royal Institution of Great Britain, Albemarle Street, London W1X 4BS, UK

1. Introduction

For over twenty years, computational methods have provided a diverse and increasingly powerful range of tools for exploring the structures and properties of matter at the atomic level. The range of applications is now very broad encompassing molecular biology, materials and surface science, mineralogy and small molecule chemistry. Moreover, atomistic modelling now links to modelling of systems with greater length and time scales, i.e. to modelling at the mesoscale.

This book focuses on the application of atomistic computer modelling techniques to materials science. In this chapter, we present a brief survey of the aims and scope of the field and short introduction to the main methodologies, which subsequent chapters will explore in much greater detail. We illustrate the current state of the art of computer modelling studies of materials by recent applications to bulk and surface properties of topical oxide materials.

1.1 Range and Scope

Computer modelling is a distinct type of scientific technique which bridges fundamental theory with experimental observation on complex systems. Its success rests first on the fact that the basic physics controlling the behaviour of matter at the atomic level has been established for over seventy years; secondly on the development over the last twenty years of approximate but effective methods for implementing this physics in complex systems; and thirdly on the ability of modern computers to handle the enormous demands regarding both processing power and memory required for realistic modelling of complex systems.

Computer simulation studies aim to provide reliable models at the atomistic level, which fulfil three main rôles: first they can provide general insight and understanding of the systems simulated; secondly they can provide models, which can directly assist the interpretation of experimental data; and thirdly they can provide accurate numerical data on important parameters, which may be either difficult to measure or entirely inaccessible to experiment. As an example of the first rôle we may cite the application of modelling techniques to the study of microporous materials [e.g. 1] where they have given extensive insights into the structural and catalytic properties of these complex materials. The second rôle is illustrated by many applications in the science of defects and solids, where simulation tools have been used together with experiment to construct models of complex defect structures [e.g. 2]. An excellent illustration of the third rôle is found in the field of high pressure science where computer modelling has enabled the accurate calculation of physical properties of materials under the enormous pressures present in the Earth's mantle and core [e.g. 3].

The main characteristic of the present status of the field is its increasingly predictive nature and the growing complexity and realism of contemporary simulations, as will be illustrated in section 3 of this chapter and in many of the subsequent chapters in this book.

2. Methodologies

Two broad classes of technique are available for modelling matter at the atomic level. The first avoids the explicit solution of the Schrödinger equation by using *interatomic potentials*, which express the energy of the system as a function of nuclear coordinates. Such methods are fast and effective within their domain of applicability and good interatomic potential functions are available for many materials. They are, however, limited as they cannot describe any properties and processes, which depend explicitly on the electronic structure of the material. In contrast, *electronic structure* calculations solve the Schrödinger equation at some level of approximation allowing direct simulation of, for example, spectroscopic properties and reaction mechanisms. We now present an introduction to interatomic potential-based methods (often referred to as atomistic simulations); comprehensive discussions of electronic structure techniques are presented in the chapters of Harrison, Abarenkov and Postnikov.

2.1 Atomistic Simulations

2.1.1 Potential Functions and their Evaluation

The potential energy function $U(r_1 \cdots r_N)$ expresses the energy of an assembly of N atoms or ions as a function of the nuclear coordinates $r_1 \cdots r_N$. The Born-Oppenheimer approximation is, of course, implicit in the use of such functions but there is no explicit inclusion of the effects of the electronic structure of the system: such effects are subsumed into the potential function. The energy zero for such functions is normally taken to be that of the component atoms (or ions) at rest at infinity, *i.e.* the self energies (electron-nuclear) of the atoms (or ions) are not included in U.

The potential function is commonly expanded as follows:

$$\mathbf{U} = \frac{1}{2}\sum_{i}^{N}\sum_{j}^{N}{}'\varphi_{ij}(\mathbf{r}_i,\mathbf{r}_j) + \frac{1}{3}\sum_{i}^{N}\sum_{j}^{\infty}\sum_{k}^{\infty}{}'\varphi_{ijk}(\mathbf{r}_i,\mathbf{r}_j,\mathbf{r}_k) + \ldots\ldots\quad (1)$$

where the φ_{ij} are 'two-body' functions which depend only on the positions of pairs of atoms i and j; φ_{ijk} are the 'three-body' terms depending explicitly on the positions of atoms i, j and k and where the prime symbol indicates that we avoid multiple counting of terms. The expansion extends to four-body terms φ_{ijkl} and higher order terms which are, however, rarely included. Indeed it has been common to approximate U by including only the two-body component φ_{ij}, which may be usefully decomposed into Coulombic and non-

Coulombic terms:

$$\varphi_{ij}(\mathbf{r}_i, \mathbf{r}_j) = \frac{\mathbf{q}_i \mathbf{q}_j}{\mathbf{r}_{ij}} + \mathbf{V}_{ij}(\mathbf{r}_{ij}) \qquad (2)$$

where q_i is the charge on the atom or ion (and is normally treated as a point entity). The Coulombic term which is purely two-body in nature is handled by the procedures described in the previous section. The non-Coulombic, V_{ij}, is then usually approximated by analytical functions which generally include both attractive and repulsive components; the latter describes the Pauli repulsion due to overlap of closed shell electron configurations, and the former attractive terms from dispersion (*i.e.* induced dipole-induced-dipole) and covalence effects. Several such functions are available, notably the Lennard-Jones potential:

$$V(r) = Ar^{-12} - Br^{-6}, \qquad (3)$$

the Morse potential:

$$\mathbf{V}_{(r)} = \mathbf{D}\{1 - \exp[-\beta(\mathbf{r} - \mathbf{r}_e)]\}^2 \qquad (4)$$

and the Buckingham potential:

$$V(r) = A\exp(-r/\rho) - Cr^{-6}. \qquad (5)$$

Lennard-Jones potentials have been used widely in modelling rare-gas and molecular crystals. Morse potentials become more appropriate when covalent systems are being studied; D may then be interpreted as the covalent bond-dissociation energy and r_e the equilibrium bond length. Buckingham potentials have been very widely used in the study of ionic and semi-ionic solids [4].

Inclusion of many-body effects may be achieved by the use of angle dependent forces. For example, Sanders *et al.* [5] employed simple 'bond-bending' terms of the type,

$$\mathbf{E}(\theta) = \frac{1}{2}\mathbf{k}_B(\theta - \theta_0)^2 \qquad (6)$$

where θ is the angle subtended by three atoms and θ_o is an equilibrium value; k_B is the appropriate force constant. The use of such terms is most appropriate in systems with directional covalent bonding, *e.g.* silicates, where θ_o will correspond to the tetrahedral angle subtended by O-Si-O bonds in the tetrahedral SiO_4 groups.

Alternative, three-body functions are provided by the 'triple-dipole' formation of Axilrod and Teller [6], which takes the form:

$$\mathbf{V}_{ijk} = \frac{\mathbf{k}_T(1 + 3\cos\theta_1\cos\theta_2\cos\theta_3}{\mathbf{r}_{ij}^3\mathbf{r}_{jk}^3\mathbf{r}_{ik}^3} \qquad (7)$$

where the angles and distances relate to the triangle defined by the three atoms i, j, and k. Such functions describe the three body contributions to the induced dipole attractive forces, and have been shown by Meath and co-workers [7] to make important contributions to the cohesive energies of rare gas crystals. Work of Baetzold *et al.* [8] has shown that such terms may be of value in constructing potential models for the silver halide crystals.

Finally we note that atomic and ionic polarisation in solids is essentially a many-body effect. Moreover in ionic materials polarisation energies may be large especially in the vicinity of charged defects and surfaces. Omission of polarisation, while obviously leading to the failure to describe high frequency dielectric properties, also results in marked inadequacies in calculated phonon dispersion curves. The simplest way of modelling polarisability is to use the point polarisable ion (PPI) model which assigns a linear constant of proportionality *i.e.* the polarisability, α, between the dipole moment μ and the field E acting on the atoms, that is:

$$\boldsymbol{\mu} = \alpha \mathbf{E} \tag{8}$$

However, as the dipole is a point entity, it cannot describe the physical basis of polarisability which is the displacement of the valence shell electron density in response to the applied field. Consequently, important physical effects, in particular the coupling of polarisability and short range repulsion, are omitted by these models which fail badly in describing both dielectric and lattice dynamical properties of crystals, as discussed in greater detail in reference [9]. These deficiencies are largely overcome by the shell model originally proposed by Dick and Overhauser [10] which describes the development of a dipole moment in terms of the displacement of a massless shell of charge Y (representing the valence shell electrons) from a core (representing the nucleus and core electrons), the core and the shell being connected by an harmonic restoring force for which the spring constant is k; the free atom polarisability, α, is then given by $\alpha = Y^2/k$. Shell model potentials have been very widely used in studies of lattice dynamical and defect properties of ionic solids. Further discussion is given in reference 9.

Having established the framework of the potential model to be used, it is next necessary to fix the variable parameters, *i.e.* those used in the description of the short range potential, V(r), the shell model constants Y and k and the effective atomic charges q (although we note that in many modelling studies of ionic oxides, halides and even silicates, these have been fixed at integral fully ionic values).

2.1.2 Parameterisation

This is undertaken by two procedures: first, *empirical methods,* in which variable parameters are adjusted, generally *via* a least squares fitting procedure to observed crystal properties. The latter must include the crystal structure (and the procedure of 'fitting' to the structure has normally been achieved by minimising the calculated forces acting on the atoms at their observed positions in the unit cell). Elastic constants should, where available, be included; and dielectric properties are required to parameterise the shell model constants. Phonon dispersion curves provide valuable information on interatomic forces; and *force constant* models (in which the variable parameters are first and second derivatives of the potential) are commonly fitted to lattice dynamical data. This has been less common in the fitting of parameters in *potential models* which are our present concern as they are required for subsequent use in simulations. However, empirically derived potential models should always be tested against phonon dispersion curves when the latter are available.

An important aspect of empirical potential parameterisation is the question of transferability. Are, for example, models derived in the study of binary oxides, transferable to ternary oxides? Considerable attention has been paid to this problem by Cormack *et al.* [11] who have examined the use of potentials in spinel oxides, *e.g.* $MgAl_2O_4$, $NiCr_2O_4$ *etc.*; in addition Parker and Price [12,13] have made a very careful study of silicates especially Mg_2SiO_4. These studies conclude that transferability works well in many cases, although systematic modifications are needed when potentials are transferred to compounds with different coordination numbers; for example the correct modelling of $MgAl_2O_4$ requires that the potential developed for MgO, in which the magnesium has octahedral coordination, be modified in view of the tetrahedral coordination of Mg in the ternary oxide. The correction factor is based on the difference Δr^c between the effective ionic radii for the different coordination numbers. If an exponential, Born-Mayer, repulsive term is used, the pre-exponential factor is modified as follows:

$$\mathbf{A}' = \mathbf{A} \exp(\Delta \mathbf{r}^C / \rho) \quad (9)$$

where the unprimed and primed pre-exponential factors refer to original and modified coordination numbers. Cormack *et al.* [11] have shown that such procedures yield potentials which are successful in modelling spinels.

There are now several sets of empirical potential parameters for a range of solids. Parameter sets are available for organic, molecular solids (see, for example, the work of Williams [14] and of Kiselev [15]). Ionic solids have been widely studied in the last twenty years, and parameters are available for ionic halides [16], oxides [17,18] and silicates [12,13].

Empirical procedures are still extensively used in developing potential parameters. They do, however, have obvious inadequacies: firstly, it is necessary to have a good range of accurate empirical data which may not be possible when new systems are being studied; secondly, there is no guarantee of the validity of the potential function when used outside the range of interatomic spacings employed in the parameterisation. For this reason there has been considerable incentive in recent years for the development of reliable theoretical procedures for calculating parameters. To date, such efforts have been directed largely towards parameterisation of effective charges and of the short range potential, V, and in a restricted number of cases, the three-body components of the potential; it is still very difficult to calculate shell model parameters. Calculations of interatomic potentials have been of two types.

2.1.3 *Electron gas methods*

These methods are based on the electron gas description of the atom which is used to calculate the interaction energies of pairs or in principle larger numbers of atoms (or ions) as a function of their nuclear coordinates. The method was established by Wedepohl [19] and by Gordon and Kim [20] and wide ranging and successful studies of ionic halides and oxides are reported by Mackrodt and co-workers [see for example references 21,22] (note also the compilation of parameters given in reference 23). In addition, modified electron gas procedures developed by Cohen and co-workers [24] have enjoyed considerable success when applied to oxides and silicates.

As the electron gas method is now very standard and has been reviewed by the present author and others (references 9 and 19,20), a more detailed discussion is not presented here; a summary of the essential features of the methodology is described in reference 25.

2.1.4 *Explicit quantum mechanical methods*

These are methods in which the Schrödinger equation is solved approximately for a cluster or periodic array of atoms for a variety of internuclear geometries; the resulting variation in the total energy is then fitted to an interatomic potential function. *Ab initio* methods are being increasingly used in such studies; examples are the early work of

Mackrodt and co-workers [26], and the more recent studies of Gale *et al.* [27] and Harrison *et al.* [28]. The study of silicate systems has been particularly active, with calculation on clusters ranging from $Si(OH)_4$ to $Si_{15}O_{16}H_{12}$ [29-31].

In using such methods care must be taken in the choice of basis sets (*i.e.* the atomic centred functions from which the LCAO-MOs are constructed). Unless good sets are used, then the resulting interatomic potential function will be inadequate due to basis-set superposition errors. This subtle effect arises from the use by each atom in the many atom calculation of basis functions on *other* atoms to reduce its own energy — an effect which invariably arises when incomplete sets are used. Correction procedures are available [32]; but in general, high quality potentials need high quality basis sets. Secondly, it is important to note that wave functions in crystals differ from those of free ions. It is necessary therefore to include in such calculations a representation of the effects of the surrounding ions on the interacting atoms, by using an appropriate embedding procedure. Thirdly, it should be recalled that the Hartree-Fock method does not include effects arising from electron correlation; more sophisticated and computationally expensive configurational interaction or perturbation methods are needed to describe correlation effects. Since, however, correlation is responsible for dispersive interactions, other methods are needed to calculate these terms. Methods based on the Density Functional Theory discussed in detail in later chapters do include directly certain of the effects of electron correlation, although they are still unable to model dispersive effects accurately. Perturbation theories following the original work of Mayer [33] may be used; but empirical methods may be needed for some time to come.

Ab initio techniques are generally far more expensive computationally than is the case for the electron gas techniques, and the cpu and disk requirements increase rapidly with the total number of electrons in the calculation. The former methods are, however, preferable, if high quality basis sets may be used, as they allow a more accurate description of atomic interactions, in particular the redistribution of electron densities consequent upon interaction. Electron gas methods clearly, however, provide a cheap, useful and generally applicable technique for deriving short range potential parameters; and as such they will have a continuing rôle in this field.

Finally we should note a valuable approach developed by Pyper and coworkers, see e.g. [34], for ionic solids in which Hartree-Fock methods are used first to obtain a set of crystal orbitals, the interactions between which are calculated as a function of distance including a full explicit evaluation of the exchange term (unlike the local density approximation used in the electron gas method). Estimates of the dispersive energy are then added to the resulting interaction energy. This approach is particularly successful for strongly ionic halides and oxides.

Having developed and parameterised potential models, the final stage before their use in a simulation study should be their *evaluation*. Non-empirically derived potentials should be evaluated by reference to their ability to predict empirical crystal properties. For empirical potentials, it is clearly necessary to use data outside the range employed in the parameterisation.We have already referred to the use of lattice dynamical data. Comparison with the results of high pressure studies, in particular the variation of structural and elastic properties with pressure, is also of great value; and in this context we should note the work of Harding and Stoneham [35] who examined several potentials for simple ionic materials by comparison with 'Hugoniots' *i.e.* pressure-volume trajectories obtained from shock wave studies. A further useful and demanding test is provided by the ability of potential models to predict crystal structures of polymorphs of a compound other than that used in the potential parameterisation.

The development of improved potential models will continue to be a central feature of simulation studies. Progress will require the increased availability of accurate crystal data and refinements in theoretical procedures for calculating potentials.

2.1.5 *Simulation Methodologies: Static Techniques*

The characteristic features of these methods is that they do not include any explicit representation of thermal motions. Energies and entropies are calculated for the static lattice; the entropy calculations normally assume the harmonic approximation. High temperature properties can be calculated using the quasi-harmonic approximation which simply evaluates energies and force constants for the high temperature lattice parameters; within this framework it is possible to predict lattice expansivity by calculating the free energy minimum as a function of lattice parameter, for a series of temperatures. Calculations may be performed on both perfect and defective lattice configurations; both will now be reviewed.

2.1.6 *Perfect Lattice Calculations*

The most basic quantity open to calculation is the lattice energy. If we omit the zero-point vibrational energy (and four-body) and higher order terms this may be written as:

$$\mathbf{E} = \mathbf{U}_C + \frac{1}{2}\sum_{i=1}^{N}\sum_{j=1}^{\infty}{}'\mathbf{V}_{ij}\mathbf{r}_{ij} + \frac{1}{3}\sum_{i=1}^{N}\sum_{j=1}^{\infty}\sum_{k=1}^{\infty}{}'\varphi_{ijk}(\mathbf{r}_i, \mathbf{r}_j, \mathbf{r}_k) \tag{10}$$

where U_c is the electrostatic energy. The summation conventions are as in Equation 1 and the terms V_{ij} and φ_{ijk} are as defined in Equation 1 and 2. A number of standard summation procedures are available for evaluating electrostatic (or Madelung) energies in crystals, of which the most widely used is the Ewald procedure which effects a partial transformation onto reciprocal space and which is discussed extensively in the earlier literature in the field, for example, reference 9. We note that the great majority of lattice energy calculations only include the two-body contribution to the short-range energy. One important matter of definition is that the lattice energy gives the *energy of the crystal with respect to component ions at infinity*. If it is desired to express the energy with respect to atoms at infinity (for which the more appropriate term is then the *cohesive* energy) then the appropriate ionisation energies and electron affinities will be added.

Lattice energy calculations are now routine, and may be carried out for very large unit cells containing several hundred atoms. The codes METAPOCS [36], THBREL [37] and GULP [38] undertake lattice energy calculations including both two– and three–body terms, using both bond–bending and triple–dipole formalisms.

Table 1: Relative Energies (per mole) of Microporous Siliceous Structures with respect to Quartz (after reference 39)

Structure	Energy (kJ / mole)
Silicalite	11.2
Mordenite	20.52
Faujasite	21.4

Lattice energy calculations provide valuable insight into the structure and stabilities of ionic and semi-ionic solids. The technique is most powerful when combined with energy minimisation procedures, which generate the structure of minimum energy. These are discussed below after the calculation of entropies have been described. The results in Table 1 give a good illustration of the value of lattice energy studies. They are the energy minimum lattice energies calculated for a number of purely siliceous microporous zeolitic structures which are compared with the lattice energy of α-SiO_2. The latter has the lowest value as would indeed be expected since the more porous structures are known to be metastable with respect to the dense α-SiO_2 polymorph. Of greater interest is the observation that of the porous structures, silicalite has the greatest stability. This accords with the fact that this polymorph can only be prepared as a highly siliceous compound unlike the case with the other zeolitic structures which are normally synthesised with high aluminium contents. The calculations which are discussed in greater detail in reference 39, suggest that this behaviour has its origin at least in part in the thermodynamic stability of the compounds. We note that more recently very similar results were obtained by Henson *et al.* [40] who also showed that the calculated values were in excellent agreement with experiment.

In addition to calculating energies, it is also possible to calculate routinely a range of crystal properties, including the lattice stability, the elastic and dielectric and piezoelectric constants, and the phonon dispersion curves. The techniques used which are quite standard require knowledge of both first and second derivatives of the energy with respect to the atomic coordinates. Indeed it is useful to describe two quantities: first the vector, g, whose components $\mathbf{g}_i^\alpha$ are defined as:

$$\mathbf{g}_i^\alpha = \left[\frac{\partial \mathbf{E}}{\partial \mathbf{x}_i^\alpha}\right] \quad (11)$$

i.e. the first derivative of the lattice energy with respect to a given Cartesian coordinate (α) of the ith atom. The second derivative matrix W has components $\mathbf{W}_{ij}^{\alpha\beta}$; defined by:

$$\mathbf{W}_{ij}^{\alpha\beta} = \left[\frac{\partial^2 \mathbf{E}}{\partial \mathbf{x}_i^\alpha \partial \mathbf{x}_j^\beta}\right] \quad (12)$$

The expressions used in calculating the properties referred to above from these derivatives are discussed in greater detail in reference 9. For more detailed discussions of the calculation of phonon dispersion curves from the second derivative or 'dynamical' matrix W, the reader should consult references 41 and 42. Finally, we note that by the term 'lattice stability' we refer to the equilibrium conditions both for the atoms within the unit cell, and for the unit cell as a whole. The former are available from the gradient vector g, while the latter are described in terms of the 6 components $\varepsilon_1....\varepsilon_1$ which define the strain matrix ε, where

$$\boldsymbol{\varepsilon} = \begin{bmatrix} \varepsilon_1 & \frac{1}{2}\varepsilon_4 & \frac{1}{2}\varepsilon_5 \\ \frac{1}{2}\varepsilon_4 & \varepsilon_2 & \frac{1}{2}\varepsilon_6 \\ \frac{1}{2}\varepsilon_5 & \frac{1}{2}\varepsilon_6 & \varepsilon_3 \end{bmatrix} \quad (13)$$

So when the unit cell as a whole is strained, we describe the modification of an arbitrary vector $\underline{r}$ in the unstrained matrix to a vector $\underline{r}'$ in the strained matrix, using the equation:

$$\mathbf{r}' = (\mathbf{1} + \varepsilon)\,\mathbf{r} \tag{14}$$

where **1** is the unit matrix. The six derivatives of energy with respect to strain, $\left[\frac{\partial E}{\partial \varepsilon_i}\right]$, therefore measure the forces acting on the unit-cell. The equilibrium condition for the crystal therefore requires that $g = 0$ and $\left[\frac{\partial E}{\partial \varepsilon_i}\right] = 0$ for all i.

2.1.7 Entropy Calculations

The entropy in a solid arises first from configuration terms which for a perfect solid are zero; while for a solid showing orientational or translational disorder configurational expressions based on the Boltzmann expression S = kln(W) may be used. In this section we shall pay more attention to the second term, which is due to the population of the vibrational degrees of freedom of the solid. Thus the entropy of a solid may be written as:

$$S_{vib} = k\int_0^Q dQ \sum_i \left\{ \frac{h\nu_i}{kT}\left[\mathbf{exp}\left[\frac{h\nu_i}{kT}\right] - 1\right]^{-1} - \mathbf{ln}\left[1 - \mathbf{exp}\left[\frac{-h\nu_i}{kT}\right]\right]\right\} \tag{15}$$

where the sum is over all phonon frequencies and the integral is over the Brillouin zone. In practice the integral is normally evaluated by sampling over the zone for which a variety of techniques are available [43]. Vibrational terms also give a contribution to the lattice energy of the crystal:

$$E_{vib} = kT\int_0^Q dQ \sum_i \left\{ \frac{h\nu_i}{2kT} + \frac{h\nu_i}{kT}\left[\mathbf{exp}\left[\frac{h\nu_i}{kT}\right] - 1\right]^{-1}\right\} \tag{16}$$

which results in the following expression for the crystal free energy with respect to ions at rest of infinity:

$$F = E + kT\int_0^Q dQ \sum_i \left\{ \frac{h\nu_i}{2kT} + \mathbf{ln}\left[1 - \mathbf{exp}\left[\frac{h\nu_i}{kT}\right]\right]\right\} \tag{17}$$

where E is the lattice energy (omitting vibrational terms) given by Equation 10.

2.1.8 Energy Minimisation

Having evaluated energies and free energies of a crystal structure it is desirable to be able to use these in an energy (or free energy) minimisation procedure. Let us consider first the simple case of minimisation to constant volume (*i.e.* within fixed cell dimensions). We write the energy of the crystal as a Taylor expansion in the displacements of the atoms, d, from that current configuration giving:

$$E(\delta) = E_0 + g\delta + \frac{1}{2}\delta W\delta + ... \tag{18}$$

If we terminate this function at the second order term and minimise E with respect to d, we obtain for the energy minimum:

$$0 = g + W\delta \qquad i.e. \quad \delta = -gW^{-1} \tag{19}$$

Displacement of the coordinates by δ as given in Equation 18 will generate the energy minimum configuration. Of course, in practice, it will not be valid to truncate the summation at the quadratic term, except when very close to the minimum. However, Equation 19 provides the basis of an effective iterative procedure for attaining the minimum. Indeed this 'Newton Raphson' method is widely used in both perfect and defect lattice energy minimisation, as it is generally rapidly convergent. Its main disadvantage is that it requires the calculation, inversion and storage of the second derivative matrix, W. Recalculation and inversion each iteration may be avoided by use of updating procedures due to Fletcher and Powell [44] which are discussed elsewhere [9,45]. The storage problem may become serious with very large structures owing to the high cpu memory requirements. Recourse may be made to gradient methods, *e.g.* the well known conjugate gradients technique, which make use only of first derivatives. Such methods are, however, more slowly converging. The increasing availability of very large cpu memories is, however, reducing the difficulties associated with the storage of the W matrix.

For evaluation of the energy minimum with respect to constant pressure (*i.e.* with variable cell dimensions), first we note that we can define the six components of the mechanical pressure acting on the solid, corresponding to the six strain components, defined in Equation 13, *i.e.*

$$P^{\varepsilon_i} = \frac{1}{V}\left[\frac{dU_i}{d\varepsilon_i}\right] \tag{20}$$

where V is the unit cell volume. The strains can then be evaluated, using Hooke's law,

$$\varepsilon = PC^{-1} \tag{21}$$

where C is the (6x6) elastic constant tensor, which may be calculated from W. Substitution of these calculated strain components into Equation 21 then yields the new cell dimensions and atomic coordinates. Again, the procedure is iterative, as it is only strictly valid in the region of applicability of the harmonic approximation. With a sensible starting point, however, only a small number of iterations (typically 2-5) is required.

The treatment above assumes that the pressure and corresponding strains are entirely mechanical in origin. However, at finite temperatures there will be a 'kinetic pressure' arising from the changes in the vibrational free energy with volume. These may be written as:

$$P_{vib}^{\varepsilon_i} = \frac{1}{V}\left[\frac{dF_{vib}}{d\varepsilon_i}\right] \tag{22}$$

where F_{vib} is the vibrational free energy. These kinetic pressures are most simply evaluated by applying small arbitrary strains to the structure and calculating the corresponding changes in F_{vib}. If P_{vib} is added to the mechanical pressure P in Equation 20, it enables us to carry out free energy minimisation. Parker and coworkers have written a general computer code, PAPAPOCS, for such calculations and the same functionality is available in the GULP code [38]. Further discussion is given in references [42] and [46] which also describe how the techniques may be used to calculate lattice expansivity, either directly or by calculating the cell dimension as a function of temperature or by calculation of the thermal Grüneisen parameter.

2.1.9 Defect Simulations

Two techniques are available for calculating the properties of defects: first we may set up a periodic array of defects; second, we may embed the defect in an infinite representation of the surrounding crystal.

2.1.10 Super Cell Methods

The simplest procedure for calculating the formation energies, entropies and hence free energies of defects is to set up a defect supercell, *i.e.* a large periodically repeating structure with the defect being at the centre of each unit cell. The techniques described in the previous section may then be employed. Thus to evaluate the defect energy we perform a calculation on the perfect lattice, which may be equilibrated either to constant volume or pressure. The lattice energy E_{PERF} is then compared with the value E_{DEF} obtained under the same conditions for the defective lattice. Thus the defect formation energy E_D is simply written as:

$$E_D = E_{DEF} - E_{PERF}. \tag{23}$$

A similar procedure is, of course, applicable for the entropies and hence free energies. The method is attractively simple, but has a number of drawbacks the principal one being the need for large supercells, especially if large complex defects are being considered. However, we note that supercells containing several hundred atoms are accessible to modern computers. A second drawback is that even with large supercells, the calculated defect energy will inevitably include a term arising from defect interaction as well as formation terms. In certain contexts, however, this may be an advantage, and indeed by calculating the defect energy for different sizes of the supercell, defect interactions may be calculated as a function of defect spacing. Such a study was performed by Cormack *et al.* [47] who investigated the interaction of the two-dimensional shear planes which form in non-stoichiometric oxides. Thirdly, there are problems in including charged defects. Either the defects must be neutralised by including a defect of opposite charge in the unit cell, in which case the isolated defect formation energy cannot be studied, or electroneutrality is achieved by adding a background neutralising charge or by making small adjustments to the charges of the other atoms in the unit cell.

Despite these difficulties, supercell methods have found considerable application in defect studies [48,49], and indeed the results compare well with those obtained using the alternative techniques described below.

2.1.11 Mott-Littleton Methods

In these calculations, the isolated defect or defect cluster is embedded in the crystal which extends to infinity and the contrast between this approach and that used in the supercell methods is illustrated diagrammatically in Figure 1. The normal procedure in a Mott-Littleton calculation is to relax all the atoms in a region of crystal surrounding the defect, containing typically 100-300 atoms, until all the atoms are at zero force. Newton-Raphson minimisation methods are generally used. The relaxation of the remainder of the crystal is then described by more approximate methods in which the polarisation, P at a point r, is calculated for crystals which have dielectric isotropy, from the expression:

$$P = \frac{1}{4\pi}\frac{qr}{r^3}\left(1 - \varepsilon_0^{-1}\right) \tag{24}$$

where q is the charge on the defect, and ε_0 is the static constant. More complex expressions are used for dielectrically anisotropic crystals. In practice, an interface region is normally used between the two regions described above; in the interface, displacements are calculated using formulae of the type given above, but are evaluated as the sum of the response to individual defects in the inner region, rather than as a response to the net charge of the defect configuration which is used for the remainder of the outer region. Short range interactions between ions in the inner region and those in the interface are also calculated explicitly.

The methods have been widely discussed and reviewed in the literature [9,50]. Several good automated computer codes are available: the original HADES2 programme written by Norgett [50] was confined to crystals of cubic symmetry; the HADES3 code [51] extended the methodology to include crystals of any symmetry. Leslie [52] developed a program CASCADE, optimised for use on the CRAY vector processing super computers; the latest versions of this program permit the inclusion of many-body terms using bond-bending or triple-dipole formalisms. The GULP code written by Gale [38] also includes a Mott-Littleton module. A large number of applications of these techniques are now reported and reviewed in the references given above. Indeed, work using these techniques has helped to establish the quantitative reliability of simulation techniques in solid state studies.

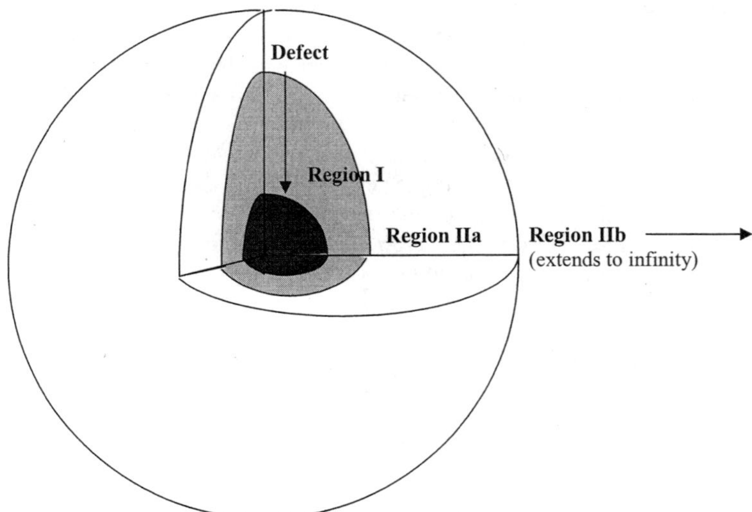

Figure 1: Schematic representation of the Mott-Littleton technique for treating relaxations around defects. (In practice an interface region IIa is included between the inner and outer regions.)

In addition to energies, defect entropies may also be calculated by related, although distinct, techniques. As with the case of the perfect lattice, defect entropies have two terms — configurational and vibrational — and the former can be calculated using simple combinatorial expressions. The latter arises from the perturbation of the vibrational frequency of the surrounding lattice atoms by the defect. The vibrational defect entropy S_{vib} is a function of the ratio of the perturbed to the unperturbed frequencies, *i.e.*

$$S_{vib} = -k\,\mathbf{ln}\frac{\prod_{i=1}^{3N'}\omega'_i}{\prod_{i=1}^{3N}\omega_i} + 3k(N' - N)\left[1 - \mathbf{ln}\left(\frac{\hbar}{kT}\right)\right] \qquad (25)$$

To evaluate the perturbed frequencies, Greens function techniques may be used. But Gillan and Jacobs [53] and Harding [54] have shown that embedded crystallite methods are more effective. Here the initial relaxation of the lattice surrounding the defect is undertaken as in the defect energy calculations; the perturbed force constant matrix for the relaxed region is then obtained. There are problems with the 'convergence' of the calculated entropy as a function of the size of the perturbed region; these may, however, be treated by the techniques discussed by Gillan and Jacobs [53].

Work of Harding and coworkers [54-56] has clearly established the value of defect entropy calculations using these methods. Moreover, we note that comparisons have been made [56] between entropies and energies calculated using supercell and embedded crystallite techniques. It is reassuring that the techniques yield the same defect parameters for large sizes of the supercell and of the crystallite.

Both energy and entropy calculations of the type described above are generally performed at constant volume, or more precisely constant lattice parameter. These may, however, be converted to constant pressure values using the following expressions:

$$h_p = u_v - V\beta T\left[\frac{df_v}{dV}\right], \qquad s_p = s_v - V\beta\left[\frac{df_v}{dV}\right] \qquad (26, 27)$$

where h_p and u_V are the constant pressure enthalpy and constant volume energy terms, s_p and s_V are the corresponding entropy terms, f_V is the constant volume free energy, the volume derivative of which may be calculated by studying the variation of u_V and s_V with lattice parameter; V is the unit cell volume and β is the expansivity of the solid. Discussion of these relationships is given in references 57 and 58. Work of Catlow *et al.* [58] on AgCl and Jackson *et al.* [59] on UO_2 has shown their value in studying high temperature, defect properties.

In concluding this section we note that defect calculations may be used to study defect mobilities as well as defect formation and interaction processes. Assuming the validity of the hopping model of defect transport then the frequency of defect jumps ν be written as:

$$\nu = \nu_0\,\mathbf{exp}\left(\frac{-g_{ACT}}{kT}\right) \qquad (28)$$

where g_{ACT} is the activation free energy for the defect migration, *i.e.* the difference between the free energies of the saddle point for the defect jump and that of its ground state. Saddle points may be identified by examining the potential energy surface for the migrating defect (although this may be difficult for crystals of low symmetry); once identified, the techniques discussed above may be used to calculate u_{ACT} and s_{ACT}.

Good discussions of the calculations of activation energies are given by Harding [56]. The validity of the hopping model, discussed in detail in reference 60, requires generally that $u_{ACT} >> kT$ and that the 'jump time' for the migration process be far less than the 'residence time' of defects at individual sites. These conditions apply to the great majority of transport processes in solids. However, for materials with very high mobilities the conditions may break down and it becomes necessary to use the dynamical simulation methods discussed in the next section.

2.1.12 Molecular Dynamics Techniques

In contrast to the static methods discussed in the previous section, molecular dynamics (MD) includes thermal energies explicitly. The method is conceptually simple: an ensemble of particles represents the system simulated and periodic boundary conditions are normally applied to generate an infinite system. The particles are given positions and velocities, the latter being assigned in accordance with a target temperature. In simulations of crystalline solids the simulation box will normally be a super cell of the basic unit cell. The system is then allowed to evolve in time by solving the classical equations of motion using a specified time step, t (which is typically 10^{-15} - 10^{-14}s). A variety of updating algorithms are available, as discussed in detail by Allen and Tildesley [61]. In the limit of an infinitesimal time step they all reduce to the simple classical equations of motion:

$$x(t+\tau) = x(t) + v(t)\tau \qquad (29)$$

$$v(t+\tau) = v(t) + \frac{f_i}{m_i}\tau \qquad (30)$$

where f_i is the force acting on the particle of mass m_i.

In the initial stages of the simulation, the ensemble equilibrates, *i.e.* it achieves an equilibrium distribution of velocities and equipartition between potential and kinetic energy. After this period, which may take several thousand time steps, the simulation is run for as long a period as is computationally feasible, and the trajectories of all particles are, if possible, stored.

MD simulations yield rich and detailed information on the system simulated. Structural information is available from radial distribution functions (RDFs), and diffusion coefficients may be calculated from the mean square displacement of the particle as a function of time, using the relationship:

$$D_\alpha = \frac{<r_\alpha^2>}{6t} \qquad (31)$$

where D_α is the diffusion coefficient of the particle of type α. Typical results are shown in Figure 2 for water in the CaF_2 and $CaCO_3$ mineral systems [62]. The increase of $<r_\alpha^2>$ with t clearly shows that diffusion is occurring in the CaF_2 system, unlike the case for $CaCO_3$ over the timescale of this simulation. Calculation of the velocity auto-correlation function (VAF) and van Hove self-correlation function yields more detailed dynamical information which may be compared with the results of inelastic neutron scattering studies; for further discussion we refer to references 57, 61 and 63.

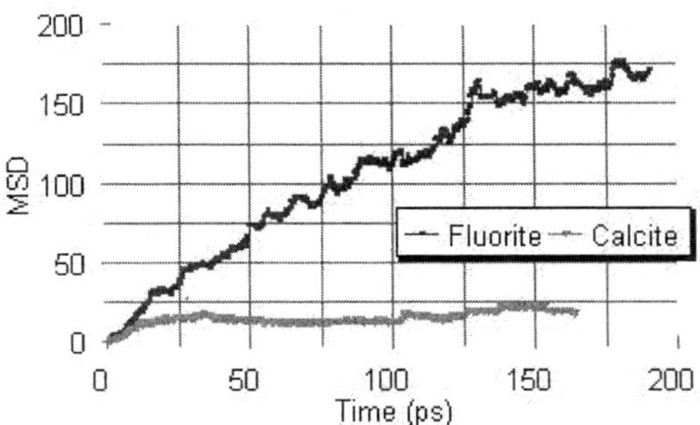

Figure 2: Mean square displacement *vs* time for H_2O at CaF_2 and $CaCO_3$ surfaces at 300K.

Despite its evident power, the technique has, however, serious limitations, the principal of which are as follows:

(i) Even with modern supercomputers it is rarely possible to run simulations for longer than 10 ns. If the probability of the events of interest (*e.g.* a defect jump) are low in a period of this length, then the simulation may be of little value. Thus in practice, with currently available computer power, diffusion can only be studied for values of the diffusion coefficient $D > 10^{-9}$ cm^2sec^{-1}.

(ii) Inclusion of polarisability in the potential greatly increases the computational requirements for the simulation. For this reason many MD simulations performed do not include this term which may be a serious omission for ionic solids.

(iii) The calculation of thermodynamic properties is difficult using MD techniques. The case of defects is especially problematic and both formation and migration energies are far more effectively calculated using static simulation techniques.

(iv) In conventional MD employing periodic boundary conditions, there are no surface effects. Thus any process involving the generation or removal of species from the surface cannot be simulated. An example is provided by the generation of Schotty disorder in crystals, which requires displacement of ions to the surface, which cannot be simulated by conventional MD studies.

Despite these limitations, the scope of MD techniques is expanding rapidly mainly due to the rapid growth in computer power, especially the development of parallel architecture systems. It is now possible to carry out routinely, simulations on several thousand particles, for periods of 500 ps – 1 ns. Moreover, the scope of the method has been expanded by a number of more specialised developments.

2.1.12.1 Constant Pressure MD

The conventional MD technique uses a fixed size for the simulation box, *i.e.* the calculation is performed under constant volume conditions. Using methods developed by Parrinello and Rahman [64] and by Berendsen and coworkers [65] it is possible to undertake constant pressure simulations by allowing cell dimensions to vary during the simulation. Detailed discussions are given in reference [61]. The most obvious field of application of this technique is to the study of phase transitions, and useful applications have been reported to the study of melting and glass formation.

2.1.12.2 MD Simulations of Glasses

Increasing use is being made of MD techniques to generate structural models for amorphous solids. The procedure used is simple: a molten system is generated; the simulation is then rapidly cooled to below the glass transition temperature. RDFs and other properties of the resulting simulated glass may then be calculated. A detailed account of this exciting field is given in the chapter by Cormack in this book.

2.1.13 Monte Carlo Techniques

Like MD, Monte Carlo (MC) methods involve the generation of successive configurations of an ensemble of particles representing the system studied; but unlike MD, there is no temporal connection between the different configurations. The aim of the technique is to generate a sufficient and representative number of configurations from which ensemble averages may then be calculated with acceptable accuracy.

A central feature of the procedure is the choice of a weighting scheme whereby configurations are included according to their probability. In particular it is desirable to weight the probability of including a configuration according to its Boltzmann factor:

$$W = \exp(-E / kT), \tag{32}$$

where E is the energy of the configuration. This may be achieved by by an ingenious and simple procedure, known as the Metropolis sampling method. Here the Boltzmann factor for a given configuration is calculated and compared with a random number P, which is generated in the range 0-1. If $W < P$, then the configuration is accepted, but if $W > P$ it is rejected. It will be seen that, over a large number of configurations, the procedure weights the probability of their inclusion according to their Boltzmann factor. It necessarily involves the generation of configurations which, after evaluation of their energy, are rejected; and a number of procedures are available to increase the efficiency of the method by reducing the number of rejected configurations [61].

The manner in which the sequence of configurations is generated depends on the system studied. For example, in a collection of atoms, a given particle may be selected at random and displaced by a small amount. For molecules, displacement and rotation may be involved. Indeed, care must be exercised in generating configurations in order to maximise the sampling efficiency. One special type of system concerns vacancy disordered compounds in which we are interested in diffusional properties. Here the vacancies are selected at random as is their direction; the Boltzmann factor for the jump is calculated (or obtained from a 'look-up' table) and compared with the random number P following the normal Metropolis procedure. Such methods have been extensively used by Murch and co-workers to study diffusion in alloys [66] and an application to oxygen diffusion in the superionic oxygen conductor Y/CeO_2 was reported by Murray and coworkers [67].

We note that a Monte Carlo 'move' can consist of the insertion or deletion of a particle, to so-called Grand Canonical Monte Carlo technique, for detailed discussion of which and of other technical aspects of M.C. simulation we refer to reference 61.

We also note an important class of applications which uses the MC method to explore complex energy surfaces and to find low energy regions, from which minima can subsequently be located by standard minimisation procedures. *Simulated annealing* methods have been used with great effect in generating crystal structures from initial random configurations of atoms or ions. The 'energy term' in simulating annealing calculations may take a variety of forms. It is commonly a simple 'cost function' based on coordination numbers and connectivity. Lattice energies, calculated from Born model potentials, may of course be employed, but this procedure becomes computationally expensive. It should, however, be noted that simulated annealing procedures employing energies calculated by electronic structure techniques are becoming feasible; although in practice it would seem to be more computationally efficient to use simpler procedures to calculate the energy (or cost function), and to refine the approximate configuration generated by simulated annealing by electronic structure methods when the use of such methods is needed.

The 'temperature' used in a simulated annealing study is normally a parameter with no real physical significance. Higher temperatures will result in the acceptance of an increasing number of high energy configurations, allowing the exploration of a wider range of the energy surface but, of course, at increased computational cost. In practice, the simulation usually starts with a high 'temperature' which is then reduced as the lower energy regions of the surface are identified. Simulated annealing like MC methods may also be adapted to the analysis of experimental data. In this case the 'energy term' becomes the deviation between the calculated data (*e.g.* the calculated X-ray intensity *vs* scattering angle for a structural model) and experiment. Such techniques are described as 'Reverse Monte Carlo' [68].

2.2 Electronic Structure Techniques

Over the last ten years there has been an enormous expansion in the applicability of explicit electronic structure methods in materials science. Hartree Fock (HF) methods still have an important rôle to play in the field but methods based on Density Functional Theory (DFT) are now very widely exploited. Detailed accounts of both approaches are presented in the chapters by Harrison, Abarenkov and Postnikov. We note that both HF and DFT Hamiltonians can be implemented in the context of periodic boundary conditions as well as molecular clusters, which in the latter case should where possible include an “embedding” procedure to represent the effects of the surrounding lattice. An example of the application of the latter approach will be presented in the next section.

3. Applications

As will be shown in subsequent chapters in this book, the range of applications of modelling techniques in materials science is now very wide. This section will attempt to give an indication of the complexity of system and problem that can now be modelled by taking three recent examples.

3.1 Case Study I: Grain boundary structures in Mantle minerals

Grain boundaries are known to be a major factor, controlling mechanical and rheological properties of materials. Detailed knowledge of their structures is, however, limited. Simulation methods have made a major contribution over the past 20 years in developing models for grain boundaries as in the work of Wolf and co-workers on a variety of metal systems [69,70] and Harding, Parker and co-workers on oxides [71-73].

Recent work has explored grain boundary properties in the Mantle mineral forsterite Mg_2SiO_4, a member of the olivine group of minerals which comprise a major proportion of the upper part of the Earth’s Mantle [74]. Knowledge of the grain boundary structure of this material is vital for developing an improved understanding of the rheology of the Mantle. Modelling boundaries in this material, however, presents substantial challenges owing to the complexity of the crystal structure.

The recent work of de Leeuw and co-workers investigated this problem using static lattice simulation techniques. They modelled the forsterite grain boundaries using the potential model for SiO_2 derived empirically by Sanders *et al.* [5], which was successfully used in many previous studies, including a recent investigation of the effect of hydration on the surface structure of quartz [75], in combination with the potential model for MgO derived by Lewis and Catlow [17] which was used previously in a number of applications, based on both static lattice calculations [76,77] and molecular dynamics simulations [78].

Atomistic simulation techniques are appropriate for these calculations because they are capable of modelling systems consisting of large numbers of ions which is necessary when modelling grain boundaries, as shown in many previous studies including the recent work of Harris *et al.* on MgO [001] tilt grain boundaries [79,80]. Energy minimisation techniques were used to investigate the structure and stability of the grain boundaries and the interactions between the lattice ions at the boundaries and adsorbed species, such as protons and dissociated water molecules, to identify the strength of interaction with specific boundary features. They employed the energy minimisation code METADISE, which is designed to model dislocations, interfaces and surfaces [79]. The detailed methods used for modelling surfaces and interfaces are treated in the chapter of de Leeuw *et al.* A grain boundary is created by fitting two surface blocks together in different orientations. In the present case, two series of tilt grain boundaries (M1 and M2, defined by the type of cation site at the surface) were created from appropriate models of stepped forsterite (010) surfaces at increasing boundary angles. Both boundary and adhesion energies were calculated, which describe the stability of the boundary with respect to the bulk material and free surfaces respectively. Results are reported in table 2 and figure 3. The atomistic models generated are illustrated in figure 4.

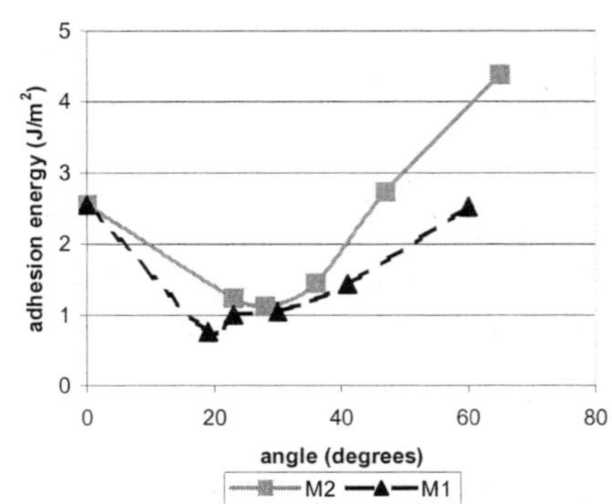

Figure 3. Adhesion energies as a function of grain boundary tilt angle

Table 2. Calculated boundary energies of (010) tilt grain boundaries in forsterite

Boundary	Boundary angle	Boundary energy (Jm^{-2})
M2	65^0	1.32
	47^0	2.72
	36^0	3.57
	28^0	3.50
	23^0	3.09
M1	60^0	2.12
	41^0	3.13
	30^0	3.19
	23^0	2.94
	19^0	2.88

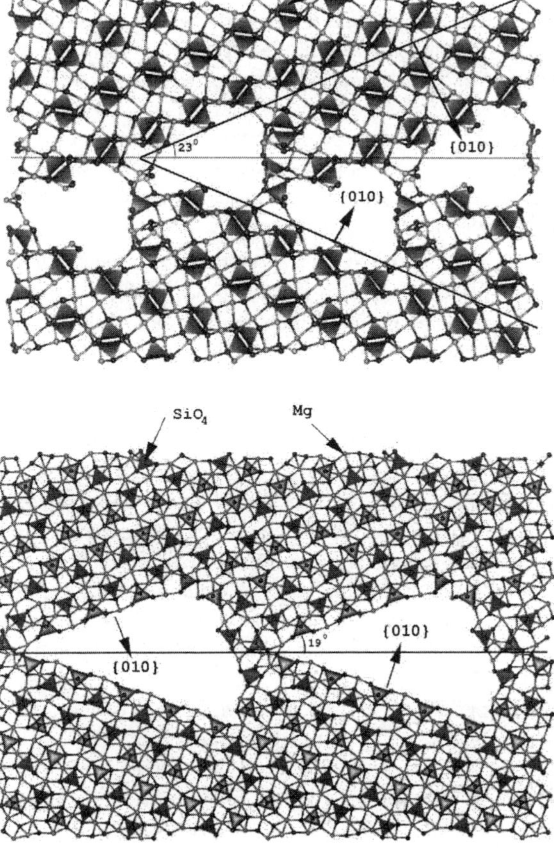

Figure 4. Relaxed structures of tilt grain boundaries with (010) mirror terraces, top (100) step wall showing two round channels per terrace, bottom (001) step wall with one triangular channel per terrace.

The larger grain boundaries do not form a continuously disordered interface but rather a series of open channels in the interfacial region with practically bulk termination of the two mirror planes (figure 4). We would expect that physical processes such as melting and diffusion of ions and molecules, e.g. oxygen or water, will be enhanced especially at the larger-terraced boundaries due to the low density of these regions compared to the bulk crystal. The minima in the adhesion energies at ϕ = ~20^0 (M1) or ~30^0 (M2) (Figure 3) indicate the boundaries which are most easily cleaved and are due to the relative stabilitities of the grain boundaries and corresponding free surfaces. Overall, the results show the ability of simulation methods to generate realistic models for these complex interfaces.

3.2 Case Study II: Reaction Mechanisms on the surfaces of Zinc Oxide

A large quantity of methanol, in excess of 22 million tonnes worldwide, is produced annually using the multicomponent $Cu/ZnO/Al_2O_3$ catalyst and feed gas, $CO_2/CO/H_2$. Many experimental studies of this process have been performed but without any definite reaction mechanism for the production of methanol being established. However, it has long been acknowledged that the important rate-determining step is the hydrogenation of adsorbed intermediates, for example the formate ion, at the active sites. Proposed mechanisms for methanol synthesis require the chemisorption of CO_2 before hydrogenation via formate to methanol. Theoretical studies of these systems have been hampered by the difficulty of modelling the catalytically active polar surfaces of ZnO, as well as the problems associated with the restrictions on the size of the system that can be modelled.

The nature of the active site for sorption/catalysis of CO_2 still remains unclear; and as a test system or model catalyst it has been proposed to use clean oxygen terminated surfaces of zincite. Temperature programmed desorption (TPD) studies have shown that the processes that occur at that particular surface are analogous to those on the industrial $Cu/ZnO/Al_2O_3$ catalyst [81,82]. It seems reasonable therefore to concentrate on the polar $(000\bar{1})$ oxygen terminated surface. In a given sample, the polar character of the surface necessitates a charge transfer between opposing polar surfaces, in order to quench surface induced polarisation of the material, which may be achieved in zincite by abstracting *ca.* 25% of oxygen ions from the surface layer, thus creating *vacant oxygen interstial surface sites*. The presence of such sites has been confirmed by optical and ESR spectroscopies [83-85], they have, moreover been suggested as the active catalytic sites for methanol synthesis and are therefore, the focus of the investigation in the work which is now described.

Recent work of French et al. [86] used novel solid-state embedding techniques in order to study important methanol precursors, including carbon dioxide, formate and methoxy ions. Their study provides insight into the mechanism of methanol synthesis by calculating binding energies, bond lengths and angles of these and related species. The hybrid QM/MM approach implemented and applied in this work follows molecular embedded cluster methods developed to treat point defects and related localised states in ionic materials [87-91].

At the heart of embedding approaches is a definition of a reference system and a single localised state, which causes only minor long-range perturbation of the reference system. In order to characterise the structural properties and interactions in the reference system it is desirable to split it into structural elements. Zincite, as with many other important metal oxide catalysts, is a polar material; and many physical properties of these materials can be described using an ionic model. In this model, the strong long-range interactions are accounted for by Coulombic forces between nearly spherical ions. These ions, particularly negatively charged anions, are highly polarisable species. Upon perturbation of the reference system by a localised defect, two major effects occur: i) dramatic short-range effects, (e.g. charge transfer and bond breaking) and ii) long-range polarisation of electron groups due to the field of the defect. Therefore, the model can be treated at two levels of approximation. In the present case, the short-range effects were dealt with fully *ab initio*, by Density Functional Theory (DFT), while the long-range polarisation effects are included through the use of the shell model of Dick and Overhauser [10].

This methodology (details of which are given in reference [86]) has been implemented within ChemShell, which is a computational chemistry environment, based on the Tcl programming language [92]. A DFT treatment, as implemented in the GAMESS-UK code [93], was used. The study employed the B97-1 exchange correlation functional [94] and a TZV2P basis set on O and Zn, which has been reoptimised for the ionic state of ZnO, along with the standard 6-311+(2p2d) basis set on all atoms of the adsorbates and a medium quality integration mesh [93]. In this work, interactions in region 2 and the interface were dealt with using our recently developed atomistic potentials [95] within the GULP code [38].

Detailed models of the $(000\bar{1})$–O terminated polar surfaces were investigated using interatomic potentials within the two-dimensional periodic slab model realised within the MARVIN code [96]. Two typical local environments were identified as terrace termination and interstitial vacancy sites; the latter has been used as the active site for the majority of calculations, as illustrated in Figure 5. The atoms highlighted in the figure indicate the surface embedded quantum region. For the calculations reported, region 1 consists of 12 atoms, the interface has 13 Zn sites, while region 2 comprises approximately 3000 atoms and 250 terminating point charges.

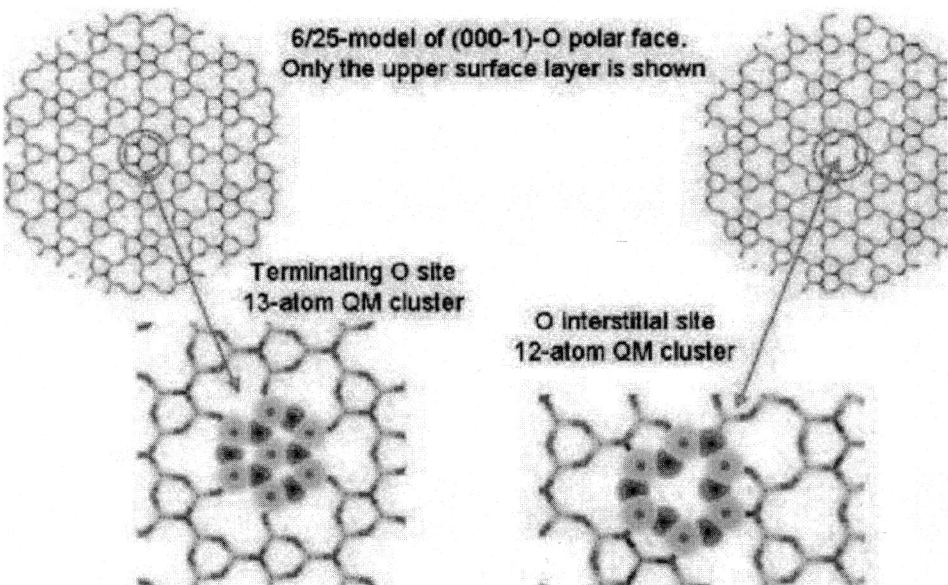

Figure 5. QM regions used for modelling of *vacant oxygen interstitial surface sites*. Only the upper surface layer is shown

A probable reaction sequence for the conversion of the feed gas to methanol has been summarised by Chinchen et al. [97], see Figure 6, where the gas phase and adsorbate species are indicated by (g) and (a) respectively. Reaction 4 is a complex process that includes multiple hydrogenation steps, which we have considered separately. The resulting species, which were not considered by Chinchen and others, were found to be stable, although, they may possibly be transient and short-lived.

The main catalytically active site that can facilitate the formation of anionic adsorbates is the *oxygen interstitial* site, as described above and illustrated in Figure 5. This site is able to trap an electron and thus adsorption can occur in two different ways; i) neutral adsorbates approach the active site at which an electron is trapped, to form a surface anionic species, ii) transfer of an electron to the neutral active site occurs simultaneously with the adsorption of neutral adsorbates.

$$\mathbf{H_2(g) \rightarrow 2H(a)}$$
$$\mathbf{CO_2(g) \rightarrow CO_2(a)}$$
$$\mathbf{CO_2(a) + H(a) \rightarrow HCO_2(a)}$$
$$\mathbf{HCO_2(a) + 2H(a) \rightarrow CH_3O(a) + O(a)}$$
$$\mathbf{CH_3O(a) + H(a) \rightarrow CH_3OH(g)}$$
$$\mathbf{CO(g) + O(a) \rightarrow CO_2(g)}$$
$$\mathbf{H_2(g) + O(a) \rightarrow H_2O(g)}$$

Figure 6. Catalytic cycle; all reactants geometries are obtained by energy minimisation of embedded clusters

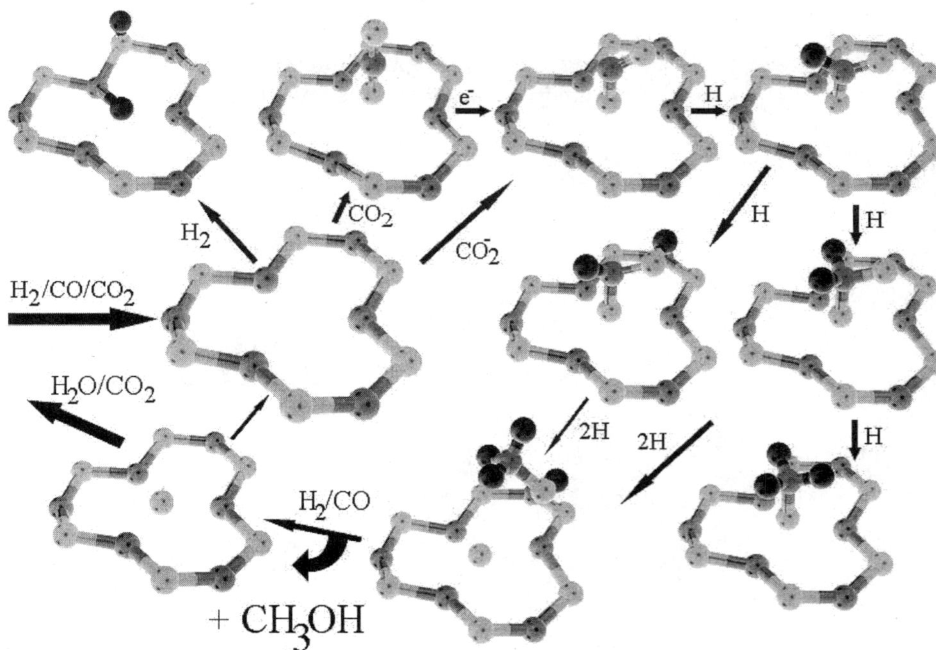

Figure 7. Proposed reaction sequence for methanol synthesis over ZnO

The proposed catalytic cycle is summarised in Figure 7, starting with the adsorption of CO_2 and H_2 (figure 6). CO_2 upon adsorption in a neutral interstial site retains the linear structure exhibited in the gas phase. Upon adding an electron the neutral CO_2 molecule bends and the extra electron populates an antibonding level, which leads to a rearrangement from an sp to an sp^2 hybridised configuration. The interaction with the surface stabilises the radical CO_2^- species.

The reaction then proceeds via the hydrogenation of the adsorbed CO_2^-, by surface hydrogen, to the formate ion. The geometry resulting from the calculations is shown in Figure 7. The adsorption of hydrogen leads to a closed shell species. Further hydrogenation can proceed either through the formation of $H_2CO_2^-$ or $HCOOH^-$ (formic acid) as shown in Figure 7. TPD data do not detect these species, which suggests a short lifetime and therefore high reactivity. Computational techniques allow us to differentiate and investigate these different scenarios. Further hydrogenation and interactions of the resulting species with the surface and possible surface defects lead to a large variety of possible intermediates. We show examples of a methoxy ion (CH_3O^-) chemisorbed to the surface and physisorbed methanol in Figure 7. To complete the catalytic cycle, methanol is removed from the surface and the active site is recycled by desorption of carbon dioxide and water.

Further details of these calculations can be found in reference [86]. They clearly show the viability of embedded cluster methodologies for investigating complex reactions on surfaces.

3.3 Case Study III: Docking of Growth Inhibitors at Mineral Surfaces

A large number of studies has been reported on the inhibition of calcite growth by organic as well as inorganic ions. The motivation for the work arises to a large extent from operational problems in the oil and water industries [98] in both of which calcite scale build-up is a major operational hazard. Some of the most widely used inhibitors are organic phosphonate ions and it is generally considered that diphosphonates and polyphosphonates are more efficient crystal growth poisons than monophosphonates. It is generally accepted that it is the phosphonate head group that interferes with the nucleation and growth process, preventing step flow and propagation, thereby inhibiting growth and augmenting the crystal morphology. Recent work [99,100] has applied computer simulation techniques to develop models for the inhibition mechanism and its dependence on the structural properties of the inhibitor molecules.

The effects of phosphonate poisons on crystal growth and morphology performed by numerous workers including Brooks *et al.* [101]. Growth inhibition processes have been identified by Hatch and Rice [102] and Reitmeier and Buehrer [103] as surface controlled events which require a very small concentration of poison to inhibit crystal formation; these observations were supported much later by Nancollas *et al.* [104] and Weigen *et al.* [105]. This surface controlled crystal growth mechanism was later shown to have a rate determining step governed by the rate of kink formation [106].

Techniques such as Atomic Force Microscopy (AFM) have been used to understand the molecular basis of the models for the inhibition mechanism [107]. Gratz *et al.* and Hillner *et al.* [108-110] observed the crystal growth inhibition process, by performing real-time, in situ AFM imaging; they found, accumulation of HEDP was found at the step-edge sites of the calcite crystal rather than terrace sites.

Although the processes involved in crystal growth inhibition are acknowledged as complex, there is a general consensus that the three main factors that influence the degree of interaction between organic phosphate containing compounds with various calcite crystal planes are charge distribution, steriochemistry and geometry as reported by Davey [111], Mann *et al.* [112], Austin *et al.* [113] and Teng *et al.* [114]. It is also clear that the inhibition is strongly determined by external factors such as pH, temperature and supersaturation.

Additionally, inhibitor efficacy can be substantially enhanced by the functional groups within the structural framework of the poison [115-119], as will be underlined by the calculations summarised here. Indeed, the phosphonates in this work were selected to investigate the extent to which electrostatic interactions between phosphonates and calcite surfaces were enhanced or decreased by functional groups, which may donate or withdraw electron density from the phosphonate oxygen.

Recent work of Ojo et al [100] examined the five phosphonates listed in Table 3. HEMP, DMP and PEMP are the mono-substituted systems. HEDP and PMP were chosen to compare with existing experimental data. HEMP exhibits a methyl and a hydroxyl group; PEMP contains two methyl groups, whilst DMP has two hydroxyl functionalities; HEDP is the diphosphonate analogue of HEMP ion whist PMP contains a phenyl group attached to the phosphonate motif.

Using surface simulation techniques as described in the chapter of de Leeuw *et al.*, phosphonate molecules were docked onto the planar calcite $\{10\bar{1}4\}$ surface. A combination of minimisation and dynamics was used to generate the most energetically favourable conformation. This procedure was repeated for the stepped $\{10\bar{1}4\}$ surface/phosphonate systems and in the calculation of replacement energies. A schematic illustration of the type of docked configuration studied is shown in figure 8.

The following reaction schemes may be proposed for the replacement by a monophosphonate ion of surface CO_3^{2-} species;

$$CO_3^{2-}\,(\text{surface}) + \text{monoPO}_3^{2-}\,(\text{solution}) \rightarrow CO_3^{2-}\,(\text{solution}) + \text{monoPO}_3^{2-}\,(\text{surface}) \qquad (33)$$

for which the corresponding replacement energy is given by:

$$E_{\text{replacement}} = E_{\text{solution}}\,(CO_3)^{2-} + E_{\text{surface}}\,(\text{monoPO}_3)^{2-} - E_{\text{solution}}\,(\text{monoPO}_3)^{2-} - E_{\text{surface}}\,(CO_3)^{2-}. \qquad (34)$$

which requires an estimate of the solvation energies, which was achieved by immersing the inhibitor molecules in a simulated droplet as discussed by Nygren et al. [99].

Table 3. Five phosphonate ions modelled in this work

HEDP	$[(PO_3)_2C(CH_3)(OH)]^{4-}$
HEMP	$[(PO_3)C(CH3)(OH)]^{2-}$
DMP	$[(PO_3)C(OH)_2]^{2-}$
PEMP	$[(PO_3)C(CH3)_2]^{2-}$
PMP	$[(PO3)Ph]^{2-}$

N.B. The charge neutral analogue of each phosphonate was used in the binding calculations.

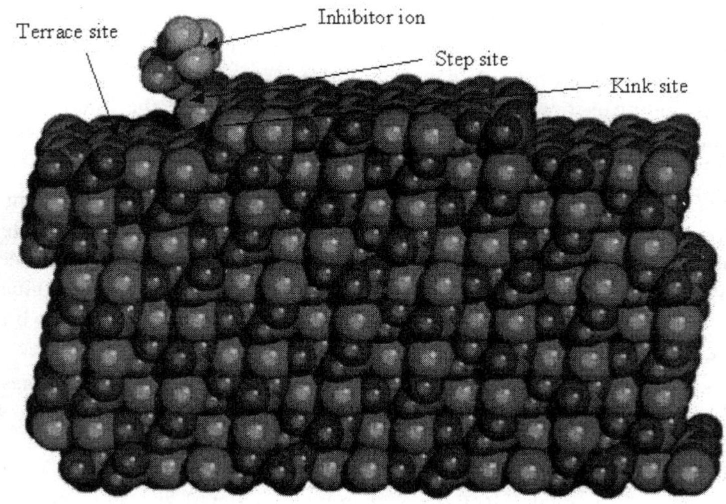

Figure 8. Schematic of the terrace, step and kink sites at a calcite $\{10\bar{1}4\}$ surface with inhibitor ion approaching kink site.

The binding energy of a neutral phosphonate species with respect to the planar $\{10\bar{1}4\}$ calcite surface which contained no defects, was calculated for all five phosphonates; the results are reported in table 4.

All calculated binding energies were favourable and their magnitude suggests that the inhibitors will physisorb rather than chemisorb to the planar surface. The negative binding energy for HEDP reinforces the concept that *binding* rather than *incorporation* is the dominant mode of interaction to the planar surface.

Table 4 Calculated binding energies ($kJmol^{-1}$) for the phosphonate ions at the planar $\{10\bar{1}4\}$ surface after 20ps NAT dynamics performed at 298K

HEDP	-85.9
HEMP	-41.5
DMP	-47.3
PEMP	-54.0
PMP	-18.4

Table 5 Calculated binding energies ($kJmol^{-1}$) for the phosphonate ions at the stepped calcite plane after 20ps NAT dynamics performed at 298K

HEDP	-276.0
HEMP	-140.9
DMP	-182.4
PEMP	-166.9
PMP	-118.7

Analysis of the simulations showed that the phosphonate ions were predominantly bound *via* electrostatic attraction between the oxygen atoms of the phosphonate group and the surface. Trends within the calculated binding energies are also consistent with the idea that surface binding is dominated by the phosphonate group; the binding energy of the diphosphonate is approximately twice that of the monophosphonates (with the exception of PMP).

Ojo *et al.* next calculated binding energies for the phosphonates at the stepped $\{10\bar{1}4\}$ calcite surface, the results of which are given in table 5. The phosphonates are more strongly bound to the stepped surface compared to the planar surface as has been predicted and observed experimentally [107]. The phosphonate group binds electrostatically with the step edge and additional stabilisation may be facilitated by the side groups binding to the step. Indeed, the calculated binding energies were approximately three-times those previously calculated on the planar $\{10\bar{1}4\}$ surface. We note that PMP has the lowest calculated binding energy since the phenyl ring is repelled by the step edge, whilst DMP and HEMP possess polar oxygen atoms within hydroxyl groups which increase the binding to calcite by approximately 20-40 $kJmol^{-1}$ (relative to PEMP).

The work of Ojo et al. [100] also examined in detail the interaction of inhibitor molecules with kink sites and the results suggest that inhibititor-kink interactions may play a significant rôle in the inhibition mechanism. Overall, the results illustrate the ability of simulation techniques to provide molecular models for crystal growth and inhibition.

4. Summary

Computer modelling techniques are now having a substantial impact on a broad range of problems in materials science. They are able to provide insight as to structures and mechanisms at the atomic level, and they are developing an increasingly predictive rôle. Advances in techniques, algorithms and computer power offer an exciting future with a great expansion in the predictive and explanatory capabilities of the methods, as will be apparent from the chapters in this book.

5. Acknowledgements

I am grateful to N.H. de Leeuw, S.C. Parker, A.A. Sokol, S.A. French, S. Ojo and B. Slater for their contributions to the work described in this chapter. I wish to thank Accelrys for provision of software.

6. References

1. C.R.A. Catlow, R.A. van Santen and B. Smit, *Modelling of Microporous Materials*, Academic Press, London 2003.
2. L. Minervini and R.W. Grimes, *J. Phys. Chem. Solids* **60** (1999) 235.
3. D. Alfe, G.D. Price and M.J. Gillan, *Phys. Rev. B* **65** (2002) 165118.
4. A.M. Stoneham and J.H. Harding, *Ann. Rev. Phys. Chem.* 37 (1986) 37, 53.
5. M.J. Sanders, M. Leslie and C.R.A. Catlow, *J. Chem. Soc. Chem. Comm.* (1984) 1271.
6. B.M. Axilrod and E. Teller, *J. Chem. Phys.* **11** (1943) 299.
7. W.J. Meath and R.A. Aziz, *Mol. Phys.* **52** (1984) 225.
8. R.C. Baetzold, C.R.A. Catlow, J. Corish, F.M. Healy, P.W.M. Jacobs, M. Leslie and Y. Tan, *J. Phys. Chem. Solids* 50 (1989) 791.
9. C.R.A. Catlow, Computer Modelling of Solids. In: C.R.A. Catlow and W.C. Mackrodt (eds.), Lecture Notes in Physics, Springer, Berlin, 1982, Vol. 166.
10. B.G. Dick and A.W. Overhauser, *Phys. Rev.* 112 (1958) 90.
11. A.N. Cormack, G.V. Lewis, S.C. Parker and C.R.A. Catlow, *J. Phys. Chem. Solids* 49 (1988) 53.

12. S.C. Parker and G.D. Price, *Phys. Chem. Miner.* 10 (1984) 209.
13. G.D. Price, A. Wall and S.C. Parker, *Phil. Trans. Roy. Soc. Lond.* A328 (1989) 391.
14. D.E. Williams, *Top. Curr. Phys.* 26 (1981) 3.
15. A.K. Kiselev, A.A. Lopatkin and A.A. Shulga, *Zeolites* 5 (1985) 261.
16. C.R.A. Catlow, K.M. Diller and M.J. Norgett, *J. Phys. C.* 10 (1977) 1395.
17. G.V. Lewis and C.R.A. Catlow, *J. Phys. C.* 18 (1985) 1149.
18. C.R.A. Catlow, C.M. Freemen, M.S. Islam, R.A. Jackson, M. Leslie, S.M. Tomlinson, Phil. Mag. A 58 (1988) 123.
19. P.T. Wedepohl, *Proc. Phys. Soc.* 92 (1967) 79.
20. R.G. Gordon and Y.S. Kim, *J. Chem. Phys.* 56 (1972) 3122.
21. W.C. Mackrodt and R.F. Stewart, *J. Phys. C: Condensed Matter* 12 (1979) 431.
22. W.C. Mackrodt and R.F. Stewart, *J. Phys. C: Condensed Matter* 12 (1979) 5015.
23. A.M. Stoneham, UKAEA Report (1981) AERE-R9598.
24. R.E. Cohen, L.L. Boyer and M.J. Mehl, *Phys. Rev.* B 35 (1987) 5749.
25. C.R.A. Catlow, In: A.K. Cheetham and P. Day (eds.) Solid State Chemistry: Techniques, Oxford University Press, 1986.
26. W.C. Mackrodt, R.F. Stewart, J.C. Cambell and I.H. Hillier, *J. Phys (Paris)* 41 (1980) C764.
27. J.D. Gale, C.R.A. Catlow and W.C. Mackrodt, *Modelling Simul. Mater. Sci. Eng,* 1 (1992) 73.
28. N. Harrison and M. Leslie, *Mol. Sim.* **9** (1992) 171.
29. K.P. Schroder and J. Sauer, *J. Phys. Chem.* **100** (1996) 11043.
30. B.W.H. van Beest, G.J. Kramer and R.A. van Santen, *Phys. Rev. Lett.* 60 (1990) 1955.
31. J. Purton, R. Jones, M. Heggie, S. Oberg and C.R.A. Catlow, *Phys. Chem. Minerals* 18 (1992) 389.
32. B. Johnson and B. Nelendar, *Chem. Phys.* 25 (1977) 263.
33. J.E. Mayer, *J. Chem. Phys.* 1 (1933) 270.
34. N.C. Pyper, In C.R.A. Catlow (ed.), Advances in Solid State Chemistry, JAI Press, 1992, Vol II.
35. J.H. Harding and A.M. Stoneham, *J. Phys. C: Cond. Matter* 17 (1984) 3401.
36. S.C. Parker, C.R.A. Catlow and A.N. Cormack, *Acta Cryst. B* 40 (1984) 200.
37. M. Leslie — Daresbury Laboratory, Warrington, WA4 4AD
38. J.D. Gale, *J. Chem. Soc. Faraday Trans. 93* (1997) 629
39. G. Ooms, R.A. van Santen, C.J.J. den Ouden, R.A. Jackson and C.R.A. Catlow, *J. Phys. C.* 92 (1988) 4462.
40. N.J. Henson, A.K. Cheetham and J.D. Gale, *Chem. Mater.* **6** (1994) 1647.
41. W. Cochran, *Crit. Rev. Solid Sci.* **2** (1971) 1.
42. S.C. Parker and G.D. Price, In: C.R.A. Catlow (ed.), Advances in Solid State Chemistry, JAI Press, 1990, Vol I.
43. G. Filippini, C.M. Gramacciolli, M. Simonetta and G.B. Suffritti, *Acta Cryst. A* **32** (1976) 259.
44. R. Fletcher and M.J.D. Powell, *Computer J.* **6** (1963) 16.
45. M.J. Norgett and R. Fletcher, *J. Phys. C.: Condensed Matter* **3** (1970) L190.
46. C.R.A. Catlow (ed.) *Computer Modelling in Inorganic Crystallography*, Academic Press, London, 1997.
47. A.N. Cormack, Rachel M. Jones, P.W. Tasker and C.R.A. Catlow, *J. Solid State Chem.* **44** (1982) 174.
48. M. Leslie and M.J. Gillan, *J. Phys.C: Condensed Matter* **18** (1985) 973.
49. N.L. Allan, W.C. Mackrodt and M. Leslie, In: C.R.A. Catlow and W.C. Mackrodt (eds.) Advances in Ceramics **23** (1987) 4257.
50. M.J. Norgett, UKAEA Report (1974) R7650.
51. C.R.A. Catlow, R. James and W.C. Mackrodt, *Phys. Rev. B: Condensed Matter* **25** (1982) 1006.
52. M. Leslie, SERC Daresbury Laboratory Report (1982) Rep. DL-SCI-TM31T.
53. M.J. Gillan and P.W.M. Jacobs, *Phys. Rev. B* **28** (1983) 759.
54. J.H. Harding, *Physica B* **131** (1985) 13.
55. J.H. Harding and A.M. Stoneham, *Phil. Mag B.* **43** (1981) 705.
56. J.H. Harding, *Rep. Prog. Phys.* **53** (1990) 1403.
57. C.R.A. Catlow, *Ann. Rev. Mater. Sci.* **16** (1986) 517.
58. C.R.A. Catlow, J. Corish, P.W.M. Jacobs and A.B. Lidiard, *J. Phys. C.* **14** (1981) L121.
59. R.A. Jackson, A.D. Murray, J.H. Harding and C.R.A. Catlow, *Philos. Mag. A.* **53** (1986) 27.
60. C.R.A. Catlow, *Solid State Ionics* **8** (1983) 89.
61. M.P. Allen and D.J. Tildesley, Computer Simulation of Liquids, Oxford University Press, 1987.
62. N.H. de Leeuw and S.C. Parker, *Mol. Simul.* **24** (2000) 71.
63. C.R.A. Catlow, *Solid State Ionics* **53** (1992) 955.
64. M. Parrinello and A. Rahman, *J. Chem. Phys.* **80** (1984) 860.
65. H.J.C. Berendsen, J.P.M. Postma, W.F. van Gunsteren, A.D. Nola and J.R. Haak, *J. Chem Phys.* **81**

(1984) 3684.
66. G.E. Murch, *Phil. Mag. A* **46** (1982) 575.
67. A.D. Murray, G.E. Murch and C.R.A. Catlow, *Solid State Ionics* **18-19** (1986) 196.
68. D.A. Keen and R.L. McGreevy, *Nature* **344** (1990) 423.
69. J.A. Jaszczak and D. Wolf, *Phys. Rev. B* **46** (1992) 2473.
70. P. Keblinski P, D. Wolf, S.R. Phillpot and H. Gleiter, *Philos. Mag. A* **79** (1999) 2735.
71. D.M. Duffy, J.H. Harding and A.M. Stoneham, *Philos. Mag. A* **67** (1993) 865.
72. D.C. Sayle, S.C. Parker and J.H. Harding, *J. Mater. Chem.* **4** (1994) 1883.
73. D.J. Harris, G.W. Watson and S.C. Parker, *Phys. Rev. B* **56** (1997) 11477.
74. N.H. de Leeuw, S.C. Parker, C.R.A. Catlow and G.D. Price, *Am. Mineral.* **85** (2000) 1143.
75. N.H. de Leeuw, F.M. Higgins and S.C. Parker, *J. Phys. Chem. B* **103** (1999) 1270.
76. N.H. deLeeuw, G.W. Watson and S.C. Parker, *J. Phys. Chem.* **99** (1995) 17219.
77. N.H. de Leeuw, G.W. Watson and S.C. Parker, *J. Chem. Soc. Faraday Trans.* **92** (1996) 2081.
78. N.H. de Leeuw and S.C. Parker, *Phys. Rev. B* **58** (1998) 13901.
79. G.W. Watson, E.T. Kelsey, N.H. deLeeuw, D.J. Harris and S.C. Parker, *J. Chem. Soc. Faraday Trans.* **92** (1996) 433.
80. D.J. Harris, G.W. Watson and S.C. Parker, *Am. Mineral.* **84** (1999) 138.
81. M. Bowker, H. Houghton and K.C. Waugh, *J. Chem. Soc. Faraday Trans. 1* **77** (1981) 3023
82. V. E. Heinrich and P. A. Cox, *The Surface Science of Metal Oxides*, Cambridge University Press, Cambridge, 1996.
83. K. Vanheusden, C.H. Seager, W.L. Warren, D.R. Tallant, J. Caruso, M.J. Hampden-Smith and T.T. Kodas, *J. Luminescence* **75** (1997) 11
84. A.R. Gonzalez-Elipe and J. Soria, *J. Chem. Soc. Faraday Trans. 1* **84** (1988) 3961
85. B. Yu, C. Zhu, F. Gan and Y. Huang, *Mater. Letts.* **33** (1998) 247
86. S.A. French, A.A. Sokol, S.T. Bromley, C.R.A. Catlow, S.C. Rogers, F. King and P. Sherwood, *Angew. Chem. Int. Ed.* **40** (2001) 4437.
87. J.H. Harding, A.H. Harker, P.B. Keegstra, R. Pandey, J.M. Vail and C. Woodward, *Physica B* **131** (1985) 151
88. A.L. Shluger, E.A. Kotomin and L.N. Kantorovich, *J. Phys. C 19* (1986) 4183
89. Z. Barandiarán and L. Seijo, *J. Chem. Phys.* **89** (1988) 5739
90. M.N. Nygren, L.G.M. Petersson, Z. Barandiaran and L. Seijo, *J. Chem. Phys.* **100** (1994) 2010.
91. P.V.Sushko, A.L.Shluger and C.R.A.Catlow, *Surf. Sci.* **450** (2000) 153.
92. P. Sherwood, A.H. Devries, S.J. Collins, S.P. Greatbanks, N.A. Burton, M.A. Vincent, I.H. Hillier, *Faraday Discuss.* **106** (1997) 79.
93. GAMESS-UK is a package of ab initio programs written by M.F. Guest, J.H. van Lenthe, J. Kendrick, K. Schoffel, and P. Sherwood, with contributions from R.D. Amos, R.J. Buenker, H.J.J. van Dam, M. Dupuis, N.C. Handy, I.H. Hillier, P.J. Knowles, V. Bonacic-Koutecky, W. von Niessen, R.J. Harrison, A.P. Rendell, V.R. Saunders, A.J. Stone, D.J. Tozer, and A.H. de Vries. The package is derived from the original GAMESS code due to M. Dupuis, D. Spangler and J. Wendoloski, NRCC Software Catalog, *Vol. 1*, Program No. QG01 (GAMESS) 1980.
94. F.H. Hamprecht, A.J. Cohen, D.J. Tozer, N.C. Handy, *J. Chem. Phys.* **109** (1998) 6264
95. L. Whitmore, A.A. Sokol, C.R.A. Catlow, *Surf. Sci.* **498** (2002) 135.
96. D.H. Gay, A.L. Rohl, *J. Chem. Soc. Faraday Trans.* **91** (1995) 92.
97. G.C. Chinchen, M.S. Spencer, K.C. Waugh, D.A. Whan, *J. Chem. Soc. Faraday Trans. I* **83** (1987) 2193.
98. A.H. Rankin and P.J.C. Sutcliffe, *Proc. Geologists Ass.* **110** (1999) 33.
99. M.A. Nygren; D.H. Gay, C.R.A. Catlow, M.P. Wilson and A.L. Rohl, *J. Chem. Soc., Faraday Trans.* **94** (1998) 3685
100. S.A. Ojo, B. Slater, C.R.A. Catlow, *Mol. Simulat.* **28** (2002) 591.
101. R. Brooks, L.M. Clark and E.F. Thurston, *Phil. Trans. R. Soc. London A* **145** (1950) 145.
102. G.B. Hatch and O Rice, *Ind. Eng. Chem.* **31** (1939) 54.
103. R.F. Reitemeier and T.F. Buehrer, *J. Phys. Chem.* **44** (1940) 535.
104. G.H. Nancollas, T.F. Kazmierczak and E Schuttringer, *Corrosion* **37** (1981) 76.
105. M.P.C. Weigen, W.G.J. Marchee and G.M. van Rosmalen, *Desalination* **47** (1983) 81.
106. J. Christoffersen and M.R. Christoffersen, *J. Cryst. Growth* **100** (1990) 100.
107. A.J. Gratz, P.E. Hillner and P.K. Hansma, *Geochim. Cosmochim. Acta.* **57** (1993) 491.
108. P.E. Hillner, A.J. Gratz and S. Manne, *Geology*, **20** (1992) 359.
109. P.E. Hillner, S. Manne, P.K. Hansma and A.J. Gratz, *Faraday Discuss.* **95** (1993) 191.
110. A.J. Gratz and P.E. Hillner, *J. Cryst. Growth* **129** (1993) 789.
111. R.J. Davey, *J. Cryst. Growth* **34** (1976) 109.
112. S. Mann, D.D Archibald, J.M. Didymus, T. Douglas, B.R. Heywood, F.C. Meldrum and N. J.

Reeves, *Science* **261** (1993) 6.
113. A.E. Austin, J.F. Miller, D.A. Vaughan and J.F. Kircher, *Desalination* **16** (1975) 345.
114. H.H. Teng, P.M. Dove, C.A. Orme and J.J. DeYoreo, *Science* **282** (1998) 5389.
115. A.L. Rohl, D.H. Gay, R.J. Davey and C.R.A. Catlow, *J. Am. Chem. Soc.* **118** (1996) 642.
116. S.N. Black, L.A. Bromley, D. Cottier, R.J. Davey, B. Dobbs and J.E. Rrout, *J. Chem. Soc. Faraday Trans.* **87** (1991) 3409.
117. R.J. Davey, S.N. Black, L.A. Bromley, D. Cottier, D.B. Dobbs and J.E. Rout, *Nature* **353** (1991) 549.
118. G.A. Falini, S Albeck, S. Weiner and L. Addadi, *Science* **271** (1996) 67.
119. J.S. Gill and R.G. Varsanik, *J. Cryst. Growth* **76** (1986) 57.

Computational Materials Science
C.R.A. Catlow and E.A. Kotomin (Eds.)
IOS Press, 2003

Methods of electronic structure calculations

A. V. Postnikov
Theoretical Low-Temperature Physics,
Gerhard Mercator University, D-47048 Duisburg, Germany

Abstract. A short overview is given of methods traditionally used in electronic structure calculations by either Hartree-Fock, or density-functional methods. Advantages and disadvantages of localized basis sets, planewave pseudopotential schemes, and several intermediate, or combined, approaches are emphasized.

1 Basis set decomposition

This contribution presents a concise overview of numerical methods used in order to achieve quantitative solution, with an acceptable accuracy, of either Hartree-Fock (HF), or Kohn-Sham (KS) equations. An introduction into the basics of these well established approaches has been provided by the lectures by N. Harrison[1] and I.Abarenkov[2] at this school. In both of these approaches, one has to solve iteratively a system of integro-differential equations on one-particle spin-orbitals $\varphi_\alpha(\mathbf{r})$:

$$\left[-\frac{\hbar^2}{2m}\nabla^2 + u(\mathbf{r}) + \int \frac{\rho(\mathbf{r}')\,d\mathbf{r}'}{|\mathbf{r}-\mathbf{r}'|}\right]\varphi_\alpha(x) - \int \frac{\rho(\mathbf{r}',\mathbf{r})\,\varphi_\alpha(\mathbf{r}')}{|\mathbf{r}-\mathbf{r}'|}d\mathbf{r}' = \varepsilon_\alpha\varphi_\alpha(\mathbf{r}) \tag{1}$$

in the HF formalism, or similarly, substituting the exchange term by an exchange-correlation potential $V_{\text{XC}}(\mathbf{r})$,

$$-\int \frac{\rho(\mathbf{r}',\mathbf{r})\,\varphi_\alpha(\mathbf{r}')}{|\mathbf{r}-\mathbf{r}'|}d\mathbf{r}' \quad \Rightarrow \quad V_{\text{XC}}(\mathbf{r})\varphi_\alpha(\mathbf{r})$$

in the KS equations. $\varphi_\alpha(\mathbf{r})$ are either the "optimal" one-electron wavefunctions providing a single-determinant representation of the multielectron wave function (in the HF formalism), or the support functions that serve for the decomposition of one-electron density $\rho(\mathbf{r})$:

$$\rho(\mathbf{r}) = \sum_{\alpha=1}^{N} \varphi_\alpha^*(\mathbf{r})\varphi_\alpha(\mathbf{r})\,.$$

The latter expression assumes summation over N (=the number of electrons in the system) states with lowest eigenvalues ε_α and holds in the HF formalism as well. $u(\mathbf{r})$ stands for the external electrostatic potential due to, e.g., atomic nuclei; the density matrix $\rho(\mathbf{r}',\mathbf{r})$ explicitly appearing in the HF equations is defined as

$$\rho(\mathbf{r}',\mathbf{r}) = \sum_{\alpha=1}^{N} \varphi_\alpha^*(\mathbf{r}')\varphi_\alpha(\mathbf{r})\,,$$

and the exchange-correlation potential in the KS equations $V_{XC}(\mathbf{r})$ is available in a closed form based on some additional assumptions – e.g., local density approximation etc. An important difference between HF and KS equations is the *locality* of $V_{XC}(\mathbf{r})$ (that is the same for all orbitals) vs. *non-locality* of the exchange potential (that itself depends on the orbital in question) in the HF formalism. Otherwise, the HF and KS equations are quite similar, and their practical solution cam be achieved by identical methods. Since the analytical solution in closed form is not possible (apart from certain model cases), an iterative treatment is a general rule. Whereas at the early stage of solving HF equations for "real" (albeit simple) systems one tended to follow general prescriptions for the solution of differential equations, an important breakthrough has been reached in 1951 by Roothaan[3] who proposed to expand the orbitals searched for over a set of appropriate basis functions:

$$\varphi_\alpha(\mathbf{r}) = \sum_{p=1}^{Q} C_{\alpha p}\chi_p(\mathbf{r})\,, \tag{2}$$

where the dimension of basis Q is reasonably larger than the number of occupied states N, and $\chi_p(\mathbf{r})$ may retain the dependence on both spatial and spin variables. Eq. (1) then reduces to

$$\sum_p C_{\alpha p}\mathcal{H}\chi_p(\mathbf{r}) = \varepsilon_\alpha \sum_p C_{\alpha p}\chi_p(\mathbf{r})\,,$$

where $\mathcal{H}$ is the Fock operator, i.e. everything that acts on a given spin-orbital in the left side of Eq. (1), or its counterpart in the KS approach. We shall refer to it as "Hamiltonian", since it is formally indeed a Hamiltonian of the reference system of independent particles, which produce the one-electron density equal to the correct one. Multiplying the above equation on the left by $\chi_q^*(\mathbf{r})$ and integrating in $\mathbf{r}$, we get

$$\sum_p C_{\alpha p}\left[\int \chi_q^*(\mathbf{r})\mathcal{H}\chi_p(\mathbf{r})d\mathbf{r} - \varepsilon_\alpha \int \chi_s^* q(\mathbf{r})\,\chi_p(\mathbf{r})\,d\mathbf{r}\right] = 0\,. \tag{3}$$

That is a system of algebraic equations on the expansion coefficients $C_{\alpha p}$, or a generalized (i.e., with overlap) diagonalization problem. The overlap matrix elements

$$S_{qp} \equiv \int \chi_q^*(\mathbf{r})\,\chi_p(\mathbf{r})\,d\mathbf{r} \tag{4}$$

need to be calculated only once (for basis functions fixed in advance). The matrix elements of the Hamiltonian

$$H_{qp} \equiv \int \chi_q^*(\mathbf{r})\left[-\frac{\hbar^2}{2m}\nabla^2 + u(\mathbf{r}) + \int \frac{\rho(\mathbf{r}')d\mathbf{r}'}{|\mathbf{r}-\mathbf{r}'|} + V_{XC}(\mathbf{r})\right]\chi_p(\mathbf{r})d\mathbf{r}$$

are actually dependent (via ρ and V_{XC}) on yet unknown coefficients $C_{\alpha p}$. Specifically, only the eigenvectors corresponding to N lowest eigenvalues ε_α must be included in the construction of the ground-state density (similarly to the *aufbau principle* in the HF formalism), so the density is

$$\rho(\mathbf{r}) = \sum_{\alpha=1}^{N} \varphi_\alpha^*(\mathbf{r})\,\varphi_\alpha(\mathbf{r}) = \sum_{pq} \underbrace{\left[\sum_{\alpha=1}^{N} C_{\alpha q}^* C_{\alpha p}\right]}_{\equiv D_{pq},\ \text{density matrix}} \chi_q^*(x)\,\chi_p(x)\,. \tag{5}$$

The density can be either diagonal in spin variables, if basis functions for both spin components remain completely decoupled, or it may retain a general form, if spin components interact due to spin-orbit interaction and/or spin non-collinearity effects being included. The internal dependency of H_{pq} on $C_{\alpha q}$ via the density matrix prevents the solution of Eq. (3) in a single matrix manipulation. However, an iterative solution is possible: one fixes a trial set of $C_{\alpha q}$ in the density matrix and constructs the Hamiltonian, then (3) becomes a system of linear equations solvable by a single diagonalization. Among the $C_{\alpha q}$ ($q = 1, \dots Q > N$) found by diagonalization one selects those corresponding to the N lowest eigenvalues, uses them to update the density matrix and proceeds till everything converges.

While S_{qp} remain constant if the basis functions do not change (that is the case, at least, in solving a static electronic-structure problem with fixed positions of atoms), H_{qp} can be separated into several terms where at least several ones (those including the density matrix and exchange-correlation potential) need to be re-calculated many times in the course of iterations:

$$\begin{aligned} H_{qp} &= \int \chi_q^*(\mathbf{r}) \left[-\frac{\hbar^2}{2m}\nabla^2 \right] \chi_p(\mathbf{r}) d\mathbf{r} && (\leq\text{2-center integrals}) \\ &+ \int \chi^*(\mathbf{r})\, u(\mathbf{r})\, \chi_p(\mathbf{r}) d\mathbf{r} && (\leq\text{3-center integrals}) \\ &+ \sum_{q'p'} \underbrace{\left[\sum_{\alpha=1}^{N} C^*_{\alpha q'}\, C_{\alpha p'} \right]}_{\text{density matrix}} \int \frac{\chi_q^*(\mathbf{r})\, \chi_{q'}^*(\mathbf{r}')\, \chi_{p'}(\mathbf{r}')\, \chi_p(\mathbf{r})}{|\mathbf{r}-\mathbf{r}'|} d\mathbf{r}\, d\mathbf{r}' && (\leq\text{4-center integrals}) \\ &+ \int \chi_q^*(\mathbf{r})\, V_{\text{XC}}(\mathbf{r})\, \chi_q(\mathbf{r}) d\mathbf{r}\,. && (6) \end{aligned}$$

The term "n-center integral" refers to the situation when the basis functions are centered somewhere in space, usually at atom sites. This is often the case – but now always, as the site-independent basis functions (e.g., plane waves) may be also used. For site-centered bases, one usually takes into account the fact that the external potential $u(x)$ contains, along with other possible contributions, the Coulomb potential from atom nuclei, either screened by core electrons or not:

$$u(x) = \sum_{\mu} \frac{eZ_{\mu}}{|\mathbf{r} - \mathbf{R}_{\mu}|} + \dots$$

Then, depending on whether both basis functions and the corresponding atom-centered component relate to the same center or not, the contributions of the form

$$\int \chi_q^{\alpha *}(\mathbf{r} - \mathbf{R}_{\alpha}) \frac{1}{|\mathbf{r} - \mathbf{R}_{\mu}|} \chi_p^{\beta}(\mathbf{r} - \mathbf{R}_{\beta})\, d\mathbf{r}$$

may be one- , two- or three-center integral. Similarly, the Hartree contribution to the Hamiltonian (the 3d line of Eq. (6) may contain one- to four-center terms. The exchange-correlation part depends on $\rho(\mathbf{r})$ and may also give rise to up to four-center integrals, as its particular form in the HF-formalism, the exchange term, would obviously do. The calculation of one-center integrals is usually trivial; two-center integrals may be simplified by an appropriate coordinate transformation (e.g., using elliptic coordinates); the three- and four-center integrals however may present a problem, and their evaluation deviates between different calculation methods.

With Eq. (3) solved, the total energy can be recovered from the calculated N lowest eigenvalues and elements of the density matrix, e.g., in the KS case:

$$\begin{aligned} E_{\text{tot}} &= \sum_{i=1}^{N} \varepsilon_i - \frac{e^2}{2} \sum_{pqrs} D_{ps}\, D_{rs} \int \chi_q^*(\mathbf{r})\, \chi_p(\mathbf{r})\, \chi_r^*(\mathbf{r}')\, \chi_s(\mathbf{r}')\, \frac{d\mathbf{r}\, d\mathbf{r}'}{|\mathbf{r}-\mathbf{r}'|} - \\ &- e \sum_{pq} D_{rq} \int \chi_q^*(\mathbf{r})\, \chi_p(\mathbf{r})\, V_{\text{XC}}(\mathbf{r})\, d\mathbf{r} + E_{\text{XC}}[\rho]\,. \end{aligned} \tag{7}$$

Response properties, e.g. forces, may follow by more or less straightforward (depending on the actual choice of basis functions) analytical differentiation; a rather systematic description on this subject has been given by Pulay[4].

2 Possible calculation setups

So far, we haven't yet specified the form of basis functions χ. As already mentioned, a quite common choice is some atom-centered functions localized in real space, but the functions localized in reciprocal space (plane waves) could be a convenient option, or some hybrid bases. The basis of choice is often related to the nature of boundary conditions used. Thus, for a simulation of a periodic solid with Bloch – von Kármán boundary conditions the use of plane wave basis may seem a natural choice, whereas non-periodic systems like molecules or clusters may seem to be more naturally treated with localized atom-centered bases. However, in both these cases an alternative choice of basis is equally possible and in effect broadly used: molecules are simulated in a repeated "simulation box" with periodic boundary conditions thus imposed and plane waves used as basis set; for periodic solids, the (k-dependent Bloch sums can be constructed by lattice summation of localized functions, as:

$$\tilde{\chi}^{\alpha}_{p\mathbf{k}}(\mathbf{r}) = \frac{1}{\sqrt{N}} \sum_{\nu} \chi^{\alpha}_{p}(\mathbf{r} - \mathbf{u}_{\alpha} - \mathbf{R}_{\nu}) \exp[i\mathbf{k}(\mathbf{R}_{\nu} + \mathbf{u}_{\alpha})]\,, \tag{8}$$

where $\mathbf{R}_\nu$ runs over lattice vectors and $\mathbf{u}_\alpha$ is the basis vector of atom α in the unit cell. Between the two extremities of plane waves and atom-centered functions as candidates for a basis set definition in a workable method, large variety of combined methods has been developed and is in use. The need for combined methods is due to the fact that none of both "pure" choices is fully satisfactory. The practical difficulty is that the electron density in real materials behave very differently near atomic cores (large fluctuations with short spatial period) and far from the cores (smooth distribution of density with possibly elevated values on the chemical bonds, i.e. away from atoms). In discussing a compromise between computational efficiency and accuracy, let us sort out advantages and disadvantages of plane waves and atom-centered basis functions.

Plane waves

(+) Ultimately complete basis; systematic augmentation of accuracy is controlled by a single parameter (planewave cutoff);

(+) Easy analytical manipulation, e.g., when calculating matrix elements of different observables, derivatives of the total energy etc.

(−) Boundary conditions can be only periodic, therefore in the course of simulating finite fragments "in the box" the spurious interactions across the box boundary are built in, and

their suppression may demand for a large box size.
(−) The number of plane waves necessary to describe fluctuations of all-electron charge density is usually beyond the reasonable computational resources. The use of pseudopotentials is a typical solution, but this is a certain approximation.
(−) For a given cutoff (i.e., the largest wavevector used in the planewave expansion), the size of basis grows very rapidly with the size of simulation cell, irrespectively on whether it contains extra atoms or not. This is a serious handicap in a simulation of large systems, but also of open systems, like surfaces or molecules.

Atom-centered functions
(+) Since all charge is physically delivered by one-electron functions centered on atoms, the basis size scales linearly with the number of atoms, irrespectively of empty space in the system. This is especially important for simulations of open systems.
(+) Boundary conditions can be either periodic or strictly "vacuum-like", with no spurious interaction between repeated fragments.
(−) The lack of systematics in gradually enhancing the completeness of basis; additional basis functions are added *ad hoc*, and no asymptotic completeness of the basis is guaranteed.
(−) Difficulties in calculating matrix elements. This is probably the most serious drawback that can be overcome by the following tricks:
– if possible, calculate in advance and store in tables for subsequent fast interpolation (good, e.g., for dynamical simulations);
– use efficient statistical scheme rather than straightforward integration;
– use localized basis functions which allow analytical (or otherwise easy) integration (Gaussian basis sets).

3 Localized basis functions; Gaussian orbitals

Among "analytical" localized basis functions, important are so-called Slater-type and Gaussian orbitals. The Slater-type orbitals (STO) imitate the asymptotic of one-electron wavefunctions, known for the hydrogen atom:

$$\chi_{\mathrm{STO}}(\mathbf{r}) \sim r^{n-1} e^{-\zeta r}\, Y_l^m(\theta,\phi)\,. \tag{9}$$

The STOs form a complete basis of solutions for the central-field problem, while the textbook solutions for the hydrogen atom are only complete if taken together with the eigenfunctions corresponding to the continuous energy spectrum. n in STOs has the meaning of principal quantum number and ζ is "effective nuclear charge", but essentially it is an adjustable (free to vary) parameter. A peculiar feature of STOs is that, differently from the hydrogen-atom solutions, they have no node structure, but a combination of several STOs makes it possible to imitate the nodes – and hence fluctuations of charge density near the atomic core. One speaks of "minimal basis", "double-zeta", "triple-zeta" etc., the latter combining several exponents with different ζ values. This terminology is also transferred to the classification of Gaussian bases discussed below.

Gaussian-type orbitals (GTO) have been proposed by Boys in 1950[5]:

$$\chi_{\mathrm{GTO}} \sim r^{2(n-1)}\, e^{-\alpha r^2}\, Y_l^m(\theta,\phi)\,. \tag{10}$$

Their big advantage is the possibility to calculate the integrals (2- to 4-centered in the construction of Hamiltonian, as well as in many cases matrix elements of other operators of interest) analytically. In GTO, only functions with $l = n - 1$ are explicitly used, i.e. $1s$, $2p$, $3d$ etc. but not $2s$, $3p$. Thus, the node structure of one-electron wavefunctions is not included in any single GTO, but can be imitated by a linear combination of several basis functions. The drawback of gaussian bases is their slower convergency as compared to STO. This follows from the fact that individual GTOs are relatively poor representatives of true one-electron wavefunctions: GTOs have wrong asymptotic in the infinity (fall down too fast) and wrong behaviour near the nucleus. The performance can be improved by using several GTO (3 – 4) to imitate a single STO. If keeping all them independent, the bottleneck is the storage of many integrals. A possible solution is to use *fixed* linear combinations of GTO in a calculation rather than individual GTOs; such combinations are referred to as *contracted* GTOs. One can fit contracted GTOs either to STOs, or to numerical solutions for single atoms or ions.

Analytical evaluation of integrals appearing in Eq. (6) becomes possible due to *Gaussian product rule*, according to which a product of two Gaussians is again a Gaussian, centered at some point on the line connecting original centers. For s-Gaussian (without polynomial in $\mathbf{r}$) in one dimension the relation is straightforward:

$$\exp(-\alpha x_A^2)\,\exp(-\beta x_B^2) = \underbrace{\exp(-q\,Q_x^2)}_{\substack{\text{constant}\\ \equiv K_{AB}}} \times \underbrace{\exp(-p\,x_P^2)}_{\substack{\text{a new}\\ \text{Gaussian}}} \tag{11}$$

$$\text{with } p = \alpha + \beta\,, \quad q = \frac{\alpha\beta}{\alpha + \beta}\,, \quad pP_x = \alpha A_x + \beta B_x\,, \quad qQ_x = A_x - B_x\,.$$

Along y (or z)-direction, the product of two original exponents would yield $\exp[-(\alpha+\beta)y^2]$ with the same exponent prefactor p as in the x-dependence, therefore the above product formula is valid in 3-dimensional case:

$$\exp(-\alpha\mathbf{r}_A^2)\,\exp(-\beta\mathbf{r}_B^2) = K_{AB}\,\exp(-p\mathbf{r}_P^2)\,.$$

In the product of two non-s Gaussians that contain powers of $(\mathbf{r} - \mathbf{A})$ and $(\mathbf{r} - \mathbf{B})$ as well, the polynomial terms can be rearranged in another polynomial of $(\mathbf{r} - \mathbf{P})$. So a product of two arbitrary Gaussians yields

$$G_{ijk}(\mathbf{r}, \alpha, \mathbf{A})\,G_{i'j'k'}(\mathbf{r}, \beta, \mathbf{B}) = \sum_{(\alpha)} C^{(\alpha)} G_{(\alpha)}(\mathbf{r}, p, \mathbf{P})$$

where (α) incorporates different possible combinations of powers of x, y and z, up to the maximal power $i{+}j{+}k{+}i'{+}j'{+}k'$. After such decomposition, the overlap matrix elements 4 reduce to the sum of integrals

$$S_{qp} = \sum_{(\alpha)} C^{(\alpha)} \int d\,\Omega Y_l^m(\theta, \phi) \int_o^\infty r^{n(\alpha)} e^{-\gamma r^2} r^2 dr$$

that are taken using (with $\gamma > 0$, $n = 0, 1, \ldots$):

$$\int_O^\infty x^{2n} e^{-\gamma x^2} dx = \frac{(2n-1)!!}{2(2\gamma)^n}\sqrt{\frac{\pi}{\gamma}}\,; \qquad \int_O^\infty x^{2n+1} e^{-\gamma x^2} dx = \frac{n!}{2\gamma^{n+1}}\,.$$

Analyzing different terms in Eq. (6), one can sort them out into
• multipole moments, $\langle G_A|\, x_C^e\, y_C^f\, z_C^g\, |G_B\rangle$ with $G_A = x_A^i\, y_A^j\, z_A^k \exp(-\alpha r_A^2)$ etc. and integration done over coordinates of one electron. Such integrals describe, for example, the effect of the external field;
• integrals $\langle G_A|\, e^{ik_C x_C}\, |G_B\rangle$ that appear, e.g., in the calculation of exchange-correlation term, when the Fourier transformation of a general-form density is used. These two types of integrals can be analytically taken by parts and are available in closed form;

• kinetic energy. Here one can make use of recurrence relations describing the differentiation of Gaussians (which are also useful for calculating the forces on atoms),

$$\frac{\partial G_n}{\partial A_x} = -\frac{\partial G_n}{\partial x} = 2\alpha G_{n+1} - n\, G_{n-1}\,. \tag{12}$$

Therefore, the matrix elements of the kinetic energy operator can be expressed via overlap integrals between GTOs of reduced and increased order, up to ± 2.

The calculation of (up to) four-center Coulomb integrals,

$$\int \chi_q^*(\mathbf{r})\, \chi_p(\mathbf{r})\, \chi_{q'}^*(\mathbf{r}')\, \chi_{p'}(\mathbf{r}')\, \frac{d\mathbf{r}\, d\mathbf{r}'}{|\mathbf{r}-\mathbf{r}'|}\,.$$

can be (by applying the Gaussian product rule) gradually reduced to a Coulomb interaction between two Gaussian distributions of charge:

$$V_{\mathbf{AB}} = \int d\mathbf{r}_1 \int d\mathbf{r}_2 \frac{\rho_{\mathbf{A}}(\mathbf{r}_A)\, \rho_{\mathbf{B}}(\mathbf{r}_B)}{|\mathbf{r}_A - \mathbf{r}_B|}\,. \tag{13}$$

For both distributions of (normalized) s-Gaussian form,

$$\rho_{\mathbf{A}}(\mathbf{r}_A) = \left(\frac{\alpha}{\pi}\right)^{3/2} e^{-\alpha r_A^2}\,; \qquad \rho_{\mathbf{B}}(\mathbf{r}_B) = \left(\frac{\beta}{\pi}\right)^{3/2} e^{-\beta r_B^2}\,; \tag{14}$$

one arrives at a closed expression reducible to the error function:

$$V_{\mathbf{AB}} = \sqrt{\frac{4\alpha\beta}{\pi}}\, \frac{1}{\sqrt{\alpha+\beta}} \int_0^1 \exp\left(-\frac{\alpha\beta}{\alpha+\beta} R_{AB}^2 t^2\right) dt\,, \tag{15}$$

with $R_{AB} \equiv R_A - R_B$, and (more complicated) similar relations can be obtained for Gaussians including non-zero powers of Cartesian coordinates – see, e.g., Ref. [6] for the guidelines of derivation.

In spite of this mathematical "simplicity" of expanding "everything" (i.e., one-electron orbitals, one-particle density...) in Gaussians centered on atoms (and, probably, in additional interstitial positions), it often turns out desirable in practical calculation schemes to allow some amount of residual general-form charge density beyond the Gaussian expansions. This serves two purposes: considerably reduces the amount of multicenter integrals to keep trace of, and somehow compensates for an unavoidable incompleteness of the Gaussian basis set. A typical structure of such decomposition of density could be:

$$\begin{aligned} \rho(\mathbf{r}) &= \rho_{\text{loc}}(\mathbf{r}) + \rho_{\text{non-loc}}(\mathbf{r}) + \rho_{\text{nucl}}(\mathbf{r}) \\ &= [\rho_{\text{loc}}(\mathbf{r}) + \rho_{\text{nucl}}(\mathbf{r}) + \xi(\mathbf{r})] + [\rho_{\text{non-loc}}(\mathbf{r}) - \xi(\mathbf{r})]\,, \end{aligned}$$

where $\rho_{\text{nucl}}(\mathbf{r})$ is a well localized near nuclei contribution of deep core states, $\rho_{\text{loc}}(\mathbf{r})$ is a valence charge density near atoms, that can be typically expanded over spherical harmonics inside a sphere of appropriate radius around each site. $\rho_{\text{non-loc}}(\mathbf{r})$ is a smoothly varying charge density in the interstitial region. The separation density $\xi(\mathbf{r})$ is chosen in such way that each square bracket remains neutral and as smooth as possible. For $\xi(\mathbf{r})$, a dual representation is used: by spherical harmonic expansion around centers in the first bracket and e.g. by plane wave expansion in the second bracket. Then the contribution to the potential follows by multipole expansion from the first bracket and by Fourier transformation from the second bracket.

4 Planewave basis set

Another extremity in the choice of the basis set, namely fully delocalized planewaves, can be normalized to the volume of the simulation cell as follows:

$$\chi_{\mathbf{k}}^{\mathbf{G}} = \frac{1}{\sqrt{\Omega}} e^{i(\mathbf{k}+\mathbf{G})} .$$

Here, $\mathbf{k}$ stands for a point in the Brillouin zone labeling a certain irreducible representation of the translation group. In other words, this is an "artefact" of the Born – von Kármán periodic boundary conditions. Infinite number of cells in a crystal maps onto infinite number $\mathbf{k}$ points, effectively a continuous distribution, over the Brillouin zone. As we scan the Brillouin zone, we study the energy dispersion. As the repeated unit cell gets larger (in one dimension, as for slabs, or in all dimensions, as for a "molecule in a box"), the Brillouin zone shrinks correspondingly, and the dispersion effectively disappears – one gets instead a sequence of discrete energy levels. In this case, it may be sufficient to use just one $\mathbf{k}$-point, $\mathbf{k} = 0$.

For each $\mathbf{k}$ (the results for different $\mathbf{k}$ do not mix other than via the construction of density), $\mathbf{G}$ runs over plane waves actually used as basis functions. $\mathbf{G}$ labels the sites of the reciprocal lattice, usually within some cutoff, $|\mathbf{k}+\mathbf{G}| \leq \mathrm{G}_{\max}$. (The actual set of $\mathbf{G}$ can be therefore slightly different for different $\mathbf{k}$). Further on, $\mathbf{G}$ will number the matrix elements in the planewave representation, e.g.,

$$S_{\mathbf{G}\mathbf{G}'} = \frac{1}{\Omega} \int_{\Omega} d\mathbf{r}\, e^{i(\mathbf{G}'-\mathbf{G})\,\mathbf{r}} = \delta_{\mathbf{G}\mathbf{G}'} .$$

The big advantage of the planewave basis is the easiness with which some matrix elements can be calculated. The momentum operator and hence the kinetic energy are diagonal in the planewave basis, yielding

$$T_{\mathbf{G}\mathbf{G}'}^{\mathbf{k}} = \frac{\hbar^2}{2m} (\mathbf{G}+\mathbf{k})^2\, \delta_{\mathbf{G}\mathbf{G}'} .$$

The presence of non-constant potential disturbs the simple parabolic band structure, and eigenfunctions are not "pure" planewaves anymore, but their linear combinations. In order to calculate them by diagonalization, one needs the matrix elements of the Hamiltonian that read

$$
\begin{aligned}
H^{\mathbf{k}}_{\mathbf{G}\mathbf{G}'} &= \frac{\hbar^2}{2m}(\mathbf{k}+\mathbf{G})^2\,\delta_{\mathbf{G}\mathbf{G}'} + \frac{e}{\Omega}\int e^{i(\mathbf{G}'-\mathbf{G})\mathbf{r}}u(\mathbf{r})d\mathbf{r} + \\
&+ \frac{e^2}{\Omega^2}\sum_{\mathbf{G}_1\mathbf{G}_2} D^{\mathbf{k}}_{\mathbf{G}_1\mathbf{G}_2}\int\!\!\int \frac{\exp[i(\mathbf{G}'-\mathbf{G})\mathbf{r}]\,\exp[i(\mathbf{G}_2-\mathbf{G}_1)\mathbf{r}']}{|\mathbf{r}-\mathbf{r}'|}\,d\mathbf{r}\,d\mathbf{r}' + \\
&+ \frac{e}{\Omega}\int e^{i(\mathbf{G}'-\mathbf{G})\mathbf{r}}\,V_{\mathrm{XC}}(\mathbf{r})d\mathbf{r}\,, \qquad (16)
\end{aligned}
$$

where $D^{\mathbf{k}}_{\mathbf{G}_1\mathbf{G}_2}$ is the density matrix as introduced in Eq. (5). The Hartree term in the second line reduces to the Fourier transform of the Coulomb interaction,

$$
\underbrace{\int d\mathbf{r}e^{i(\mathbf{G}'-\mathbf{G}+\mathbf{G}_2-\mathbf{G}_1)\mathbf{r}}}_{\Omega\,\delta(\mathbf{G}'-\mathbf{G}+\mathbf{G}_2-\mathbf{G}_1)}\int d(\mathbf{r}'-\mathbf{r})\frac{\exp[i(\mathbf{G}_2-\mathbf{G}_1)(\mathbf{r}'-\mathbf{r})]}{|\mathbf{r}'-\mathbf{r}|}\,,
$$

whereas external potential and $V_{\mathrm{XC}}(\mathbf{r})$ contribute via their Fourier transforms:

$$
\langle\mathbf{G}|V_{\mathrm{XC}}|\mathbf{G}'\rangle = \sum_{\mathbf{Q}} V_{\mathrm{XC}}(\mathbf{Q})\,\delta(\mathbf{Q}+\mathbf{G}'-\mathbf{G})\,.
$$

Only these last terms are actually problematic in a planewave calculation, because they may be strongly localized in real space, so that their Fourier expansions do not converge up to a prohibitively large **G**-cutoff. This can be understood as a consequence of the fact that these terms provide a strong perturbation to the empty lattice case and hence mix the basis states far separated in energy. Or, a deep potential in the vicinity of atomic cores results in rapidly fluctuating density and hence high **G** cutoff for its proper description. Such problems did not arise in tight-binding-type schemes, where the basis functions could be tailored to account for such fluctuations.

The only possibility to save a purely planewave basis for practical calculations is to discard core states, considering them as "frozen", with the justification that they hopefully do not participate in the chemical bonding. This simplification gives rise to a family of pseudopotential methods, which are contrasted to "all-electron methods" (i.e., those where all electrons, valence and core, are treated on equal footing). Modern pseudopotentials in use are largely of *ab initio* type. They are "cooked" (with the use of certain approximations and criteria, to be discussed further on) from "true" (all-electron; in a sense that all electrons participate in the solution of the Kohn–Sham equations) solutions for free atoms or ions.

The construction of a pseudopotential typically starts with the choice of an appropriate reference configuration (e.g., Fe$3d^7 4s^1$) and "pseudoization radii" r_{C} that can be different for different l-channels. As a rule, the following conditions are imposed:

- the pseudofunction must have no nodes (in order to avoid wiggles that would demand for higher cutoff);
- the pseudofunction matches the all-electron one beyond the cutoff radius;
- norm conservation, meaning that the charge contained within the pseudoization radius is the same for the pseudofunction and the all-electron one. (This condition however can be relaxed; deviations from this rule give rise to ultrasoft pseudopotentials, to be specially discussed below);
- the eigenvalues corresponding to pseudofunctions must be equal to those of the all-electron

solution – at least for the reference configuration, and in the ideal case over a range of reasonable configurations, meaning the pseudopotential is *transferable*.

In practice, one typically starts from a smooth pseudofunction – there is a certain freedom in choosing it – and then *inverts* the radial Schrödinger equation to get the corresponding pseudopotential:

$$V_l^{\text{pseudo}}{}_{\text{screened}}(\mathbf{r}) = \varepsilon_l - \frac{l(l+1)}{r^2} + \frac{1}{2r R_l^{\text{PS}}(r)} \frac{d^2}{dr^2}\left[r\, R_l^{\text{PS}}(r)\right] . \quad (17)$$

This gives a *screened* pseudopotential, which incorporates the Hartree and exchange-correlation terms related to the pseudocharge only, i.e. to the charge distribution in a pseudo-wavefunction. When using the pseudopotential in the following, we would need to re-calculate these terms according to how the pseudo-WF evolves; so we subtract the corresponding terms and get an *unscreened* pseudopotential,

$$V_l^{\text{pseudo}}{}_{\text{unscreened}}(\mathbf{r}) = V_l^{\text{pseudo}}{}_{\text{screened}}(\mathbf{r}) - V_l^{\text{pseudo-H}}(\mathbf{r}) - V_l^{\text{pseudo-XC}}(\mathbf{r}) .$$

A substantial progress in the use of pseudopotential methods was achieved along the line of implementing the following priorities:

- the PP should be as soft as possible;
- it should be as transferable as possible (i.e., the PP generated for a given atomic configuration should reproduce others accurately), that is essential for applications dealing with simulations of chemical bonding;
- The pseudo-charge density should reproduce the "true" valence charge density as accurately as possible.

These goals are essentially conflicting, for the reasons to be discussed below. In the search for a compromise, the concept of **norm conservation** remained for a long time generally adopted. The condition of norm-conservation reads

$$\int_0^{r_c} r^2\, dr\, \left|\varphi_v^{\text{PS}}(\mathbf{r})\right|^2 = \int_0^{r_c} r^2\, dr\, \left|\varphi_v^{\text{all-electron}}(\mathbf{r})\right|^2 . \quad (18)$$

that must hold for all valence pseudofunctions. Since all-electron (single-atom) wavefunctions and eigenvalues ε_v are different for different angular momentum values l, so would be cutoff radii and pseudopotentials. Such (l-dependent) pseudopotentials are called *semi-local*.

It is good for the transferability of pseudopotential if the slope of the pseudofunction equals that of the true wavefunction at the pseudoization radius. Taken together with the norm-conservation, this formulates the condition on the logarithmic derivatives:

$$\frac{1}{\varphi^{\text{PS}}(r_c, E)} \left.\frac{d\,\varphi^{\text{PS}}(r_c, E)}{d\,r}\right|_{r=r_c} = \frac{1}{\varphi^{\text{all-el.}}(r_c, E)} \left.\frac{d\,\varphi^{\text{all-el.}}(r_c, E)}{d\,r}\right|_{r=r_c} . \quad (19)$$

The larger the energy "window" within which the logarithmic derivatives match approximately, the better the transferability.

In 1990, Vanderbilt[7] suggested to abandon the norm-conservation condition (18), that would allow to make the pseudoization radius r_c large – essentially, limited only by the condition that the spheres of this radius centered on different atoms must not overlap in a simulation. A big advantage would be that pseudopotentials generated with larger r_c are much softer

and hence much lower planewave cutoff would be needed. There are also disadvantages:
• the formalism becomes less transparent and more mathematically involved, because the functions get a non-trivial overlap; charge density is not $\sum|\varphi^{\rm PS}(\mathbf{r})|^2$ anymore but includes contributions from "augmentation functions";
• a pseudopotential becomes less transferable.

The latter disadvantage is not very serious because one can typically afford generating the pseudopotential anew several times, as the calculation converges and the final charge configuration becomes apparent. The augmentation functions are non-zero only inside the pseudoization radius associated with each atom, but they may turn out to be strongly localized there, therefore the planewave expansion of the augmentation density (necessary, e.g., for solving the Poisson equation by fast Fourier transform) may have quite high energy cutoff. The success of the ultrasoft PP scheme is due to the possibility to search, by varying the tolerated deviation from norm-conservation, for an optimal balance between computational loads related, on one side, to the planewave expansions of the pseudo-wave functions and, on the other side, to the additional treatment of somehow localized augmentation functions.

In this sense, the ultrasoft PP method is an example of an approach attempting to combine the advantages of both spatially localized and fully delocalized basis constructions.

5 Dual representation of the wavefunction; LAPW and LMTO methods

Essentially a very close idea, to use different description of one-electron orbitals and the charge density constructed from them, near the nuclei and away from them, was put forward by Slater in his formulation of the augmented planewave method as early as in 1937. Although his original *muffin-tin approximation* (spherical symmetric potential in the non-overlapping atomic spheres, constant potential in the interstitial) is hardly in use anymore, the *muffin-tin geometry* (central expansion of electron orbitals and potential over spherical harmonics inside each atomic sphere of, say, radius S; general-shape but smoothly varying functions in the interstitial region I) is used in state-of-art high-precision electronic structure codes. The basic idea is the dual representation for the wavefunction

$$\psi_{\mathbf{k}}(\mathbf{r}) = \begin{cases} \frac{1}{\sqrt{\Omega}}\sum\limits_{\mathbf{G}} C_{\mathbf{G}} e^{i(\mathbf{k}+\mathbf{G})\mathbf{r}}\,, & \mathbf{r}\in \mathrm{I} \\ \sum\limits_{lm} A_{lm}^{\alpha\mathbf{k}} u_l^{\alpha}(r) Y_l^m(\hat{\mathbf{r}})\,, & r<S \end{cases} \tag{20}$$

$u_l(r)$ is a regular (at the origin) solution of radial Schrödinger equation, that with a spherically symmetric potential yields:

$$\left[-\frac{d^2}{dr^2}+\frac{l(l+1)}{r^2}+V(r)-\varepsilon_l\right] r u_l(r) = 0\,. \tag{21}$$

ε_l is a parameter, variable in order to achieve the continuity of the wavefunction at the sphere boundary. The continuity of the dual representation (20) is necessary in order to have well defined kinetic energy. The matching condition, relating $C_{\rm G}$ and A_{lm}, follows from the planewave expansion

$$e^{i\mathbf{K}\mathbf{r}} = 4\pi e^{i\mathbf{K}\mathbf{R}_\alpha}\sum_{lm} i^l j_l\left(K|\mathbf{r}-\mathbf{R}_\alpha|\right) Y_l^{m*}(\hat{\mathbf{K}}) Y_l^m(\mathbf{r}\hat{-}\mathbf{R}_\alpha) \tag{22}$$

around an arbitrary cite $\mathbf{R}_\alpha$. One can achieve an augmentation of any planewave with $\mathbf{K} = \mathbf{k} + \mathbf{G}$ to numerical solutions inside each sphere, taken at *a priori* chosen energy parameter ε_l. However, the augmented function constructed in such way cannot be used as a fixed basis function χ_q of Eq. 3, because the resulting eigenvalue ε_α would not necessarily coincide with ε_l. Instead, one has to scan the values of the parameter ε that fulfills the condition

$$\|\langle \mathbf{G}_1, \varepsilon | \mathcal{H} - \varepsilon \mathcal{S} | \mathbf{G}_2, \varepsilon \rangle\| \equiv D(\varepsilon) = 0\,, \tag{23}$$

where $|\mathbf{G}\varepsilon\rangle$ stands for a plane wave $\mathbf{G}$ augmented to the (regular at the origin) radial solution inside each atomic sphere for energy ε. Eq. 23 is not a single diagonalization problem; it is typically solved by tracing zeros of the determinant.

A related calculation scheme, that is however derived from quite different starting point, is the **Korringa–Kohn–Rostoker** (KKR) method. It uses the fact that in the interstitial, i.e. in the region with constant potential V_0, the Schrödinger equation degenerates into the Helmholz equation

$$\left(-\frac{\hbar^2}{2m}\nabla^2 + V_0 - \epsilon\right)\psi = 0$$

with a solution

$$\psi = \exp\left[i\sqrt{\frac{2m}{\hbar^2}(\varepsilon - V_0)}\, r\right] \equiv e^{i\kappa r}$$

that can be expanded in spherical harmonics with $j_l(\kappa^2, r)$ – regular at the origin and $n_l(\kappa^2, r)$ – singular at the origin radial functions. Such functions to be used in the following are, up to a normalization, spherical Bessel and Neumann functions:

$$j_l(\kappa^2, r) \equiv j_l(\kappa r)\frac{(2l-1)!!}{2(\kappa S)^l}\,; \qquad n_l(\kappa^2, r) \equiv -n_l(\kappa r)\frac{(\kappa S)^{l+1}}{(2l-1)!!}\,. \tag{24}$$

These plane waves are *scattered* on individual atomic potentials, that gives rise to phase shifts η. The phase shift shows in which relation regular and singular solutions are present in the scattered wave. In the most direct way the relation between J_l and n_l is accounted for by the *potential function* of scattering problem $P(\epsilon, \kappa^2)$. In terms of it, the wavefunction can be expanded over spherical harmonics inside and outside the sphere with scattering potential in the following way:

$$\psi(\varepsilon, \mathbf{r}) = \sum_{lm} Y_l^m(\hat{\mathbf{r}}) \times \begin{cases} u_l(\varepsilon, r)\,, & r < S \\ n_l(\kappa^2, r) - P_l(\varepsilon, \kappa^2)\, j_l(\kappa^2, r) & r > S \end{cases} \tag{25}$$

The potential function can be obtained from the condition that both the radial wavefunction $u_l(\varepsilon, r)$ and its first derivative match a combination of $n_l(\kappa^2, r)$ and $j_l(\kappa^2, r)$:

$$P(\varepsilon, \kappa^2) = \frac{\{u_l(\varepsilon), n_l(\kappa^2)\}}{\{u_l(\varepsilon), j_l(\kappa^2)\}}\,, \tag{26}$$

where the Wronskian of two radial functions, taken at the sphere boundary, is defined as

$$\{f, g\} \equiv S^2[f(S)g'(S) - f'(S)g(S)]$$

that gives specifically for the functions (24)

$$\{n_l, j_l\} = \frac{S}{2}\,.$$

The so calculated potential function relates to the phase shift η_l in the following way:

$$P_l(\varepsilon,\kappa^2) = -\cot\eta_l(\varepsilon,\kappa)\frac{2(S\kappa)^{2l+1}}{[(2l-1)!!]^2}\,. \tag{27}$$

Indeed, for the case of no scattering (the potential equal to V_0 everywhere inside the sphere) $u_l = j_l$, $P(\varepsilon,\kappa^2) = \infty$ and $\eta = 0$. It follows from Eq. (27) that $P_l(\varepsilon,\kappa^2)$ is an always increasing function of ε. Adding $P_l(\varepsilon,\kappa^2)j_l(\kappa^2,r)$ to the wavefunction representation (25) results in *muffin-tin orbital* (MTO) χ_l:

$$\chi_{lm}(\varepsilon,\kappa^2,\mathbf{r}) = Y_l^m(\hat{\mathbf{r}}) \times \begin{cases} u_l(\varepsilon,r) + P_l(\varepsilon,\kappa^2)j_l(\kappa^2,r)\,, & r < S \\ n_l(\kappa^2,r)\,. & r > S \end{cases} \tag{28}$$

One could in principle fix ε and use muffin-tin orbitals as localized basis functions, but the quality of this basis would be quite bad for any general ε, for the same reasons as discussed above for augmented plane waves. Here also, instead of performing a diagonalization one can, as a more practical scheme, scan the integral of energies of interest and search for those values of ε at which the matching conditions are satisfied at all sphere boundaries simultaneously. This is achieved by taking into account the expansion of non-regular solution, taken at any center $\mathbf{R}_\alpha$, around any other center $\mathbf{R}_\beta$, that has the following form:

$$\begin{aligned} n_{lm}(\kappa^2,|\mathbf{r}-\mathbf{R}_\alpha|) &= \sum_{l'm'} j_{l'm'}(\kappa^2,|\mathbf{r}-\mathbf{R}_\beta|) \times \\ &\times \sum_{l''} 4\pi\, C_{l\;l'\;l''}^{m\,m'\,m''}\, i^{-l+l'-l''}\, \frac{2(\kappa R_\alpha)^{l'+l-l''}(2l''-1)!!}{(2l'-1)!!\,(2l-1)!!}\, n^*_{l''(m'-m)}(\kappa^2,|\mathbf{R}_\alpha-\mathbf{R}_\beta|) \\ &\equiv \sum_{l'm'} j_{l'm'}(\kappa^2,|\mathbf{r}-\mathbf{R}_\beta|)\, S_{\mathbf{R}_\beta l'm',\mathbf{R}_\alpha lm}(\kappa^2)\,. \end{aligned} \tag{29}$$

The condition that a linear combination of MTOs is the solution in all space reads

$$\sum_{\mathbf{R}lm}\left[P_{\mathbf{R}'l'}(\varepsilon,\kappa^2)\,\delta_{\mathbf{R}\mathbf{R}'}\,\delta_{ll'}\,\delta_{mm'} - S_{\mathbf{R}'l'm',\mathbf{R}lm}(\kappa^2)\right]C_{\mathbf{R}lm} = 0\,. \tag{30}$$

The zeros of the determinant of this equation

$$||\mathbf{P}(\varepsilon,\kappa^2) - \mathbf{S}(\kappa^2)|| = 0$$

determine the solutions of the scattering problem.

Comparing the performance of the KKR method with the augmented plane wave scheme one can note the following advantages:

• a much more compact basis set can be typically used. Whereas a high **G**-cutoff in the augmented plane wave method may typically result in ~ 1000 plane waves per atom, the size of the basis set in the KKR is essentially limited by maximal l in the expansion of structure constants. The latter converge quite fast, with $l_{\text{max.}} \approx 2-3$ being typically acceptable, resulting

in $(l+1)^2 \sim 16$ functions per atom (for the description of valence states);
• radial solutions and functions in the interstitial are allowed to match *along with their derivatives.*
• The structure constants do not depend on scattering potentials and they are scaled in a simple way as the lattice constant changes; therefore they do not need to be recalculated in the course of electronic-structure iterations and can be stored in advance (as functions of κ^2) for any given crystal structure.

Still the problem preventing to use a single diagonalization in order to obtain all eigenvalues at once is insufficient variational freedom of numerical solutions inside the spheres, taken at certain fixed ε. This can be helped by *linearization*, due to an observation by Andersen[8] that $u_l(\varepsilon, r)$ as functions of *energy* is with good accuracy linear:

$$u_l(\varepsilon, r) \approx u_l(\varepsilon_\nu, r) + (\varepsilon - \varepsilon_\nu)\dot{u}_l(\varepsilon_\nu, r)\,. \tag{31}$$

The linearization of augmented plane waves leads to the LAPW method, the linearization of scattering-wave formalism – to the LMTO method. The advantage of the former is a relatively easy generalization over non-spherical potentials, the advantage of the latter – compact and highly efficient basis; the inclusion of potentials of general form is also possible but somehow more mathematically involved.

Differentiating Eq. (21) in energy yields:

$$\left[-\frac{d^2}{dr^2} + \frac{l(l+1)}{r^2} + V(r) - \varepsilon_l\right] r\,\dot{u}_l(r, \varepsilon_l) = r\,u_l(r, \varepsilon_l)\,, \tag{32}$$

with

$$\dot{u}_l \equiv \frac{d}{d\varepsilon} u_l(r, \varepsilon)\,.$$

In the LAPW formalism, both functions u_l and $\dot{u}_l$ are used to match the plane wave on the sphere boundary, whereby the wavefunction expansion (20) takes a form:

$$\psi_{\mathbf{k}}(\mathbf{r}) = \begin{cases} \frac{1}{\sqrt{\Omega}} \sum\limits_{\mathbf{G}} C_{\mathbf{G}} e^{i(\mathbf{k}+\mathbf{G})\mathbf{r}}\,, & \mathbf{r} \in \mathrm{I} \\ \sum\limits_{lm} \left[A^{\mathbf{k}}_{lm} u_l(r) + B^{\mathbf{k}}_{lm} \dot{u}_l(r)\right] Y_l^m(\hat{\mathbf{r}})\,, & r < S\,. \end{cases} \tag{33}$$

This combination now has sufficient variational freedom: if band energy deviates from an in advance fixed value ε_ν, the shape of the wavefunction inside the sphere can still be recovered using the approximation (31). One can then use as basis functions $\psi^{\mathbf{G}}(\mathbf{r})$, corresponding to different $\mathbf{G}$ in the dual representation (33), with the coefficients A_{lm} and B_{lm} obtained from matching conditions there. The basis functions are then *energy independent*, and the electronic-structure problem can be solved by matrix diagonalization, as in any method with fixed basis. The error in the band energy due to linearization is $\sim (\varepsilon - \varepsilon_\nu)^4$. Typically, it is negligible within an energy window of 1–1.5 Ry.

With this, the secular equation becomes:

$$\left[H^{\mathbf{k}}_{\mathbf{GG}'} - \varepsilon\, S_{\mathbf{GG}'}\right] C^{\mathbf{k}}_{\mathbf{G}'} = 0\,. \tag{34}$$

The expression for the overlap matrix

$$S_{\mathbf{GG}'} = \frac{1}{\Omega} \int d\mathbf{r} e^{i(\mathbf{G}-\mathbf{G}')\mathbf{r}} \Theta(\mathbf{r}) + \sum_{\mathbf{R}_\alpha} S_\alpha(\mathbf{G}, \mathbf{G}') \tag{35}$$

contains a step function, $\Theta(\mathbf{r})$, that is $= 0$ inside any muffin-tin sphere, and $S_\alpha(\mathbf{G},\mathbf{G}')$, contributions to overlap from a sphere centered at $\mathbf{R}_\alpha$. The Hamiltonian matrix

$$\begin{aligned} H^{\mathbf{k}}_{\mathbf{GG}'} &= \frac{1}{\Omega}\int d\mathbf{r}e^{-i(\mathbf{G}+\mathbf{k})\mathbf{r}}\,\Theta(\mathbf{r})\left[\hat{T}+\hat{V}_{\rm PW}\right]e^{i(\mathbf{G}'+\mathbf{k})\mathbf{r}} + \\ &+\sum_{\mathbf{R}_\alpha}\left[H_\alpha(\mathbf{G},\mathbf{G}')+V^{\rm NS}_\alpha(\mathbf{G},\mathbf{G}')\right] \end{aligned} \qquad (36)$$

has $H_\alpha(\mathbf{G},\mathbf{G}')$ – spherical and $V^{\rm NS}_\alpha(\mathbf{G},\mathbf{G}')$ – non-spherical (i.e., from $l \neq 0$) contributions from muffin-tin sphere at R_α. The details of practical implementations of full-potential LAPW method are well described in a book by D. Singh.[9]

The full-potential LAPW method arguably provides superior accuracy, as compared to other practical implementations of the density functional theory, in evaluation of band energies, total energies, forces and other properties derivable from the band structure. Its results are sometimes referred to as "DFT-truth". The reason for such exceptional standing is the practical lack of other approximations than the linearization 31, whose error in determining the eigenvalues is usually quite tolerable. The ambiguity in the choice of exchange-correlation potential is shared by all methods, and it is usually decoupled from the algorithm to solve the Kohn-Sham equations as implemented in any computer code in question. Apart from this, the augmentation of the basis size and hence the convergency of results is controlled by a few cutoff parameters and can be achieved for any system of interest, if necessary computational resources are available. The price paid for this comfortable feature is usually a large size of matrices to be diagonalized, and the increase in the number of basis functions with the size of the unit cell rather than with the number of atoms. The LMTO method that operates with more compact atom-centered basis set may be more favourable in the latter aspects, but it sometimes faces problems in describing the charge density distributed over the unit cell with sufficient accuracy. Most of the practically working full-potential LMTO schemes use finite-grid representation of (at least part of) the charge density and solves the Poisson equation by fast Fourier transform.

References

[1] N.M. Harrison. This volume, p. 45

[2] I.V. Abarenkov. This volume, p. 71

[3] C. C. J. Roothaan. New Developments in Molecular Orbital Theory, *Rev. Mod. Phys.* **23**, 69 (1951).

[4] Peter Pulay. Analytical Derivative Techniques and the Calculation of Vibrational Spectra, in: *Modern Electronic Structure Theory, Part II*, ed. by David R. Yarkony, World Scientific, Singapore (1995), pp. 1191–1240.

[5] S. F. Boys. Electronic wave functions. I. A general method of calculation for the stationary states of any molecular system, *Proc. Roy. Soc. (London)* **A200**, 542 (1950).

[6] T. Helgaker and P. R. Taylor. "Caussian Basis Sets and Molecular Integrals", in: "Modern Electronic Structure Theory", Ed. D. R. Yarkony, World Scientific, Singapore 1995.

[7] David Vanderbilt. Soft self-consistent pseudopotentials in a generalized eigenvalue formalism, Phys. Rev. B **41**, 7892–7895 (1990).

[8] O. Krogh Andersen. Linear methods in band theory, *Phys. Rev. B* **12**, 3060–3083 (1975).

[9] D. J. Singh. *Planewaves, pseudopotentials and the LAPW method* (Kluwer Academic Publishers, Boston, 1994).

Computational Materials Science
C.R.A. Catlow and E.A. Kotomin (Eds.
IOS Press, 2003

An Introduction to Density Functional Theory

N. M. Harrison

Department of Chemistry, Imperial College of Science Technology and Medicine, SW7 2AY, London and

CLRC, Daresbury Laboratory, Daresbury, Warrington, WA4 4AD

For the past 30 years density functional theory has been the dominant method for the quantum mechanical simulation of periodic systems. In recent years it has also been adopted by quantum chemists and is now very widely used for the simulation of energy surfaces in molecules. In this lecture we introduce the basic concepts underlying density functional theory and outline the features that have lead to its wide spread adoption. Recent developments in exchange correlation functionals are introduced and the performance of families of functionals reviewed.

The lecture is intended for a researcher with little or no experience of quantum mechanical simulations but with a basic (undergraduate) knowledge of quantum mechanics. We hope to provide sufficient background to enable informed judgements on the applicability of a particular implementation of density functional theory to a specific problem in materials simulation.

For those who wish to go more deeply into the formalism of density functional theory there are a number of reviews and books aimed at intermediate and advanced levels available in the literature [1,2,3]. Where appropriate source articles are referred to in the text.

1. The Solution of the Schrödinger Equation

During the course of this lecture we will be primarily concerned with the calculation of the ground state energy of a collection of atoms. The energy may be computed by solution of the Schrödinger equation – which, in the time independent, non-relativistic, Born-Oppenheimer approximation is[1];

$$\hat{H}\Psi(\mathbf{r}_1,\mathbf{r}_2,\ldots,\mathbf{r}_N) = E\Psi(\mathbf{r}_1,\mathbf{r}_2,\ldots,\mathbf{r}_N)$$

Equation 1

[1] Atomic units are used throughout.

The Hamiltonian operator, H, consists of a sum of three terms; the kinetic energy, the interaction with the external potential (V_{ext}) and the electron-electron interaction (V_{ee}). That is;

$$\hat{H} = -\frac{1}{2}\sum_{i}^{N}\nabla_i^2 + \hat{V}_{ext} + \sum_{i<j}^{N}\frac{1}{|\mathbf{r}_i - \mathbf{r}_j|}$$

Equation 2

In materials simulation the external potential of interest is simply the interaction of the electrons with the atomic nuclei;

$$\hat{V}_{ext} = -\sum_{\alpha}^{N_{at}}\frac{Z_\alpha}{|\mathbf{r}_i - \mathbf{R}_\alpha|}$$

Equation 3

Here, $\mathbf{r}_i$ is the coordinate of electron *i* and the charge on the nucleus at $\mathbf{R}_\alpha$ is Z_α. Note that in order to simplify the notation and to focus the discussion on the main features of DFT the spin coordinate is omitted here and throughout this article. Equation 1 is solved for a set of Ψ subject to the constraint that that the Ψ are anti-symmetric – they change sign if the coordinates of any two electrons are interchanged. The lowest energy eigenvalue, E_0, is the ground state energy and the probability density of finding an electron with any particular set of coordinates $\{\mathbf{r}_i\}$ is $|\Psi_0|^2$.

The average total energy for a state specified by a particular Ψ, not necessarily one of the eigenfunctions of Equation 1, is the expectation value of H, that is;

$$E[\Psi] = \int \Psi^* \hat{H} \Psi d\mathbf{r} \equiv \left\langle \Psi \middle| \hat{H} \middle| \Psi \right\rangle$$

Equation 4

The notation $[\Psi]$ emphasises the fact that the energy is a *functional* of the wavefunction. The energy is higher than that of the ground state unless Ψ corresponds to Ψ_0 – which is the *variational theorem*;

$$E[\Psi] \geq E_0$$

Equation 5

The ground state wavefunction and energy may be found by searching all possible wavefunctions for the one that minimises the total energy. Hartree-Fock theory consists of an ansatz for the structure of Ψ - it is assumed to be an antisymmetric product of functions (ϕ_i) each of which depends in the coordinates of a single electron, that is;

$$\Psi_{HF} = \frac{1}{\sqrt{N!}} \det\left[\phi_1 \phi_2 \phi_3 \ldots \phi_N\right]$$

Equation 6

where, det indicates a matrix determinant [4]. Substitution of this ansatz for Ψ into the Schrödinger equation results in an expression for the Hartree Fock energy;

$$\begin{aligned} E_{HF} = & \int \phi_i^*(\mathbf{r}) \left(-\frac{1}{2} \sum_i^N \nabla_i^2 + V_{ext} \right) \phi_i(\mathbf{r}) d\mathbf{r} \\ & + \frac{1}{2} \sum_{i,j}^N \int \frac{\phi_i^*(\mathbf{r}_1) \phi_i(\mathbf{r}_1) \phi_j^*(\mathbf{r}_2) \phi_j(\mathbf{r}_2)}{|\mathbf{r}_i - \mathbf{r}_j|} d\mathbf{r}_1 d\mathbf{r}_2 \\ & - \frac{1}{2} \sum_{i,j}^N \int \frac{\phi_i^*(\mathbf{r}_1) \phi_j(\mathbf{r}_1) \phi_i(\mathbf{r}_2) \phi_j^*(\mathbf{r}_2)}{|\mathbf{r}_i - \mathbf{r}_j|} d\mathbf{r}_1 d\mathbf{r}_2 \end{aligned}$$

Equation 7

The second term is simply the classical Coulomb energy written in terms of the orbitals and the third term is the *exchange* energy. The ground state orbitals are determined by applying the variation theorem to this energy expression under the constraint that the orbitals are orthonormal. This leads to the Hartree-Fock (or SCF) equations;

$$\left[-\frac{1}{2} \nabla^2 + v_{ext}(\mathbf{r}) + \int \frac{\rho(\mathbf{r}')}{|\mathbf{r} - \mathbf{r}'|} d\mathbf{r}' \right] \phi_i(\mathbf{r}) + \int v_X(\mathbf{r}, \mathbf{r}') \phi_i(\mathbf{r}') d\mathbf{r}' = \varepsilon_i \phi_i(\mathbf{r})$$

Equation 8

Where the non-local exchange potential, v_X, is such that:

$$\int v_X(\mathbf{r},\mathbf{r}')\phi_i(\mathbf{r}')d\mathbf{r}' = -\sum_j^N \int \frac{\phi_j(\mathbf{r})\phi_j^*(\mathbf{r}')}{|\mathbf{r}-\mathbf{r}'|}\phi_i(\mathbf{r}')\,d\mathbf{r}'$$

Equation 9

The Hartree-Fock equations describe non-interacting electrons under the influence of a mean field potential consisting of the classical Coulomb potential and a *non-local* exchange potential.

From this starting point better approximations (correlated methods) for Ψ and E_0 are readily obtained but the computational cost of such improvements is very high and scales prohibitively quickly with the number of electrons treated (for an excellent introduction see ref. [4]). In addition, accurate solutions require a very flexible description of the wavefunction's spatial variation, i.e. a large and basis set is required which also adds to the expense for practical calculations. Many correlated methods have been developed for molecular calculations [4]. The cost of the most commonly used methods, MP2, MP3, MP4, CISD, CCSD, CCSD(T) formally scales with the number of electrons raised to the power of 5,6,7,6,6,7 respectively. In most cases CCSD(T) calculations are of sufficient accuracy to determine the chemical properties of systems to sufficient accuracy to predict chemical properties (stability, reaction rates …) however, due to the computational expense the routine application of such methods to realistic models of systems of interest is not practical and not likely to become so despite rapid advances in computer technology. In this context we mention recent advances in the solution of the Schrödinger equation using variational quantum Monte Carlo approach [5].

The discussion above has established that direct solution of the Schrödinger equation is not currently feasible for systems of interest in condensed matter science – this is a major motivation for the development and use of density functional theory.

The question that arises is – *Is it necessary to solve the Schrödinger equation and determine the 3N dimensional wavefunction in order to compute the ground state energy ?*

2. Avoiding the Solution of the Schrödinger Equation

The Hamiltonian operator (Equation 2) consists of single electron and bi-electronic interactions – i.e. operators that involve on the coordinates of one or two electrons only. In order to compute the total energy we do not need to know the 3N dimensional wavefunction. Knowledge of the two-particle probability density – that is, the probability of finding an electron at $\mathbf{r}_1$ and an electron at $\mathbf{r}_2$ is sufficient.

A quantity of great use in analysing the energy expression is the second order density matrix, which is defined as:

$$P_2(\mathbf{r}_1',\mathbf{r}_2';\mathbf{r}_1,\mathbf{r}_2)=\frac{N(N-1)}{2}\int\Psi^*(\mathbf{r}_1',\mathbf{r}_1',...,\mathbf{r}_N')\Psi(\mathbf{r}_1,\mathbf{r}_2,...,\mathbf{r}_N)d\mathbf{r}_3d\mathbf{r}_4...d\mathbf{r}_N$$

Equation 10

The diagonal elements of P_2, often referred to as the *two-particle density matrix* or *pair density*, are;

$$P_2(\mathbf{r}_1,\mathbf{r}_2)=P_2(\mathbf{r}_1,\mathbf{r}_2;\mathbf{r}_1,\mathbf{r}_2)$$

This is the required two electron probability function and completely determines all two particle operators. The first order density matrix is defined in a similar manner and may be written in terms of P_2 as;

$$P_1(\mathbf{r}_1';\mathbf{r}_1)=\frac{2}{N-1}\int P_2(\mathbf{r}_1',\mathbf{r}_2;\mathbf{r}_1,\mathbf{r}_2)d\mathbf{r}_2$$

Equation 11

Given P_1 and P_2 the total energy is determined exactly;

$$E=tr\left(\hat{H}\hat{P}\right)=\int\left[\left(-\frac{1}{2}\nabla_1^2-\sum_{\alpha}^{N_{at}}\frac{Z_\alpha}{|\mathbf{r}_1-\mathbf{R}_\alpha|}\right)P_1(\mathbf{r}_1',\mathbf{r}_1)\right]_{\mathbf{r}_1=\mathbf{r}_1'}d\mathbf{r}_1+\int\frac{1}{|\mathbf{r}_1-\mathbf{r}_2|}P_2(\mathbf{r}_1,\mathbf{r}_2)d\mathbf{r}_1d\mathbf{r}_2$$

Equation 12

We conclude that the diagonal elements of the first and second order density matrices completely determine the total energy. This appears to vastly simplify the task in hand. The solution of the full Schrödinger equation for Ψ is not required – it is sufficient to determine P_1 and P_2 - and the problem in a space of 3N coordinates has been reduced to a problem in a 6 dimensional space.

Approaches based on the direct minimisation of $E(P_1,P_2)$ suffer from the specific problem of ensuring that the density matrices are legal – that is, they must be constructible from an antisymmetric Ψ. Imposing this constraint is non trivial and is currently an unsolved problem [6,7]. In view of this we conclude that Equation 12 does not lead immediately to a reliable method for computing the total energy without calculating the many body wavefunction.

The observation which underpins density functional theory is that we do not even require P_2 to find E – *the ground state energy is completely determined by the diagonal elements of the first order density matrix – the charge density.*

3. The Hohenburg-Kohn Theorems

In 1964 Hohenburg and Kohn proved the two theorems [8]. The first theorem may be stated as follows;

The electron density determines the external potential (to within an additive constant).

If this statement is true then it immediately follows that the *electron density uniquely determines the Hamiltonian operator* (Equation 2). This follows as the Hamiltonian is specified by the external potential and the total number of electrons, N, which can be computed from the density simply by integration over all space. Thus, in principle, given the charge density, the Hamiltonian operator could be uniquely determined and this the wave functions Ψ (of all states) and all material properties computed.

Hohenburg and Kohn [8] gave a straightforward proof of this theorem, which was generalised to include systems with degenerate states in proof given by Levy in 1979 [9]. It is said that the theoretical spectroscopist E. B. Wilson put forward a very straightforward proof of this theorem during a meeting in 1965 at which it was being introduced. Wilson's observation is that the electron density uniquely determines the positions and charges of the nuclei and thus trivially determines the Hamiltonian. This proof is both transparent and elegant – it is based on the fact that the electron density has a cusp at the nucleus, such that;

$$Z_\alpha = \frac{-1}{2\bar{\rho}(0)}\left[\frac{\partial\bar{\rho}(r_\alpha)}{\partial r_\alpha}\right]_{r_\alpha=0}$$

where $\bar{\rho}(\mathbf{r})$ is the spherical average of ρ and so a sufficiently careful examination of the charge density uniquely determines the external potential and thus the Hamiltonian.

Although less general than the Levy proof this observation establishes the theorem for the case of interest – electrons interacting with nuclei. The first theorem may be summarised by saying that the energy is a functional of the density – E[ρ].

The second theorem establishes a variational principle;

For any positive definite trial density, ρ_t, such that $\int \rho_t(\mathbf{r})d\mathbf{r} = N$ *then* $E[\rho_t] \geq E_0$

The proof of this theorem is straightforward. From the first theorem we know that the trial density determines a unique trial Hamiltonian (H_t) and thus wavefunction (Ψ_t); $E[\rho_t] = \langle \Psi_t \mid H \mid \Psi_t \rangle \geq E_0$ follows immediately from the variational theorem of the Schrödinger equation (Equation 5). This theorem restricts density functional theory to studies of the ground state. A slight extension allows variation to excited states that can be guaranteed orthogonal to the ground state but in order to achieve this knowledge of the exact ground state wavefunction is required.

The two theorems lead to the fundamental statement of density functional theory;

$$\delta\left[E[\rho] - \mu\left(\int \rho(\mathbf{r})d\mathbf{r} - N\right)\right] = 0$$

Equation 13

The ground state energy and density correspond to the minimum of some functional $E[\rho]$ subject to the constraint that the density contains the correct number of electrons. The Lagrange multiplier of this constraint is the electronic *chemical potential* μ.

The above discussion establishes the remarkable fact that there is a *universal* functional $E[\rho]$ (i.e. it does not depend on the external potential which represents the particular system of interest) which, if we knew its form, could be inserted into the above equation and minimised to obtain the *exact* ground state density and energy.

4. The Energy Functional

From the form of the Schrödinger equation (Equation 1) we can see that the energy functional contains three terms – the kinetic energy, the interaction with the external potential and the electron-electron interaction and so we may write the functional as;

$$E[\rho] = T[\rho] + V_{ext}[\rho] + V_{ee}[\rho]$$

The interaction with the external potential is trivial;

$$V_{ext}[\rho] = \int \hat{V}_{ext}\rho(\mathbf{r})d\mathbf{r}$$

The kinetic and electron-electron functionals are unknown. If good approximations to these functionals could be found direct minimisation of the energy would be possible; this possibility is the subject of much current research - see for instance Ref. [10].

Kohn and Sham proposed the following approach to approximating the kinetic and electron-electron functionals [11]. They introduced a fictitious system of N non-interacting electrons to be described by a single determinant wavefunction in N "orbitals" ϕ_i. In this system the kinetic energy and electron density are known exactly from the orbitals;

$$T_s[\rho] = -\frac{1}{2}\sum_i^N \langle \phi_i | \nabla^2 | \phi_i \rangle$$

Here the suffix emphasises that this is not the true kinetic energy but is that of a system of non-interacting electrons, which reproduce the true ground state density;

$$\rho(\mathbf{r}) = \sum_i^N |\phi_i|^2$$

Equation 14

The construction of the density explicitly from a set of orbitals ensures that it is legal – it can be constructed from an asymmetric wavefunction.

If we also note that a significant component of the electron-electron interaction will be the classical Coulomb interaction – or Hartree energy (this is simply the second term of Equation 7 written in terms of the density);

$$V_H[\rho] = \frac{1}{2}\int \frac{\rho(\mathbf{r}_1)\rho(\mathbf{r}_2)}{|\mathbf{r}_1 - \mathbf{r}_2|} d\mathbf{r}_1 d\mathbf{r}_2$$

the energy functional can be rearranged as;

$$E[\rho] = T_s[\rho] + V_{ext}[\rho] + V_H[\rho] + E_{xc}[\rho]$$

Equation 15

Where we have introduced the *exchange-correlation functional;*

$$E_{xc}[\rho] = (T\,[\rho] - T_s[\rho]) + (V_{ee}[\rho] - V_H[\rho])$$

E_{xc} is simply the sum of the error made in using a non-interacting kinetic energy and the error made in treating the electron-electron interaction classically. Writing the functional (Equation 15) explicitly in terms of the density built from non-interacting orbitals (Equation 14) and applying the variational theorem (Equation 13) we find that the orbitals, which minimise the energy, satisfy the following set of equations;

$$\left[-\frac{1}{2}\nabla^2 + v_{ext}(\mathbf{r}) + \int\frac{\rho(\mathbf{r}')}{|\mathbf{r}-\mathbf{r}'|}d\mathbf{r}' + v_{xc}(\mathbf{r})\right]\phi_i(\mathbf{r}) = \varepsilon_i\phi_i(\mathbf{r})$$

Equation 16

In which we have introduced a local multiplicative potential, which is the functional derivative of the exchange correlation energy with respect to the density,

$$v_{xc}(\mathbf{r}) = \frac{\delta E_{xc}[\rho]}{\delta\rho}$$

Equation 17

This set of non-linear equations (the Kohn-Sham equations) describes the behaviour of non-interacting "electrons" in an effective local potential. For the exact functional, and thus exact local potential, the "orbitals" yield the exact ground state density via Equation 14 and exact ground state energy via Equation 15.

These Kohn-Sham equations have the same structure as the Hartree-Fock equations (Equation 8) with the non-local exchange potential replaced by the local *exchange-correlation potential* v_{xc}. We note at this point that the nomenclature in general use and reproduced here is very misleading. As stated above E_{xc} contains an element of the kinetic energy and is not the sum of the exchange and correlation energies as they are understood in Hartree-Fock and correlated wavefunction theories.

The Kohn-Sham approach achieves an exact correspondence of the density and ground state energy of a system consisting of non-interacting Fermions and the "real" many body system described by the Schrödinger equation. A cartoon representing this relationship is displayed in Figure 1.

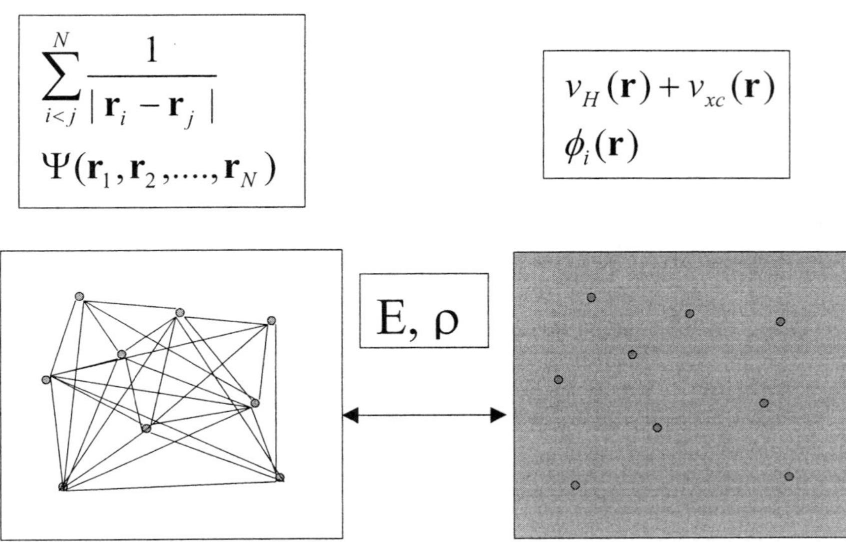

Figure 1 A cartoon representing the relationship between the "real" many body system (left hand side) and the non-interacting system of Kohn Sham density functional theory (right hand side).

The correspondence of the charge density and energy of the many-body and the non-interacting system is only exact if the exact functional is known. In this sense Kohn-Sham density functional theory is an *empirical* methodology – we do not know (and have no way of systematically approaching) the exact functional. However, the functional is universal – it does not depend on the materials being studied. For any particular system we could, in principle, solve the Schrödinger equation exactly and determine the energy functional and its associated potential. This, of course, involves a greater effort than a direct solution for the energy. Nevertheless, the ability to determine exact properties of the universal functional in a number of systems allows excellent approximations to the functional to be developed and used in *unbiased* and thus *predictive* studies of a wide range of materials – a property usually associated with an *ab initio* theory. For this reason the approximations to density functional theory discussed below are often referred to as *ab initio* or *first principles* methods.

The computational cost of solving the Kohn Sham equations (Equation 16) scales formally as N^3 (due to the need to maintain the orthogonality of N orbitals) but in current practice is dropping towards N^1 through the exploitation of the locality of the orbitals.

For calculations in which the energy surface is the quantity of primary interest DFT offers a practical and potential highly accurate alternative to the wavefunction methods discussed above. In practice, the utility of the theory rests on the approximation used for $E_{xc}[\rho]$.

5. The Local Density Approximation for $E_{xc}[\rho]$

The generation of approximations for E_{xc} has lead to a large and still rapidly expanding field of research. There are now many different flavours of functional available which are more or less appropriate for any particular study. Ultimately such judgments must be made in terms of results (i.e.: the direct comparison with more accurate theory or experimental data, which will be discussed below) but knowledge of the derivation and structure of functionals is very valuable when selecting which to use in any particular study.

The early thinking that lead to practical implementations of density functional theory was dominated by one particular system for which near exact results could be obtained – the homogeneous electron gas. In this system the electrons are subject to a constant external potential and thus the charge density is constant. The system is thus specified by a single number - the value of the constant electron density $\rho=N/V$.

Thomas and Fermi studied the homogeneous electron gas in the early 1920's [12]. The orbitals of the system are, by symmetry, plane waves. If the electron-electron interaction is approximated by the classical Hartree potential (that is exchange and correlation effects are neglected) then the total energy functional can be readily computed [12]. Under these conditions the dependence of the kinetic and exchange energy (Equation 7) on the density of the electron gas can be extracted (Dirac [13,1,14]) and expressed in terms of a *local* functions of the density. This suggests that in the inhomogeneous system we might approximate the *functional* as an integral over a local *function* of the charge density. Using the kinetic and exchange energy densities of the non-interacting homogeneous electron gas this leads to;

$$T[\rho] = 2.87 \int \rho^{5/3}(\mathbf{r})d\mathbf{r}$$

and,

$$E_x[\rho] = 0.74 \int \rho^{4/3}(\mathbf{r})d\mathbf{r}$$

Equation 18

These results are highly suggestive of a representation for E_{xc} in an inhomogeneous system. The local exchange correlation energy per electron might be approximated as a simple function of the local charge density (say, $\varepsilon_{xc}(\rho)$). That is, an approximation of the form;

$$E_{xc}[\rho] \approx \int \rho(\mathbf{r})\varepsilon_{xc}(\rho(\mathbf{r}))d\mathbf{r}$$

Equation 19

An obvious choice is then to take $\varepsilon_{xc}(\rho)$ to be the exchange and correlation energy density of the uniform electron gas of density ρ - this is the *local density approximation* (LDA). Within the LDA $\varepsilon_{xc}(\rho)$ is a *function* of only the local value of the density. It can be separated into exchange and correlation contributions;

$$\varepsilon_{xc}(\rho) = \varepsilon_x(\rho) + \varepsilon_c(\rho)$$

Equation 20

The Dirac form can be used for ε_x (Equation 18);

$$\varepsilon_x(\rho) = -C\rho^{1/3}$$

Equation 21

Where for generality a free constant, C, has been introduced rather than that determined for the homogeneous electron gas. This functional form is much more widely applicable than is implied from its derivation and can be established from scaling arguments [1]. The functional form for the correlation energy density, ε_c, is unknown and has been simulated for the homogeneous electron gas in numerical quantum Monte Carlo calculations which yield essentially exact results [15]. The resultant exchange correlation energy has been fitted by a number of analytic forms [16,17,18] all of which yield similar results in practice and are collectively referred to as LDA functionals.

A cartoon of the principles underlying the LDA is displayed in Figure 2. The energy density for the material represented by the inhomogeneous density of the left hand panel, is assigned a value from the known density variation of the exchange-correlation energy density of the homogeneous electron gas (right hand panel). This value is assigned with no regard to location or variations of the inhomogeneous density.

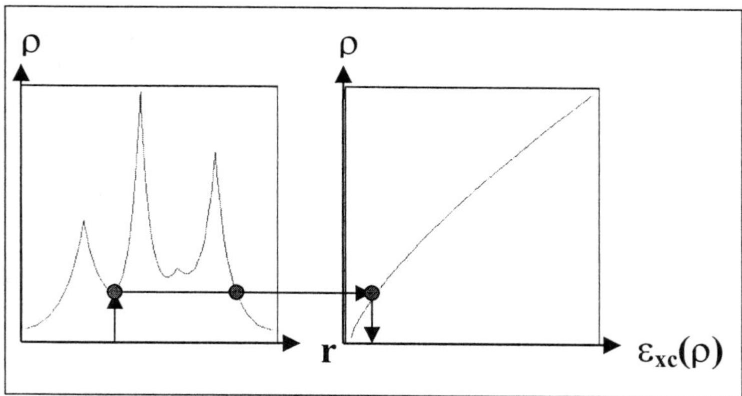

Figure 2 A cartoon depicting the use of the local density approximation to determine the exchange correlation energy of an inhomogeneous system. The exchange energy density of the homogeneous electron gas is depicted on the right and the density of the inhomogeneous system on the left [19].

The LDA has proven to be a remarkably fruitful approximation. Properties such as structure, vibrational frequencies, elastic moduli and phase stability (of similar structures) are described reliably for many systems (explicit cases will be discussed below). However, in computing energy differences between rather different structures the LDA can have significant errors. For instance, the binding energy of many systems is overestimated (typically by 20-30 %) and energy barriers in diffusion or chemical reactions may be too small or absent. Nevertheless, the remarkable fact is that the LDA works as well as it does given the reduction of the energy functional to a simple local function of the density.

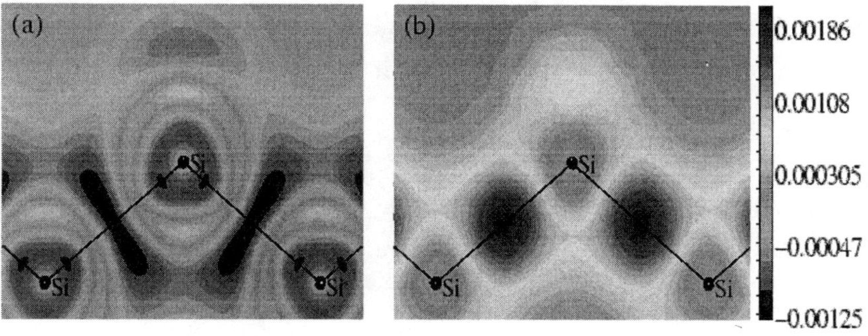

Figure 3 The difference between the exchange (left) and correlation (right) energy densities computed using variation quantum Monte Carlo and the local density approximation in bulk Silicon[20].

The magnitude of the errors in the LDA energy densities has recently been estimated by computing the energy density of bulk silicon with variational quantum Monte Carlo (V-QMC) calculations. In Figure 3 the difference between LDA densities and the, V-QMC result is plotted. There are very significant errors in the exchange and correlation energies but, as the exchange energy is generally underestimated and the correlation energy overestimated, these errors tend to cancel. The success of the LDA appears to be in part due to this cancellation of errors.

For further insight into the reasons for the success of the LDA we return to our discussion of the density matrices. As demonstrated above (Equation 12) the first and second order density matrices are sufficient to determine the exact total energy. Insight into the behaviour of functionals can be obtained by examining how well they approximate P_2. A commonly used device is to convert P_2 - the probability of finding an electron at $\mathbf{r}_1$ and an electron at $\mathbf{r}_2$ – into the conditional probability of finding an electron at $\mathbf{r}_2$ given that there is an electron at $\mathbf{r}_1$; this quantity is the exchange correlation hole;

$$P_{xc}(\mathbf{r}_1,\mathbf{r}_2) = \frac{P_2(\mathbf{r}_1,\mathbf{r}_2)}{\rho(\mathbf{r}_1)} - \rho(\mathbf{r}_2)$$

Equation 22

We may think of P_{xc} as the hole an electron placed at $\mathbf{r}_1$ digs for itself in the surrounding density. There are a number of exact properties of the hole which one would hope to reproduce in approximations. For instance, for any point $\mathbf{r}_1$ the reduction in the surrounding density should be one-electron, that is;

$$\int P_{xc}(\mathbf{r}_1,\mathbf{r}_2)d\mathbf{r}_2 = -1$$

Equation 23

This result follows immediately if P_2 of Equation 22 is inserted in Equation 11.

The exact variation of the exchange hole for a neon atom is plotted as a function of $\mathbf{r}_2$ for $\mathbf{r}_1$=0.09a_0 and 0.4 a_0 Figure 4 and compared to that computed within the LDA (from ref. [21]). It is clear that the LDA is a very poor approximation to P_2.

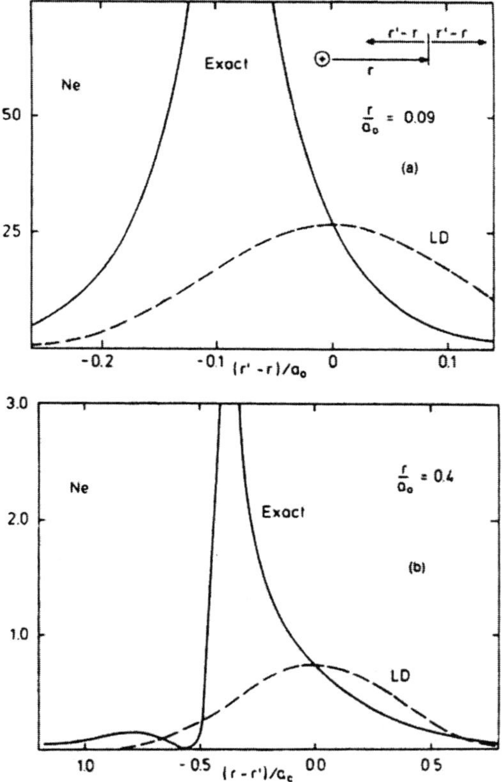

Figure 4 The exchange hole, P_x(r,r'), for a neon atom comparing the exact result (continuous line) to that of the LDA (dashed line) [21]. The top panel is for r=0.09a$_0$ and the lower for r=0.4a$_0$

We are faced with the question – how can the LDA produce such reasonable energetics if the pair correlation function is so poorly described? The answer is based on the structure of the Coulomb operator. We remember from Equation 12 that the electron-electron interaction can be written in terms of P_2 as;

$$V_{ee} = \frac{1}{2}\int \frac{1}{|\mathbf{r}_1 - \mathbf{r}_2|} P_2(\mathbf{r}_1, \mathbf{r}_2) d\mathbf{r}_1 d\mathbf{r}_2$$

Equation 24

From this it seems readily apparent that a poor approximation to P_2 leads directly to a poor estimate of the electron-electron interaction. However, the Coulomb operator depends only on the magnitude of the separation of $\mathbf{r}_1$ and $\mathbf{r}_2$ Substitution of $\mathbf{u} = \mathbf{r}_1 - \mathbf{r}_2$ yields;

$$V_{ee} = \frac{1}{2}\int\left[\int \frac{1}{u} P_2(\mathbf{r}_1, \mathbf{r}_1 + \mathbf{u}) d\mathbf{r}_1\right] d\mathbf{u}$$
$$= \frac{1}{2}\int_0^\infty 4\pi u^2 . \left[\frac{\int P_2(\mathbf{r}_1, \mathbf{r}_1 + \mathbf{u}) d\mathbf{r}_1}{u} \frac{d\Omega_u}{4\pi}\right] du$$

Equation 25

Thus the electron-electron interaction depends only on the *spherical average* of the pair density - P(u);

$$.P(u) = \int P_2(\mathbf{r}_1, \mathbf{r}_1 + \mathbf{u}) d\mathbf{r}_1 \frac{d\Omega_u}{4\pi}$$

Equation 26

The exact P(u) for the neon atom is compared to that resulting from the LDA in Figure 5 for the same positions of **r** used in Figure 4. The LDA makes a reasonable approximation to the spherically averaged hole and preserves the normalisation of the hole to –1 (Equation 23). This observation explains in part the success of the LDA.

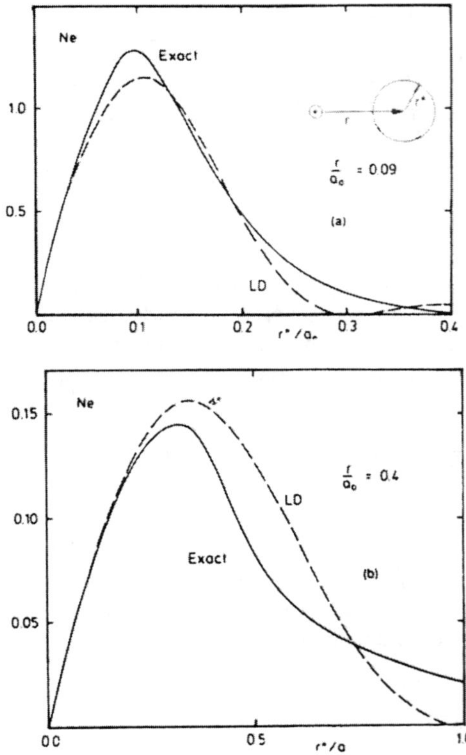

Figure 5 The spherical average of the exchange hole in a neon atom. comparing the exact result (continuous line) with that computed in the LDA (dashed line) [21].

We can conclude that the remarkable performance of the LDA is a consequence of its reasonable description of the spherically averaged exchange correlation hole coupled with the tendency for errors in the exchange energy density to be cancelled by errors in the correlation energy density. An understanding of these features is an important pre-requisite to developing functionals that seek to improve on the LDA

6. Beyond the Local Density Approximation

In many systems the exchange contribution to the energy is dominant over the correlation energy. Thus, at first sight, a very natural extension of the LDA is to compute the non-local exchange potential exactly as in Hartree Fock theory (Equation 7) whilst approximating the correlation potential with a local functional. This would yield a functional of the form:

$$E_{xc} \approx E_{Fock} + E_c^{LDA}$$

The greater complexity [22] associated with the calculation of the non-local exchange potential would be offset by potentially significantly greater accuracy. However, the performance of the LDA is, in part, based on rather delicate cancellations between the exchange and correlation interactions and, in general, the use of the exact exchange interaction produces rather poor results.

In the homogeneous electron gas the non-local exchange potential has effectively infinite range and its contribution to the electron-electron interaction diverges at the Fermi surface [23]. In metals we conclude that the non-local exchange potential does not yield the correct physics – indeed this behaviour was one of the main motivating factors in the early work of Thomas and Fermi [12], which formed the basis of density functional theory.

Further evidence for the poor behaviour of the non-local exchange interaction in systems where the band gap (HOMO-LUMO gap) is small is obtained from studies of the H_2 molecule at extended bond lengths. The exchange correlation hole for this system is displayed in Figure 6.

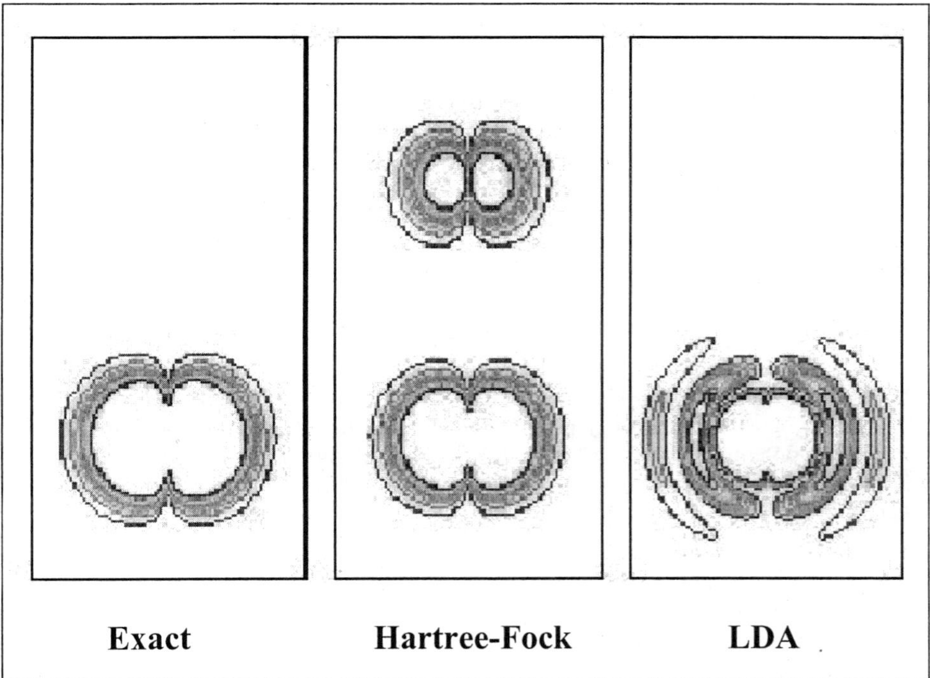

Figure 6 The exchange correlation hole in H_2 at an extended bond lengths computed exactly, with the non-local exchange potential of Hartree-Fock theory and in the LDA. The hole density is plotted as a function of r' with r centred on the H_2 bond [19].

The Hartree-Fock potential produces a very reasonable local and semi-local description – which is superior to that of the LDA – but introduces a pathological non-local feature.

We can conclude that in order to improve on the LDA approximation semi-local theories which incorporate some of the features of the exact exchange interaction are required and that theories which preserve the analytic properties of the exchange correlation hole are likely to be successful.

7. The Generalised Gradient Approximation

The local density approximation can be considered to be the zeroth order approximation to the semi-classical expansion of the density matrix in terms of the density and its derivatives [24]. A natural progression beyond the LDA is thus to the gradient expansion approximation (GEA) in which first order gradient terms in the expansion are included. This results in an approximation for the exchange hole [24] which has a number of unphysical properties; it does not normalise to –1, it is not negative definite and it contains oscillations at large u [25].

In the generalised gradient approximation (GGA) a functional form is adopted which ensures the normalisation condition and that the exchange hole is negative definite [26,27]. This leads to an energy functional that depends on both the density and its gradient but retains the analytic properties of the exchange correlation hole inherent in the LDA.

The typical form for a GGA functional is;

$$E_{xc} \approx \int \rho(r)\varepsilon_{xc}(\rho, \nabla\rho)d\mathbf{r}$$

Equation 27

As we will see below the GGA improves significantly on the LDA's description of the binding energy of molecules – it was this feature which lead to the very wide spread acceptance of DFT in the chemistry community during the early 1990's. A number of functionals within the GGA family [28,26,27,29,30,31] have been developed. The performance of these functionals will be discussed below.

8. Meta-GGA functionals

Recently functionals that depend explicitly on the semi-local information in the Laplacian of the spin density or of the local kinetic energy density have been developed [32,33,34]. Such functionals are generally referred to as meta-GGA functionals.

The form of the functional is typically;

$$E_{xc} \approx \int \rho(r)\varepsilon_{xc}(\rho, |\nabla\rho|, \nabla^2\rho, \tau)d\mathbf{r}$$

Equation 28

Where the kinetic energy density τ is;

$$\tau = \frac{1}{2}\sum_i |\nabla\varphi_i|^2$$

Equation 29

9. Hybrid Exchange Functionals

There is an *exact* connection between the non-interacting density functional system and the fully interacting many body system via the integration of the work done in gradually turning on the electron-electron interactions. This adiabatic connection approach [35] allows the exact functional to be formally written as:

$$E_{xc}[\rho]=\frac{1}{2}\int d\vec{\mathbf{r}}d\vec{\mathbf{r}}'\int_{\lambda=0}^{1}d\lambda\frac{\lambda e^{2}}{|\vec{\mathbf{r}}-\vec{\mathbf{r}}'|}[<\rho(\vec{\mathbf{r}})\rho(\vec{\mathbf{r}}')>_{\rho,\lambda}-\rho(\vec{\mathbf{r}})\delta(\vec{\mathbf{r}}-\vec{\mathbf{r}}')]$$

Equation 30

where the expectation value $<...>_{\rho,\lambda}$ is the density-density correlation function and is computed at density $\rho(\mathbf{r})$ for a system described by the effective potential;

$$V_{eff}=V_{en}+\frac{1}{2}\sum_{i\neq j}\frac{\lambda e^{2}}{|\vec{\mathbf{r}}_{i}-\vec{\mathbf{r}}_{j}|}$$

Equation 31

Thus the exact energy could be computed *if* one knew the variation of the density-density correlation function with the coupling constant, λ. The LDA is recovered by replacing the pair correlation function with that for the homogeneous electron gas.

The adiabatic integration approach suggests a different approximation for the exchange-correlation functional. At λ=0 the non-interacting system corresponds identically to the Hartree-Fock ansatz, while the LDA and GGA functionals are constructed to be excellent approximations for the fully interacting homogeneous electron gas – that is, a system with λ=1. It is therefore not unreasonable to approximate the integral over the coupling constant as a weighted sum of the end points – that is, we might set:

$$E_{xc}\approx aE_{Fock}+bE_{xc}^{GGA}$$

with the coefficients are to be determined by reference to a system for which the exact result is known. Becke adopted this approach [36] in the definition of a new functional with coefficients determined by a fit to the observed atomisation energies, ionisation potentials, proton affinities and total atomic energies for a number of small molecules [36]. The resultant (three parameter) energy functional is,

$$E_{xc} = E_{xc}^{LDA} + 0.2(E_X^{Fock} - E_X^{LDA}) + 0.72\Delta E_X^{B88} + 0.81\Delta E_c^{PW91}$$

Here ΔE_X^{B88} and ΔE_c^{PW91} are widely used GGA corrections [37,38] to the LDA exchange and correlation energies respectively.

Hybrid functionals of this type are now very widely used in chemical applications with the B3LYP functional (in which the parameterisation is as given above but with a different GGA treatment of correlation [39]) being the most notable. Computed binding energies, geometries and frequencies are systematically more reliable than the best GGA functionals.

10. The Performance of Various Functionals

The functionals currently used in density functional simulations form a natural hierarchy. Although it cannot be claimed that there is a systematic approach to the exact functional it is clear that improvements are being made in the underlying functional form and that the description of ground state properties is improving. The most notable recent advances being those in which the non-local nature of the exchange potential is introduced in one form or another. The current hierarchy is summarized in Table 1.

Dependencies	**Family**
Exact exchange, $\|\nabla\rho\|$, ρ	Hybrid
$\nabla^2\rho$, τ	Meta-GGA
$\|\nabla\rho\|$	GGA
ρ	LDA

Table 1 The current hierarchy of exchange correlation functionals.

In producing functionals there are two broad schools of thought, which may be summarized as follows;

- Adopt an appropriate functional form and introduce parameters to be determined by reference to experimental data or data from explicitly correlated calculations. This is largely an empirical approach.
- Use the exact properties of the functional to determine both its structure and the parameters in its functional form.

There are, of course, a number of functionals that rather cross the boundary between the two schools but, as we will see below, the distinction is often valuable in assessing the likely accuracy of a particular functional in a new application. Unfortunately in order to discuss a variety of functionals they must be labeled with a mnemonic – a selection of which are listed in Table 2 along with an indication of the family they come from and the number of empirically determined parameters which enter the functional. The parameters are determined in practice from fitting to the properties of large training sets of molecular properties.

Mnemonic	Family	Parameters
LDA	Local	-
BLYP	GGA	Light
PBE	GGA	-
HCTH	GGA	18
VS98	Meta-GGA	21
PKZB	Meta-GGA	1
Hybrid	Hybrid-exchange	3

Table 2 Mnemonics used to define density functionals of various families and the number of external parameters fitted when defining the functional.

Apart from the LDA the functionals listed are the GGA functionals BLYP [29, 39], PBE [31], HCTH [40] the meta-GGA functionals VS98 [41], PKZB [34] and the Hybrid functional of Becke [36]. The degree of empirical parameterisation of these functionals varies considerably with the BLYP, PBE and PKZB functionals aiming to be virtually *ab-initio* and parameter free while the HCTH and VS98 functionals are very heavily parameterised with reference to large molecular training sets.

The performance on these functionals in calculations of a number of molecular and material properties have recently been tabulated by Kurth *et. al.* [42] and Adamo *et. al.* [43]. Here we summarise some of the key data from these studies.

Atomisation Energies		
	M.R.E %	**M.A.E. (max) kcal/mol**
LDA	22 %	-
BLYP	5 %	-
PBE	7 %	17 (51)
HCTH	3 %	-
VS98	2 %	3 (12)

PKZB	3 %	5 (38)
Hybrid	-	3 (20)

Table 3 Atomisation Energies - the mean relative error (M.R.E) for a collection of 20 molecules [42] **and the mean absolute error (M.A.E.) in kcal/mol for a collection of 148 molecules** [43] **with the maximum absolute error given in brackets.**

In Table 3 errors in computed molecular atomisation energies are tabulated. The tendency of the LDA to over bind by 20-30% discussed above is seen clearly in this data. The GGA functionals yield very significant improvements with relative errors in the range 3-7% and an average absolute error of 17 kcal/mol. The highly parameterised HCTH functional performs somewhat better than the BLYP and PBE functionals. The meta-GGA functionals yield relative errors of 2-3% and average absolute errors of 3-5 kcal/mol. Again the highly parameterised functional – VS98 - performs slightly better than PKZB. The meta-GGA functionals achieve a distinct improvement over the GGA functionals. The relatively lightly parameterised hybrid functional performs as well as the best meta-GGA. For many chemical reactions an accuracy of 1-2 kcal/mol is sufficient for predictive modelling of the thermochemistry. It is clear that average errors yielded by the meta-GGA and hybrid functionals are approaching this limit. Performance is not systematically at this level; maximum errors for ‘difficult’ systems are still in the range 12-38 kcal/mol.

Structures		
	M.R.E. % [42]	**M.A.E. (max) Angstrom [43]**
LDA	5%	-
BLYP	8%	-
PBE	4%	0.011 (0.064)
HCTH	6%	-
VS98	8%	0.008 (0.08)
PKZB	3%	0.019 (0.111)
Hybrid	-	0.007 (0.062)

Table 4 Mean relative error (M.R.E.) in the unit cell volumes of 12 crystals [42] **and mean absolute errors (M.A.E.) in the bond lengths of 23 molecules** [43] **with the maximum absolute error given in brackets.**

Results for unit cell volumes of 12 different crystal structures and for the bond lengths in 23 molecules are summarised in Table 4. The LDA is a remarkably good approximation in bulk crystals and its error for volumes (5%) compares favourably to that produced by GGA functionals (4-8%) and meta-GGA functionals (3-8%). In molecular systems the highly parameterised meta-GGA and the hybrid functional are somewhat more accurate than the PBE GGA functional. It is interesting to note that although the VS98 functional is somewhat more accurate than the PKZB in molecules this situation is reversed for calculations on crystals. This is part of a general trend in that lightly parameterised functionals do not contain a bias towards molecular systems and tend to be transferable to materials applications while highly parameterised functionals are not.

Bulk Modulus and Vibrational Frequencies		
	Bulk Modulus **M.R.E. %**	**Vibrational Frequencies** **M.A.E. (max) (cm^{-1})**
LDA	19%	-
BLYP	22%	-
PBE	10%	65 (-194)
HCTH	20%	-
VS98	29%	33 (-109)
PKZB	9%	72 (+144)
Hybrid	-	40 (-209)

Table 5 Mean relative error (M.R.E) in the bulk moduli of 12 crystals [42] **and mean absolute error (M.A.E.) in the vibrational frequencies of 55 molecules** [43] **with the maximum absolute error given in brackets.**

Results for the bulk moduli of 12 different crystal structures and for the vibrational frequencies in a set of 55 molecules are summarised in Table 5. For the calculation of bulk moduli the GGA functionals offer a small but non-systematic improvement over the 19% error of the LDA. The PKZB meta-GGA functional produces a significantly smaller error of 9%. For vibrational frequencies in molecules the VS98 and hybrid exchange functionals perform somewhat better than the PBE GGA functional or the PKZB meta-GGA functional. As for computed structures there is a tendency for the highly parameterised VS98 meta-GGA functional to perform well in molecules but rather poorly in crystalline solids.

11. Conclusions

Density functional theory provides us with a relatively efficient and *unbiased* tool with which to compute the ground state energy in realistic models of bulk materials and their surfaces. The reliability of such calculations depends on the development of approximations for the exchange-correlation energy functional. Significant advances have been made in recent years in the quality of exchange correlation functionals as dependence on local density gradients, semi-local measures of the density and non-local exchange functionals have been introduced.

The local density approximation is a very simple and remarkably reliable for the structure, elastic moduli, relative phase stability of many materials but is less accurate for binding energies and details of the energy surface away from equilibrium geometries – eg: transition states. The GGA family of functionals improves binding energies to average errors of 20 kcal/mol and relative errors of 3-7% while meta-GGA and hybrid-exchange functionals reduce these errors to 3-5 kcal/mol and 2-3%. This is close to the accuracy required for predictive simulations of thermochemical properties. The GGA, meta-GGA and hybrid functionals retain, and somewhat improve on, the LDA's excellent description of bonds lengths with typical errors in the region of 1-2 milli-Angstrom. Using these functionals elastic moduli are reproduced to within 10% and vibrational frequencies to ~40cm^{-1}.

There is a distinct tendency for functionals that are highly parameterised and fitted to the properties of molecular systems to perform somewhat better than lightly parameterised functions for molecules but to perform relatively poorly in simulations on periodic materials.

References

1. R. G. Parr, and W. Yang, Density-Functional Theory of Atoms and Molecules, OUP, Oxford, (1989)
2. N. H. March (ed.), Electron Correlation in the Solid State, ICP, London, (1999)
3. J. Callaway and N. H. March, Density Functional Methods: Theory and Applications, *Solid State Physics*, **38**, 135 (1984)
4. A. Szabo and N. S. Ostlund, *Modern Quantum Chemistry*, (Macmillan, New York, 1982)
5. B. L. Hammond and W. A. Lester Jr. and P.J. Reynolds, *Monte Carlo Methods in Ab Initio Quantum Chemistry*, (World Scientific, Singapore, 1994).
6. J. Coleman, Calculation of the first- and second-order density matrices, in, The Force Concept in Chemistry, Ed. B. M. Deb, (Van Nostrand Reinhold, New York, 1981)
7. R. Erdahl and V. H. Smith Jr. Eds. *Density matrices and Density Functionals*, (Reidel, Dordrecht 1987).
8. P. Hohenburg and W. Kohn, *Phys. Rev.* **136** B864 (1964)
9. M. Levy, *Proc. Natl. Acad. Sci. USA* **76**, 6062 (1979); M. Levy, *Phys. Rev. A* **26**, 1200 (1982); M. Levy and J. P. Perdew, *The Constrained Search Formalism for Density Functional Theory*, in *Density Functional Methods in Physics*, Ed. R. M. Dreisler and J. da Providencia, (Plenum, New York 1985)
10. M. Foley and P. A. Madden, *Phys. Rev.* B **53**, 10589 (1996)
11. W. Kohn and L. J. Sham, *Phys. Rev.* **140** A1133 (1956)
12. E. Fermi *Z. Phys.* **48** 73 (1928); L. H. Thomas, *Proc. Camb. Phil. Soc.* **23** 542 (1927); these articles are reproduced in N. H. March, Self Consistent Fields in Atoms, Plenum, Oxford, (1975)
13. P. A. M. Dirac, *Proc. Camb. Phil. Soc.* **26** 376 (1930)
14. E. H. Lieb, *Rev. Mod. Phys.*, **53** 603 1981
15. D. M. Ceperley and B. J. Alder, *Phys. Rev. Lett.*, **45** 566 1980
16. J. P. Perdew and A. Zunger, *Phys. Rev.* B **23** 5048 (1981)
17 U. von Barth and L. Hedin, *J. Phys.* C **5** 1629 (1972)
18. S. H. Vosko, L. Wilk and M. Nusair, Accurate spin-dependent electron liquid correlation energies of local spin density calculations: a critical analysis", *Can. J. Phys.* **58**, 1200 (1980)
19. This image is used with the kind permission of A. Savin (unpublished work)
20. R. Q. Hood, M. Y. Chou, A. J. Williamson, R. Rajagopal, R.J. Needs and W. M. C. Foulkes, *Phys. Rev. Lett.* **78** 3350 1997.
21. O. Gunnarsson, M. Jonson and B. I. Lundqvist, *Phys. Rev.* B **20** 3136 (1979).
22. The exact exchange potential can be computed rather readily for periodic materials when a basis set Gaussian type orbitals is used as in the CRYSTAL software: R. Dovesi, V. R. Saunders, C. Roetti, M. Causà, N. M. Harrison, R. Orlando, C. M. Zicovich-Wilson, CRYSTAL98 User's Manual, University of Torino, Torino, 1998; (*http://www.cse.dl.ac.uk/Activity/CRYSTAL*).
23. See, for instance, the discussion in Chapter 17 of, *Solid State Physics*, N. W. Ashcroft and N. D. Mermin, second edition, (HoltSaunders, Philadelphia, 1976)
24. R. M. Dreizler and E. K. U. Gross, *Density Functional Theory* (Springer Verlag, Berlin, 1990)
25. Y. Yang, J. P. Perdew, J. A. Cevary, L. D. Macdonald and S. H. Vosko, *Phys. Rev.* A **41**, 78 (1990)
26. J. P. Perdew and Y. Wang, *Phys. Rev.* B **33**, 8800, (1986); Ibid. E **34**, 7406, (1986)
27. J. P. Perdew, in *Electronic Structure of Solids 91*, Ed. P. Ziesche and H. Eschrig (Akademie Verlag, Berlin, 1991)
28. D. C. Langreth, M. J. Mehl, *Phys. Rev.* B, **28**, 1809 (1983)
29. A. D. Becke, *Phys. Rev.* A **38** 3098 1988

30. C. Lee, W. Yang, R. G. Parr, *Phys. Rev.* B **37** 785 (1988)
31. J. P. Perdew, K. Burke, M. Ernzerhof, *Phys. Rev. Lett.* **77**, 3865, (1996); Ibid. E **78,** 1396 (1997)
32. V. Tschinke, T. Zieglar, *Can. J. Chem.* **67**, 460, (1989)
33. R. Neumann, N. C. Handy, Chem. Phys. Lett.,**266**, 16 (1997)
34. J. P. Perdew, S. Kurth, A. Zupan and P. Blaha, *Phys. Rev. Lett.* **82**, 2544, (1999); Ibid E **82**, 5179 (1999)
35. D. Pines and P. Nozières, *The Theory of Quantum Liquids* (Benjamin, New York, 1966); J. Harris and R. O. Jones, *J. Phys.* F **4,** 1170 (1974), O. Gunnarsson and B. I. Lundqvist, *Phys. Rev.* B **13**, 4274 (1976).
36. A. D. Becke, *J. Chem. Phys.*, **98**, 1372 (1993); A.D. Becke, Ibid. **98,** 5648 (1993)
37. A. D. Becke, *J. Chem. Phys.,* **88**, 1053 (1988)
38. J. P. Perdew and Y. Wang, *Phys. Rev.* B **45,** 13244 (1992)
39. C. Lee, W. Yang and R. G. Parr, *Phys. Rev.* B **37**, 785 (1988)
40. F. A. Hamprecht, A. J. Cohen, D. J. Tozer and N. C. Handy, *J. Chem. Phys.* **109**, 6264 (1998)
41. T. Van Voorhis and G. E. Scuseria, *J. Chem. Phys.* **109**, 400, (1998)
42. S. Kurth, J. P. Perdew and P. Blaha, *Int. J. of Quant. Chem.* **75** 889 (1999)
43. C. Adamo, M. Ernzerhof and G. E. Scuseria, *J. Chem. Phys.*, **112**, 2643 (2000)

Computational Materials Science
C.R.A. Catlow and E.A. Kotomin (Eds.)
IOS Press, 2003

Hartree-Fock method

I.V.Abarenkov

St.Petersburg State University, St.Petersburg, RUSSIA

Abstract. In this paper the Hartree-Fock (HF) theory and its application to the electronic structure calculations of solids are discussed.

In the first part few exact results for the many-electron systems concerning the reduced density matrixes, electron density, correlation function, spin correlations and space correlations are recollected. Then the general HF theory is considered starting from the simplest one-determinant case. For this case the reduced density matrixes, the energy functional, and HF equations are obtained. The analysis of self-consistent solution procedure, canonical HF orbitals, Koopmans theorem, and population rule are made. Then non-canonical HF orbitals are introduced and the localized and delocalized orbitals are discussed. Next, the general case of several determinants is considered. The symmetry aspects (many-electron symmetry, one-electron symmetry, configuration, closed shells, open shells, translation symmetry) are examined.

The second part is devoted to the HF electronic structure calculations of ionic crystals where the HF method are expected to be the best applicable. The localized orbitals description is adopted in this section. The non-canonical localized orbitals, total energy, cluster expansion, pair interaction, three-ion interaction, one-electron energies, the problem of the band gap are discussed.

1 Introduction

The Hartree-Fock (HF) method [1, 2] has a long history and there exists an extensive literature on the HF method, including excellent monographies [3]–[9]. The HF method was first applied to the atomic structure calculations, then to molecules, and finally to solids. In this paper the main features of the HF method will be discussed and a non-conventional application to the solid state electronic structure calculations will be considered. The results of particular calculation will be used for the illustration only.

2 Density matrixes and correlation function

The wave function of a n-electron system $\Psi(x_1, x_2, \cdots, x_n)$ is a function of n variables x_i, where x is a collection of a space variable $\mathbf{r}$ and spin variable σ. The wave function contains an enormous amount of information most of which is necessary to calculate the function itself (even approximate). At the same time we are usually interested in the properties of the system and not in the wave function. To calculate a one-electron property (kinetic energy, interaction with the external field energy, dipole moment etc.) or a two-electron property (electron-electron interaction energy etc.) the much simple constructions are needed. These are the first-order (one-particle) $\rho(x_1|\,x_1')$ and the second order (two-particles) $\rho(x_1, x_2, |\,x_1', x_2')$ *reduced density matrixes* (see [10])

$$\rho(x_1|\,x_1') \;=\; n \int \Psi(x_1, x_2, \cdots, x_n)\, \Psi^*(x_1', x_2, \cdots, x_n)\, dx_2 \,\cdots\, dx_n,$$

$$\rho(x_1, x_2, |\,x_1', x_2') \;=$$
$$= n(n-1) \int \Psi(x_1, x_2, x_3, \cdots, x_n)\, \Psi^*(x_1', x_2', x_3, \cdots, x_n)\, dx_3 \,\cdots\, dx_n.$$

For the one-electron property

$$\widehat{W}_1(x_1, \cdots, x_n) \;=\; \sum_{k=1}^{n} \widehat{w}_1(x_k)$$

one has

$$\langle\Psi|\widehat{W}_1|\Psi\rangle \;=\; \int [\,\widehat{w}_1\,(x)\,\rho\,(x\,|\,x')\,]_{x'=x}\, dx. \tag{2.1}$$

For the two-electron property

$$\widehat{W}_2(x_1, \cdots, x_n) \;=\; \sum_{k<\ell=1}^{n} \widehat{w}_2(x_k, x_\ell)$$

one similarly has

$$\langle\Psi|\widehat{W}_2|\Psi\rangle \;=\; \frac{1}{2} \int [\,\widehat{w}_2(x_1, x_2)\,\rho(x_1, x_2\,|\,x_1', x_2')\,]_{x_1'=x_1,\, x_2'=x_2}\, dx_1 dx_2. \tag{2.2}$$

The reduced density matrixes have an immediate physical interpretation themselves. From the physical interpretation of the wave function it follows that $\rho(x|x)$ is the probability density to find an electron at the point x (at the point $\mathbf{r}$ with spin σ), in other words $\rho(x|x)$ is the electron density of the system (normalized to the number of electrons). Similarly $\rho(x_1, x_2|x_1, x_2)$ is the probability density to find simultaneously one electron at the point $\mathbf{r}_1$ with spin σ_1 and another electron at the point $\mathbf{r}_2$ with spin σ_2.

An important feature of a many-electron system are the electron correlations. Unfortunately, there is no unique definition of correlations in the many electron theory. In general, correlations are the difference between the actual motion of particles and the motion of similar but independent particles. It is the notion of independent particles which is difficult to define. In quantum theory one can consider the non-interacting particles simply by omitting the interaction term from the Hamiltonian. However, contrary to the classical particles the non-interacting quantum particles are not necessarily independent. For example, the non-interacting identical fermions are not independent because of the Pauli principle and hence there are correlations even in the system of non-interacting electrons. Sometimes such correlations are called the "spin correlations" or "exchange correlations". The interacting electrons experience an additional correlations because they avoid each other due to their electrostatic repulsion. Another,

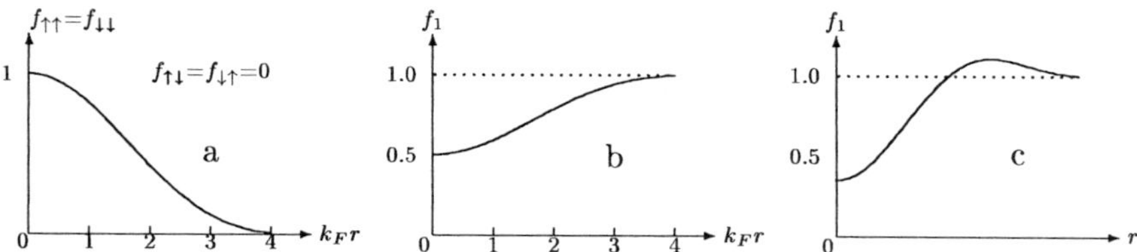

Figure 1: Correlation functions and correlation hole.
a,b - homogeneous gas of non-interacting electrons, $r = |\mathbf{r}_1 - \mathbf{r}_2|$
c -sketch of the correlation hole for interacting electrons (general case)

customary way is to consider as independent the electrons subject to the Hartree-Fock equations. This is also not a unique definition because there are many variants of the Hartree-Fock theory which can differ substantially.

In this instance the reduced density matrixes enable one to define correlations using the notion of statistically independent particles. For statistically independent electrons the probability to find simultaneously one electron at a point x_1 and another electron at a point x_2 irrespective of where are all other electrons is equal to the product of the probability to find one electron at the point x_1 irrespective of where all other electrons are and the probability to find one electron at x_2 irrespective of where all other electrons are. For the correlated electrons this equality does not hold and using the physical interpretation of the reduced density matrixes one can write

$$\rho(x_1, x_2|x_1, x_2) = \rho(x_1|x_1)\,\rho(x_2|x_2)\,\Big(1 - f(x_1, x_2)\Big) \tag{2.3}$$

If $f = 0$ the electrons are not correlated i.e. they are (statistically) independent. The function $f(x_1, x_2)$ could be referred to as the pair correlation function. Due to the wave function antisymmetry one has $\rho(x_1, x_2|x_1, x_2) = -\rho(x_2, x_1|x_1, x_2)$. Therefore $\rho(x_1, x_2|x_1, x_2)$ tends to zero when x_1 tends to x_2 while the product $\rho(x_1|x_1)\rho(x_2|x_2)$ tends to a finite value. So the correlation function $f(x_1, x_2)$ tends to 1 when x_1 tends to x_2. At the same time in most cases $f(x1, x2)$ tends to zero when electrons move far apart. Figuratively speaking there is a *correlation hole* around each particle. This is true even for the non-interacting electrons with the same spin and the corresponding hole is known as an exchange hole or spin hole.

To show the correlation hole explicitly the spatial correlation function $f_1(\mathbf{r}_1, \mathbf{r}_2)$ can be defined as follows

$$\rho(\mathbf{r}_1, \mathbf{r}_2) = \rho(\mathbf{r}_1)\rho(\mathbf{r}_2)f_1(\mathbf{r}_1, \mathbf{r}_2), \tag{2.4}$$

with the help of spatial densities

$$\rho(\mathbf{r}) = \int \rho(\mathbf{r}, \sigma|\mathbf{r}, \sigma)d\sigma, \qquad \rho(\mathbf{r}_1, \mathbf{r}_2) = \int \rho(\mathbf{r}_1, \sigma_1, \mathbf{r}_2, \sigma_2|\mathbf{r}_1, \sigma_1, \mathbf{r}_2, \sigma_2)d\sigma_1 d\sigma_2$$

Three examples of correlation functions are shown in Figure 1.

3 Hartree-Fock theory

3.1 Introduction

We consider a stationary state $\Psi(x_1, x_2, \cdots, x_n)$ of a n-electron system

$$\widehat{H}\Psi(x_1, x_2, \cdots, x_n) = E\,\Psi(x_1, x_2, \cdots, x_n)$$

assuming that the Hamiltonian

$$\widehat{H} = \widehat{H}(\mathbf{r}_1, \cdots, \mathbf{r}_n) = \sum_{k=1}^{n} \widehat{h}(\mathbf{r}_k) + \sum_{k<\ell=1}^{n} \widehat{g}(\mathbf{r}_k, \mathbf{r}_\ell) \tag{3.1}$$

does not depend on spin variables. Here one-electron operator

$$\widehat{h}(\mathbf{r}) = -\frac{\hbar^2}{2m}\nabla^2 + V(\mathbf{r}) \tag{3.2}$$

is the sum of the electron kinetic energy and its potential energy in the external field, the system's nucleus field in particular. The two-electron operator

$$\widehat{g}(\mathbf{r}_1, \mathbf{r}_2) = \frac{e^2}{|\mathbf{r}_1 - \mathbf{r}_2|}$$

is the electron-electron interaction.

The total wave function $\Psi(x_1, x_2, \cdots, x_n)$ cas always be constructed with the help of one-electron functions $\psi_j(x)$ which are usually referred to as *spin-orbitals*, and spin-orbitals can be constructed from one-electron functions $\phi_j(\mathbf{r})$ depending on space variables only

$$\psi_j(x) = \phi_j(\mathbf{r})\alpha(\sigma), \quad \text{or} \quad \psi_j(x) = \phi_j(\mathbf{r})\beta(\sigma),$$

for spin up and spin down states correspondingly. The functions $\phi_j(\mathbf{r})$ are referred to as *orbitals*. With any complete and hence infinite set of spin-orbitals the wave function could be reproduced exactly. In practical methods the number of spin-orbitals is finite, the total wave function is reproduced only approximately, and the particular choice of spin-orbitals become important to the quality of the approximation. In this respect there are two features of the Hartree-Fock (HF) method which distinguish it from all other methods.

1. The approximate total wave function is constructed with the smallest possible number of spin-orbitals necessary to reproduce some symmetry properties, spin and space, of the exact wave function.

2. The orbitals are selected to bring the minimum (in general, extremum) to the energy functional of the system.

In different variants of the HF method all, none, or some symmetry properties of the total wave function, apart from its antisymmetry, are taken into account, and different particular forms of the energy functional are used.

Besides, especially for the excited states, one should specify not only the number of orbitals but the particular choice of orbitals to be used for the total wave function construction. Because usually an energy is attributed in one way or other to any orbital the specification of orbitals to be used is expressed as a one-electron levels population rule.

We consider the spinless Hamiltonian (3.1) which commutes with the total spin operator $\widehat{\mathbf{S}}$. Therefore to take into account the spin symmetry one can impose a condition that the approximate wave function is an eigenfunction of $\widehat{S^2}$ and $\widehat{S}_z$ operators. If the system has a space symmetry as well, a condition that the approximate wave function is a member of the symmetry group irreducible representation basis could also be imposed. For the exact wave function these are the necessary conditions if the total energy level is not degenerate and only sufficient conditions in the degenerate case.

Any imposed condition is usually considered as a restriction, even if the condition serves to make the better approximation. Therefore if all mentioned above conditions are imposed on the approximate wave function the corresponding variant is referred to as the Restricted Hartree-Fock (RHF) method. Different energy functional are used in different variants of RHF method. If none of the conditions are imposed the method is referred to as the Unrestricted Hartree-Fock (UHF) method, the energy functional being the mean value of the Hamiltonian calculated with the approximate wave function. In UHF method only the antisymmetry of the wave function is taken into account and because of it the smallest possible number of spin-orbitals is equal to the number of electrons n. The UHF wave function has the simplest form. It is a single Slater determinant built of n different spin-orbitals. Usually, every spin-orbital is taken to be the spin up or spin down state. Suppose there are n_α spin-orbitals with spin up and $n_\beta - n_\alpha$ orbitals with spin down, and $n_\alpha \geq n_\beta$. Then the single Slater determinant build ot the following spin-orbitals

$$\begin{array}{rclrcl}
\psi_1(\mathbf{r},\sigma) & = & \phi_1(\mathbf{r})\alpha(\sigma), & \psi_{n_\alpha+1}(\mathbf{r},\sigma) & = & \phi_{n_\alpha+1}(\mathbf{r})\beta(\sigma),\\
\psi_2(\mathbf{r},\sigma) & = & \phi_2(\mathbf{r})\alpha(\sigma), & \psi_{n_\alpha+2}(\mathbf{r},\sigma) & = & \phi_{n_\alpha+2}(\mathbf{r})\beta(\sigma),\\
 & \cdots & & & \cdots & \\
\psi_{n_\beta}(\mathbf{r},\sigma) & = & \phi_{n_\beta}(\mathbf{r})\alpha(\sigma), & \psi_{n_\alpha+n_\beta}(\mathbf{r},\sigma) & = & \phi_{n_\beta}(\mathbf{r})\beta(\sigma),\\
\psi_{n_\beta+1}(\mathbf{r},\sigma) & = & \phi_{n_\beta+1}(\mathbf{r})\alpha(\sigma), & & & \\
 & \cdots & & & & \\
\psi_{n_\alpha}(\mathbf{r},\sigma) & = & \phi_{n_\alpha}(\mathbf{r})\alpha(\sigma). & & &
\end{array}$$

is automatically the eigenfunction of $\widehat{S}_z$ operator with eigenvalue equal (in atomic units) to half the difference between the number of spin up and spin down orbitals. At the same time this determinant is not, in general, the eigenfunction of $\widehat{S^2}$ operator.

The single Slater determinant could be the eigenfunction of $\widehat{S^2}$ only if orbitals $\phi_{n_\alpha+1}, \cdots, \phi_{n_\alpha+n_\beta}$ are (linearly independent) leanear combinations of orbitals $\phi_1, \cdots, \phi_{n_\alpha}$, in particular

$$\begin{array}{rcl}
\phi_{n_\alpha+1}(\mathbf{r}) & = & \phi_1(\mathbf{r}),\\
\phi_{n_\alpha+2}(\mathbf{r}) & = & \phi_2(\mathbf{r}),\\
\cdots & & \cdots\\
\phi_{n_\alpha+n_\beta}(\mathbf{r}) & = & \phi_{n_\beta}(\mathbf{r})
\end{array} \tag{3.3}$$

The orbitals ϕ_i with $i \leq n_\beta$ are referred to as doubly occupied and those with $n_\beta < i \leq n_\alpha$ as singly occupied. The Slater determinant built of spin-orbitals (3.3) is the eigenfunction of $\widehat{S^2}$ with eigenvalue $S(S+1)$ and the eigenfunction of $\widehat{S}z$ with eigenvalue S, where $S = (n_\alpha - n_\beta)/2$. Hence if the system has no spatial symmetry the single Slater determinant is the RHF wave function. The energy functional in the one-determinant RHF method is the mean value of the Hamiltonian. It should be stressed here that the RHF wave function is a single determinant only for the maximal (S) or minimal ($-S$) z-projection of the total spin. For all other projection the RHF wave function is a linear combination of several determinants.

Because the single Slater determinant is the simplest wave functions we will consider this case in more details in the next two sections.

3.2 One-determinant case

In this section we consider the ground state of a n-electron system and assume that the approximate wave function is a single Slater determinant

$$\Psi(x_1, x_2, \cdots, x_n) = \frac{1}{\sqrt{n!}} \begin{vmatrix} \psi_1(x_1), & \psi_1(x_2) & \cdots & \psi_1(x_n) \\ \psi_2(x_1), & \psi_2(x_2) & \cdots & \psi_2(x_n) \\ \cdots & \cdots & \cdots & \cdots \\ \psi_n(x_1), & \psi_n(x_2) & \cdots & \psi_n(x_n) \end{vmatrix}, \tag{3.4}$$

where first n_α spin-orbitals are the spin up states and the last $n_\beta = n - n_\alpha$ spin-orbitals are spin down states

$$\psi_i(x) = \phi_i(\mathbf{r})\alpha(\sigma),\ i = 1, \ldots, n_\alpha, \qquad \psi_i(x) = \phi_i(\mathbf{r})\beta(\sigma),\ i = n_\alpha + 1, \ldots, n.$$

Because of the determinant properties we can assume without any loss of generality that spin-orbitals are orthonormal

$$\langle \psi_i | \psi_j \rangle = \delta_{ij}. \tag{3.5}$$

Then the wave function (3.4) will be normalized to 1. Besides, under any unitary transformation of spin-orbitals the wave function Ψ will acquire the phase factor only. The first order reduced density matrixes in this case is

$$\rho(x|x') = \sum_{i=1}^{n} \psi_i(x)\psi_i^*(x') = \rho_\alpha(\mathbf{r}|\mathbf{r}')\alpha(\sigma)\alpha(\sigma') + \rho_\beta(\mathbf{r}|\mathbf{r}')\beta(\sigma)\beta(\sigma'), \tag{3.6}$$

where

$$\rho_\alpha(\mathbf{r}|\mathbf{r}') = \sum_{i=1}^{n_\alpha} \phi_i(\mathbf{r})\phi_i^*(\mathbf{r}'), \qquad \rho_\beta(\mathbf{r}|\mathbf{r}') = \sum_{i=n_\alpha+1}^{n} \phi_i(\mathbf{r})\phi_i^*(\mathbf{r}')$$

could be considered as the spin up and spin down spatial density matrixes respectively. The second order reduced density matrix is

$$\begin{aligned} \rho(x_1, x_2|x_1', x_2') = \sum_{i,j=1}^{n}{}' \big[& \psi_i(x_1)\psi_j(x_2)\psi_i^*(x_1')\psi_j^*(x_2') - \\ & - \psi_i(x_1)\psi_j(x_2)\psi_j^*(x_1')\psi_i^*(x_2') \big], \end{aligned} \tag{3.7}$$

where the prime by the sum means that $i \neq j$. At the same time the prime could be omitted because the term in square brackets vanishes when $i = j$. Then one can write

$$\rho(x_1, x_2 \,|\, x_1', x_2') = \rho(x_1 \,|\, x_1')\rho(x_2 \,|\, x_2') - \rho(x_1 \,|\, x_2')\rho(x_2 \,|\, x_1'). \tag{3.8}$$

The equation (3.8) is the necessary and sufficient condition for the wave function to be the single Slater determinant. The reduced density matrixes are invariant under any unitary transformation of spin-orbitals. Another important property of the first order reduced density matrix (3.6) is

$$\int \rho(x_1|x')\,\rho(x'|x_2)dx' \;=\; \rho(x_1|x_2).$$

It shows, in particular, that $\rho(\mathbf{r}_1, \sigma_1|\mathbf{r}_2, \sigma_2)$ tends to zero with increasing $|\mathbf{r}_1 - \mathbf{r}_2|$.

It is instructive to calculate the correlation function in the one-determinant case. From (2.3) and (3.8) we have

$$f(x_1, x_2) \;=\; \frac{|\rho(x_1|x_2)|^2}{\rho(x_1|x_1)\,\rho(x_2|x_2)}. \tag{3.9}$$

The general behavior of this function is similar to that shown in Figure 1(a) for the non-interacting homogeneous electron gas. If $\sigma_1 \neq \sigma_2$ it is equal to zero, so electrons with opposite spins are not correlated. If $\sigma_1 = \sigma_2$ the function f is always positive, does not exceed 1, it is equal to 1 if $x_1 = x_2$, and, in general, $f(x_1, x_2)$ tends to zero when particles moves far apart. For the function describing the correlation hole the equations (2.4), (3.6), (3.8) give

$$f_1(\mathbf{r}_1, \mathbf{r}_2) \;=\; 1 \;-\; \frac{|\rho_\alpha(\mathbf{r}_1|\mathbf{r}_2)|^2 + |\rho_\beta(\mathbf{r}_1|\mathbf{r}_2)|^2}{(\rho_\alpha(\mathbf{r}_1|\mathbf{r}_1) + \rho_\beta(\mathbf{r}_1|\mathbf{r}_1))\,(\rho_\alpha(\mathbf{r}_2|\mathbf{r}_2) + \rho_\beta(\mathbf{r}_2|\mathbf{r}_2))}. \tag{3.10}$$

The correlation hole is similar to that shown in Figure 1(b), except the value at $\mathbf{r}_1 = \mathbf{r}_2$, which is generally less than 0.5, being equal to 0.5 if $\rho_\alpha = \rho_\beta$. The equation (3.9) and (3.10) for the correlation functions are valid for any n-electron system in one-determinant approximation. The particular shape of the correlation function f for parallel spins and the shape of the correlation hole is of course different for different systems.

The energy functional

$$E \;=\; \frac{\langle\Psi|\widehat{H}|\Psi\rangle}{\langle\Psi|\Psi\rangle} \tag{3.11}$$

with the help of equations (3.1), (2.1), (2.2), (3.6), and (3.7) can be written as

$$E \;=\; \sum_{i=1}^{n} \acute{h}_{ii} \;+\; \frac{1}{2}\sum_{i,j=1}^{n}{}' \acute{J}_{ij} \;-\; \frac{1}{2}\sum_{i,j=1}^{n}{}' \acute{K}_{ij}, \tag{3.12}$$

where

$$\begin{aligned} \acute{h}_{ii} \;&=\; \langle\psi_i(x)\,|\,\widehat{h}(\mathbf{r})\,|\,\psi_i(x)\rangle; \\ \acute{J}_{ij} \;&=\; \langle\psi_i(x_1)\psi_j(x_2)\,|\,\widehat{g}(\mathbf{r}_1, \mathbf{r}_2)\,|\,\psi_i(x_1)\psi_j(x_2)\rangle; \\ \acute{K}_{ij} \;&=\; \langle\psi_i(x_1)\psi_j(x_2)\,|\,\widehat{g}(\mathbf{r}_1, \mathbf{r}_2)\,|\,\psi_j(x_1)\psi_i(x_2)\rangle \end{aligned} \tag{3.13}$$

are the one-electron integrals and the two-electron *Coulomb integrals* and *exchange integrals* respectively. The acute sign ´ shows that these integrals are calculated with spin-orbitals. It will distinguish them from similar integrals but calculated with orbitals. The first sum in (3.12) is the n-electron system kinetic energy and its energy in the external field, the second sum is the *Coulomb energy* and the third sum is the *exchange energy* of the n-electron system. The partition of the electron-electron interaction energy into Coulomb and exchange contributions could be done in another way. From (3.13) it follows that $J_{ii} = K_{ii}$. This term is referred to as the state ψ_i *self-energy*. From (3.12) it follows that the total energy does not change if the self-energy is simultaneously added to the Coulomb energy and subtracted from the exchange energy. Hence the total energy can be written as

$$E = \sum_{i=1}^{n} \hat{h}_{ii} + \frac{1}{2}\int \frac{\rho(x_1|x_1)\rho(x_2|x_2)}{|\mathbf{r}_1 - \mathbf{r}_2|} dx_1 dx_2 - \frac{1}{2}\int \frac{\rho(x_1|x_2)\rho(x_2|x_1)}{|\mathbf{r}_1 - \mathbf{r}_2|} dx_1 dx_2. \quad (3.14)$$

The second and third terms in the right hand side of (3.14) are also called the Coulomb and exchange energies respectively. However, their values differ from those of (3.12) by the total self-energy which could be quite substantial. It should be noted here that the trick with the self-energy allows one to express both the Coulomb and exchange energies in terms of the first order reduced density matrix only if the wave function is a single Slater determinant.

According to the second feature of the HF method the HF spin-orbitals are those which bring the minimum to the energy E under conditions imposed on spin-orbitals. In the UHF method the only condition is the orthonormality requirements (3.5). In the RHF method there are additional conditions (3.3). Therefore the UHF and RHF cases should be treated separately.

In the UHF case we have

$$\frac{\delta}{\delta\psi_i}\left(E - \sum_{i,j=1}^{n} \lambda_{ij}\langle\psi_i|\psi_j\rangle\right) = 0, \qquad i = 1, \ldots, n,$$

where λ_{ij} are the Lagrange parameters, which are to take account of the orthonormality conditions (3.5) and which comprise the Hermitian matrix

$$\lambda_{ij} = \lambda_{ji}^{*}$$

because the functional must be real. Taking the functional derivatives we arrive at the following system of equations for spin-orbitals

$$\widehat{F}(x)\psi_i(x) = \sum_{j=1}^{n} \lambda_{ij}\psi_j(x), \qquad i = 1, \ldots, n, \quad (3.15)$$

where

$$\widehat{F}(x) = \hat{h}(\mathbf{r}) + \widehat{J}(x) - \widehat{K}(x)$$

is the Fock operator and $\widehat{J}(x)$ and $\widehat{K}(x)$

$$\widehat{J}(x)f(x) = \int g(\mathbf{r},\mathbf{r}')\rho(x'|x')dx' f(x), \quad \widehat{K}(x)f(x) = \int g(\mathbf{r},\mathbf{r}')\rho(x|x')f(x')dx'$$

are the Coulomb and exchange operators. All these operators are invariant under any unitary transformation of spin-orbitals. The non-diagonal Lagrange parameters $\lambda_{ij}, i \neq j$ usually complicate significantly the solution of the system of equations. However, in this case the problem is easy. To obtain the orthogonal spin orbitals one can simply set the non-diagonal parameters equal to zero and replace the system of equations (3.15) by

$$\widehat{F}(x)\psi_i(x) = \epsilon_i\psi_i(x), \qquad i = 1, \dots, n. \tag{3.16}$$

All spin-orbitals will be orthogonal to each other as eigenfunctions of the same Hermitian operator $\widehat{F}$. The (3.16) is the *canonical* UHF system of equation and its solutions $\psi_i(x)$ are the UHF *canonical* spin-orbitals.

In the RHF case we should first separate the spin variables and write the energy as a functional of orbitals and not of spin-orbitals. Calculating integrals over the spin variables in (3.12) and taking into account the relation (3.3) between spin-orbitals and orbitals we arrive at the equation

$$E = \sum_{i=1}^{n_\beta} 2h_{ii} + \sum_{i=n_\beta+1}^{n_\alpha} h_{ii} + \sum_{i=1}^{n_\beta}\sum_{j=1}^{n_\alpha}(2J_{ij} - K_{ij}) + \frac{1}{2}\sum_{i,j=n_\beta+1}^{n_\alpha}(J_{ij} - K_{ij}) \tag{3.17}$$

where

$$\begin{aligned} h_{ii} &= \langle\phi_i(\mathbf{r})|\widehat{h}(\mathbf{r})|\phi_i(\mathbf{r})\rangle; \\ J_{ij} &= \langle\phi_i(\mathbf{r}_1)\phi_j(\mathbf{r}_2)|\widehat{g}(\mathbf{r}_1,\mathbf{r}_2)|\phi_i(\mathbf{r}_1)\phi_j(\mathbf{r}_2)\rangle; \\ K_{ij} &= \langle\phi_i(\mathbf{r}_1)\phi_j(\mathbf{r}_2)|\widehat{g}(\mathbf{r}_1,\mathbf{r}_2)|\phi_j(\mathbf{r}_1)\phi_i(\mathbf{r}_2)\rangle \end{aligned} \tag{3.18}$$

are the one-electron and two-electron Coulomb and exchange integrals respectively calculated with orbitals. In the RHF case the equations for orbitals are

$$\frac{\delta}{\delta\phi_i}\left(E - \sum_{i,j=1}^{n_\alpha}\lambda_{ij}\langle\phi_i|\phi_j\rangle\right) = 0, \qquad i = 1, \dots, n_\alpha,$$

Taking the functional derivative we obtain the RHF system of equations for one-determinant case

$$\begin{aligned} 2\widehat{F}_d(\mathbf{r})\phi_i(\mathbf{r}) &= \sum_{j=1}^{n_\alpha}\lambda_{ij}\phi_j(\mathbf{r}), \qquad i = 1, \dots, n_\beta \\ \widehat{F}_s(\mathbf{r})\phi_i(\mathbf{r}) &= \sum_{j=1}^{n_\alpha}\lambda_{ij}\phi_j(\mathbf{r}), \qquad i = n_\beta + 1, \dots, n_\alpha \end{aligned} \tag{3.19}$$

In this system there are two different Fock operators

$$\begin{aligned} \widehat{F}_d(\mathbf{r}) &= \widehat{h}(\mathbf{r}) + \widehat{J}_\alpha(\mathbf{r}) + \widehat{J}_\beta(\mathbf{r}) - \frac{1}{2}\left(\widehat{K}_\alpha(\mathbf{r}) + \widehat{K}_\beta(\mathbf{r})\right), \\ \widehat{F}_s(\mathbf{r}) &= \widehat{h}(\mathbf{r}) + \widehat{J}_\alpha(\mathbf{r}) + \widehat{J}_\beta(\mathbf{r}) - \widehat{K}_\alpha(\mathbf{r}) \end{aligned}$$

for the doubly ($\widehat{F}_d$) and singly ($\widehat{F}_s$) occupied orbitals. Here

$$\widehat{J}_\alpha(\mathbf{r})f(\mathbf{r}) = \int\frac{\rho_\alpha(\mathbf{r}'|\mathbf{r}')}{|\mathbf{r}-\mathbf{r}'|}d\mathbf{r}'f(\mathbf{r}), \qquad \widehat{K}_\alpha(\mathbf{r})f(\mathbf{r}) = \int\frac{\rho_\alpha(\mathbf{r}|\mathbf{r}')}{|\mathbf{r}-\mathbf{r}'|}f(\mathbf{r}')d\mathbf{r}',$$

and similar equations are valid for $\widehat{J}_\beta$ and $\widehat{K}_\beta$. The non-diagonal Lagrange parameters in (3.19) contrary to the UHF equations (3.15) could not be set equal to zero because then orbitals ϕ_i with $i \leq n_\beta$ and ϕ_j with $j > n_\beta$ would not be orthogonal to each other and hence spin-orbitals ψ_i with $i \leq n_\beta$ and ψ_j with $n_\beta < j \leq n_\alpha$ also would not be orthogonal. The non-diagonal parameters need special treatment and so the RHF equations are more complicated than the UHF equations.

Both RHF and UHF equations have the transparent physical interpretation: each electron moves in the averaged field (local Coulomb and non-local exchange) of all other electrons and not in field due to the electrons instantaneous positions. (It is easy to verify that the self-interaction terms in the Coulomb and exchange potentials cancel each other.) This is what one means considering the electrons obeying the HF equations as independent.

Comparing the UHF and the single determinant RHF methods we see that UHF equations are simpler and are easier to solve than the RHF equations. However, the UHF wave function is not the eigenfunction of $\widehat{S^2}$. Hence the solution of UHF equations corresponds not to a pure spin state but to a linear combination of several pure spin states. Therefore calculated with it properties of the system are the averaged over several states values. On the other hand, there are n different orbitals in the UHF comparing with n_α orbitals in RHF, hence the UHF wave function is somewhat more flexible, which is an advantage for the variational method. (The pure spin state wave function with n different orbitals appears in the spin-extended HF method, which is many determinant method and more complicated than the one determinant RHF method.) In its turn the RHF solution corresponds to one actual state of the system and calculated with it properties correspond to one state. However, the RHF equations and more complicated than the UHF ones and they are more difficult to solve. Still, if the package to calculate the RHF equations is made it is expedient to use it, the more so because the rigorous way to improve the RHF wave function and to incorporate the correlation effects is known, at least in principle.

3.3 The one determinant RHF method for the singlet state

In one case the one-determinant RHF equation simplifies considerably. This is the case of even number of electrons with $n_\alpha = n_\beta$, i.e. this a singlet $S = 0$ state case. Such state of a system is rather important because the ground state of many solids belongs to it. In this state all orbitals are doubly occupied

$$\psi_i(x) = \phi_i(\mathbf{r})\alpha(\sigma), \quad \psi_{n/2+i}(x) = \phi_i(\mathbf{r})\beta(\sigma), \quad i = 1, \ldots, n/2,$$

ρ_α coincide with ρ_β

$$\rho_\alpha(\mathbf{r}|\mathbf{r}') = \rho_\beta(\mathbf{r}|\mathbf{r}') = \rho(\mathbf{r}|\mathbf{r}') = \sum_{i=1}^{n/2} \phi_i(\mathbf{r})\phi_i^*(\mathbf{r}'), \tag{3.20}$$

and the system (3.19) reduces to the first equation. Only one Fock operator enters this equation and therefore all non-diagonal Lagrange parameters could be set equal to zero similarly to what has been done in the UHF method. In result the RHF system of equations can be written as

$$\widehat{F}(\mathbf{r})\phi_i(\mathbf{r}) = \epsilon_i\phi_i(\mathbf{r}), \qquad i = 1, \ldots, n/2. \tag{3.21}$$

Here

$$\widehat{F}(\mathbf{r}) = \widehat{h}(\mathbf{r}) + 2\widehat{J}(\mathbf{r}) - \widehat{K}(\mathbf{r}), \tag{3.22}$$

and

$$\widehat{J}(\mathbf{r})f(\mathbf{r}) = \int \frac{\rho(\mathbf{r}'|\mathbf{r}')}{|\mathbf{r}-\mathbf{r}'|}d\mathbf{r}' f(\mathbf{r}), \qquad \widehat{K}(\mathbf{r})f(\mathbf{r}) = \int \frac{\rho(\mathbf{r}|\mathbf{r}')}{|\mathbf{r}-\mathbf{r}'|}f(\mathbf{r}')d\mathbf{r}'. \tag{3.23}$$

The (3.21) is the RHF *canonical* system of equations and its solutions $\phi_i(\mathbf{r})$ are the RHF *canonical* orbitals.

Both (3.16) and (3.21) are systems of non-linear equations, because Fock operators depend on one-electron functions (orbitals or spin-orbitals). This make a serious problem. There is no unique method of solution the non-linear system of HF equations. The most widely used one is the iteration process where the orbitals obtained at the m-th step are employed to calculate the next step density matrix and Fock operator and the eigenfunctions of the latter are used (immediately or with modifications) as orbitals of the next, $(m+1)$-th step. The iterations proceed until the difference between the input and output orbitals become smaller than a preselected threshold. This iteration process applied to the HF equations gave rise to the name *self-consistent field* equations. The direct iterations, when the eigenfunction ϕ_i' of Fock operator obtained at the m-th step are used immediately as the one-electron functions $\phi_i^{(m+1)}$ of the next step, seldom converge. The convergence depends on the relative strength of two interactions, the electron-electron interaction and the electron-nucleus interaction, the greater strength of electron-nucleus interaction the better convergence. The direct iterations show good convergence for highly ionized systems and they never converge for the negative ions. The usual receipt to achieve convergence, or to accelerate the poor convergence, is to take the next approximation as follows

$$\phi_i^{(m+1)} = \phi_i^{(m)} + \tau(\phi_i' - \phi_i^{(m)}) \tag{3.24}$$

where τ is either the preselected parameter or is calculated with orbitals of m-th, $(m-1)$-th, and sometimes few further back steps, such as in the δ^2-process of Aitken [11], or Wegstein process [12]. The value $\tau = 1$ corresponds to the direct iterations and sometimes the process (3.24) with $\tau = 0.1$ shows reasonable convergence.

Another problem arise because Fock operator has an infinite number of eigenfunctions and eigenvalues and there should be a receipt how to select the particular set of eigenfunctions to be used as the next approximation to orbitals. Usually the receipt is based on the physical interpretation of the Fock operator eigenvalue ϵ_k. From (3.21) and (3.18) it follows

$$\epsilon_k = \langle\phi_k|\widehat{F}|\phi_k\rangle = h_{kk} + \sum_{j=1}^{n/2}(2J_{kj} - K_{kj}). \tag{3.25}$$

At the same time with the help of (3.17) one can write the equation for energies $E^{(n)}$ and $E_k^{(n-1)}$ of two states. The $E^{(n)}$ is the energy of the n-electron system singlet state with doubly occupied orbitals $\phi_i, i = 1, \ldots, n/2$ and corresponding spin-orbitals. The $E_k^{(n-1)}$ is the energy of the $(n-1)$-electron system doublet state with the same spin-orbitals, one spin-orbital $\phi_k\alpha$ or $\phi_k\beta$ being omitted. Then it is easy to show that

$$\epsilon_k = E^{(n)} - E_k^{(n-1)}. \tag{3.26}$$

The equation (3.26) shows that the eigenvalue ϵ_k of canonical RHF equation is an approximation to the negative of the ionization potential from the one-electron state ϕ_k. This statement is known as Koopmans theorem [13]. The approximation is the neglect of the ion relaxation, i.e. the assumption that spin-orbitals of the ionized state are the same as the corresponding spin-orbitals of the initial state. The Koopmans theorem (3.26) is also valid for the canonical UHF equation, the ϵ_k being corresponded to the spin orbital ψ_k, $k = 1\ldots, n$. Because of the Koopmans theorem the ϵ_k is usually referred to as the one-electron energy levels. (Note, that the actual energy of the electron in the state ψ_k differs from ϵ_k because the former contains half the interaction energy with all other electrons and not the whole interaction energy as in (3.25)). One usually assumes that in the ground state the electrons should occupy all lowest one-electron energy levels and therefore n spin-orbitals corresponding to the lowest ϵ_k should be selected to make the one-determinant wave function of the ground state. The deficiency of this commonly used occupation rule is the neglect of relaxation effects. For example, in compact systems like atoms the relaxation effects often make the n-electron states with the said population to have larger energy than those where some one-electron states with higher ϵ_k are occupied while some other with lover ϵ_k are empty. The simplest example is the ground state of the transition metal atom where 4s state is occupied while some 3d states are empty although $\epsilon_{3d} < \epsilon_{4s}$. Besides, the said occupation rule applied during the iteration process can result in divergence when the highest occupied and lowest unoccupied levels reverse their positions in each successive iteration.

The Fock operator eigenfunctions which are included into the one-determinant wave function are referred to as *occupied* (canonical) orbitals and all other eigenfunctions are referred to as *virtual* orbitals.

The canonical RHF orbitals of a molecule or a solid are usually spread over the whole system. These are the so called *delocalized* orbitals comparing with *localized* orbitals which are spread over one or few atoms only. In many cases, especially for large systems it is an advantage to employ the localized orbitals. However, the canonical are not the only orbitals which could be used in the one-determinant case. Indeed, the one-determinant wave function (3.3) built of any $n/2$ linearly independent combinations

$$\varphi_i(\mathbf{r}) = \sum_{k=1}^{n/2} C_{ik}\phi_k(\mathbf{r}) \tag{3.27}$$

of occupied canonical orbitals ϕ_k and that built of ϕ_k themselves differ by the normalization factor only. Therefore the linear combinations (3.27) with the non-singular matrix C could be used as orbitals also and these are referred to as the *non-canonical* orbitals. The occupied canonical and non-canonical orbitals span the same functional space Ω.

The non-canonical orbitals are usually normalized to 1, although it is not necessary, and, in general, they are not orthogonal. The matrixes $\mathbf{T}$ of the scalar products

$$T_{ij} = \langle\varphi_i(\mathbf{r})|\varphi_j(\mathbf{r})\rangle \tag{3.28}$$

is the *metric* matrix and $\mathbf{S} = \mathbf{T} - \mathbf{I}$ is the *overlap* matrix. In terms of non-canonical orbitals the first order reduced density matrix is

$$\rho(x|x') = \rho(\mathbf{r}|\mathbf{r}')\left(\alpha(\sigma)\alpha(\sigma') + \beta(\sigma)\beta(\sigma')\right), \tag{3.29}$$

where

$$\rho(\mathbf{r}|\mathbf{r}') = \sum_{i,j=1}^{n/2} \varphi_i(\mathbf{r}) \left\{\mathbf{T}^{-1}\right\}_{ij} \varphi_j^*(\mathbf{r}'). \tag{3.30}$$

The second order reduced density matrix is still given by the equation (3.8). The density matrixes (3.29) and (3.30) are invariant under any linear non-singular transformation of orbitals φ_i and not only under the unitary transformations as (3.20). So the density matrix (3.30) is defined by the functional space Ω and not by the particular choice of orbitals in it.

The equation for non-canonical orbitals could be obtained with the variational principle applied to the energy functional E (3.11) without any orthonormality conditions imposed on orbitals φ_i [14]. After simple but tedious calculations one arrives at the equation

$$\widehat{F}\,\varphi_i(\mathbf{r}) = \widehat{\rho}\,\widehat{F}\,\widehat{\rho}\,\varphi_i(\mathbf{r}), \qquad i = 1, \ldots, n/2, \tag{3.31}$$

where $\widehat{F}$ is the Fock operator (3.22), (3.23) with the density (3.30), and $\widehat{\rho}$

$$\widehat{\rho} f(\mathbf{r}) = \int \rho(\mathbf{r}|\mathbf{r}') f(\mathbf{r}') d\mathbf{r}'$$

is the projector onto the functional space Ω. The equation (3.31) could also be obtained directly from (3.27) taking into account the equation (3.21) for the canonical orbitals.

The equation (3.31) is the general equation for the occupied non-canonical (and canonical) orbitals as any orbital belonging to the space Ω satisfy the equation (3.31). It enables one to develop an eigenfunction and eigenvalue problem for non-canonical orbitals with specific properties. From (3.27) it follows that $\widehat{F}\varphi_i \in \Omega$ and $\widehat{\rho}\varphi_i \in \Omega$. Therefore for any Hermitian operator $\widehat{A}$ the eigenfunctions of the equation

$$\left(\widehat{F} + \widehat{\rho}\,\widehat{A}\,\widehat{\rho}\right)\varphi_i = \lambda_i \varphi_i, \qquad i = 1, \ldots, n/2 \tag{3.32}$$

belong to Ω and therefore they are the non-canonical orbitals. The equations (3.32) are known as the Adams-Gilbert equations [15, 16].

In (3.32) all non-canonical orbitals φ_i are eigenfunctions of the same Hermitian operator and therefore they are orthogonal to each other. To develop equations for non-orthogonal orbitals it is necessary to divide all occupied orbitals into N groups (atoms, ions, bonds, etc.) with $n_p, p = 1, N$ orbitals in each group, the total number of orbitals being equal to $n/2$. Then for each group one can set a particular operator $\widehat{A}_p$ to obtain the system of equations

$$\left(\widehat{F} + \widehat{\rho}\,\widehat{A}_p\,\widehat{\rho}\right)\varphi_{pi} = \lambda_{pi}\varphi_{pi}, \qquad p = 1, \ldots, N, \quad i = 1, \ldots, n_p. \tag{3.33}$$

Here orbitals of different groups are not orthogonal to each other because they are eigenfunctions of different operators. With proper choice of $\widehat{A}_p$ the non-canonical orbitals φ_{pi} of the group p will be localized. Besides, if we introduce Fock operator $\widehat{F}_p$ and operator $\widehat{U}_p = \widehat{F} - \widehat{F}_p$ for every group then the equation (3.33) can be written as

$$\left(\widehat{F}_p + \widehat{U}_p + \widehat{\rho}\,\widehat{A}_p\,\widehat{\rho}\right)\varphi_{pi} = \lambda_{pi}\varphi_{pi}, \qquad p = 1, \ldots, N, \quad i = 1, \ldots, n_p.$$

These have the form of equations for group p (atom, bond, etc.) orbitals in the field of all other groups which open the vast field for modelling. With minor modification the Adams-Gilbert equations are valid in the UHF case also.

3.4 Several determinants case

The one-determinant is the simplest approximate wave function. However, not every state can be described by the one-determinant function and an example is the excited singlet state. For the sake of simplicity we consider the two-electron system. In the excited state there should be two different orbitals $\phi_1(\mathbf{r})$ and $\phi_2(\mathbf{r})$ so there are two determinants with zero projection of the total spin

$$\Psi_1(x_1, x_2) = \frac{1}{\sqrt{2}} \begin{vmatrix} \phi_1(\mathbf{r}_1)\alpha(\sigma_1), & \phi_1(\mathbf{r}_2)\alpha(\sigma_2) \\ \phi_2(\mathbf{r}_1)\beta(\sigma_1), & \phi_2(\mathbf{r}_2)\beta(\sigma_2) \end{vmatrix},$$

$$\Psi_2(x_1, x_2) = \frac{1}{\sqrt{2}} \begin{vmatrix} \phi_2(\mathbf{r}_1)\alpha(\sigma_1), & \phi_2(\mathbf{r}_2)\alpha(\sigma_2) \\ \phi_1(\mathbf{r}_1)\beta(\sigma_1), & \phi_1(\mathbf{r}_2)\beta(\sigma_2) \end{vmatrix}.$$

These determinants are the two UHF wave functions, corresponding to the same energy (in the absence of the magnetic field) and they give rise to two RHF wave functions, singlet and triplet, with zero spin projection

$$\Psi_{00} = \frac{1}{\sqrt{2}}(\Psi_1 + \Psi_2), \qquad \Psi_{10} = \frac{1}{\sqrt{2}}(\Psi_1 - \Psi_2).$$

They are the two-determinant wave functions, and any UHF one-determinant wave function with zero spin projection is the linear combination of the singlet and triplet states.

In general, the RHF wave function is composed of several determinants in the open shells case, which is better to consider for systems with spatial symmetry. However, for the many-electron system there are several similar but still different terms associated with its symmetry, and some of them are not as transparent as they seem.

The "symmetry of the many-electron system" is the symmetry of the external field in which electrons are moving, the symmetry of the potential due to the nuclei of the system in particular. To specify the symmetry of the system it is necessary to indicate the group of symmetry operations under which the Hamiltonian (3.1) is invariant.

The expressions "the many-electron state has symmetry" or "the many-electron wave function has symmetry" means that the many-electron wave function is a member of the basis set of an irreducible representation of the symmetry group of the system.

The expression "the orbital (or spin-orbital) has symmetry" means that the orbital is a member of the basis set of an irreducible representation of some symmetry group, and the expression "the one-electron problem has a symmetry" means that such group does exist. Contrary to the many-electron symmetry group, the one-electron symmetry group is not easy to recognize because in general the one-electron functions are not the eigenfunctions of one operator but are solutions to the system of equations, the operators in different equations being different. However, the most important is the non-linearity of equations, because the linearity lies at the very base of the group theory. For example, if two solutions of the non-linear equation correspond to the same energy, their linear combination is not necessarily the solution of the equation, and even if it is a solution it does not necessarily correspond to the same energy. In the case of HF equations the energy functional should be carefully chosen to preserve the assumed symmetry of the one-electron problem.

Two points should be stressed here. First, if the nucleus field has a spatial symmetry the ground state many-electron wave function does not necessarily belong to the identical representation. It is very well known for atoms, where the nuclei field is spherically symmetrical and only a small number of atoms have spherically symmetrical many-electron wave function in the ground state. Second, the symmetry group of one-electron functions should not necessarily be that of the many-electron function. The simplest example is the Heitler-London wave function of the ground state hydrogen molecule. In this case the reflection in the plane perpendicular to the bond at its midpoint is the symmetry operation for the two-electron wave function, and it is not the symmetry operation for any of two one-electron functions. Only the set of two one-electron functions taken together makes the reducible representation basis. Sometimes when the state has a lower symmetry the term "broken symmetry" is used.

Now we assume that the electron problem has a symmetry and this symmetry coincides with the many-electron symmetry. Then every orbital belongs to the irreducible representation. We will use the index Γ for the irreducible representation, $N(\Gamma)$ for its dimension, the index $\gamma = 1, \ldots, N(\Gamma)$ to enumerate the basis functions, and the index n to distinguish functions with the same pair of indexes Γ and γ. From each orbital $\phi_{n\Gamma\gamma}$ we construct two spin-orbitals

$$\psi_{n\Gamma\gamma\uparrow}(x) = \phi_{n\Gamma\gamma}(\mathbf{r})\alpha(\sigma), \qquad \psi_{n\Gamma\gamma\downarrow}(x) = \phi_{n\Gamma\gamma}(\mathbf{r})\beta(\sigma).$$

The functional space spanned by spin-orbitals with fixed n, Γ is referred to as the $n\Gamma$ *(electronic) shell*. Suppose we have t shells $n_j, \Gamma_j, j = 1, \ldots, t$ and t integers g_j such that

$$0 < g_j \leq 2N(\Gamma_j), \qquad j = 1, \ldots, t, \qquad \sum_{j=1}^{t} g_j = n_e.$$

Here temporarily we denoted by n_e the number of electrons to distinguish it from the index n of the electron shell. Let us construct all linearly independent Slater determinants containing g_1 spin-orbitals from the shell n_1, Γ_1, g_2 spin-orbitals from the shell $n_2\Gamma_2$ etc. These determinants span the linear space of n_e-electron functions, which is referred to as the *configuration* and denoted as

$$(n_1\Gamma_1)^{g_1}\,(n_2\Gamma_2)^{g_2}\,\cdots,(n_t\Gamma_t)^{g_t}. \tag{3.34}$$

The integer g_j is usually referred to as the occupation number of $n_j\Gamma_j$ shell in the configuration (3.34). If $g_j = 2N(\Gamma_i)$ then the shell $n_j\Gamma_j$ is referred to as *closed*, if $g_j = 0$ then the shell is referred to as *empty*, and in any intermediate case the shell is referred to as *open*. We assume that any configuration under consideration contains no empty shells.

In the RHF method the approximate wave function i) is the linear combination of Slater determinants belong to one configuration and ii) this approximate wave function is an eigenfunction of $\widehat{S^2}$ and $\widehat{S}_z$ operators, and is γ-th basis vector of Γ-th representation. To ensure that the one-electron problem symmetry coincides with the many-electron system symmetry the energy functional should contain only invariants

$$\sum_{\gamma} \phi_{n\Gamma\gamma}(\mathbf{r})\phi^*_{n\Gamma\gamma}(\mathbf{r}'). \tag{3.35}$$

In general it could not be achieved if one takes the mean value of the Hamiltonian in one particular many-electron state for an energy functional. An averaging over several states should be used instead. In the case of spherically symmetrical field the averaging over S_z and γ with constant S and Γ results in the necessary form of the functional. This averaging is not an approximation as the exact total energy does not depend on S_z and γ. In the case of lower symmetry an additional averaging over states with different total energies may be necessary for the energy functional to contain only invariants (3.35). The general RHF equations are rather complicated and will not be presented here. There is one important and at the same time simple case in which configuration contains closed shells only. In this case the RHF wave function is a single Slater determinant, the mean value of the Hamiltonian calculated with one-determinant wave function contains only invariants (3.35), so in this case the one-electron and many-electron symmetries coincide. The ground state of most insulating solids corresponds to the closed shells configuration. This one-determinant case was discussed in the previous section.

In the solid state theory especially important is the translation symmetry. The one-electron problem can have the translation symmetry if the configuration contains only closed shells. For the translation symmetry the closed shell means that if the orbital $\phi_i(\mathbf{k}_m, \mathbf{r})$ is included into the one-determinant wave function, then functions with all $\mathbf{k}_{m'}$ belonging to the same **k**-star are also included. In this closed shell case the Fock operator (3.22) in the canonical RHF equations (3.21) is invariant under any translation. So in the closed shells case the RHF canonical equations for crystal is a band structure problem with the exact exchange. The band structure is a well known problem. With different approximation for the potential it was applied to various solids and there is extensive literature devoted to it. For referencies, especially concerning the HF calculations, see [8].

In the next section of the present paper another, less known approch based on localized orbitals will be considered in application to the non-metallic solids.

4 Local orbitals Hartree-Fock method for ionic solids

4.1 Localized orbitals and non-conventional variational approach

It is well established for ionic solids that they could be described as a collection of overlapping ions, the amount of overlap being correspond to the covalence admixture. We will consider here ionic solids with closed shells and large enough gap between the top of the valence band and the bottom of the conduction band. The ions in these solids also contain closed shells only. For these solids it is expedient to use the local orbital description as was shown in the pioneering work of Löwdin [17], where the cohesive properties of alkali-halide crystals were successfully calculated with the help of free ions orbitals.

The canonical HF orbitals of the perfect ionic crystal are the completely delocalized Bloch functions. They are unique apart from the arbitrary k-dependent phase factor. The Bloch functions Fourier transform are Wannier functions which are the orthogonal and normalized to unity non-canonical orbitals. Wannier functions are not unique as their shape depends on the Bloch functions phase factor. Any Wannier function is localized, as it is normalized to 1, and the said phase factor can be selected so as to provide the best possible localization. Still even these functions are not perfectly localized because of the oscillating tails necessary to provide the different Wannier functions orthogonality. Better localization can be achieved when normalized and not orthogonal linear combinations of Wannier functions are considered. However, the direct calculation of localized orbitals from Bloch functions is technically very complicated procedure, especially in the case of overlapping energy bands. Much better is to calculate the localized nonorthogonal non-canonical crystal orbitals directly with the help, for example, the Adams-Gilbert equations (3.33).

However, instead of solving the HF equations a special direct variational approach with an appropriate energy functional could be employed for calculation the non-canonical localized orbitals. In the conventional approach an analytic expression for orbitals with variational parameters is assumed so that the energy functional become a function of variational parameters. Unfortunately, with non-orthogonal orbitals this function is not a polynomial and this makes the problem practically untractable except some special simple cases. There exists another approach employing the variational parameters in another way. As ions are stable in the crystal one can assume that each localized orbital is the solution to the HF equations for an ion in an external potential, which is a confining potential for negative ions and stretching potential for positive ions. An analytic expression containing variational parameters is assumed for these external potentials and not for orbitals. The variational approach is a direct one: select values for the variational parameters, calculate orbitals for every ion, calculate the energy of the crystal, change values of the parameters and proceed until the energy minimum is achieved. Contrary to the HF potential of the whole system the external potential for an ion is not unique. Different potentials could be equally valid corresponding to different localized orbitals build up from the same canonical HF orbitals.

4.2 Energy functional

With the help of localized orbitals it is easy to find the energy functional for an infinite crystal with translation symmetry. We start with a finite crystal containing N closed shells ions and assume that all nucleus are situated at the corresponding lattice sites and that the local orbitals of a given type ions are the same irrespective of where they are situated, in the middle of the finite crystal or at its border. Each ion we label by the index g and orbitals we label by two indexes g and k accordingly

$$\varphi_{g,k}(\mathbf{r}), \qquad g = 1,2,\cdots,N, \qquad k = 1,2,\cdots,n_g,$$

where n_g is the number of orbitals localized on the ion g. The charge and the position vector of the ion g nucleus we denote Z_g and $\mathbf{R}_g$. The metric matrix $\mathbf{T}$ (3.28) is

$$\mathbf{T}_{g,k;g',k'} = \langle\phi_{g,k}|\phi_{g',k'}\rangle,$$

The orbitals of the same ion are assumed to be orthonormal, so for the overlap matrix $\mathbf{S} = \mathbf{T} - \mathbf{I}$ one has

$$\mathbf{S}_{g,k;g,k'} = 0.$$

Hence $\mathbf{S}$ describes the overlap of different ions.

In the RHF approximation the adiabatic potential (electron energy plus the nuclei-nuclei repulsion energy) of the finite crystal with closed shells is

$$W = 2\sum_{g=1}^{N}\sum_{k=1}^{n_g}\int \varphi_{g,k}^{*}(\mathbf{r})\left(-\frac{1}{2}\nabla^2 - \sum_{g'=1}^{N}\frac{Z_{g'}}{|\mathbf{r}-\mathbf{R}_{g'}|}\right)\varphi_{g,k}(\mathbf{r})d\mathbf{r} + \\ + \int\frac{2\rho(\mathbf{r}|\mathbf{r})\rho(\mathbf{r}'|\mathbf{r}') - \rho(\mathbf{r}|\mathbf{r}')\rho(\mathbf{r}'|\mathbf{r})}{|\mathbf{r}-\mathbf{r}'|}d\mathbf{r}d\mathbf{r}' + \sum_{g<g'=1}^{N}\frac{Z_g Z_{g'}}{|\mathbf{R}_g - \mathbf{R}_{g'}|} \tag{4.1}$$

where the electron density matrix $\rho(\mathbf{r}|\mathbf{r}')$ (3.30) is

$$\rho(\mathbf{r}|\mathbf{r}') = \sum_{g,g'=1}^{N}\sum_{k=1}^{n_g}\sum_{k'=1}^{n_{g'}}\varphi_{g,k}(\mathbf{r})\left\{\mathbf{T}^{-1}\right\}_{g,k;g',k'}\varphi_{g',k'}^{*}(\mathbf{r}').$$

Atomic units will be used unless stated overwise.

Let us rearrange the equation for the density matrix. The sum

$$\rho_g(\mathbf{r}|\mathbf{r}') = \sum_{k=1}^{n_g}\varphi_{g,k}(\mathbf{r})\,\varphi_{g,k}^{*}(\mathbf{r}')$$

can be thought of as the density matrix of the ion g in the crystal. This equation is similar to the density matrix of the single ion but with non-canonical orbitals of the crystal localized on the given ion. The density matrix ρ_{g_1,g_2} of the two ions g_1 and g_2 considering as a two-ion cluster is

$$\rho_{g_1,g_2}(\mathbf{r}|\mathbf{r}') = \sum_{g,g'\in g_1,g_2}\sum_{k=1}^{n_g}\sum_{k'=1}^{n_{g'}}\varphi_{g,k}(\mathbf{r})\left\{\mathbf{T}^{-1}\right\}_{g,k;g',k'}\varphi_{g',k'}^{*}(\mathbf{r}').$$

With density matrixes for both ions g_1 and g_2 this equation can be written as

$$\rho_{g_1,g_2} = \rho_{g_1} + \rho_{g_2} + \widetilde{\rho}_{g_1,g_2} \tag{4.2}$$

where $\widetilde{\rho}_{g_1,g_2}$ is the two-ion constituent of the density matrix thus defined for each pair of ions. Similarly the density of the cluster consisting of three ions g_1, g_2, g_3 can be written as

$$\rho_{g_1,g_2,g_3} = \rho_{g_1} + \rho_{g_2} + \rho_{g_3} + \widetilde{\rho}_{g_1,g_2} + \widetilde{\rho}_{g_1,g_3} + \widetilde{\rho}_{g_2,g_3} + \widetilde{\rho}_{g_1,g_2,g_3}, \tag{4.3}$$

where $\widetilde{\rho}_{g_1,g_2,g_3}$ is the three-ion constituent of the density matrix. Continuing the process the density matrix of the finite crystal can be written as

$$\rho(\mathbf{r}\,|\,\mathbf{r}') = \sum_{n=1}^{N}\sum_{g_1<g_2\cdots<g_n=1}^{N}\widetilde{\rho}_{g_1,g_2,\cdots,g_n}(\mathbf{r}\,|\,\mathbf{r}'). \tag{4.4}$$

where $\widetilde{\rho}_{g_1,g_2,\cdots,g_n}$ is the n-ion constituent of the density matrix. In (4.4)

$$\widetilde{\rho}_g(\mathbf{r}\,|\,\mathbf{r}') = \rho_g(\mathbf{r}\,|\,\mathbf{r}').$$

We will refer the equation (4.4) to as the cluster expansion for the density [18, 19]. The indexes $g_1, g_2, \cdots, g_n$ can be arranged not only in increasing order as in (4.4) but in arbitrary order providing that all clusters are accumulated in the sum and none of the clusters is counted twice.

The n-ion constituents of the density can be found by reversing the equations (4.2), (4.3), etc. However it is possible to write the equation for the n-ion constituent in (4.4) in the closed form [19] as follows

$$\tilde{\rho}_{g_1,\cdots,g_n}(\mathbf{r}\,|\,\mathbf{r}') = -\sum_{g,k}\sum_{g',k'} \varphi_{g,k}(\mathbf{r}) \left(\mathbf{Q}[g_1,\cdots,g_n]\right)_{g,k;g',k'} \varphi^*_{g',k'}(\mathbf{r}'),$$

Here the sum is over all the ions $g_1, g_2, \cdots, g_n$ of the cluster and over all orbitals of each ion. The matrix $\mathbf{Q}$ is the following

$$\mathbf{Q}[g_1,\cdots,g_n] = \sum_{m=1}^{n} \sum_{\alpha_1<\cdots<\alpha_m}^{g_1<\cdots<g_n} (-)^{m+n} \left(\mathbf{I} - (\mathbf{I}+\mathbf{S}[\alpha_1,\cdots,\alpha_m])^{-1}\right). \tag{4.5}$$

In this equation the sum is over all subsets $\alpha_1, \cdots, \alpha_m$ of the set $g_1, \cdots, g_n$. In over words the sum is over all subclusters $\alpha_1, \cdots, \alpha_m$ of the cluster $g_1, \cdots, g_n$. In the equation (4.5) all matrixes are of the same size, the overlap matrix $\mathbf{S}[\alpha_1, \cdots, \alpha_m]$ of the subcluster being completed with zeroes as follows

$$\left(\mathbf{S}[\alpha_1,\cdots,\alpha_m]\right)_{g,k;g',k'} = \begin{cases} S_{g,k;g',k'}, & \text{if } g, g' \in \alpha_1, \alpha_2, \cdots, \alpha_m, \\ 0, & \text{overwise} \end{cases}.$$

The cluster expansion for the energy follows immediately when the density matrix is substituted by cluster expansion (4.4) in the equation (4.1) for the energy

$$\begin{aligned} W &= \sum_{n=1}^{N} \sum_{g_1<g_2<\cdots<g_n=1}^{N} \widetilde{W}[g_1,g_2,\cdots,g_n] = \\ &= \sum_{g=1}^{N} \widetilde{W}[g] + \sum_{g_1<g_2=1}^{N} \widetilde{W}[g_1,g_2] + \sum_{g_1<g_2<g_3=1}^{N} \widetilde{W}[g_1,g_2,g_3] + \cdots . \end{aligned} \tag{4.6}$$

Here $\widetilde{W}[g]$ is the energy of the ion g, $\widetilde{W}[g_1, g_1]$ is the interaction energy of ions g_1 and g_2, and $\widetilde{W}[g_1, \cdots, g_n]$ is the n-ion interaction energy. The explicit equation for $\widetilde{W}[g_1, \cdots, g_n]$ could be found in [20], it is too lengthy to be presented here. The cluster expansion (4.6) of the energy appeals by its simple form corresponding to our intuitive notion of an ionic molecule or cluster as a collection of interacting ions, deformed in the process of the cluster formation but still ions. The many-body interactions between ions are native to the simulation models of crystals.

Both cluster expansions (4.4) for the electron density and (4.6) for the energy were obtained for finite clusters and the assumption about the similarity of orbitals has not been used. In (4.4) (4.6) all sums are finite and the energy W is also finite. When these expansions are applied to the infinite crystal the finite sums become an infinite series and the energy W itself becomes infinite. Let us consider the n-ion contribution to the perfect crystal energy

$$W^{(n)} = \sum_{g_1<g_2<\cdots<g_n=1} \widetilde{W}[g_1,g_2,\cdots,g_n] = \frac{1}{n!} \sum_{g_1,g_2\cdots,g_n}{}' \widetilde{W}[g_1,g_2,\cdots,g_n], \tag{4.7}$$

The prime by the second sum means that none two of the indexes are equal. In this equation we took into account that by definition the value of $\widetilde{W}[g_1, g_2, \cdots, g_n]$ does not depend on its indexes order. In the perfect crystal case the index g is a complex of two indexes, ℓ numbering the elementary cells, and t numbering the ions in the elementary cell. Now we employ the assumption that the shape of localized orbitals does not depend on ℓ, i.e. $\phi_{t\ell k}(\mathbf{r}) = \phi_{tk}(\mathbf{r} - \mathbf{R}_\ell)$. Then the value of the n-ion interaction $\widetilde{W}[g_1, g_2, \cdots, g_n]$ does not change when the cluster is translated as a whole by some lattice vector. Consequently, for the perfect crystal one can define the energy per unit cell and for the n-ion contribution to this energy obtain

$$w^{(n)} = \frac{1}{n!} \sum_{g_1} \sum_{g_2 \cdots g_n}{}' \widetilde{W}[g_1, g_2, \cdots, g_n], \qquad g_1 = t, 0. \tag{4.8}$$

In the equation (4.8) the index g_1 runs over the ions of a certain unit cell, which we will refer to as the central unit cell and will assign the value $\ell = 0$ to it, and indexes $g_2, \cdots, g_n$ run over the whole lattice including the central unit cell. The prime by the sum shows that all indexes $g_1, g_2, \cdots, g_n$ are different. The sum in (4.8) is carried over the infinite lattice. It can be shown that $w^{(n)}$ can be expressed as a combinations of conventional lattice sums and a few absolutely and rapidly convergent series.

Consequently the energy of the perfect crystal per unit cell can be written as

$$w = w^{(1)} + w^{(2)} + w^{(3)} + \cdots. \tag{4.9}$$

This is also an infinite series. The convergence of this series is a complicated problem depending on the overlap. For example, in the case of zero overlap (orthogonal orbitals) this series reduces to the sum of the first and second terms only. The numerical calculations for some alkali halide crystals with non-zero overlap show that one can retain only few first members in this series.

4.3 The energy of the crystal with NaCl lattice

The equation for the n-ion contribution to the crystal energy per unit cell can be simplified if the crystal has a point symmetry. It can be shown that if a cluster $g_1, \cdots, g_n$ being subjected to the point symmetry operation with respect to the lattice site g_1 coincide with another cluster $g_1, g_2', \cdots, g_n'$ then the n-ion interaction energies of these clusters also coincide

$$\widetilde{W}[g_1, g_2, \cdots, g_n] = \widetilde{W}[g_1, g_2', \cdots, g_n'].$$

Therefore the sum (4.8) can be written as

$$w^{(n)} = \frac{1}{n!} \sum_{g_1} \sum_{g_2 \cdots g_n}{}'' N[g_1, g_2, \cdots, g_n] \widetilde{W}[g_1, g_2, \cdots, g_n], \qquad g_1 = t, 0.$$

The double prime by the sum shows that the summation is over the symmetry nonequivalent clusters only and $N[g_1, g_2, \cdots, g_n]$ is the number of clusters equivalent to the cluster $g_1, g_2, \cdots, g_n$.

Let us consider a cubic ionic crystal with NaCl lattice. The unit cell contains two ions, cation and anion, and we assign indexes $t = 1$ and $t = 2$ to them correspondingly.

We put the frame of reference origin at some lattice site and denote e_0 the charge of the ion at this site. The position vector of every ion g is $\mathbf{R}_g = a\mathbf{n}_g$, where a is the nearest neighbors distance and $\mathbf{n}_g$ is a vector with integer Cartesian components. The equilibrium value of a we will denote a_0. The charge e_g of an ion g is either e_0 or $-e_0$ depending on whether the sum $n_{gx} + n_{gy} + n_{gz}$ is even or odd.

The simplest is the one-ion contribution to the crystal energy which is the sum

$$w^{(1)} = \widetilde{W}[g_c] + \widetilde{W}[g_a], \qquad g_c = 1,0, \quad g_a = 2,0.$$

of cation and anion energies.

The two-ion contribution in the NaCl lattice is

$$w^{(2)} = \frac{1}{2}\sum_{t=1}^{2}\sum_{g}{}'' N[g_1,g]\widetilde{W}[g_1,g] = \frac{1}{2}\sum_{t=1}^{2}\sum_{0\le n_{gx}\le n_{gy}\le n_{gz}} N[g_1,g]\widetilde{W}[g_1,g], \quad g_1 = t,0.$$

and the two-ion interaction energy can be written as a sum of the long-range and short-range components [20]

$$\widetilde{W}[g_1,g_2] = W_{lr}^{(2)}[g_1,g_2] + W_{sr}^{(2)}[g_1,g_2].$$

The long-range component

$$W_{lr}^{(2)}[g_1,g_2] = \frac{e_{g_1}e_{g_2}}{|\mathbf{R}_{g_1} - \mathbf{R}_{g_2}|}$$

is the point charges interaction energy and its contribution to the energy per unit cell is just the Madelung energy

$$w_{lr}^{(2)} = -e_0^2\frac{M}{a_0},$$

where M is the Madelung constant (for NaCl lattice M=1.7475645946). The short-range interaction decreases exponentially with increasing the distance between ions and the corresponding summation

$$w_{sr}^{(2)} = \frac{1}{2}\sum_{t=1}^{2}\sum_{0\le n_{gx}\le n_{gy}\le n_{gz}} N[g_1,g]W_{sr}^{(2)}[g_1,g], \qquad g_1 = t,0$$

can be carried out immediately.

The three-ion interaction energy can also be written as a sum of the long-range and short-range components [20]

$$\widetilde{W}[g_1,g_2,g_3] = W_{lr}[g_1,g_2,g_3] + W_{sr}[g_1,g_2,g_3].$$

The three-ion long-range interaction is

$$W_{lr}[g_1,g_2,g_3] = t(g_1,g_2|g_3) + t(g_1,g_3|g_2) + t(g_2,g_3|g_1),$$

where

$$t(g_1,g_2|g_3) = -2e_{g_3}\int\frac{\widetilde{\rho}_{g_1,g_2}(\mathbf{r}|\mathbf{r})}{|\mathbf{r} - \mathbf{R}_{g_3}|}d\mathbf{r}.$$

It decays exponentially with increasing the distance between ions g_1 and g_2, and slowly (as $1/R^2$, because $\int \widetilde{\rho}_{g_1,g_2}(\mathbf{r}|\mathbf{r})d\mathbf{r} = 0$) with increasing the distance R between the ion g_3 and the cluster g_1, g_2. To calculate its contribution to the cohesive energy one can start from the equation (4.7) to obtain

$$W_{lr}^{(3)} = \frac{1}{3!} \sum_{g_1,g_2,g_3}{}' W_{lr}[g_1, g_2, g_3] = \frac{1}{2} \sum_{g_1,g_2,g_3}{}' t(g_1, g_2|g_3).$$

If we denote

$$E_{lr}^{(3)}[g_1, g_2] = \sum_{\substack{g_3 \\ g_3 \neq g_1, g_2}} t(g_1, g_2|g_3).$$

then we see that the equation for the three-ion long-range contribution to the cohesive energy is similar to the two-ion short-range contribution, the $W_{sr}^{(2)}[g_1, g_2]$ being changed for $E_{lr}^{(3)}[g_1, g_2]$. Hence

$$w_{lr}^{(3)} = \frac{1}{2} \sum_{t=1}^{2} \sum_{0 \leq n_{gx} \leq n_{gy} \leq n_{gz}} N[g_1, g] E_{lr}^{(3)}[g_1, g], \qquad g_1 = t, 0.$$

The three-ion short-range interaction decays exponentially when the distance between any two of three ions increases. So the sum for the three-ion short-range contribution to the cohesive energy

$$w_{sr}^{(3)} = \frac{1}{6} \sum_{t=1}^{2} \sum_{g_2,g_3}{}'' N[g_1, g_2, g_3] W_{sr}[g_1, g_2, g_3], \qquad g_1 = t, 0$$

can be calculated immediately.

The same analysis can be performed for any n-ion contribution to the cohesive energy.

4.4 Confining potential

As we had already discussed the localized orbitals φ_{tk} are chosen to be the solution of HF equations for the corresponding ion in a potential $V_t(\mathbf{r})$ simulating the crystal environment of the ion

$$\left(\widehat{F}_t + V_t(\mathbf{r})\right) \varphi_{t,k} = \varepsilon_{t,k} \varphi_{t,k}, \qquad t = 1, 2.$$

Here $\widehat{F}_t$ is the Fock operator of the single ion t but calculated with the density matrix

$$\rho_t(\mathbf{r}|\mathbf{r}') = \sum_k \varphi_{tk}(\mathbf{r}) \varphi_{tk}^*(\mathbf{r}')$$

of the ion t in the crystal.

In ionic solids the influence of the crystal environment on cation is less important because cation is a tightly bounded formation compared with anion. Therefore we will consider the crystal environment potential for anion and we will refer to it as a *confining potential*. Similar arguments with an appropriate change of sign are applicable to the crystal environment potential for cation.

The particular form of the potential V_t is not crucial and quite different potentials could do the same job producing the same total electron density of the crystal, each potential corresponding to the particular local orbitals as a combinations of the same Bloch functions of the crystal. It is however important for the potential to be flexible enough and not imposing the unnecessarily crude limitations. When both ions contain only s- and p-orbitals a spherically symmetrical potential could be employed. The spherical symmetry is not so severe limitation as it seems at first. There are at least two reasons for this. First, the environment of any ion in this crystal has cubic symmetry. Therefore the admixture to s-orbitals of the ion begins with g-orbitals, and the admixture to p-orbitals begins with f-orbitals. Because of the energy considerations these admixtures should be comparatively small. Second, we consider the potential in the equations for non-orthogonal orbitals. After the orthogonalization to its neighbors the orbital of the ion g will acquire the linear combination of neighboring s- and p-orbitals admixture, which in respect to the ion g is a linear combination of orbitals with all possible values of the angular quantum number. As to the particular form of the potential, we took into account that most calculations are performed with the Gaussian basis set. Therefore it is expedient to construct the potential of Gaussian functions also.

The following equation could be suggested for the confining potential

$$V_t(\mathbf{r}) = V_{t0}\left(1 - \left(1 - e^{-\alpha_t r^2}\right)^{n_t}\right), \tag{4.10}$$

with parameters V_{t0}, α_t, and n_t controlling the depth, the radius, and the slope of the well respectively. As an example the confining potential for O^{-2} ion in MgO crystal is shown in Figure 2. Instead the parameter α_t it is suitable to use parameter r_t directly connected with the radius of the potential well and defined as the value of r where the potential is equal to one half of its value at $r = 0$

$$\left(1 - e^{-\alpha_t r_t^2}\right)^{n_t} = \frac{1}{2}.$$

All three parameters are considered as variational. The reason why the parameter V_{t0} of the potential should be taken as a variational parameter and not be fixed to coincide with the Madelung potential at the corresponding lattice site is the following. Consider an electron of a regular ion of the crystal. This electron is an extra electron with respect to any other ion. Therefore the potential influencing the electron at large distance from its ion can be expressed as a sum of two terms. First is the potential of the perfect crystal influencing an extra electron, that is the potential influencing the electron at the bottom of the conduction band. Second is the point charge +1 potential which is the difference at large distances between the potential influencing the electron of the same ion and the potential influencing an extra to the ion electron. In the HF approximation the potential for an electron at the bottom of the conduction band is the zero Fourier component of the crystal potential. As the potential simulating the crystal environment in the equation (4.10) is equal to zero at the infinity the parameter V_{t0} must be the sum of the Madelung potential at the lattice site and the zero Fourier component of the crystal potential, and the latter is not equal to zero. Indeed, the zero Fourier components of the point ion parts of the cations and anions potentials cancel each other and those of the non-point parts do not.

Among three variational parameters two of them, α_t and V_{t0}, are of principle importance and the third one, n_t, is of secondary importance as the slope of a well is less

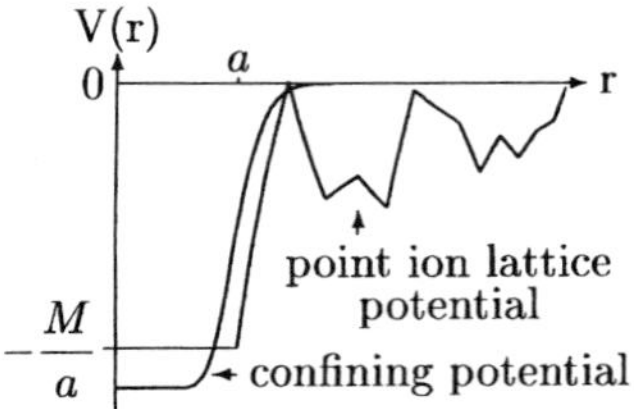

Figure 2: Confining potential and the point ion lattice potential for O^{-2} anion in MgO crystal

important compared with its depth or radius. This conclusion had been confirmed by the results of particular crystal calculations.

4.5 Application to the MgO crystal

The described above method was applied to the MgO crystal, where the O^{2-} ion is stable only because of the crystal environment field. Some of the results are presented here. For more complete data see [20]. (see also [21, 22] for comparison with earlier works). In MgO crystal the cation is very tightly bounded, so only the confining potential for the anion was employed. First, the special GTO basis (24s17p) for Mg^{2+} and (24s19p) for O^{2-} were generated to produce results close to the HF limit for ions, as it can be seen from the Table 1.

Table 1: GTO and numerical results.

	O^{2-} in crystal		Mg^{2+} single	
	Numerical	GTO	Numerical	GTO
Total energy	-84.21682	-84.21680	-198.83081	-198.83078
ϵ_{1s}	-20.61662	-20.61661	-49.76871	-49.76870
ϵ_{2s}	-1.30631	-1.30630	-4.48295	-4.48294
ϵ_{2p}	-0.59585	-0.59585	-3.00617	-3.00617

Employing the generated basis the MgO crystal was considered with the nearest neighbor distance $a = 4.0$ close to the experimental value $a_{exp} = 3.96$. The orbitals of the single Mg^{2+} ion (t=1) and of the O^{2-} ion (t=2) in the confining potential (4.10) were employed. One-ion, two-ion, and three-ion contributions were retained in the energy per unit cell w (4.9). The minimum of w with respect to all three parameters r_2, V_{20}, n_2 of the potential (4.10) was obtained as follows. For several values of n_2 the minimum of w with respect to parameters r_2 and V_{20} was found. This minimum was rather sharp. Then considering the resulting energy as a function of n_2 the minimum was found at $n_2 = 16$. This minimum is very shallow indicating that the parameter n_2 is of minor importance. With the obtained orbitals the convergence of various series in w was examined.

Table 2: Convergence of the two-ion short-range and three-ion long-range series.

n	t_1–t_2	m_{12}	$U_{sr}^{(2)}[g_1, g_2]$	$U_{lr}^{(3)}[g_1, g_2]$
1	a–c	1	0.23339	-0.04389
2	a–c	3	0.00022	-0.00008
3	a–a	2	0.02184	0.02888
4	a–a	4	0.00038	0.00028
5	a–a	6	0.00003	-0.00001

In the case of spherically symmetrical (overlapping) ions the selection of independent clusters in the sum over the lattice is easy. Two two-ion clusters are equivalent if they have the same integer square distance

$$m_{12} = (n_{x1} - n_{x2})^2 + (n_{y1} - n_{y2})^2 + (n_{z1} - n_{z2})^2$$

and one ion of the first cluster is similar to an ion of the second cluster. Two three-ion clusters are equivalent if their triangles are congruent and ions in one pair of the corresponding vertices are similar. The triangles are congruent if they have the same values of m_{12}, m_{13}, and m_{23}. The convergence of lattice series for MgO is demonstrated in Tables 2 and 3. In the Table 2 are shown the values $U_{sr}^{(2)}[g_1, g_2]$ and $U_{lr}^{(3)}[g_1, g_2]$ which are correspondingly the $W_{sr}^{(2)}[g_1, g_2]$ and $E_{lr}^{(3)}[g_1, g_2]$ multiplied by the number of equivalent clusters $[g_1, g_2]$. For clusters not shown in the Table 2 the interaction energy is equal to zero with the precision used in this Table. Such is, for example, cluster made entirely of cations even if it contains the nearest cations only. One can see from the Table 2 that the convergence in both series is rather fast.

The convergence of the series for the short-range three-ion interaction is shown in the Table 3. In this table are shown the values $U_{sr}^{(3)}[g_1, g_2, g_3]$ and $U^{(3)}[g_1, g_2, g_3]$ which are correspondingly the $W_{sr}^{(3)}[g_1, g_2, g_3]$ and $W^{(3)}[g_1, g_2, g_3]$ multiplied by the number of equivalent clusters $[g_1, g_2, g_3]$. All three-ion clusters with ions up to the 9-th neighbors including were considered. In the upper half of the table the clusters are given for which $U_{sr}^{(3)}$ is not equal to zero with the precision used in the table. They are arranged in decreasing the absolute value of $U_{sr}^{(3)}$ order. The convergence of the series for the short-range three-ion interactions is good although as many as 12 clusters was necessary to take into account. In the last column the values of $U^{(3)}$ corresponding to the total three-ion interaction energy are given, 10 clusters being added with the largest $|U^{(3)}|$ from the rest. The data of the last column show that the convergence of the series for the total three-ion interaction energy is very poor and this is obviously due to the long-range component. Hence we came at the conclusion that to achieve convergence the short-range and the long-range interactions should be separated and summed independently.

Next, four-ion contribution to the energy was considered. For this several four-ion clusters had been calculated and their energy was decomposed into the total one-ion

Table 3: Convergence of the three-ion series.

n	t_1-t_2-t_3	m_{12}	m_{13}	m_{23}	$P[g_1,g_2,g_3]$	$U_{sr}^{(3)}[g_1,g_2,g_3]$	$U^{(3)}[g_1,g_2,g_3]$
1	a–c–a	1	2	1	12	-0.03132	-0.00702
2	a–c–a	1	2	5	24	0.00312	0.00385
3	a–c–c	1	1	4	3	0.00124	-0.01727
4	a–c–c	1	1	2	12	0.00086	-0.02414
5	a–a–a	2	2	2	8	0.00085	-0.03659
6	a–a–a	2	2	6	24	0.00082	0.01612
7	a–c–a	1	2	3	24	0.00051	0.02117
8	a–c–a	1	4	1	3	-0.00050	-0.00859
9	a–a–a	2	2	8	6	0.00030	0.01140
10	a–a–a	2	2	4	12	0.00016	-0.01074
11	a–c–a	1	6	3	24	-0.00001	-0.01704
12	a–c–a	1	4	5	24	0.00001	0.00357
13	a–c–c	1	5	2	24	0.00000	0.02377
14	a–c–c	1	3	6	24	0.00000	-0.01991
15	a–c–c	1	3	2	24	0.00000	0.01952
16	a–c–a	1	6	9	48	0.00000	0.01608
17	a–a–c	2	3	5	48	0.00000	0.01588
18	a–a–c	2	5	5	36	0.00000	0.01270
19	a–c–a	1	6	5	48	0.00000	-0.01207
20	a–a–c	2	3	9	24	0.00000	-0.01037
21	a–c–c	1	5	8	24	0.00000	-0.01027
22	a–c–a	1	8	5	24	0.00000	-0.01023

$E^{(1)}$ energy, total two-ion long-range $E_{lr}^{(2)}$ interaction energy (point ions interaction), total two-ion short-range $E_{sr}^{(2)}$, total three-ion $E^{(3)}$ interaction energies and the four-ion $E^{(4)}$ interaction energy. The calculated clusters and their energies are given in the Table 4. As it can be seen from this table the four-ion interaction is considerably smaller than the three-ion interaction and we will neglect the four-ion and all higher n-ion interactions.

Consequently, in the present calculations both lattice series and the cluster expansion series could be considered as summed up to convergence. Therefore all deviations of the calculated data from the experiment could be attributed to the HF method and not to some additional approximations made in the calculations.

However shallow the energy minimum is, the corresponding to it potential $V_2(\mathbf{r})$ reasonably well describes the influence of the crystal environment on the oxygen anion. Indeed, the position of 2p energy level ε_{2p} corresponds to the energy gap of MgO crystal in the HF approximation. The latter is known to be 0.63 au (17 eV) [23, 24]. The value 0.60 au obtained in the present work is close enough to it especially considering that the zero Fourier component of the crystal potential and the position of 2p energy

Table 4: Four-ion clusters

	a–a–a–a	c–a–a–a	a–c–c–a	a–c–c–c
$E^{(1)}$	-297.32343	-421.82335	-546.32327	-670.82319
$E^{(2)}_{lr}$	4.24264	-0.87868	-2.58579	-0.87868
$E^{(2)}_{sr}$	0.02184	0.12761	0.15923	0.11669
$E^{(3)}$	-0.01830	-0.00633	-0.00519	-0.00604
$E^{(4)}$	0.00060	0.00031	-0.00003	-0.00000

level do not exactly coincide with the HF bottom of the conduction band and the top of the valence band correspondingly but are only near to them. The HF value of the energy gap is two times greater than the experimental value 7.8 eV [25, 26]. Better approximation to the crystal energy gap is the excitation energy of the O^{2-} ion in the potential $V_2(\mathbf{r})$. This could be considered as a first step of the multiconfiguration self-consistent field approach to the crystal in the nonorthogonal Wannier representation. As only one excited configuration is considered here and the correlation effects in MgO are known to be greater in the excited state compared with the ground state, the excitation energy of the oxygen ion in the crystal potential overestimates the crystal energy gap. Accordingly, the value 9.5 eV for the oxygen excitation energy obtained in the present work is greater than the experimental value 7.8 eV of the crystal energy gap but is not too far from it.

Next the cohesive properties of MgO crystal were considered. At first, the oxygen orbitals in the potential optimal for $a = 4.0$ were taken as fixed and the energy w was calculated as a function of interatomic distance a. The position of the energy minimum gives the equilibrium interatomic distance, the value of the energy minimum gives the cohesive energy of the crystal, and the second derivative of w with respect to a gives the compressibility of the crystal. The equation for w contains series over the crystal. Often these series are truncated and either the first neighbors or the first and second neighbors are retained only. These approximations were tested in the present paper and were compared with calculations where the series in the equation for w were summed up to convergence, within the precision adopted. The results presented in Table 5 show that although in the two-ion approximation the first and second neighbors produce almost converged result, the three-ion contribution can not be neglected. At the same time the convergency of the three-ion series is rather slow, so the first and second neighbors approximation is not an adequate.

As it can be seen from the Table 5, where the periodic HF [27] and experimental values of the cohesive properties are given for comparison, the fixed orbitals approximation produce an acceptable result for the interatomic distance but the inverse compressibility is too big. Better approximation can be obtained performing the variational calculations for every value of a independently. The results of such calculations are shown in the Table 5 as flexible orbitals variant. They are in a very good agreement with the periodic HF calculation results. The coincidence of the results of calculations within the same HF approximation but employing quite different methods of solution the HF equations ensures us that the obtained in the present paper localized orbitals are the

Table 5: Cohesive properties of MgO.
Elementary cube edge d (Å), cohesive energy E_{coh} (kJ/mol),
inverse compressibility $1/\beta$ (GPa)

variant			d	E_{coh}	$1/\beta$
fixed orbitals	first neighbours	two-ion	4.31	2775	243
	fisrst and second neighbours	two-ion	4.35	2724	227
		+ three-ion	4.02	2917	230
	sum to convergency	two-ion	4.36	2723	229
		+ three-ion	4.21	2815	236
flexible orbitals	sum to convergency	two-ion + three-ion	4.19	2815	183
intraionic correlations included			4.23	2980	170
periodic HF calculation [27]			4.19	—	181
experiment (0°K)			4.19[a]	3030[a]	165[b]

[a] Reference [28]
[b] Reference [29, 30]

non-canonical HF orbitals of the crystal.

The intraionic correlation effects in O^{2-} crystal ion were calculated in the present paper. For this the numerical configuration interaction program [31, 32] was employed. In this program the excited configurations are constructed with Sturm orbitals thus considerably reducing the necessary configuration space size, so that even few hundred configurations can recover up to 85% of the correlation energy. The results of test calculations performed for the isoelectronic single Ne atom, single Na^+, F^- ions, and O^{2-} ion in the potential (4.10) with parameters $V_2 = -1.05$, $r_2 = 4.0$, $n_2 = 16$ are given in the Table 6. The radial 1s, 2s, 2p orbitals are the atomic orbitals of the ground state $1s^2\,2s^2\,2p^6$ configuration and all others are Sturm functions adjusted to the ground state orbitals. The calculations were of SD-CI quality. The results of calculation with different number of electron shells taken into account show good convergence. The experimental values of the correlation energies were estimated as follows. First, the experimental values of the single atom energy for F, Ne, and Na were taken from the paper [33] and with the help of experimental values for the electron affinity for F atom [34] and ionization potential for Na atom the experimental energies of F^- and Na^+ ions were obtained. Next, the HF energies for F^-, Ne, and Na^+ were calculated numerically. Corrections (mass correction, relativistic correction, and Lamb correction) to the HF energies for atoms were taken from [33] and added to the HF energies of ions, the difference between corrections for an ion and for the corresponding atom being

Table 6: The correlation energy of the isoelectronic ions.
Na^+, Ne, F^- — single ions, O^{2-} — ion in the confining potential

		Na^+	Ne	F^-	O^{2-}
E_{HF}		-161.6770	-128.5471	-99.4595	-84.2168
r_{1s}		0.143	0.158	0.176	0.199
r_{2s}		0.779	0.892	1.036	1.237
r_{2p}		0.796	0.965	1.256	1.512
shells	N_{conf}	E_{corr}			
1s 2s 2p 3s 3p 3d 4s	42	-0.1942	-0.1856	-0.1921	-0.1922
+4p 4d 4f	100	-0.2670	-0.2595	-0.2597	-0.2345
+5s 5p 5d 5f	222	-0.3100	-0.3045	-0.3022	-0.2981
+6s 6p 6d 6f	392	-0.3296	-0.3254	-0.3229	-0.3152
+5g	444	-0.3326	-0.3284	-0.3259	-0.3189
experiment[a]		-0.390	-0.390	-0.397	

[a] see text for explanations.

neglected. The difference between the experimental and corrected HF energies was taken for the experimental value of the correlation energy shown in Table 6.

The correlation energy in O^{2-} ion obtained in the present paper (0.32 au calculated value and 0.4 au estimation, higher excitations being included) is close to the value obtained in the paper [35] (0.35 au, which is the sum of 0.10 au reported in [35] as the difference between correlation energies in O^{2-} ion and O atom, and 0.25 au correlation energy in O atom [36]). It is normal that these intraionic correlation energies do not coincide. Their difference, apart from the computational details, results from the different localization potentials used in [35] and in the present paper, which causes the difference in the total density distribution between ions and consequently different partition of the correlation energy between intraionc and interionic components.

The Table 6 shows that the HF total energies and radii of 1s, 2s, and 2p states are quite different for ions considered. Still, the correlation energy is approximately the same. This implies that the intraionic correlations contribute to the value of the cohesive energy itself but their influence on the equilibrium interatomic distance and the compressibility is small. It is really so as can be seen from the Table 5 where the "itraionic correlations included" variant means that in the cohesive properties calculation the intraionc correlation energy was calculated for every interatomic distance and was added to the crystal energy w calculated with flexible orbitals.

The fact that the intraionic correlations increase the equilibrium distance a_0 is qualitatively correct because the interatomic correlations are known to correspond to the attraction between ions thus diminishing a_0. To calculate the interatomic correlations it is necessary to construct the localizing potential for the ion pairs, which must be consistent with the developed localizing potential for one ion. The construction of such potential is in progress now.

The RHF method with the special variational procedure, the variational parameters being transferred from the wave function into the confining potential, enables us to calculate the non-canonical localized Hartree-Fock orbitals of O^{2-} ion in MgO crystal. This procedure makes it possible to incorporate in the natural way the intraionic correlation effects. The calculated cohesive properties of MgO are in good agreement with the experiment.

5 Concluding remarks

The Hartree-Fock method with localized orbitals could be successfully applied to ionic solids. This method is very useful to incorporate the correlation effects and to investigate solids with impurities. The same approach can be applied to the covalent solids as well. However, the orbitals localized on bonds are not so well defined as orbitals localized on ions. Besides, a single bonding diagram is not always sufficient. Still, this approach seems promising.

References

[1] D.R.Hartree, Proc.Cambr.Phil.Soc. **24** (1928) 89.

[2] V.A.Fock, Z.Phys. **61** (1930) 126-148.

[3] D.R.Hartree, The calculation of atomic structures. John Wiley & Sons, NY, 1957.

[4] J.C.Slater, Quantum theory of molecules and solids. Volume 1. Quantum theory of molecules. McGraw-Hill Book Company, NY, 1963.

[5] C.C.J. Roothaan and P. Bagus, in Methods in: Computational Physics, v.2, Academic Press, NY, 1963, pp. 47-94.

[6] J.C.Slater, Quantum theory of molecules and solids. Volume 4. The Self-Consistent Field for Molecules and Solids, McGraw-Hill Book Company, NY 1974

[7] C. Froese Fisher, The Hartree-Fock method for atoms, John Wiley and Sons, NY 1977

[8] C.Pisani, R.Dovesi, C.Roetti, Hartree-Fock an initio treatment of crystalline systems. Lecture notes in chemistry, No 48, ISBN: 3-540-19317-0. Springer-Verlag, Berlin, 1988.

[9] L.Szasz, The Electronic Structure of Atoms, John Wiley and Sons, NY, 1992.

[10] R. McWeeny, Methods of molecular quantum mechanics, Academic Press, NY, 1989.

[11] A.C.Aitken, Proc.Roy.Soc.Edinburgh **57** (1937) 269

[12] J.H.Wegstein, Comm.Assoc.Comput.Math. **1** (1958) 9

[13] T.A.Koopmans, Physica **1** (1933) 104

[14] I.V.Abarenkov, V.F.Bratsev, A.V.Tulub, Quantum chemistry fundamentals. Vushaja Shkola, Moskow, 1989 (in Russian)

[15] W.H.Adams, J.Chem.Phys. **37** (1962) 2009–2018

[16] T.L.Gilbert "Self-consistent equations for localized orbitals in polyatomic systems", in Molecular orbitals in chemistry, physics and biology, Acad.Press, N-Y, (1964).

[17] Lowdin P.O., *A theoretical investigation into some properties of ionic crystals*, Thesis, Uppsala, 1948.

[18] I.V.Abarenkov, I.M. Antonova, Physica status solidi, **38** (1970) 783-797

[19] I.V.Abarenkov, I.M. Antonova, Sov.Phys.:Solid State, **20** (1978) 326-328

[20] I.V.Abarenkov, I.M.Antonova, Russian J.Phys.Chem. **74**, suppl.2 (2000) 253-266

[21] I.V.Abarenkov, I.M. Antonova, Physica status solidi (b) **92** (1979) 389-396

[22] N.C. Pyper, Phil.Trans.R.Soc.Lond.A **352** (1995) 89-124

[23] S.T. Pantelides, D.J. Michish, A.B. Kunz, Phys.Rev.B, **10** (1974) 5203–5212

[24] R.Pandey, J.E.Jaffe, A.B.Kunz, Phys.Rev.B **43** (1991) 9228–9237

[25] D.M.Roessler, W.C.Walker, Phys.Rev. **159** (1967) 733–738

[26] M.L.Bortz, R.H.French, D.J.Jones, R.V.Kasowski, F.S.Ohuchi, Physica Scripta **41** (1990) 537-541

[27] M.I.McCarthy, N.M.Harrison, Phys.Rev.B **49** (1994) 8574-8582

[28] A.J.Cohen, R.G.Gordon, Phys.Rev.B **14** (1976) 4593-4605

[29] O.L.Anderson, P.Andreatch Jr., J.Am.Ceram.Soc. **49** (1966) 404-409

[30] K.Marklund, S.A.Mahmoud, Physica Scripta **3** (1971) 75-76

[31] S.A.Kotochigova, I.I.Tupitsyn, Optics and Spectroscopy **61** (1986) 1161-1166 (in Russian)

[32] I.I.Tupitsyn, Radiochemistry (1987) No 4 478-481 (in Russian)

[33] A.Veilard, E.Clementi, J.Chem.Phys. **49** (1968) 2415-2421

[34] R.A.Kendall, T.H.Dunning Jr., R.J.Harrison, J.Chem.Phys. **96** (1992) 6796-6806

[35] K.Doll, M.Dolg, P.Fulde, Phys.Rev.B **52** (1995) 4842-4848

[36] D.Feller, E.R.Davidson, J.Chem.Phys. **90** (1989) 1024-1030

Computational Materials Science
C.R.A. Catlow and E.A. Kotomin (Eds.)
IOS Press, 2003

On the performance of various Hamiltonians in the study of crystalline compounds. The case of open shell systems

G. Mallia[1], R. Orlando[2], M. Llunell[1] and R. Dovesi[1,3]

[1] *Dipartimento di Chimica IFM, Università di Torino, via Giuria 5, I-10125 Torino, Italy*

[2] *Dipartimento di Scienze e Tecnologie Avanzate, Università del Piemonte Orientale "Amedeo Avogadro", c.so Borsalino 54, I-15100 Alessandria, Italy*

[3] *Unità INFM di Torino, sezione F via Giuria 5, I-10125 Torino, Italy*

Abstract. The Hartree-Fock, Density Functional (in its local and non local or gradient corrected variants) and "hybrid" schemes, such as B3LYP, are applied to the study of three typical *spin polarised infinite systems*, namely F-centres in alkali halides, trapped hole centres in alkaline earth oxides and transition metal insulators. As a rule, DFT functionals tend to delocalise the unpaired electron much more than HF, LDA providing the most widely delocalised solution and "hybrid" schemes being intermediate. In some cases, as for the F-center in LiF, this different behaviour involves only a different degree of localisation of the unpaired electron, however, in other cases (trapped electron holes), DFT functionals delocalise the unpaired electron completely and unphysically.

1 Introduction

In the years 80-95, the near totality of the electronic structure calculations of crystalline compounds were performed within the Local Density Approximation (LDA), whereas more recently the Generalized Gradient Approximation (GGA) has become the method of choice. In molecular quantum chemistry, on the contrary, Hartree-Fock (HF) has been the reference method for many decades and only in the last years the "hybrid" B3LYP scheme has become very popular, as being able to provide quite accurate results for atomization energies, equilibrium geometries and vibrational frequencies. HF and B3LYP have been applied to crystalline compounds very seldom and, as a matter of fact, at the moment being only one computer program, CRYSTAL98 [1], permits the calculation of the electronic structure of crystalline compounds at both the HF and B3LYP levels. Hence, one question we will try to answer in this paper concerns the performance of HF and B3LYP in the study of crystalline compounds as compared to LDA and GGA. Preliminary applications to silicates [2, 3] show that the same considerations about the merits and the limits of the various hamiltonians in molecular calculations [4, 5] apply to covalent infinite systems. More surprising is the observation that B3LYP seems to perform fairly well also in situations [6] which are much closer to the pure ionic than to the pure covalent limit, *i.e.* rather far from the class of molecules used in the parametrization of the method [7]. In fact, the binding energy of $Mg(OH)_2$ and $Ca(OH)_2$, as well as the interlayer interaction energy, the oxide hydration ($MO+H_2O \rightleftharpoons M(OH)_2$) energy and the Raman vibrational frequencies obtained by Baranek *et al.* [6] confirm the limits of both HF (underestimation of binding energy, overestimation of frequencies) and LDA (overestimation of binding energy, interlayer and hydration energies; underestimation of frequencies), whereas B3LYP on the average performs better than GGA.

In this paper we extend this analysis to the large class of spin polarised systems. In particular, we investigate two typical situations in solid state, namely:

- paramagnetic defects in ionic systems
- first row transition metal oxides and fluorides

More specifically, the following cases are considered:

- An F-center in LiF. In this very typical defect [8] an anion vacancy is created in a fully ionic lattice. The extra electron, which maintains electroneutrality, interacts with the electrostatic field at the vacancy and is accomodated in it forming a bound state. Accurate Electron Paramagnetic Resonance (EPR) and Electron Nuclear Double Resonance (ENDOR) experimental data are available for comparison [9].
- Trapped electron hole centers in MgO [10, 11, 12, 13, 14, 15, 16]. This defect, which has been investigated by EPR and ENDOR extensively, is characterised by the presence of an electron hole localised at one oxygen ion in a p-type orbital. The electrostatic field at the ionic sites is much stronger and the unpaired electron is more strongly localised than in the F-center in LiF.
- $KMnF_3$ [17] and NiO [18, 19] as prototypes of transition metal compounds, the former exhibiting a more pronounced ionic character than the latter, as containing the more electronegative F anions. The discussion will involve the capabilities of the various hamiltonians to describe unpaired electrons in d-type orbitals, the relative stability of ferro- and antiferro-magnetic phases and the degree of polarization of the anions as a key to the interpretation of the superexchange mechanism.

2 Computational details

The present calculations have been performed with the most recent implementation of the periodic CRYSTAL code [1] at the Unrestricted Hartree-Fock (UHF), Kohn-Sham (KS) and "hybrid" levels. One Local Spin Density Approximation (LSDA) functional (Dirac-Slater [20] parameterization of exchange and Perdew-Zunger parameterization of the correlation part [21] (SPZ)) and two GGA exchange-correlation functionals (Becke [22]—Lee-Yang-Parr (BLYP) [23] and Perdew-Burke-Ernzerhof [24] (PBE)) have been considered. As regards the hybrids, the popular B3LYP [7, 23], based on a mixed Hartree-Fock and DFT exchange functionals, has been tested.

As regards the basis set, Bloch functions are built from localised functions, namely Gaussian type functions (GTFs) given by the product of a Gaussian and a real solid spherical harmonic [25], or, more generally, by a linear combination (contraction) of GTFs to be indicated in the following as Atomic Orbitals, AOs.

For the F-center in LiF the same basis set as in previous work [26, 27] has been adopted: for Li and F, 5 and 13 AOs have been used (resulting from a contraction of 6 and 1 GTFs for the $1s$ and $2sp$ shells of Li, and a contractions of 7, 3, 1 and 1 GTFs for the $1s$, $2sp$, $3sp$

and $4sp$ shells of F, respectively). In order to provide an accurate description of the unpaired electron at the vacancy, a basis set has been added at the center of the vacancy, *i.e.* an sp shell (a Gaussian with exponent $\alpha_{sp} = 0.093\ a_0{}^{-2}$, optimized at the UHF level [26]).

With respect to previous work on MgO [28, 29, 30, 31], richer and more accurate basis sets have been used [27, 32]. Mg and O are described by 13 AOs (resulting from a contraction of 8, 5 , 1 and 1 GTFs for the $1s$, $2sp$, $3sp$ and $4sp$ shells of Mg and contractions of 8, 4, 1 and 1 GTFs for the $1s$, $2sp$, $3sp$ and $4sp$ shells of O, respectively).

As regards H and Li, the same basis sets as in previous works have been used: for the former [30, 31] three s functions (21-1 contraction) and one p shell (the exponent of the most diffuse functions are $\alpha_s = 0.14$ and $\alpha_p = 1.0\ a_0{}^{-2}$); for the latter [28, 29] two s functions and 1 p shell are used, the most diffuse gaussians being those with $\alpha_s = 0.5$ and $\alpha_p = 0.6\ a_0{}^{-2}$,

The basis sets used for $KMnF_3$ and NiO are reported in Refs. [17, 27]. They consist of 27 AOs for Mn and Ni, 17 AOs for K, 13 AOs for O and F. The basis sets of Mn and Ni result from the contraction of 8, 6, 4, 1, 1, 4, 1 gaussian functions into $1s$, $2sp$, $3sp$, $4sp$, $5sp$, $3d$, $4d$ shells; that of K from 8, 6, 5, 1, 1 gaussian functions into $1s$, $2sp$, $3sp$, $4sp$, $5sp$ shells; that of O(F) from 8(7), 4(3), 1, 1 gaussian functions into $1s$, $2sp$, $3sp$, $4sp$ shells.

The exchange–correlation potential is expanded into an auxiliary basis set of symmetrised atom-centred Gaussian type functions, with even–tempered exponents [33]. They can be found in Ref. [34].

3 Results and discussion

3.1 F-center in LiF

The isolated defect situation is simulated with the supercell scheme [26]: a supercell containing 32 atoms has been adopted, as it provides converged results for the properties investigated here.

The results obtained with the various functionals are summarised in Table 1 and Figures 1 and 2. The charge q attributed to the vacancy according to Mulliken partition and the spin

Table 1: Mulliken charge $q^{\alpha+\beta}$ (electrons) and spin density $\rho^{\alpha-\beta}$ ($10^{-2}\ a_0{}^{-3}$) at the vacancy site (F-center) and hyperfine coupling constants a and b (MHz) at the nearest Li and F nuclei around the vacancy.

	F-center		Li		F	
	q	$\rho^{\alpha-\beta}$	a	b	a	b
UHF	1.05	2.08	39.23	3.20	77.16	11.30
SPZ	0.89	1.66	42.07	2.93	120.81	13.46
BLYP	1.06	1.86	55.13	3.11	100.80	15.93
PBE	0.93	1.73	48.26	2.93	113.35	11.62
B3LYP	1.03	1.89	53.30	3.13	99.96	15.15
EXP.[35]			39.06	3.2	105.94	14.96

density $\rho^{\alpha-\beta}$ at the centre of the anion vacancy are reported in the second and third columns in Table 1. q varies from 0.86 electrons (LSDA) to 1.06 electrons (BLYP) and $\rho^{\alpha-\beta}$ from 0.0166 $a_0{}^{-3}$ (LSDA) to 0.0208 (UHF), where in the latter case GGA and hybrid functionals

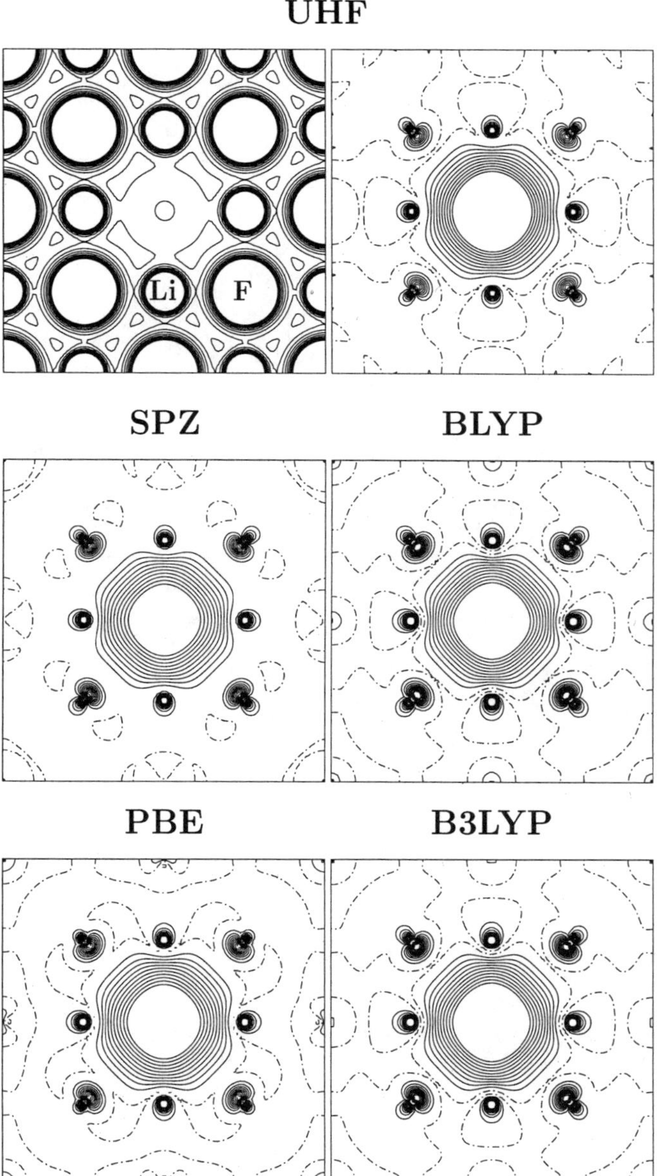

Figure 1: Spin density maps for an F–center in LiF as obtained by using different hamiltonians. The HF total charge density is also reported (top-left) to locate the ions. The section is parallel to the (100) plane through the defect. The unpaired electron is at the center of the map, whose side is 7.98 Å long. The separation between contiguous isodensity curves is 0.01 and 0.001 $a_0{}^{-3}$, for the electron charge and spin density, respectively. The density range is 0.01÷0.1 (charge) and $-0.01 \div 0.01$ $a_0{}^{-3}$ (spin maps). Continuous, dashed and dot-dashed lines denote positive, negative and zero values.

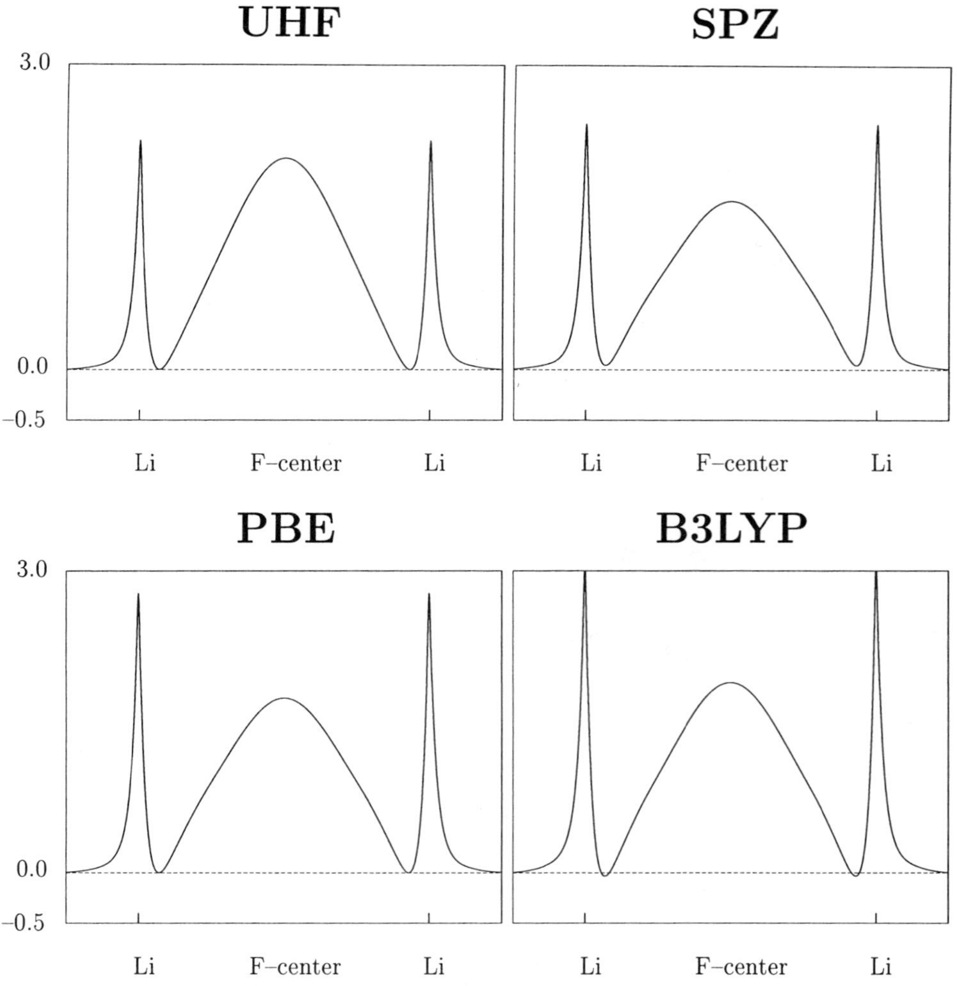

Figure 2: Spin density profile (in $10^{-2}\ a_0{}^{-3}$) along a line connecting the F–center to opposite nearest neighbouring Li ions. The path is 5.985 Å long; ticks indicate nuclear positions.

provide intermediate values. The general picture emerging from the table (the unpaired electron is nearly totally localised at the vacancy site; the various functionals provide a similar description, with differences of the order of 20% between the two extremes) is confirmed by spin density maps (Figure 1) and profiles (Figure 2). The interesting feature is the spin polarisation of the Li^+ and F^- ions surrounding the defect. As both ^7Li and ^{19}F have nuclear spin ($\frac{3}{2}$ and $\frac{1}{2}$, respectively [36]) it has been possible to measure the interaction between the unpaired electron and nuclei as far as the seventh neighbours [9]. In a previous study, an overall comparison of the UHF and SPZ results has been performed [26]. In Table 1 the isotropic a and anisotropic b coupling constants calculated at the nearest F and Li nuclei with the various hamiltonians are compared with the experiment. The values of a and b are more scattered than q and $\rho^{\alpha-\beta}$, differing by as much as 40%. In all cases, UHF provides the lowest value of a (confirming that the spin density is mostly localised at the vacancy), LSDA provides the highest value of a at F, the GGA and hybrid functionals being intermediate. b values, that describe the anisotropy of the hyperfine coupling tensor, do not show any similar trend and present a lower dispersion. When looking at farther neighbours, it results very clearly that the value of a from UHF is always smaller than from SPZ [26], whereas those obtained with gradient corrected functionals and hybrid methods are intermediate (unpublished work). In summary, in the present case, where the unpaired electron occupies a very diffuse s-like state, the various functionals provide similar descriptions.

3.2 *Trapped-hole in MgO*

Despite its wide and robust applicability, DFT has betrayed some restricted and insufficient nature.

The electron trapped holes that are formed in alkaline earth oxides upon irradiation [10, 37] are the second class of defects we will discuss. Let us consider magnesium oxide, when one Mg atom is replaced with a H or a Li atom and an electron hole is created. Extremely accurate EPR and ENDOR results are available concerning the coupling between the unpaired electron and the H [38, 12, 16] or Li [10, 14, 39] nuclear spin ($I_z = \frac{1}{2}, \frac{3}{2}$, respectively) that allow to characterize the electronic structure of the defect very accurately: in both cases experimental data [40] are compatible with a model where the unpaired electron is strongly localised at one of the oxygen ions around the substituted atom (see Figures 3,4).

These defects have been investigated in the approximation of five different hamiltonians (see Tables 2-4) and with a 32 atom supercell.

Table 2: Energy difference (ΔE in mhartree) between the relaxed geometry (only Li or H and O_1 are allowed to relax) and the non-relaxed geometry. d_H, d_{Li} and d_{O_1} (in Å) indicate the displacement of H, Li and O_1 from the perfect lattice position. d_{H-O_2} and d_{Li-O_1} are the resulting distances between the indicated atoms. a_0 is the conventional cell lattice parameter optimized for each hamiltonian (experimental value 4.195 Å).

		MgO:[H]^0				MgO:[Li]^0			
$\mathcal{H}$	a_0	d_{H-O_2}	d_H	d_{O_1}	ΔE	d_{Li-O_1}	d_{Li}	d_{O_1}	ΔE
UHF	4.21	1.003	1.1	–0.1	–140.1	2.503	0.3	–0.1	–11.5
SPZ	4.17	1.085	1.0	–0.2	–66.9	2.185	0.0	–0.1	–0.9
BLYP	4.29	1.045	1.1	–0.2	–87.8	2.345	0.1	–0.1	–2.2
PBE	4.26	1.030	1.1	–0.1	–81.9	2.230	0.0	–0.1	–1.8
B3LYP	4.25	1.025	1.1	–0.1	–97.6	2.425	0.2	–0.1	–2.7

The equilibrium lattice parameters have been redetermined for each hamiltonian. Previous studies at the UHF level [28, 30] had shown that only the subtitutional atom and the O ion with the electron hole are affected by large nuclear relaxation, the displacements of all the other atoms from their perfect lattice sites being much smaller. Hence, only these atoms have been involved in the geometry optimization process. In Table 2 the equilibrium data for H (Li) and O_1 and the corresponding relaxation energy are reported, whereas Table 3 provides an analysis of charge and spin density in terms of Mulliken populations. Spin density maps are given in Figures 3 and 4, showing that different hamiltonians represent a very different

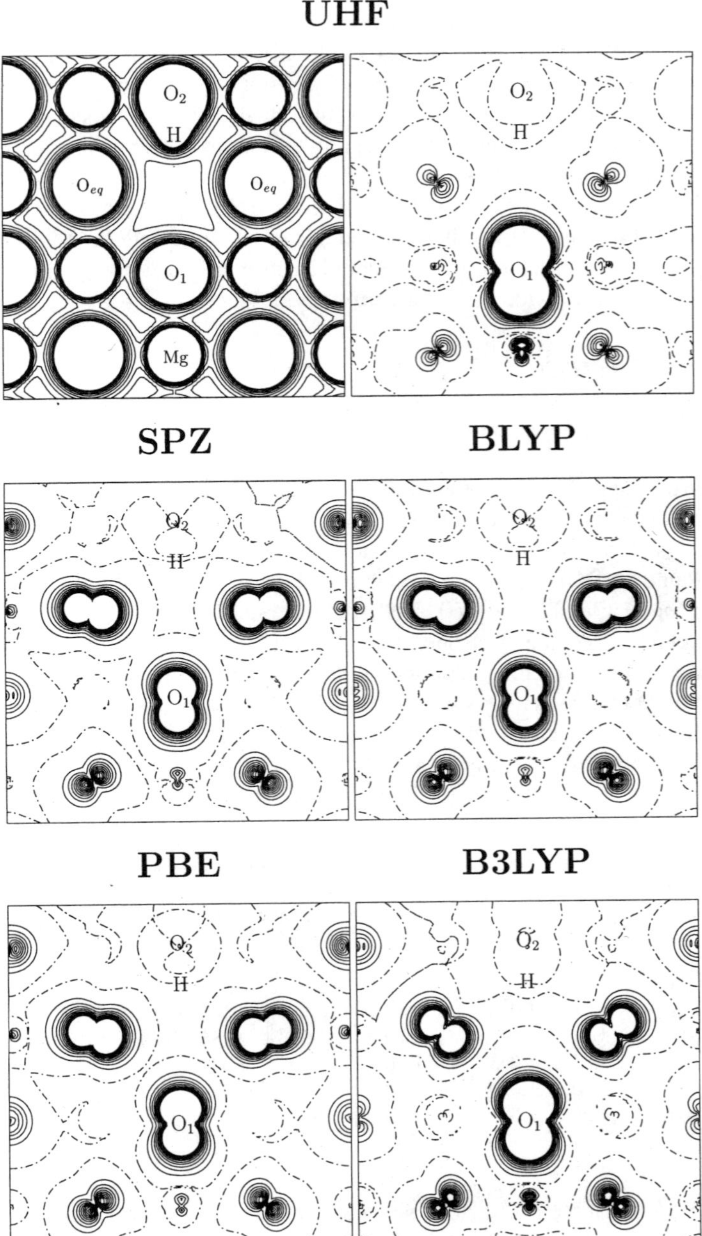

Figure 3: Spin density maps for MgO:$[H]^0$ as obtained by using different hamiltonians. The UHF total charge density is also reported (top-left) to locate the ions. The section is parallel to the (100) plane through the defect. Map sides measure twice the lattice parameter (see Table 2). Scales and symbols as in Figure 1.

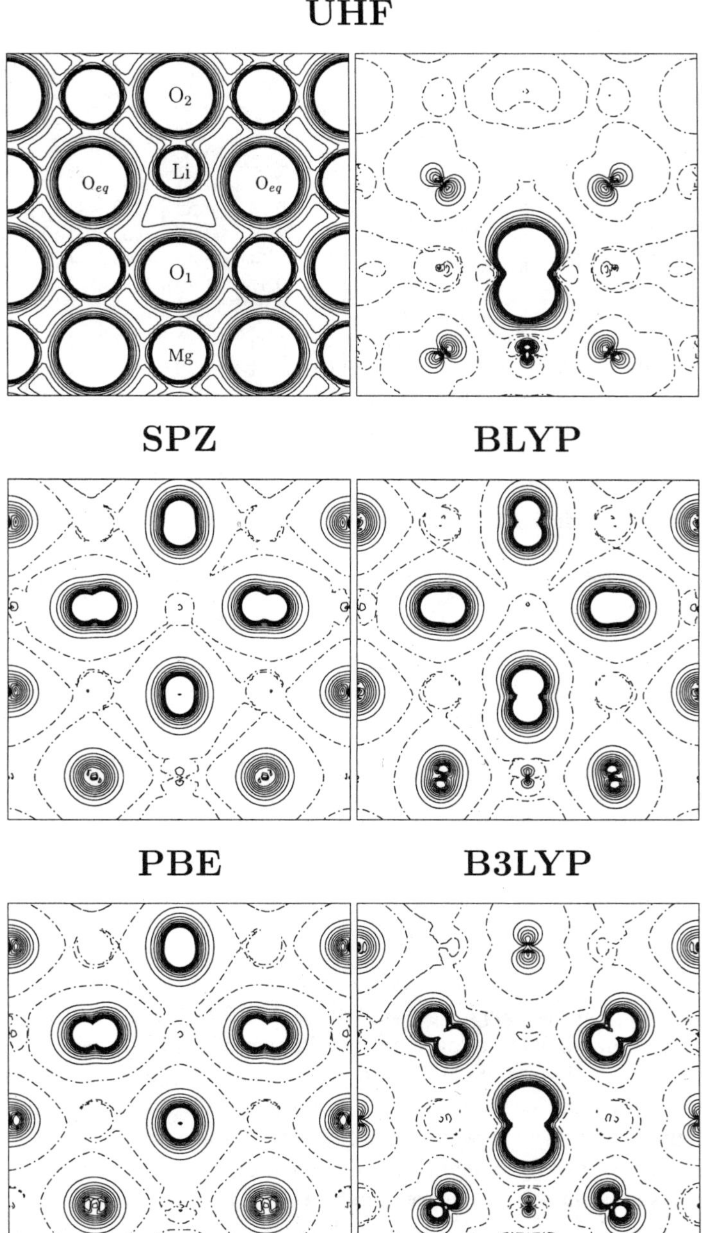

Figure 4: Spin density maps for $MgO:[Li]^0$ as obtained by using different hamiltonians. The UHF total charge density is also reported (top-left) to locate the ions. See Figure 3 for details.

Table 3: Net atomic charges (q) and spin moments (μ) evaluated according to a Mulliken partition of charge and spin densities. Labels of atoms as in Figure 3. M stands for H (top) or Li (bottom). q and μ as number of electrons.

		M		O_1		O_2		O_{eq}	
	$\hat{\mathcal{H}}$	q	μ	q	μ	q	μ	q	μ
MgO:[H]^0	UHF	+0.59	−0.00	−0.99	0.97	−1.55	0.00	−1.55	0.00
	SPZ	+0.45	−0.00	−1.56	0.20	−1.52	0.00	−1.61	0.15
	BLYP	+0.53	−0.00	−1.57	0.19	−1.47	0.00	−1.61	0.14
	PBE	+0.51	−0.00	−1.53	0.24	−1.49	0.00	−1.62	0.15
	B3LYP	+0.55	0.00	−1.34	0.51	−1.49	0.00	−1.68	0.10
MgO:[Li]^0	UHF	+0.99	0.00	−1.03	0.97	−1.91	0.00	−1.88	0.01
	SPZ	+0.98	0.00	−1.71	0.09	−1.69	0.12	−1.69	0.11
	BLYP	+0.97	0.00	−1.66	0.14	−1.70	0.10	−1.68	0.11
	PBE	+0.98	0.00	−1.74	0.08	−1.71	0.10	−1.71	0.10
	B3LYP	+0.98	−0.00	−1.48	0.41	−1.80	0.01	−1.72	0.10

Table 4: Calculated and experimental hyperfine isotropic (a) and anisotropic (b) coupling constants and nuclear quadrupole (P) coupling constant (in MHz).

	MgO:[H]^0		MgO:[Li]^0		
	H		^{7}Li		
$\hat{\mathcal{H}}$	a	b	a	b	P
UHF	0.073	2.366	−2.393	2.258	−0.017
SPZ	−0.256	−0.189	−3.691	−0.129	-0.003
BLYP	−0.694	−0.043	−3.078	0.092	−0.003
PBE	−0.379	0.115	−3.979	−0.202	−0.010
B3LYP	−0.074	1.142	−3.339	0.480	−0.004
EXP.	0.044[a]	2.376[a]	−4.539[b]	2.313[b]	−0.014[b]
	0.101[c]	2.371[c]			
	0.070[d]	2.360[d]			

[a]Ref. [12, 13] ; [b]Ref. [14, 15] ; [c]Ref. [16] ; [d]Ref. [11].

degree of localization of the unpaired electron, as quantified by the spin moment of O_1 in Table 3, which varies in the range from 0.97-0.98 (UHF) to 0.2-0.1 (SPZ, BLYP). This is generally true for spin moments, as different hamiltonians provide substantially different pictures, although the distribution of electron charge density is indeed not very sensitive to the choice of the hamiltonian (see the values of q in Table 3). In particular, in the UHF approximation the unpaired electron is not spread over the nearest and next nearest neighbouring O ions as is observed with all the other hamiltonians. The different degree of localization of the electron hole has important consequences on the extension of the displacement of the small Li ion, which is maximum for UHF, since in this case the electrostatic interaction with O_1 is minimum (q_{O_1} is approximately -1), whereas the displacement is minimum for LSDA and PBE. On the contrary, all the hamiltonians predict very similar relaxation for O_1, because O_1 always interacts with the same Li ion bearing the same charge (+1) in all cases and, despite the variabilty of its own electron charge in the different cases, its displacement is sterically hindered.

At this point one might wonder whether in the case of LSDA, GGA and B3LYP some lower energy state corresponding to a localised unpaired electron exists that we were unable to find. However, several attempts were done at driving the Self Consistent Field (SCF) cycle to a localised solution, including the use of the UHF localised solution as a starting guess, but they all failed and the same fully delocalised solution was found at each time. One might also object that the experimental evidence of a localised electron hole is not strong enough to state its existence unambiguously. In fact, the size and shape of the orbital describing the unpaired electron is not obtained from the spectral analysis directly, but through the use of an interpreting model which contains some arbitrariness. For this reason, we calculated the values of the isotropic (a) and anisotropic (b) hyperfine coupling constants and the nuclear quadrupole coupling constant (P), that compare to the parameters fitted on the experimental spectra straightforwardly. These are reported in Table 4. It is evident that the UHF results do not only compare to the experimental values qualitatively, but are also in quantitative agreement in all cases except for a in [MgO]:Li, whereas most DFT results differ in magnitude and, in some cases, even for their sign. The disagreement of the DFT data with experiment is particularly striking for b and even B3LYP performs very poorly in this case, so that an accurate description of the nonlocal exchange interaction appears to be very important for the success of UHF and the self-interaction error is probably the main source of error in DFT. Nevertheless, it must be mentioned that, when heavier ions are considered, the UHF values of a and b are much less satisfactory [41], probably because the spin polarization of the inner electrons needs to be described at a more sophisticated level of approximation, where correlation effects are correctly taken into account (unfortunately, in this case, the DFT functionals we used are equally ineffective).

3.3 $KMnF_3$

As a third example, let us consider $KMnF_3$, an ionic perovskite [17, 42], whose structure is shown in Fig. 5. Each Mn ion is at the center of a regular F^- octahedron. Table 5 provides the equilibrium lattice parameter as resulting from the various hamiltonians. As expected LSDA underestimates it (-1.5%) and UHF overestimates it (+2%) with respect to experiment. BLYP, PBE and B3LYP give results quite close to UHF.

In Figure 5 three different magnetic structures are reported. The structure on the left represents the ferromagnetic (FM) phase, where all the Mn ions have the same spin, say α.

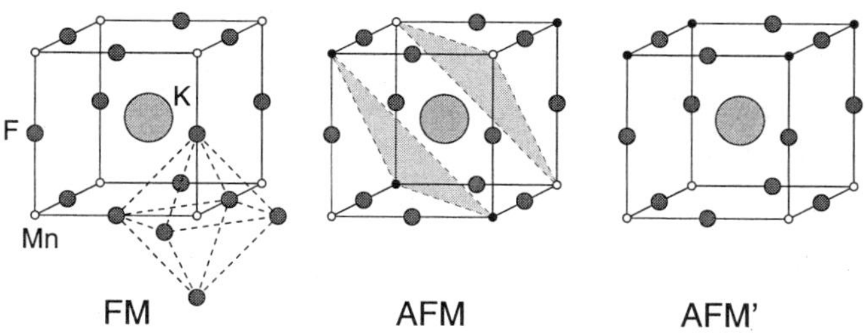

Figure 5: Cubic unit cell of $KMnF_3$, different magnetic phases: the FM (left) and two possible AFM (AFM, center and AFM', right) structures are shown. Black and white small spheres represent spin up and down Mn ions. In the central figure, the dashed lines connect Mn ions with the same spin in (111) planes.

Table 5: Comparison between the lattice parameter **a** calculated using different hamiltonians and experimental value [43].

$KMnF_3$ – Lattice Parameter		
$\hat{\mathcal{H}}$	**a** (Å)	Δ %
UHF	4.28	+2.21
SPZ	4.13	–1.42
BLYP	4.31	+2.86
PBE	4.28	+2.12
B3LYP	4.27	+1.98
EXP.	4.19	

The picture in the middle corresponds to an antiferromagnetic (AFM) phase, the one where each α-spin Mn ion is surrounded by six β-spin nearest neighbours. In this arrangement (111) planes entirely of α-Mn and others entirely of β-Mn are stacked alternatively. This AFM magnetic cell must be twice as large as the FM unit cell. On the right, another possible AFM phase is shown that is formed from the stacking of alternating α-Mn and β-Mn (001) planes.

In a fully ionic picture, each Mn ion should have five electrons in d-type orbitals with parallel spin. This picture is actually not far from the quantum mechanical solution, as is shown in Table 6, where charge and spin density data as resulting from Mulliken population analysis are reported. Net charges (q) are very close to their formal values (+2, -1 and +1 for Mn, F and K), the largest difference being about 0.2 electrons for Mn between UHF and LSDA.

Table 6: Net atomic charges (q) and spin moments (μ) evaluated according to a Mulliken partition of the spin densities. q and μ in $|e|$.

$KMnF_3$												
	FM						AFM					
	Mn		F		K		Mn		F		K	
$\hat{\mathcal{H}}$	q	μ	q	μ	q	μ	q	μ	q	μ	q	μ
UHF	1.77	4.94	–0.92	0.02	0.99	0.00	1.77	4.94	–0.92	0.00	0.99	0.00
SPZ	1.56	4.76	–0.85	0.08	0.98	–0.02	1.54	4.68	–0.84	0.00	0.98	0.00
BLYP	1.58	4.77	–0.85	0.08	0.98	–0.02	1.56	4.70	–0.85	0.00	0.98	0.00
PBE	1.59	4.79	–0.86	0.07	0.98	–0.01	1.58	4.73	–0.85	0.00	0.98	0.00
B3LYP	1.64	4.84	–0.88	0.05	0.99	–0.01	1.64	4.81	–0.87	0.00	0.99	0.00

As regards spin moment, the five unpaired electrons are almost fully localised at Mn, although a small polarization of F (and K) is observed (see Figure 6). It is, however, to be noticed that this small polarization is responsible for the superexchange interaction along the Mn-F-Mn path [17]. For symmetry reasons it is nearly null in the AFM solution and, as a consequence, the AFM phase is slightly more stable than FM $KMnF_3$ [17]. In fact, the AFM-FM energy difference is approximately proportional to the difference in the spin moment of F. This quantity is four times larger for LSDA than for UHF (as can be seen also in Figure 6 from the number of isolines present at the F sites), whereas B3LYP is intermediate and BLYP and PBE are very close to LSDA (note that the corresponding calculated values of the lattice parameter were very close to the UHF value).

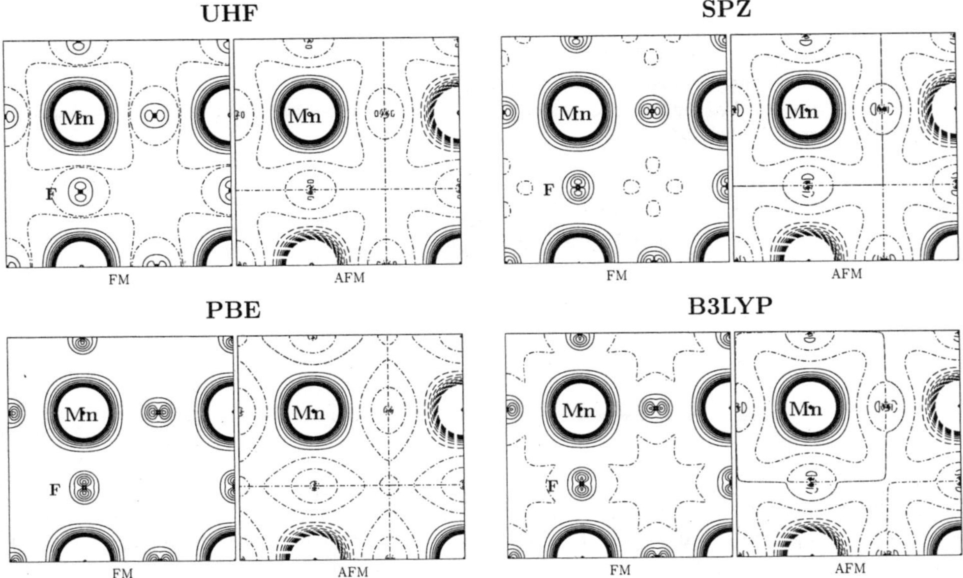

Figure 6: Spin density maps obtained with different hamiltonians for the FM and AFM solutions of $KMnF_3$ in the (001) plane through the Mn anf F atoms. The separation between contiguous isodensity lines is $0.01a_0^{-3}$; the function is truncated in the core region at $\pm 0.1a_0^{-3}$. Continuous, dashed and dot–dashed lines correspond to up-, down- and zero-spin density, respectively.

The experimental information concerning the magnetic interactions is usually interpreted in terms of the superexchange constant J, that can be related to the energy difference between the AFM and FM phases through the equation:

$$\Delta E = \frac{2zJS_z^2}{k} \tag{1}$$

where ΔE refers to the double cell containing 2 Mn ions, z is the number of bonds relating atoms within the cell and their neighbors with opposite spin (2 atoms · 6 neighors /2 = 6 in the present case, where the division by two is due to the fact that each bond is shared by two atoms), S_z is its conventional spin (5/2 for Mn) and $k = 3.1577 \cdot 10^5$ is the ratio between the conversion factor from hartree to joule (1 hartree = $4.359748 \cdot 10^{-18}$ J) and the Boltzmann constant ($1.3806558 \cdot 10^{-23}$ JK^{-1}). In Table 7 the total energy of the FM and AFM phases (E), their difference (ΔE), the superexchange coupling constant (J) and the ratio of the calculated to the experimental [44] value of J are reported. The UHF value of J is about $\frac{1}{3}$ the experimental J, whereas all the DFT functionals provide values that are larger than experiment, LDA overestimating J by a factor 4.5 and B3LYP by a factor 2. Another interesting property that has been investigated experimentally is the dependence of J (or, equivalently, ΔE) on the Mn–Mn distance. Experimentally, this behaviour is analysed by replacing the K ion with heavier (Rb, Tl) ions, with a consequent increase of the lattice parameter and in the hypothesis that the monovalent cation has no influence on the magnetic structure. Alternatively, Mn ions are inserted in other fluorides in order to device specific Mn–Mn distances.

Table 7: Total energy (in hartree) of the ferromagnetic (FM) and antiferromagnetic (AFM) phases of $KMnF_3$ as obtained with various hamiltonians. Both cells contains 10 atoms. $\Delta E = E_{\rm AFM} - E_{\rm FM}$ in mhartree. J is the superexchange coupling constant (in Kelvin). Experimental values from Ref. [44]; they have however been multiplied by 2, because the Hamiltonian used by L. J. De Jong and R. Block contains a factor two that is absent from the Hamiltonian leading to eq. 1 ($\hat{H} = -\sum_{i,j} J\hat{S_{z_i}}\hat{S_{z_j}}$).

$KMnF_3$					
$\mathcal{H}$	$E_{\rm FM}$	$E_{\rm AFM}$	ΔE	$J_{calc.}$	$\frac{J_{calc.}}{J_{exp.}}$
UHF	–4095.286164	–4095.286758	0.594	2.50	0.34
SPZ	–4089.532113	–4089.540105	7.992	33.64	4.58
BLYP	–4101.523522	–4101.531254	7.732	32.56	4.43
PBE	–4100.020100	–4100.026538	6.438	27.10	3.69
B3LYP	–4101.056636	–4101.060127	3.491	14.70	2.00
EXP.				7.30 7.40	

In the theoretical model we simply modify the lattice parameter. The experimental and theoretical results are shown in Figure 7, where both the experimental and calculated points are fitted to the power of the Mn–Mn distance. In this case the various functionals provide values within the error bar of the experimental value, UHF being at the top of the range and the DFT hamiltonians at the bottom.

3.4 NiO

NiO is a high spin antiferromagnetic insulator in the rock-salt face-centered-cubic structure [45]. It is slightly less ionic than $KMnF_3$ and a comparative analysis of these two compounds can be instructive.

In NiO the spin polarization of the anion is five to six times larger than in $KMnF_3$ (compare the anion spin moments in Tables 6 and 8), so that the spin density distribution is quite different according to the different Hamiltonians. At opposite limits of the variability range

Table 8: Net atomic charges (q) and spin moments (μ) in NiO evaluated according to a Mulliken partition of the charge and spin densities for the FM phase of NiO. μ_{e_g} and $\mu_{t_{2g}}$ are the spin population of the $3d$ orbitals ($d_{2z^2-x^2-y^2} + d_{x^2-y^2}$ and $d_{xy} + d_{xz} + d_{yz}$, respectively).

	Ni				O	
$\mathcal{H}$	q	μ	μ_{e_g}	$\mu_{t_{2g}}$	q	μ
UHF	1.87	1.92	1.918	0.003	−1.87	0.08
SPZ	1.55	1.56	1.362	0.195	−1.55	0.44
BLYP	1.56	1.60	1.526	0.060	−1.56	0.41
PBE	1.59	1.61	1.530	0.066	−1.59	0.39
B3LYP	1.68	1.73	1.718	0.006	−1.68	0.27

there are LSDA and UHF: at the LSDA level, only 1.5 electrons are localised at the cation, as almost a half of the two unpaired electrons is at the anion, whereas UHF assigns 1.92 electrons to Ni and 0.08 to O (see also the spin density maps in Figure 8).

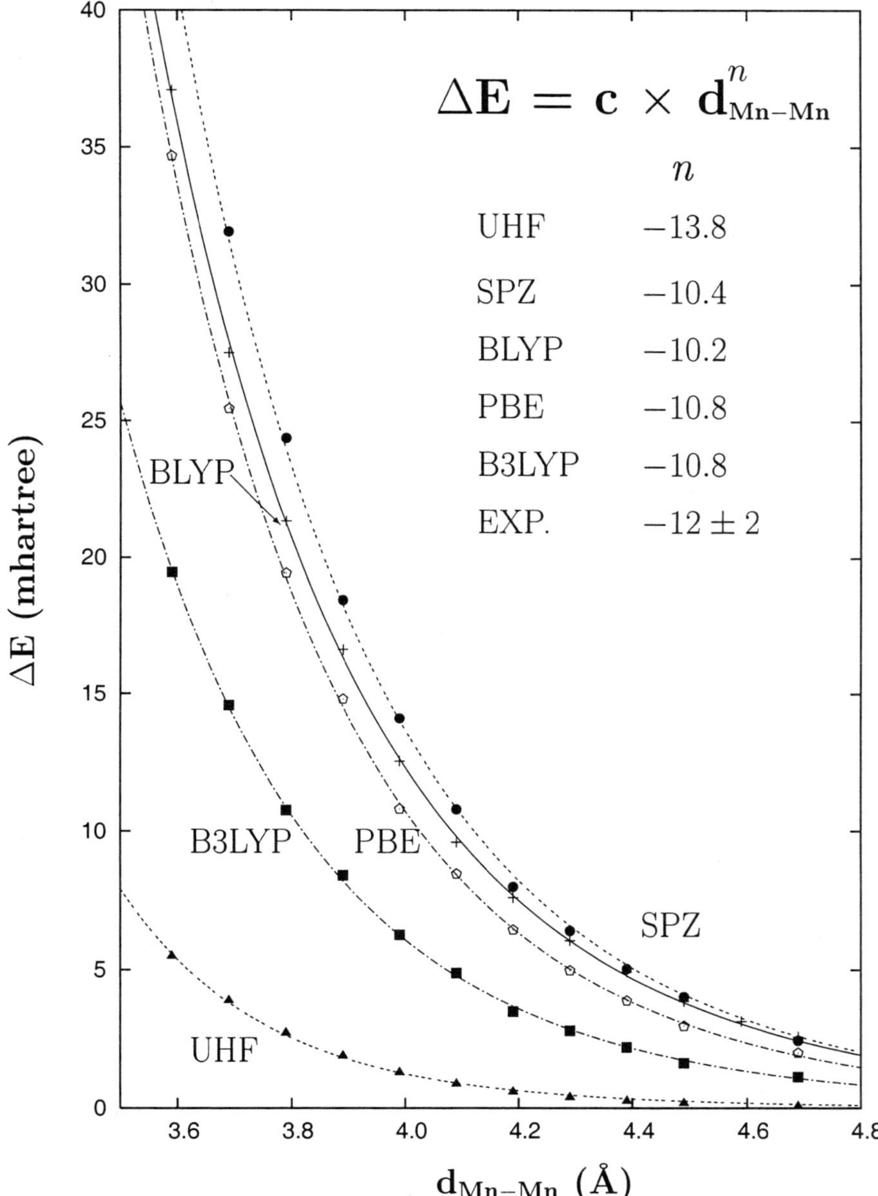

Figure 7: Energy difference (ΔE in mhartree) per formula unit between the $KMnF_3$ FM and AFM phases, obtained with different hamiltonians, as a function of the Mn–Mn distance d_{Mn-Mn}.

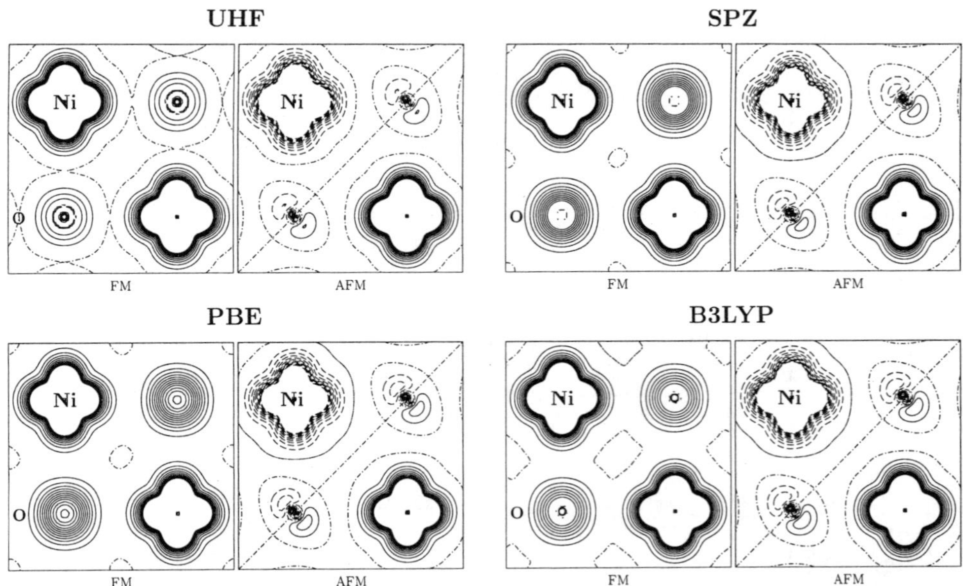

Figure 8: Spin density maps obtained with different hamiltonians for the FM and AFM solutions of NiO in the (001) plane through the Ni and O atoms. Symbols and scales as in Figure 6.

The different characterization of the electronic structure of NiO, as results from UHF and DFT hamiltonians, is also reflected in the values of the magnetic coupling constants J_1 and J_2, which refer mainly to the interaction between nearest and next nearest Ni ions of opposite spin, respectively. The trend of J_1 and J_2, as compared to the experimental values, is very similar to that observed in the case of $KMnF_3$. For example, J_2(SPZ) is 16 times larger than J_2(UHF) (Table 9) (see also Ref. [46]).

Another major difference in the outcome from different Hamiltonians concerns the population of the d-type AOs. The Mulliken populations of e_g and t_{2g} components reported in Table 8 (we remind that Ni^{2+} d electron configuration is d^8, and the simple ideal model of a d^8 ion in an octahedral cage predicts a $(t_{2g}^{\alpha})^3$ $(t_{2g}^{\beta})^3$ $(e_g^{\alpha})^2$ configuration) differ substantially in the different cases: UHF and B3LYP assign very little spin density to t_{2g}^{β} orbitals, whereas with PBE, BLYP and particularly SPZ t_{2g}^{β} orbitals do have some amount of spin density. This is a consequence of the metallic nature of the LSDA and GGA solutions, as shown in Figure 9, where there is a strong mixing of the various bands. On the contrary, UHF and B3LYP characterize NiO as an insulator, though with a relatively small band gap. Also in $KMnF_3$ (Figure 10) the band gap decreases drastically when changing from UHF to LSDA, but the band gap is larger and $KMnF_3$ was characterized as an insulating system in all cases. We only mention here that this problem of the mixing up of levels that are very close in energy, when using DFT hamiltonians, becomes very critical when Jahn-Teller distorsion is expected to take place, as happens in the case of $KFeF_3$ [17], $KCoF_3$ [17] and $KCuF_3$ [49], for which unbalancing the occupation within the t_{2g} and e_g subsets becomes impossible.

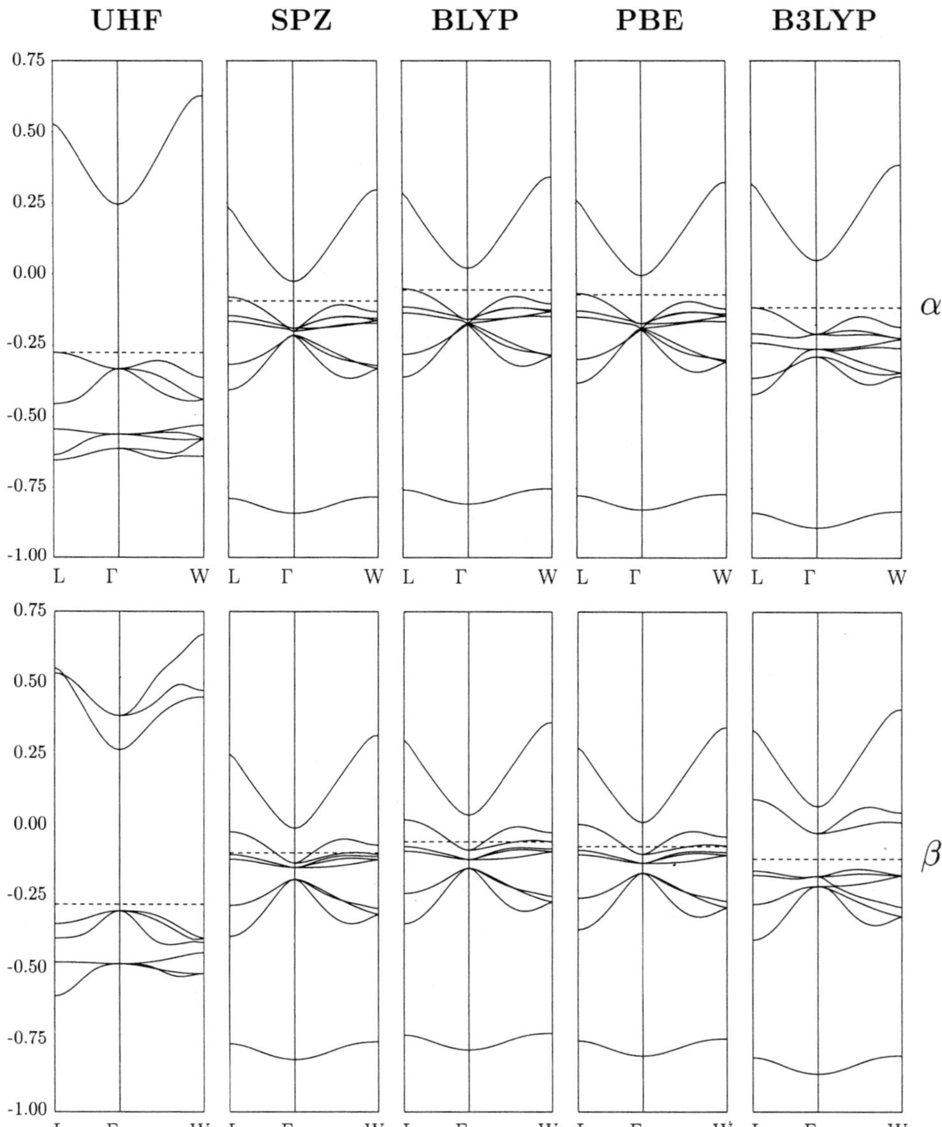

Figure 9: Bottom conduction (above the dashed line) and valence bands of the ferromagnetic phase of NiO. α and β indicate the majority and minority spin states, respectively. The dashed horizontal line corresponds to the Fermi level.

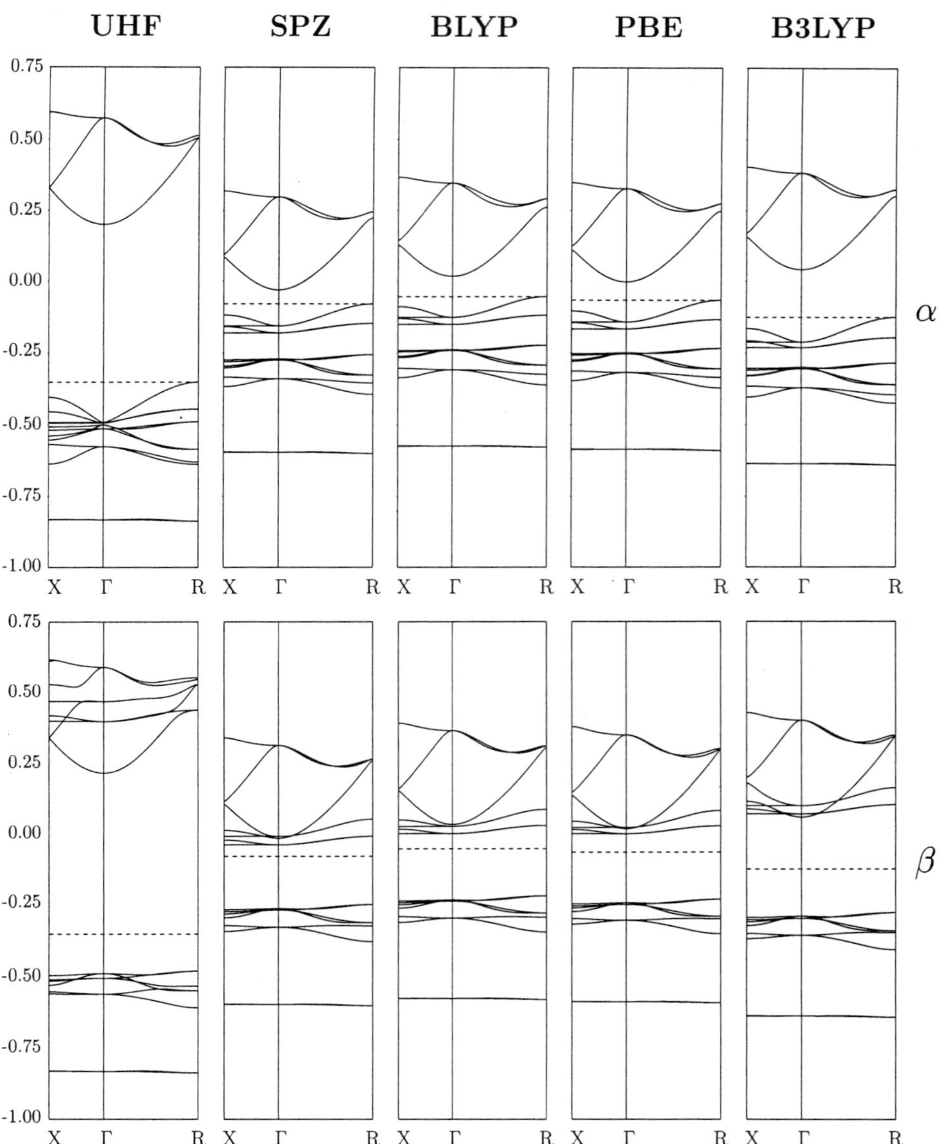

Figure 10: Bottom conduction (above the dashed line) and valence bands of the ferromagnetic phase of $KMnF_3$. α and β indicate the majority and minority spin states, respectively.

Table 9: Total energy (in hartree) of the ferromagnetic (FM) phases of NiO as obtained with various hamiltonians for a cell containing 10 atoms. J_1 and J_2 are the magnetic constants (in Kelvin). The ratios are calculated with respect to the most recent data.

NiO					
$\hat{\mathcal{H}}$	E_{FM}	J_1	$\frac{J_{1\ calc.}}{J_{1\ exp.}}$	J_2	$\frac{J_{2\ calc.}}{J_{2\ exp.}}$
UHF	-3163.609403	9.58	0.6	-51.84	0.3
SPZ	-3160.891377	122.11	7.5	-826.56	4.1
BLYP	-3167.347608	51.19	3.2	-622.59	3.1
PBE	-3166.580646	48.07	3.0	-609.82	3.0
B3LYP	-3167.013654	27.27	1.7	-278.05	1.4
EXP.		$+15.9^a$, $+16^b$		-221^a,-201^b	

[a]Ref. [47] ; [b]Ref. [48]

4 Conclusion

Various DFT functionals have been used in the study of four different systems involving unpaired electrons. In the first case (F-center in LiF) the unpaired electron is localised in a quite diffuse s-type function, in the second case ([MgO]:H and [MgO]:Li) p-type orbitals are involved, whereas in the last two cases ($KMnF_3$ and NiO) the localised d-type AOs of Mn are involved. In all cases UHF and LSDA provide the most localised and delocalised solution, respectively, and GGA and hybrid functionals give intermediate results. The difference in the results provided by the various functionals is minimal in the F-center case. It becomes larger for $KMnF_3$, where the small (but crucial for magnetism) polarization of the F ion is four times larger with LSDA than UHF, with a consequent much larger value of the superexchange coupling constant. In the case of [MgO]:H and [MgO]:Li the difference is not only quantitative, but also qualitative: at the UHF level the unpaired electron is strongly localised at a single O ion in a p-type orbital, in agreement with experiment, whereas DFT (both LSDA and GGA) and hybrid functionals distribute the unpaired electron to several O ions. Electron self-interaction is certainly responsible for the more widely spread LSDA density and it is expected to be less important when the unpaired electron itself is quite diffuse. On the contrary, it can affect results significantly in those cases where the orbitals occupied by the unpaired electron are more contracted, as in the case of [MgO]:Li (p-type orbital), $KMnF_3$ and NiO (d-type orbitals). However, in the former case the one unpaired electron in the infinite crystal can be distributed over many anionic sites, while in the latter cases such a large redistribution of the unpaired electrons is inhibited by the presence of the other Mn (Ni) ions (there is one Mn or Ni ion per unit cell) and the very stable configuration of the F ions (complete octet). The present results indicate that the various functionals should be used with caution. In particular, for strongly localised unpaired electrons the evaluation of the exchange interaction needs to be as accurate as possible. In this case the choice of UHF and, in part, B3LYP is much more appropriate.

References

[1] V.R. Saunders, R. Dovesi, C. Roetti, M. Causà, N. M. Harrison, R. Orlando and C.M. Zicovich-Wilson, *CRYSTAL98 User's Manual*, Università di Torino (Torino,1998).

[2] P. Ugliengo, B. Civalleri, C.M. Zicovich-Wilson and R. Dovesi, *Chem. Phys. Lett.* **318**, 247–255 (2000).

[3] M. Catti, B. Civalleri and P. Ugliengo, *J. Phys. Chem. B* **104**, 7259 (2000).

[4] M. O. Sinnokrot and C. D. Sherrill, *J. Chem. Phys.* **115**, 2439 (2001).

[5] B. G. Johnson, P. M. W. Gill and J. A. Pople, *J. Chem. Phys.* **98**, 5612 (1993).

[6] Ph. Baranek, A. Lichanot, R. Orlando, R. Dovesi, *Chem. Phys. Letters* **240**, 362 (2001).

[7] A. D. Becke, *J. Chem. Phys.* **98**, 5648 (1993).

[8] B. Henderson, *Defect in crystalline solids*, Edward Arnold,Ltd., London (1972).

[9] W.C. Holton and H. Blum, *Phys. Rev.* **125**, 89 (1962).

[10] O.F. Schirmer, *J. Phys. Chem. Solids* **32**, 499 (1971).

[11] B. Henderson and J.E. Wertz, "Defects in alkaline earth oxides", *Adv. Phys.* **17**, 786 (1969).

[12] P.W. Kirklin, P. Auzins and J. E. Wertz, *J. Phys. Chem. Solids* **26**, 1067 (1965).

[13] W.C. O'Mara and J.E. Wertz, *Solid State Commun.* **8**, 807 (1970).

[14] M. M. Abraham,W. P. Unruh and Y. Chen, *Phys. Rev. B* **10**, 3540 (1974).

[15] Y. Chen and M. M. Abraham, *J. Phys. Chem. Solids* **51**, 747 (1990).

[16] W.P. Unruh, Y. Chen and M.M. Abraham, *J. Chem. Phys.* **59**, 3284 (1973).

[17] R. Dovesi, F. Freyria Fava, C. Roetti and V.R. Saunders, *Faraday Discuss.* **106**, 173 (1997).

[18] W.C. Mackrodt, N.M Harrison, V.R. Saunders, N.L.Allan, M.D. Towler, E. Aprà and R. Dovesi, *Philos. Magaz. A* **68**, 653 (1993).

[19] M.D. Towler, N.L. Allan, N.M. Harrison, V.R. Saunders, W.C. Mackrodt and E. Aprà, *Phys. Rev. B* **50**, 5041 (1994).

[20] P. A. M. Dirac, *Proc. Cambridge Phil. Soc.* **26**, 376 (1930).

[21] J. P. Perdew and A. Zunger, *Phys. Rev. B* **23**, 5048 (1981).

[22] A. D. Becke, *Phys. Rev. A* **38**, 3098 (1988).

[23] C. Lee, W. Yang, and R. G. Parr, *Phys. Rev. B* **37**, 785 (1988).

[24] J. P. Perdew, K. Burke, and M. Ernzerhof, *Phys. Rev. Lett.* **77**, 3865 (1996).

[25] C. Pisani, R. Dovesi and C. Roetti, *Hartree-Fock ab initio Treatment of Crystalline Systems*, *Lecture Notes in Chemistry*, volume 48, Springer Verlag, Heidelberg (1988).

[26] G. Mallia, R. Orlando, C. Roetti, P. Ugliengo, R. Dovesi, *Phys. Rev. B* **63**, 235102 (2001).

[27] *http://www.ch.unito.it/ifm/teorica/Basis_Sets*.

[28] A. Lichanot, C. Larrieu, R. Orlando and R. Dovesi, *J. Phys. Chem. Solids* **59**, 7 (1998).

[29] A. Lichanot, C. Larrieu, C. Zicovich-Wilson, C. Roetti, R. Orlando and R. Dovesi, *J. Phys. Chem. Solids* **59**, 1119 (1998).

[30] A. Lichanot, R. Orlando, G. Mallia, M. Merawa and R. Dovesi, *Chem. Phys. Lett.* **318**, 240 (2000).

[31] A. Lichanot, Ph. Baranek, M. Merawa, R. Orlando and R. Dovesi, *Phys. Rev. B* **62**, 12812 (2000).

[32] M.I. McCarthy and N.M. Harrison, *Physical Review B* **49**, 8574 (1994).

[33] M.D. Towler, A. Zupan and M. Causà, *Comp. Phys. Commun.* **98**, 181 (1996).

[34] *http://www.ch.unito.it/ifm/teorica/AuxB_Sets.*

[35] H. Seidel, H.C. Wolf, *The Physics of Colour Centers.*, Ed. Fowler, New York (1968).

[36] J.A. Weil, J.R. Bolton and J.E. Wertz, *Electron Paramagnetic Resonance*, JOHN WILEY & SONS, INC, 605 Third Avenue, New York (1994).

[37] G. Rius, R. Cox, R. Picard and C. Santier, *C.r. Acad. Sci. Paris* **271**, 824 (1970).

[38] J.J. Davies, *Phys. Lett. A* **28**, 9 (1968).

[39] B. Henderson, *J. Phys. C:Solid State Phys.* **9**, 579 (1976).

[40] G. Rius and A. Hervé, *Solid State Communications* **15**, 399 (1974).

[41] P. Baranek, G. Pinarello, C. Pisani and R. Dovesi, *Phys. Chem. Chem. Phys.* **2**, 3893 (2000).

[42] N.M. Harrison, V.R. Saunders, R. Dovesi and W.C. Mackrodt, *Phil. Trans. R. Soc. Lond. A* **356**, 75 (1998).

[43] Madelung, *Landolt-Börnstein New Series*, Springer Verlag, Berlin (1982).

[44] L.J. de Jongh and R. Block, *Physica* **79B**, 568 (1975).

[45] P.A. Cox, *The transition metal oxides*, Oxford University Press, Oxford (1992).

[46] I. de P.R.Moreira, F. Illas and R. L. Martin, *Phys. Rev. B* (submitted).

[47] M. T. Hutchings and E. J. Samuelsen, *Phys. Rev. B* **6**, 3447 (1972).

[48] R. Shanker and R. A. Singh, *Phys. Rev. B* **7**, 5000 (1973).

[49] M.D. Towler, R. Dovesi and V. R. Saunders, *Phys. Rev. B* **52**, 10150 (1995).

Computational Materials Science
C.R.A. Catlow and E.A. Kotomin (Eds.)
IOS Press, 2003

Translation Symmetry in Imperfect Crystal Modelling

Robert A. Evarestov
Quantum Chemistry Department, St.Petersburg State University,
Stary Peterghof,University Prospect,2, 198504, St.Petersburg,Russia, evarest@hm.csa.ru

Abstract. There are considered two models of defective crystalline solids based on use of periodic boundary conditions and translation symmetry:supercell model (SCM) and cyclic cluster model (CCM). The symmetry of both models is analyzed in terms of the symmetry properties of a perfect bulk crystal. There are given the principles of supercell and cyclic cluster shape and size choice in defective crystals calculations. To illustrate the general consideration there are discussed the results of LCAO Hartree-Fock calculations of interstitial oxygen atom in MgO crystal in SCM and perfect rutile TiO_2 in CCM. For comparison results of MSINDO CCM calculations of rutile are included.

1 Introduction

In the theory of point defects in crystalline solids two models are applied which make use of supercells and the Born-von Karman periodic boundary conditions (PBC): the supercell model (SCM) and the cyclic cluster model (CCM) [1, 2, 3]. The quasimolecular large unit cell (QLUC) approach [4] is in fact connected with the calculation of a cyclic cluster embedded in the Madelung field of the surrounding crystal. The embedded CCM and the SCM have both similarities and differences [5]. Similar is that in both models the direct lattice translation vectors are transformed. A large unit cell (supercell) is introduced for the perfect host crystal in such a way that the point symmetry of the corresponding Bravais lattice is maintained. The difference between the approaches concerns the crystal region for which PBC are introduced: in the CCM it is made for the supercell itself, in the SCM the PBC are introduced for the *main region* of a crystal. The latter consists of $N = N_1 \times N_2 \times N_3$ primitive unit cells (N_i primitive unit cells in the direction of basic translation vector $\mathbf{a}_i$, $i = 1, 2, 3$; N_i are very large numbers).

In the SCM the loss of periodicity in a crystal due to the presence of defects is restored and the momentum space description can be applied so that the methods developed for band structure calculations of perfect crystals can be directly extended to the SCM [6, 7, 8, 9, 10, 11, 12, 13, 14, 15, 16, 17].

As a disadvantage of the SCM one receives the artificially introduced periodicity of point defects and therefore the necessity to use such a $\mathbf{k}$ set for the defective crystal that minimizes the interaction of defect wave functions in the superlattice [6] and reduces the dispersion of the defect levels between various $\mathbf{k}$ points. Additional problems of the SCM are connected with the consideration of charged point defects and host atom relaxation around defects. The SCM is most often applied to covalent and weakly ionic systems [1].

The CCM avoids some problems of the SCM as it excludes direct defect-defect interaction and at the same time has advantages compared with the molecular cluster model (MCM). The latter is realized by simply cutting out of the crystal some portion consisting of the point defect and surrounding host atoms, followed either by a saturation of dangling bonds of atoms at the cluster boundaries with hydrogen atoms or pseudo atoms, or by an embedding in a representation of the Madelung field of the surrounding lattice.

In the MCM the space symmetry of the host crystal is reduced. An artificial inequivalence of sites of the host crystal appears and the symmetry connection between the MCM and the **k** space one-electron states is lost.

It is well known that the CCM, as an intermediate between the SCM and the MCM, unifies the advantages of the SCM (correct symmetries of states, absence of boundary effects, reasonable description of both occupied and virtual states) and MCM (direct space approach, possibility of simple extension, applicability of molecular quantum chemistry techniques, absence of unwanted defect-defect interaction) [1]. The computational realization of the CCM is connected with the modification of interatomic distance and directional cosine matrices of each atom in order to satisfy the PBC. The point symmetry is maintained in the CCM by an averaging over matrix elements of those atoms which appear on the surface of the corresponding Wigner-Seitz supercell [1].

The above mentioned modifications were realized in semiempirical versions of the Hartree-Fock (HF) LCAO method, at CNDO level [18, 19] and at INDO level [20, 21, 22, 23, 24, 25], and in the local density approximation (LDA) of the density-functional theory (DFT) [26]. An implementation of the CCM at the non-empirical HF LCAO level encounters the problems connected with a large number of two-electron integrals for crystals and can be made basing on HF LCAO computer codes for band calculations [27].

2 Cyclic Clusters and Their Symmetry Groups

Our consideration does not depend neither on the dimension n of the crystal space, nor on the dimension $m \leq n$ of the translational periodicity of the crystal. More than three dimensional crystal spaces are useful for describing the so-called incommensurate crystal phases [28, 29].

Let G be a space symmetry group of an infinite crystal and $T = T_{\mathbf{a}}$ be its invariant translational subgroup of infinite order. The factor group $G/T_{\mathbf{a}}$ is isomorphic to a crystallographic point group(crystal class) F of the order n_F. We use the Seitz's notation $g = (R|\mathbf{v}_R + \mathbf{a})$ for a general element $g \in G$ of the group where $R \in F$, $\mathbf{v}_R$ and $\mathbf{a}$ are improper and lattice translations, respectively.

To determine a finite cyclic system $C^{(\mathbf{A})}$(cyclic cluster) and its symmetry group let us make a linear transformation of the basic translation vectors $\mathbf{a}_i$ $(i = 1, 2, ..., m)$ of the translation subgroup T of the space group G of the model of an infinite crystal

$$\mathbf{A}_j = \sum_i l_{ji} \cdot \mathbf{a}_i, \qquad j = 1, 2, ...m, \qquad L = |\det l| \succeq 1, \tag{1}$$

with integer coefficients l_{ij} forming the matrix l.

The transformation (1) for $L > 1$ determines a large unit cell (LUC) of the direct lattice. It may be taken in the form of a parallelepiped with the vectors $\mathbf{A}_j$ as edges or in the form of a corresponding Wigner-Seitz unit cell.

The particular form of the matrix l in (1) defines both the LUC volume which is L times larger than that of a primitive unit cell and the symmetry of the direct lattice based on the translation group $T_{\mathbf{A}}$ of the translations

$$\mathbf{A}_n = \sum_j n_j \cdot \mathbf{A}_j, \qquad n_j \text{ are arbitrary integers.} \tag{2}$$

The group $T_{\mathbf{A}}$ is an invariant subgroup of T so that the cosets in the decomposition

$$T = \sum_{s=1}^{L} (E|\mathbf{a}'_s) \cdot T_{\mathbf{A}} \tag{3}$$

form the factor group $T/T_{\mathbf{A}} = \widetilde{T}^{(\mathbf{A})}$ of order L. Inner translations of the LUC are supposed to be chosen as $\mathbf{a}'_s$ in (3).

A cyclic cluster $C^{(\mathbf{A})}$ is defined as a LUC with identical opposite faces, or, in another words, the cyclic boundary conditions are introduced for the LUC. The translation symmetry of $C^{(\mathbf{A})}$ is characterized by the group $T^{(\mathbf{A})}$ of order L with the elements $(E|\mathbf{a}'_s)$ from (3) and by the multiplication law where all the translations over vectors (2) are considered as the unit element (multiplication modulo $T_{\mathbf{A}}$). The group $T^{(\mathbf{A})}$ is isomorphic to the group $\widetilde{T}^{(\mathbf{A})}$.

Let $l^{(\mathbf{a})}(g)$ be the transformation matrix of the translation vectors $\mathbf{a}_i$ under the point symmetry operation $R \in F$,

$$R \cdot \mathbf{a}_i = \sum_{i'} l^{(\mathbf{a})}_{i'i}(R) \cdot \mathbf{a}_{i'}. \tag{4}$$

As the operations $g \in G$ are compatible with the translation symmetry of the lattice, the vectors $R \cdot \mathbf{a}_i$ are lattice translations (integer linear combinations of basic translation vectors $\mathbf{a}_i$), i.e. the matrix elements $l^{(\mathbf{a})}_{i'i}(R)$ of the matrix $l^{(\mathbf{a})}(R)$ are integers for all $R \in F$.

The corresponding transformation matrix for the translation vectors $\mathbf{A}_j$ in (1) is

$$l^{(\mathbf{A})}(R) = l \cdot l^{(\mathbf{a})}(R) \cdot l^{-1}, \qquad R\,\mathbf{A}_j = \sum_{j'} l^{(\mathbf{A})}_{j'j}(R) \cdot \mathbf{A}_{j'}. \tag{5}$$

We choose the matrix l so that all the matrix elements $l^{(\mathbf{A})}_{i'i}(R)$ of the matrix $l^{(\mathbf{A})}(R)$ be integers for all $R \in F$. In this case the point symmetry of the lattice composed of LUC's is the same as the point symmetry of the initial lattice generated by the basic translation vectors $\mathbf{a}_i$, and the transformation (1) is called symmetrical [30].

For symmetrical transformation the coset representatives $(R|\mathbf{v}_R + \mathbf{a}'_s)$ in decomposition

$$G = \sum_{R \in F} \sum_{s=1}^{L} (R|\mathbf{v}_R + \mathbf{a}'_s) \cdot T_{\mathbf{A}} \tag{6}$$

form a group $G^{(\mathbf{A})}$ of order $n_F \cdot L$ if a multiplication modulo $T_{\mathbf{A}}$ is accepted as a group multiplication law for its elements. The group $G^{(\mathbf{A})}$ is isomorphic to the factor group $G/T_{\mathbf{A}}$. It is obvious that this group describes the symmetry of the cyclic system.

The value L for transformation (1) is equal to the number of unit cells in a given cyclic cluster, i.e. it determines its size. One can generate an infinite number of cyclic systems of different size. The cyclic cluster defined for a very large L-value is often introduced in the solid state theory and called main region of a crystal. Usually no attention is paid to the fact that the main region possesses the point symmetry of an infinite crystal only if the corresponding transformation (1) is symmetrical. The smallest possible cyclic cluster $C^{(\mathbf{A})}$ is a primitive unit cell with the symmetry group $G^{(\mathbf{A})}$ of order n_F isomorphic to point group F of a crystal.

For symmetric transformation (1) the group of symmetry $G^{(\mathbf{A})}$ of a cyclic cluster is homomorphic to the symmetry group G of the model of an infinite crystal with the kernel of homomorphism $T_{\mathbf{A}}$.

The space groups $G^{(\mathbf{A})}$ and G have different multiplication laws for their elements. Group $G^{(\mathbf{A})}$ is finite, group G is infinite. But they have the same point symmetry (a crystal class), the same system of basic translations and the same structure. The latter means that the both groups $G^{(\mathbf{A})}$ and G have the invariant subgroups of translations $T^{(\mathbf{A})}$ and T, respectively, and have the same coset decomposition with respect to the groups $T^{(\mathbf{A})}$ and T:

$$G^{(\mathbf{A})} = \sum_{R\in F}(R|\mathbf{v}_R)\cdot T^{(\mathbf{A})}, \qquad G = \sum_{R\in F}(R|\mathbf{v}_R)\cdot T. \tag{7}$$

The groups $G^{(\mathbf{A})}$ are in fact space groups with finite subgroups of translations.

The finite group $G^{(\mathbf{A})}$ has the same structure as the infinite space group G, therefore its irreps may be generated by the same way as for the group G. As for the space group G, the irreps of $G^{(\mathbf{A})}$ may be characterized by irreducible $\mathbf{k}$-stars and the index α that numbers nonequivalent irreps with the same irreducible $\mathbf{k}$-star. Only the number of irreducible $\mathbf{k}$-stars in the case of the group $G^{(\mathbf{A})}$ is finite because its invariant translational subgroup $T^{(\mathbf{A})} \lhd G^{(\mathbf{A})}$ is also finite.

To find the correspondence between all the irreps of the group $G^{(\mathbf{A})}$ and some irreps of the group G we use the homomorphism $G \to G^{(\mathbf{A})}$ ($T_{\mathbf{A}}$ is the kernel of homomorphism). Every irrep $D^{({}^*\mathbf{k},\alpha)}((R|\mathbf{v}_R+\mathbf{a}'_s)\cdot T_{\mathbf{A}})$ of the factor group $G/T_{\mathbf{A}}$ engenders some irrep of G. In these irreps of G all the elements $(R|\mathbf{v}_R+\mathbf{a}'_s+R\cdot\mathbf{A}_\mathbf{n})$ of the coset $(R|\mathbf{v}_R+\mathbf{a}'_s)\cdot T_{\mathbf{A}}$ in decomposition (6) are mapped by the same matrix $D^{({}^*\mathbf{k},\alpha)}((R|\mathbf{v}_R+\mathbf{a}'_s+R\cdot\mathbf{A}_\mathbf{n})) = D^{({}^*\mathbf{k},\alpha)}((R|\mathbf{v}_R+\mathbf{a}'_s))$. The latter is the matrix of the irrep $D^{({}^*\mathbf{k},\alpha)}((R|\mathbf{v}_R+\mathbf{a}'_s)\cdot T_{\mathbf{A}})$ corresponding to this coset of the factor group $G/T_{\mathbf{A}}$ or to the element $(R|\mathbf{v}_R+\mathbf{a}'_s) \in G^{(\mathbf{A})}$ as $G^{(\mathbf{A})} \longleftrightarrow G/T_{\mathbf{A}}$. Thus to obtain all the irreps of the group $G^{(\mathbf{A})}$, if the irreps of the space group G are known, is sufficient to pick out those irreps of G in which all the translations $(E|\mathbf{A}_\mathbf{n}) \in T_{\mathbf{A}}$ are mapped by unit matrix and to attribute the matrices of these irreps to the elements of the group $G^{(\mathbf{A})}$ according to the homomorphism $G \to G^{(\mathbf{A})}$. In Bloch basis the matrices $D^{({}^*\mathbf{k},\alpha)}((E|\mathbf{A}_\mathbf{n}))$ are diagonal with $\exp(-i\cdot\mathbf{k}_t\cdot\mathbf{A}_\mathbf{n})$ on the main diagonal. For the irreps of G under consideration the relation

$$\exp(-i\cdot\mathbf{k}_t\cdot\mathbf{A}_\mathbf{n}) = 1, \qquad \mathbf{k}_t \in^* \mathbf{k}, \tag{8}$$

has to be valid for all vectors $\mathbf{A}_\mathbf{n}$. This relation is fulfilled for the set $\widetilde{\mathbf{K}}^{(\mathbf{A})}$ of L points $\mathbf{k}_t$ in the Brillouin zone (BZ):

$$\mathbf{k}_t = \sum_{j=1}^{m} q_{tj}\cdot\mathbf{b}_j, \qquad t = 1, 2, \ldots, t, \tag{9}$$

where integers q_{tj} assure the position of point $\mathbf{k}_t$ in the BZ, and

$$\mathbf{b}_j = \sum_i (l^{-1})_{ij}\mathbf{B}_i \qquad (l \text{ is matrix from (1)}). \tag{10}$$

The vectors $\mathbf{b}_j$ and $\mathbf{B}_i$ define the transformed and initial reciprocal lattices:

$$(\mathbf{a}_i\cdot\mathbf{B}_{i'}) = 2\pi\cdot\delta_{ii'}, \qquad (\mathbf{A}_j\cdot\mathbf{b}_{j'}) = 2\pi\cdot\delta_{jj'}. \tag{11}$$

For symmetrical transformation (1), the set $\widetilde{\mathbf{K}}^{(\mathbf{A})}$ contains vectors $\mathbf{k}_t$ by the whole k-stars (with respect to the point group F), as $R \cdot \mathbf{k}_t \in \widetilde{\mathbf{K}}^{(\mathbf{A})}$ $(R \in F)$, if $\mathbf{k}_t \in \widetilde{\mathbf{K}}^{(\mathbf{A})}$. The irreps of the group G with irreducible stars $^*\mathbf{k} \in \widetilde{\mathbf{K}}^{(\mathbf{A})}$ determine all the irreps of the group $G^{(\mathbf{A})}$. Therefore all the irreps of a cyclic system $C^{(\mathbf{A})}$ symmetry group are in a simple way unambiguously related to some irreps of space symmetry group G of the model of an infinite crystal.

This consideration is equally applicable both to single-valued and double-valued reps of the space groups $G^{(\mathbf{A})}$ and G. An additional degeneracy connected with the time-reversal symmetry can easily be taken into account for the irreps of the group $G^{(\mathbf{A})}$ as this degeneracy is tabulated for all the space groups G.

3 Symmetry of the Supercell Model of a Defective Crystal

Let a point defect (impurity atom or molecule, one or few vacancies) occupy a position q in perfect crystal with a space group G (crystal class F). A site symmetry group of the defect in the crystal S_D consists of common elements of two groups: of the symmetry group $S_D^{(o)}$ of the defect itself and the site symmetry group $S_\mathbf{q}$ of the position q in the crystal occupied by the defect

$$S_D = S_D^{(o)} \cap S_\mathbf{q}. \tag{12}$$

The group S_D depends on the mutual orientation of the symmetry elements of the groups $S_D^{(o)}$ and $S_\mathbf{q}$.

Let us consider now a periodical structure that arises when embedded defect is periodically repeated and occupies in this structure some positon $\mathbf{Q}$ with site symmetry group $\tilde{S}_\mathbf{Q} = S_D$ (supercell model). The symmetry of the obtained periodical structure is characterized by some space group G_D. One seeks for all the possible space groups $G_D = T_\mathbf{A} F_D$ of the defective crystal for a given host crystal (space group G) and for a given point defect in crystal (point group S_D). Some symmetry operations disappear when one goes from the space group G to the space group G_D. Therefore the group G_D has to be a subgroup of G

$$G_D \subseteq G \tag{13}$$

with a crystal class satisfying the condition

$$S_D \subseteq F_D \subseteq F. \tag{14}$$

The possible space groups G_D of the crystal classe F_D are those which have the group S_D as a site subgroup. By this way, one obtains the list of space groups $\tilde{G}_D$ that may describe the symmetry of the supercell model of a defective crystal for a given site symmetry S_D of the embedded defect and space group G of a host crystal. In the frame of the same space group G_D the periodical structures may differ by the length of basic translation vectors $\mathbf{A}_1, \mathbf{A}_2, \mathbf{A}_3$ for the same type of the Bravais lattice. The vectors $\mathbf{A}_1, \mathbf{A}_2, \mathbf{A}_3$ have a definite orientation with respect to the symmetry elements of the group $\tilde{G}_D$, and sometimes there is a definite relation between their lengths. Besides they have to be an integer linear combinations

$$\mathbf{A}_j = \sum_{i=1}^{3} l_{ji}\mathbf{a}_i, \quad L = |\det l| > 1, \;\; i,j = 1,2,3 \tag{15}$$

of basic translation vectors $\mathbf{a}_1, \mathbf{a}_2, \mathbf{a}_3$ of the host perfect crystal. The supercell size is defined by the number of primitive unit cells L in the supercell. Let the transformation (15) be a symmetric one.

These conditions determine the matrix l (or vectors $\mathbf{A}_1, \mathbf{A}_2, \mathbf{A}_3$), i.e. the translational subgroup $T_{\mathbf{A}}$ of the symmetry group G_D. At last it is necessary to verify if the obtained group $\tilde{G}_D$ is a subgroup of the symmetry group of the host crystal (condition (13). This condition assures that all the atoms of the system transform into themselves under the operatioons of the group G_D. If $\tilde{G}_D \subset G$, then $G_D = \tilde{G}_D$; if $G_D \not\subset G$, then the group $\tilde{G}_D$ has to be excluded from the list of the possible groups G_D. The Wyckoff position $\mathbf{q}$ of the host crystal occupied by a defect may be multiple: $\mathbf{q}_i$ $(i = 1, 2, \ldots, n_{\mathbf{q}})$. A defect may occupy one or several positions $m \leq n_{\mathbf{q}}$ in the unit cell of a host crystal. On the other hand, in the group G_D the position $\mathbf{Q}$ may be also multiple: $\mathbf{Q}_j$ $(j = 1, 2, \ldots, n_{\mathbf{Q}})$. The points

$$\mathbf{q}_i + \sum_{i'} m_{i'}\mathbf{a}_{i'} \qquad (i = 1, 2, \ldots, n_{\mathbf{q}};\ m_{i'} \text{ are integers}) \tag{16}$$

of a host crystal have the same symmetry. Point defects in the crystal with a periodic defect may occupy some of these positions but cannot occupy other positions of the host crystal. In particular all the points $\mathbf{Q}_j$ have to be contained in the set (16). We assume that the coordinate systems K_G and K_D are chosen for the groups G and G_D in the same way as in [33] and that the points $\mathbf{Q}_1$ and $\mathbf{q}_1$ represent the same point of the host and imperfect crystals. All the points $\mathbf{Q}_j$ are contained in the set (16), if

$$m_1^{(ij)}\mathbf{a}_1 + m_2^{(ij)}\mathbf{a}_2 + m_3^{(ij)}\mathbf{a}_3 = \mathbf{Q}_j - \mathbf{Q}_1 - (\mathbf{q}_i - \mathbf{q}_1). \tag{17}$$

Relation (17) has to be satisfied for any $j = 1, 2, \ldots, n_{\mathbf{Q}}$ and at least for one $i = 1, 2, \ldots, n_{\mathbf{q}}$. When using relations (15) and $\mathbf{q}_i = \sum_{i'} \alpha_{ii'}\mathbf{a}_{i'}$, $\mathbf{Q}_j = \sum_{j'} \beta_{jj'}\mathbf{A}_{j'}$, equation (17) takes the form

$$m_{i'}^{(ij)} = \sum_{j'} (\beta_{jj'} - \beta_{1j'})l_{j'i'} - (\alpha_{ii'} - \alpha_{1i'}). \tag{18}$$

As $m_{i'}$ must be integers, relation (17) or (18) may impose additional restrictions on the matrices l defining the possible supercells.

A group G_D may have several independent sets of positions $\mathbf{Q}_{jp}$ $(j = 1, 2, ..., n_{\mathbf{Q}}, p = 1, 2, \ldots)$ with the same site symmetry group S_D. The location of the defects in one of the sets of positions $\mathbf{Q}_{jp}$ with $p \neq 1$ (instead of the positions $\mathbf{Q}_{j1}$) does not give a new structure as all the sets are symmetry equivalent.

If there are several noneqivalent Wyckoff positions with the same symmetry S_D, the location of the defects in several sets of positions simultaneously gives either new structures with a periodic defect or structures having already appeared earlier. In the latter case new translational and point symmetry elements may appear and the symmetry of the structure seems to be higher (the change of crystal class and/or lattice type).

If a defect has a principal symmetry axis, the number of structures with a periodic defect increases as there are different crystal structures with the same space group but with different orientation of a molecular axis relative to the symmetry elements of the group S_D (see an example in Section 4.2).

In particular, when the space group G_D has the lowest possible point symmetry, the group S_D is the group of the highest possible site symmetry in the group G_D and is isomorphous with its crystal class $S_D \leftrightarrow F_D$. In this case the defect occupies the Wyckoff position a

($\mathbf{Q}_a = (0,0,0)$) with a site symmetry $S_D = S_a$. The group G_D itself is symmorphic and may have any of the Bravais lattice types possible for a given crystal class . The symmetrical transformation (15) is mostly used for application of the supercell model for the investigation of physical properties of a single defect in a crystal but for a given L the symmetrical transformation not always assures the largest possible distance between the nearest defects in the defect sublattice(defect period).

Finally, the generation of the possible space groups with a periodical defect consists of the following steps [31]:

1. Determination of the site group S_D of the defect in the crystal;
2. Determination of the crystal classes satisfying the condition (14);
3. Determination in these classes the space groups having the site symmetry subgroups S_D;
4. Generation of the matrix l and vectors $\mathbf{A}_j$ (15);
5. Verification of the condition (13).

The groups G_D of a supercell model with one point defect for a supercell are symmorphic, belong to the crystal class $F_D = S_D$ with all the possible for this crystal class types of crystal lattices.

4 Point defects in MgO crystal

We apply the approach developed for point defects in MgO as a host crystal [31]. Its space group is $O_h^5(Fm\overline{3}m)$ with the basic translation vectors

$$\mathbf{a}_1 = (0,1/2,1/2), \quad \mathbf{a}_2 = (1/2,0,1/2), \quad \mathbf{a}_3 = (1/2,1/2,0) \tag{19}$$

(the length of the elementary cube edge a is assumed to be unity). Let the origin of the coordinate system K_G be at the oxygen site. The oxygen atom occupies in the primitive unit cell the position a ($\mathbf{q}_a = (0,0,0)$) with the site symmetry group O_h, and magnesium atom occupies the position b ($\mathbf{q}_b = (1/2,1/2,1/2)$) with the same site symmetry. Let us consider the neutral defects – an oxygen interstitial atom and an oxygen molecule O_2^{2-} in oxygen site of MgO crystal.

4.1 Interstitial oxygen atom

The proper symmetry of a free oxygen atom is described by the full orthogonal group $O(3)$. There are different sites for embedding of an oxygen atom in MgO crystal. Three of them are the most symmetrical ones: positions c ($S_c = T_d$), d ($S_d = D_{2h}$) and e ($S_e = C_{4v}$). These positions correspond to the volume-centred (v), face-centred (f) and edge-centred (e) interstitials. We consider only position c, other two are considered in [31].

The positions ($\mathbf{q}_{c1} = (1/4,1/4,1/4)$, $\mathbf{q}_{c2} = (3/4,3/4,3/4)$) of the group O_h^5 have the site symmetry T_d. The symmetry of the defect in a crystal is $S_D = O(3) \cap T_d = T_d$. Let the origin of the coordinate system K_{PD} be in the interstitial oxygen atom (position $\mathbf{q}_{c1}$) and the axes be parallel to the axes of the K_G system. The possible space groups of the crystal

Table 1: Structures with interstitial oxygen in position c

Structure	$l^{(i)}$	L	$D(N_n)$	N_d	Remark
$\mathrm{Oc}T_d^1a$	$l^{(1)}$	$4n^3$	$n\,(6)$	1	
$\mathrm{Oc}T_d^2a$	$l^{(2)}$	n^3	$n/\sqrt{2}\,(12)$	1	
$\mathrm{Oc}T_d^3a$	$l^{(3)}$	$16n^3$	$\sqrt{3}n\,(8)$	1	
$\mathrm{Oc}O_h^4a$	$l^{(1)}$	$4n^3$	$\sqrt{3}n/2\,(8)$	2	n odd
$\mathrm{Oc}O_h^5c$	$l^{(2)}$	n^3	$n/2\,(6)$	2	n odd
$\mathrm{Oc}O_h^7a$	$l^{(2)}$	n^3	$\sqrt{3}n/4\,(4)$	2	n even
$\mathrm{Oc}T_d^2abc$	$l^{(2)}$	n^3	$\sqrt{3}n/4\,(4, 8)$	3	n even

$$l^{(1)} = \begin{pmatrix} -n & n & n \\ n & -n & n \\ n & n & -n \end{pmatrix} \quad l^{(2)} = \begin{pmatrix} n & 0 & 0 \\ 0 & n & 0 \\ 0 & 0 & n \end{pmatrix} \quad l^{(3)} = \begin{pmatrix} 3n & -n & -n \\ -n & 3n & -n \\ -n & -n & 3n \end{pmatrix}$$

with a periodic defect belong to the crystal classes T_d and O_h (one and two interstitials in a supercell, respectively). At first let us consider all the possible space groups of the crystal class T_d: T_d^1, T_d^2, T_d^3. The basic translation vectors $\mathbf{A}_1\,(p,0,0)$, $\mathbf{A}_2\,(0,p,0)$, $\mathbf{A}_3\,(0,0,p)$ for the group T_d^1 have equal lengths and are directed along Cartesian axes (form a simple cubic lattice). Let us rewrite the vector relation (15) in Cartesian components,

$$\begin{aligned} &(1/2)(l_{12}+l_{13}) = p, \quad (1/2)(l_{11}+l_{13}) = 0, \quad (1/2)(l_{11}+l_{12}) = 0, \\ &(1/2)(l_{22}+l_{23}) = 0, \quad (1/2)(l_{21}+l_{23}) = p, \quad (1/2)(l_{21}+l_{22}) = 0, \\ &(1/2)(l_{32}+l_{33}) = 0, \quad (1/2)(l_{31}+l_{33}) = 0, \quad (1/2)(l_{31}+l_{32}) = p. \end{aligned}$$

The solution of this system is the following:

$$l_{11} = l_{22} = l_{33} = -p, \quad l_{12} = l_{13} = l_{21} = l_{23} = l_{31} = l_{32} = p.$$

Therefore $p = n$ is an integer. In this way we obtain the first structure $\mathrm{Oc}T_d^1a$ (see Table 1) of a point defect in supercell model. The symbol $\mathrm{Oc}T_d^1a$ means that the oxygen interstitial in position c of a host crystal in the supercell model is described by the space group T_d^1 and occupies the Wyckoff position a in the imperfect crystal with this space group. In the same way we obtain the structures $\mathrm{Oc}T_d^2a$ and $\mathrm{Oc}T_d^3a$. Table 1 gives the main parameters of the different possible structures in the supercell model: the symbol of the structure, the matrix l in the tranformation (15), the augmentation L of the primitive unit cell of the host crystal, the distance D between the nearest neighbours and their number N_n in the defect sublattice (the numbers N_n are given in parentheses) and the number N_d of defects in a supercell.

In crystal class O_h there are only three space groups having positions with site symmetry T_d: O_h^4 (position a), O_h^5 (c) and O_h^7 (a and b). In the space group O_h^4 two poins $\mathbf{Q}_{a1} = (0,0,0)$ and $\mathbf{Q}_{a2} = (n/2, n/2, n/2)$ of the position a have the site symmetry T_d. For any n they correspond to the position c in the group O_h^5 of the host crystal. Indeed, for the group O_h^4 (simple cubic lattice) the matrix l of the transformation (15) is $l^{(1)}$, and according to equation (18) we have

$$\begin{aligned} &m_1^{(11)} = m_2^{(11)} = m_3^{(11)} = 0, \qquad m_1^{(21)} = m_2^{(21)} = m_3^{(21)} = -1/2, \\ &m_1^{(12)} = m_2^{(12)} = m_3^{(12)} = n/2, \quad m_1^{(22)} = m_2^{(22)} = m_3^{(22)} = (n-1)/2. \end{aligned}$$

Thus, the position $\mathbf{Q}_{a1}$ of the crystal with periodic defect is translationally equivalent for any n to the position $\mathbf{q}_{c1}$ of the host crystal; the position $\mathbf{Q}_{a2}$ is translationally equivalent to the position $\mathbf{q}_{c1}$ for even n and to $\mathbf{q}_{c2}$ for odd n. For even n the system has an additional symmetry element (translation vector $\mathbf{Q}_{a2} = (n/2, n/2, n/2)$), which transforms a simple cubic lattice into the body-centered cubic one . Besides, the inversion (in O_h^4 it is accompanied by an improper translation $\mathbf{Q}_{a2}$) replaces the oxygen and magnesium atoms by each other and, therefore, is not a symmetry operation of the system in consideration. In this case the structure obtained is described by the space group of the crystal class T_d with a body-centered crystal lattice – T_d^3 and coincides with the structure $\mathrm{Oc}T_d^3a$. For odd n we obtain a new structure $\mathrm{Oc}O_h^4a$ (see Table 1).

Analogously, two more structures with a periodic defect are obtained: $\mathrm{Oc}O_h^5c$ for odd n (for even n this structure coincides with $\mathrm{Oc}T_d^1a$) and $\mathrm{Oc}O_h^7a$ for even n (when interstitial oxygen occupies the positions a or b). In the latter case, both positions being occupied, the structure coincides with $\mathrm{Oc}O_h^4a$.

Let us return to the imperfect crystal with the space group of the crystal class T_d. In space group T_d^1 the two positions a and b have the site symmetry T_d: $\mathbf{Q}_a = (0,0,0)$, $\mathbf{Q}_b = (n/2, n/2, n/2)$. The latter corresponds to the position c of the group O_h^5 of the host crystal for any n, so that embedding the defect in position b does not give a new structure with a periodic defect. When placing the defects in both a and b positions one does not obtain a new structure. The occupation of symmetry nonequivalent positions in the group T_d^1 leads to additional symmetry elements in the system: translation for even n and inversion with an improper translation for odd n. This structure coincides with $\mathrm{Oc}T_d^3a$ for even n and with $\mathrm{Oc}O_h^4a$ for odd n.

In the space group T_d^2 there are three more positions with the site symmetry T_d: b ($\mathbf{Q}_b = (n/2, n/2, n/2)$), c ($\mathbf{Q}_c = (n/4, n/4, n/4)$), d ($\mathbf{Q}_d = (3n/4, 3n/4, 3n/4)$). As we have seen, the position b for any n corresponds to positions c in the group O_h^5 of the host crystal. In the same way it may be proved that the positions c and d correspond to the position c in the group O_h^5 of the host crystal only for even n. A new structure with a periodic defect may be obtained only when all the three positions a, b and c are occupied simultaneously (structure $\mathrm{Oc}T_d^2abc$). Defects in positions a and b give the structure $\mathrm{Oc}O_h^5c$, in positions a and c – $\mathrm{Oc}O_h^7a$, and in positions a, b, c and d – $\mathrm{Oc}O_h^4a$. The rest of possible combinations is simply reduced to those mentioned above.

In space group T_d^3 there is only one position (a) with site symmetry T_d.

4.2 Oxygen molecule in oxygen site

Let us describe the peculiarity of a molecular type point defect. We consider an electroneutral (with regards to the host crystal) molecular ion O_2^{2-} (in symmetry consideration we call it for brevity molecule O_2). The O_2 molecule has a proper symmetry $S_D^{(o)} = D_{\infty h}$. Let us place it in the oxygen site of the MgO crystal having the site symmetry $G_a = O_h$ (position a of the space group O_h^5). Different site symmetry groups S_D of the periodic defect in the crystal arise for different orientation of the molecular axis relatively to the elements of the site symmetry group O_h. There are three possible orientations giving the groups S_D of the highest possible symmetry: [111] ($S_D = D_{3d}$), [110] ($S_D = D_{2h}$) and [001] ($S_D = D_{4h}$).

Let $S_D = D_{3d}$. The space groups with site symmetry D_{3d} are the following: D_{3d}^1 (a, b), D_{3d}^3 (a, b), D_{3d}^5 (a, b), O_h^4 (c), O_h^7 (c, d), O_h^9 (c). Table 2 gives the space groups of the structures with a periodic defect in the crystal class D_{3d}.

Table 2: Structures with a molecular oxygen in position a of a host crystal and with [111] orientation

Structure	$l^{(i)}$	L	$D(N_n)$	N_d	Remark
$O_2aD^1_{3d}a$	$l^{(1)}$	$9n_1^2n_2$	$\sqrt{3/2}n_1$ (6), $\sqrt{3}n_2$ (2)	1	
$O_2aD^3_{3d}a$	$l^{(2)}$	$3n_1^2n_2$	$n_1/\sqrt{2}$ (6), $\sqrt{3}n_2$ (2)	1	
$O_2aD^5_{3d}a$	$l^{(3)}$	$2n_1^3+n_2^3$ $-n_1^2n_2$	$\sqrt{(3n_1^2+n_2^2+2n_1n_2)/2}$ (6), $\mid n_1-n_2\mid/\sqrt{2}$ (6), $(n_1+n_2)/\sqrt{2}$ (2)	1	$n_1\neq n_2$

$$l^{(1)}=\begin{pmatrix} n_1 & n_1 & -2n_1\\ -2n_1 & n_1 & n_1\\ n_2 & n_2 & n_2\end{pmatrix}\quad l^{(2)}=\begin{pmatrix} -n_1 & n_1 & 0\\ 0 & -n_1 & n_1\\ n_2 & n_2 & n_2\end{pmatrix}\quad l^{(3)}=\begin{pmatrix} n_1 & n_1 & n_2\\ n_1 & n_2 & n_1\\ n_2 & n_1 & n_1\end{pmatrix}$$

5 Electron structure of MgO crystal with a molecular defect

To illustrate the application of the SCM of a crystal with a periodic defect we give in Table 3 the results of the electronic structure calculations of molecular ion O_2^{2-} in oxygen site of MgO crystal oriented along the direction [110]. For these Hartree-Fock calculations we used the program Crystal-92 [32] and atomic functions of Gaussian type which were fitted to obtain the lattice constant and the bonding energy of a perfect crystal in good agreement with the experiment.

The accuracy of the calculation is determined by the following computational parameters: a) the five threshold parameters that control the accuracy of the bielectronic series (the good quality [11] set 10^{-5}, 10^{-5}, 10^{-5}, 10^{-5}, 10^{-11} was adopted); b) the self-consistent convergence thresholds on eigenvalues and total energy (the values 10^{-7} a.u. and 10^{-5} a.u. were adopted); c) the sampling **k**-point net for the integration in the reciprocal space when calculating the Fermi energy and density matrix (the points belonging to Monkhorst set forming a reciprocal lattice with basis vectors $\mathbf{b}_i/s$ $(i=1,2,3)$ were taken with shrinking factor $s=2$ for supercells with $L=16$ and $s=4$ for supercells with $L=4,8$). To investigate the convergence of the calculation results with increasing supercell, the symmetrical transformation with matrices $l^{(5)}$ $(L=4)$, $l^{(9)}$ $(L=8)$ and $l^{(8)}$ $(L=16)$ was used to generate supercells. The defect-defect distance D appears to be a, $a\sqrt{2}$ and $a\sqrt{3}$ respectively (a=4,19 Å is the lattice constant of the perfect crystal). The interatomic distance d in O_2^{2-} ion was taken to be 1.35 Å for all the calculations in discussion (this value is close that found for different supercells by minimizing the total energy).

In Table 3 there are given: the defect energy E_d defined as the difference of the total energies of supercells with the same size for imperfect (with O_2^{2-} ion at anion site) and perfect crystals, the Fermi-energy E_F, the charge on the oxygen atoms in the molecule O_2^{2-} and the bandwidth ΔE_d of the highest occupied defect state energy band. The convergence of the results with increasing distance D between the nearest neighbours in the defect sublattice (the numbers N_n are given in parentheses) is seen.

In Table 4 the same results are given for different choices of the supercell with fixed volume $(L=16)$. The results obtained show that for the fixed supercell size the best choice of the transformation matrix corresponds to the symmetrical transformation of the host lattice basic translation vectors (space group D^{25}_{2h}): the distance D is the largest one, so that the bandwidth ΔE_d is the smallest one. This choice of the supercell ensures also the minimum value of the total energy. The charge distribution seems to be practically independent on

Table 3: The convergence of the results of the Hartree-Fock LCAO calculations of the molecular ion O_2^{2-} in oxygen site of MgO crystal

Space group	$D_{2h}^{19}(a)$	D_{2h}^{25}	D_{2h}^{23}
L	4	8	16
(n_1, n_2, n_3)	(2,0,1)	(2,2,2)	(2,2,1)
$D(N_n)$	a (6)	$a\sqrt{2}$ (12)	$a\sqrt{3}$ (8)
E_d (a.u.)	-74.5442	-74.5455	-74.5457
E_F (a.u.)	-0.2589	-0.2772	-0.2962
$q(e)$	-1.059	-1.069	-1.065
ΔE_d (eV)	0.66	0.55	0.35

choice of the the supercell.

Table 4: The results of the Hartree-Fock LCAO calculations of the molecular ion O_2^{2-} in oxygen site of MgO crystal in the supercell model ($L = 16$, 33 atoms in unit cell, $a = 4.19$ Å

Space group	$D_{2h}^{19}(a')$	$D_{2h}^{19}(a'')$	$D_{2h}^{19}(a)$	D_{2h}^{1}	D_{2h}^{23}	D_{2h}^{25}
(n_1, n_2, n_3)	(2,2,1)	(1,1,4)	(3,1,2)	(2,2,2)	(2,2,4)	(2,2,1)
$D(N_n)$	$a/\sqrt{2}$ (2)	$a\sqrt{3/2}$ (4)	$a\sqrt{2}$ (2)	$a\sqrt{2}$ (4)	$a\sqrt{2}$ (4)	$a\sqrt{3}$ (8)
E_d (a.u.)	-74.4146	-74.5435	-74.5452	-74.5434	-74.5434	-74.5457
E_F (a.u.)	-0.2799	-0.2901	-0.2950	-0.2915	-0.2931	-0.2962
$q(e)$	-1.022	-1.067	-1.065	-1.068	-1.068	-1.065
ΔE_d (eV)	1.16	0.76	0.38	0.51	0.46	0.35

6 The symmetry of the Supercell Model of Metal-Doped TiO_2 Structures

The titanium atom has different point symmetry groups F_D in two TiO_2 structures under consideration.

In rutile structure $G = D_{4h}^{14}$, $F_D = D_{2h}$, the basic translation vectors $\mathbf{a}_1 = (a, 0, 0)$, $\mathbf{a}_2 = (0, a, 0)$, $\mathbf{a}_3 = (0, 0, c)$ define a simple tetragonal lattice. There are two titanium atoms in the primitive unit cell occupying the two-site position a with coordinates $(0, 0, 0)$ and $(a/2, a/2, c/2)$.

In anatase structure $G = D_{4h}^{19}$, $F_D = D_{2d}$, the basic translation vectors $\mathbf{a}_1 = (-a/2, a/2, c/2)$, $\mathbf{a}_2 = (a/2, -a/2, c/2)$, $\mathbf{a}_3 = (a/2, a/2, -c/2)$ define a body-centered tetragonal lattice. There are also two titanium atoms in the primitive unit cell occupying the two-site position a with coordinates $(0, 0, 0)$ and $(0, a/2, c/4)$.

6.1 Metal-doped rutile structure

The site symmetry group $S_{\mathbf{q}} = S_a = D_{2h}$ consists of the elements:

$$E, U_{xy}, U_{\bar{x}y}, C_{2z}, I, \sigma_{xy}, \sigma_{\bar{x}y}, \sigma_{2z}. \tag{22}$$

An impurity vanadium atom V in rutile structure, replacing an atom Ti, has the symmetry $S_D = D_{2h}$. The possible crystal classes F_D for defective crystals in the supercell model are D_{2h} and D_{4h} (one or two impurity atoms for a supercell, respectively). Positions with the site symmetry D_{2h} are contained in the following symmorphic space groups of crystal class D_{2h}: $D^1_{2h} = Pmmm$ (positions from a to h), $D^{19}_{2h} = Cmmm$ (positions from a to d), $D^{23}_{2h} = Fmmm$ (positions a and b), $D^{25}_{2h} = Immm$ (positions from a to d). The site symmetry group $S_{\mathbf{Q}} = S_a = D_{2h}$ in any of these groups in standard setting of [33] consists of the elements:

$$E, U_x, U_y, C_{2z}, I, \sigma_x, \sigma_y, \sigma_{2z}, \tag{23}$$

i.e. the symmetry elements of the group $S_{\mathbf{Q}}$ are rotated through the angle $-\pi/4$ around the Z-axis relatively of the symmetry elements of the group $S_{\mathbf{q}}$ (or the coordinate system of the space group D^{14}_{4h}). To combine the site symmetry group (23) with the site symmetry group (22) let us rotate the spaces of the groups G_D through an angle $\pi/4$ around the axe Z. Then the symmetry elements of $S_D = \tilde{S}_a = D_{2h}$ take the form (22), and the basic translation vectors $\mathbf{A}_j$ in a new setting will have with respect to the coordinate system related to the space group D^{14}_{4h} new components.

To receive the transformation matrix l defining the supercell one uses the relation (15) taking the rotated vector $\mathbf{A}_j$ in a new setting [31]. In particular, for defective crystal space group D^{19}_{2h} it gives $l_{11} = -l_{22} = p_1 - p_2$; $l_{12} = -l_{21} = p_1 + p_2$; $l_{13} = l_{23} = l_{31} = l_{32} = 0$; $l_{33} = p_3$. As the matrix elements of l have to be integers $p_1 - p_2 = n_2, p_1 + p_2 = n_1, p_3 = n_3$.

Thus the transformation matrix l found connects the host crystal and the supercell model translation vectors for the defective crystal space group $G_D = D^{19}_{2h}$:

$$l = \begin{pmatrix} n_2 & n_1 & 0 \\ -n_1 & -n_2 & 0 \\ 0 & 0 & n_3 \end{pmatrix}, \quad L = (n_1^2 - n_2^2)n_3. \tag{24}$$

Matrices l for other space groups G_D are found by the same way:

$$G_{PD} = D^1_{2h}, \quad l = \begin{pmatrix} n_1 & n_1 & 0 \\ -n_2 & n_2 & 0 \\ 0 & 0 & n_3 \end{pmatrix}, \quad L = 2n_1n_2n_3; \tag{25}$$

$$G_{PD} = D^{23}_{2h}, \quad l = \begin{pmatrix} -n_2 & n_2 & n_3 \\ n_1 & n_1 & n_3 \\ n_1 - n_2 & n_1 + n_2 & 0 \end{pmatrix}, \quad L = 4n_1n_2n_3; \tag{26}$$

$$G_{PD} = D^{25}_{2h}, \quad l = \begin{pmatrix} -n_1 & -n_2 & n_3 \\ n_1 & n_2 & n_3 \\ n_2 & n_1 & -n_3 \end{pmatrix}, \quad L = 2(n_1^2 - n_2^2)n_3. \tag{27}$$

As a particular case one obtains the matrices of the symmetric transformation of basic translations vectors in the simple tetragonal lattice of the host crystal: $n_1 = n_2$ for (22) and (25), $n_2 = 0$ for (23) and (24). In published supercell calculations of Li,Na-doped rutile [34] only the matrices (24) for $n_1 = n_2 = n_3$ were used, so that the convergence of the results with supercell increasing was not investigated.

Let now two titanium atoms in the supercell of rutile are substituted by the impurity atoms of the same chemical element (in practice such a model may appear to be useful for the investigatioon of the dependence of the electronic structure changes on the percentage of doping). In this case the heighest possible point symmetry of the defective crystal is $F_D = D_{4h}$ (the same as for the host crystal). Following the procedure given in Section 3 one finds the space groups corresponding to the point symmetry D_{4h} and at the same time generating the site symmetry D_{2h} of impurity atom in the supercell model: D^1_{4h}, D^5_{4h}, D^9_{4h}, D^{10}_{4h}, D^{14}_{4h}, D^{17}_{4h}, D^{18}_{4h}. Checking the condition (13) one finds that only the space group D^{14}_{4h} is acceptable as the defective crystal group G_D. The matrix l in this case is diagonal

$$G_{PD} = D^{14}_{4h}, \quad l = \begin{pmatrix} 2n_1 & 0 & 0 \\ 0 & 2n_1 & 0 \\ 0 & 0 & n_3 \end{pmatrix}, \quad L = 4n_1^2 n_3. \tag{28}$$

6.2 Metal-doped anatase structure

The titanium atom site symmetry group in anatase structure is D_{2d} (the second order axes as in the case of rutile are oriented along xy, $\overline{x}y$ and z directions). The following four symmorphic space groups of D_{2d} point symmetry are compatible with the site symmetry group D_{2d}: D^1_{2d}, D^5_{2d}, D^9_{2d}, D^{11}_{2d}.

For two of them (D^5_{2d}, D^9_{2d}) in standard setting [33] the orientation of the basic translation vectors relatively symmetry elements is the same as in the anatase space group D^{19}_{4h}. For rest two space groups (D^1_{2d}, D^{11}_{2d}) the same rotation of the system is necessary that was mentioned for rutile structure.

Using (15) one finds the following matrices l defining the supercells:

$$G_{PD} = D^1_{2d}, \quad l = \begin{pmatrix} -n_1 & n_1 & 0 \\ n_1 & n_1 & 2n_1 \\ n_2 & n_2 & 0 \end{pmatrix}, \quad L = 4n_1^2 n_2; \tag{29}$$

$$G_{PD} = D^5_{2d}, \quad l = \begin{pmatrix} 0 & n_1 & n_1 \\ n_1 & 0 & n_1 \\ n_2 & n_2 & 0 \end{pmatrix}, \quad L = 2n_1^2 n_2; \tag{30}$$

$$G_{PD} = D^9_{2d}, \quad l = \begin{pmatrix} n_1 & -n_2 & 0 \\ -n_2 & n_1 & 0 \\ n_2 & n_2 & n_1 + n_2 \end{pmatrix}, \tag{31}$$

$$L = (n_1 + n_2)(n_1^2 - n_2^2);$$

$$G_{PD} = D^{11}_{2d}, \quad l = \begin{pmatrix} n_2 & n_2 - n_1 & -n_1 \\ n_2 & n_1 + n_2 & n_1 \\ n_1 - n_2 & -n_2 & n_1 \end{pmatrix}, \quad L = 4n_1^2 n_2. \tag{32}$$

All the matrices (27)-(30) correspond to the symmetric transformations of basic translation vectors of the body-centered tetragonal lattice of the host crystal.

If two titanium atoms in the supercell model of anatase structure are substituted by the impurity atoms only one (D^{19}_{4h}) of the tetragonal space groups satisfies to the relation (15), and therefore can be taken as G_D:

$$G_{PD} = D^{19}_{4h}, \quad l = \begin{pmatrix} 2n_1 + 1 & 0 & 0 \\ 0 & 2n_1 + 1 & 0 \\ 0 & 0 & 4n_2 + 1 \end{pmatrix}, \tag{33}$$

$$L = (2n_1 + 1)^2 (4n_2 + 1).$$

Hartree-Fock calculations of V-doped rutile were made [31] basing on the SCM symmetry consideration of this section.

7 Density matrix of crystalline systems in the Hartree-Fock approximation

The implementation of the CCM in the HF LCAO method may be realized based on the corresponding periodic approaches and requires some modifications[27]. Such a possibility is connected with the fundamental property of idempotency for the one-particle density matrix (DM) of the crystalline system. The idempotency of the DM is a consequence of the orthonormality of the one-electron Bloch functions that are the basis of the crystalline orbitals. In the CCM the DM has to be idempotent when a one-determinant wave function is used and convergence with cluster size is achieved. The infinite DM of a solid formally does not have this property, but the infinite solid itself is modeled by a very large cyclic cluster describing the main region of the crystal when Born-von Karman PBC are introduced to ensure macroscopic periodicity of a crystal model.

The extension of the HF method to models of infinite crystals leads to difficulties connected with the one-electron Bloch functions (BF) behavior since they do not fall off to zero at infinity and therefore cannot be normalized to unity in space [35, 36]. To overcome this difficulty, PBC are introduced for the BF normalized in the main region of volume $V_N = NV_a$ of the crystal containing a large number N of primitive unit cells of volume V_a.

The BF are labeled by the wave vector $\mathbf{k}$, assuming N values in the BZ. From these BF the one-determinant wave function of a crystal is constructed and the HF equations for the main region of volume V_N are obtained by means of the variational procedure:

$$\left\{-\frac{1}{2}\Delta + V(\mathbf{r}) + \int \frac{\rho_{\mathbf{r}'\mathbf{r}'}}{|\mathbf{r}-\mathbf{r}'|} d^3\mathbf{r}'\right\} \psi_{n\mathbf{k}}(\mathbf{r}) - \frac{1}{2}\int \frac{\rho_{\mathbf{r}\mathbf{r}'}}{|\mathbf{r}-\mathbf{r}'|} \psi_{n\mathbf{k}}(\mathbf{r}') d^3\mathbf{r}' = E_{n\mathbf{k}}\psi_{n\mathbf{k}}(\mathbf{r}) \quad (34)$$

In eqn. (34) n denotes the number of an energy band, $V(\mathbf{r})$ is the potential of an electron in the periodic field of nuclei, $E_{n\mathbf{k}}$ are one-electron band energies, and $\psi_{n\mathbf{k}}$ are the one-electron BF. The density matrix for crystals with filled energy bands

$$\rho_{\mathbf{r}\mathbf{r}'} = \frac{1}{N}\sum_{\mathbf{k}} P_{\mathbf{r}\mathbf{r}'}(\mathbf{k}) = 2\sum_{n}\sum_{\mathbf{k}} \psi^*_{n\mathbf{k}}(\mathbf{r})\psi_{n\mathbf{k}}(\mathbf{r}') \quad (35)$$

is calculated by summing over occupied energy bands and N values of $\mathbf{k}$ vector in the BZ.

In eqn. (34) the local Coulomb operator is expressed through the charge density $\rho_{\mathbf{r}\mathbf{r}}$. The exchange operator is non-local and depends on the non-diagonal matrix elements $\rho_{\mathbf{r}\mathbf{r}'}$.

The DM is normalized to the total number of NN_e electrons in the main region of the crystal (N_e is the number of electrons in the primitive unit cell):

$$\begin{aligned} \int_{V_a} \rho_{\mathbf{r}\mathbf{r}} d^3\mathbf{r} &= N_e \\ \int_{NV_a} \rho_{\mathbf{r}\mathbf{r}} d^3\mathbf{r} &= NN_e \end{aligned} \quad (36)$$

For the DM in $\mathbf{k}$ space one receives, using eqns. (35) and (36):

$$\begin{aligned} P_{\mathbf{r}\mathbf{r}'}(\mathbf{k}) &= 2N\sum_{n} \psi^*_{n\mathbf{k}}(\mathbf{r})\psi_{n\mathbf{k}}(\mathbf{r}') \\ \int_{V_B} P_{\mathbf{r}\mathbf{r}'}(\mathbf{k}) d^3\mathbf{k} &= N_e \end{aligned} \quad (37)$$

where V_B is the BZ volume. As follows from eqn. (37) the DM in $\mathbf{k}$ space is normalized on N_e independently of the choice of N in the direct lattice. In the limit of the infinite crystal the summation over $\mathbf{k}$ in eqn. (35) is replaced by an integral over the BZ

$$\rho_{\mathbf{rr}'} = \frac{1}{V_B}\int P_{\mathbf{rr}'}(\mathbf{k})d^3\mathbf{k} \tag{38}$$

The one-particle DM of crystalline systems, its symmetry properties, long-range behavior, and related computational problems have been considered before [37]. Here, we discuss the idempotency property of the DM connected with the orthonormality of the BF.

Using the periodicity of the DM in direct lattice $\rho_{\mathbf{r}+\mathbf{a}_n,\mathbf{r}'+\mathbf{a}_n} = \rho_{\mathbf{rr}'}$ ($\mathbf{a}_n$ being a translation vector of the Bravais lattice) one can write $\rho_{\mathbf{r},\mathbf{r}'} = \rho_{\mathbf{r},\mathbf{r}'}(\mathbf{a}_n)$ where the arguments $\mathbf{r},\mathbf{r}'$ are electron coordinates in the primitive unit cell ($\mathbf{a}_n = 0$). As for BF the relation $\psi_{n\mathbf{k}}(\mathbf{r}+\mathbf{a_n}) = e^{i\mathbf{k}\mathbf{a}_n}\psi_{n\mathbf{k}}(\mathbf{r})$ is satisfied, one receives from eqn. (37) the relation

$$\rho_{\mathbf{rr}'}(\mathbf{a}_n) = \frac{1}{N}\sum_{\mathbf{k}} e^{-i\mathbf{k}\mathbf{a}_n}P_{\mathbf{rr}'}(\mathbf{k}) \tag{39}$$

connecting the DM $\rho_{\mathbf{rr}'}(\mathbf{a}_n)$ in the main region of the direct lattice and the DM $P_{\mathbf{rr}'}(\mathbf{k})$ at N points of the BZ.

The introduction of PBC for the main region of a crystal leads to a one-to-one correspondence between the number of primitive unit cells considered in the direct lattice and the number of $\mathbf{k}$ points in the BZ. The DM $\rho_{\mathbf{rr}'}(\mathbf{a}_n)$ and $P_{\mathbf{rr}'}(\mathbf{k})$ are connected by relation (39) and by the following relation

$$P_{\mathbf{rr}'}(\mathbf{k}) = \sum_{\mathbf{a}_n}\rho_{\mathbf{rr}'}(\mathbf{a}_n)e^{i\mathbf{k}\mathbf{a}_n} \tag{40}$$

The DM $P_{\mathbf{rr}'}(\mathbf{k})$ is periodic in reciprocal space, so that relation (40) is a Fourier series for $P_{\mathbf{rr}'}(\mathbf{k})$ with the Fourier expansion coefficients $\rho_{\mathbf{rr}'}(\mathbf{a}_n)$ from eqn. (39).

The relations (39) and (40) may be written not only for the main region of a crystal but for any cyclic cluster of smaller size in the direct lattice (defined by the translation vectors $\mathbf{A}_j$), received by the following Large Unit Cell (LUC) transformation of primitive lattice vectors $\mathbf{a}_i$ ($i = 1, 2, 3$) 15 The choice of l defines both size and point symmetry of the cyclic cluster. The symmetry of the CCM has been considered in section 2.

The space group G of an infinite crystal is defined as the limit of the symmetry groups of a sequence of cyclic clusters with increasing size [38].

The introduction of PBC for a cyclic cluster, defined by a symmetric transformation (15), means that for any LUC translation vector $\mathbf{A_m} = \sum_i m_i\mathbf{A}_i (i = 1, 2, 3)$ the one-electron BF's satisfy the equation

$$\psi_{n\mathbf{k}}(\mathbf{r}+\mathbf{A_m}) = \psi_{n\mathbf{k}}(\mathbf{r}) = e^{i\mathbf{k}\mathbf{A_m}}\psi_{n\mathbf{k}}(\mathbf{r}) \tag{41}$$

Eqn. (41) is satisfied only for those BF's of the infinite crystal which transform according to the unit representation of the translation group $T_\mathbf{A} \subset T$ defined by LUC translations $\mathbf{A_m}$.

Therefore all the irreducible representations of $G^\mathbf{A}$ may be found from those space group G irreducible representations which satisfy the relation

$$e^{i\mathbf{k}\mathbf{A_m}} = 1 \tag{42}$$

For a symmetric transformation (15) the whole stars of wave vectors $\mathbf{k}$ satisfy eqn. (42).

Thus the irreducible representations of the cyclic cluster symmetry group $G^{\mathbf{A}}$ may be easily found from irreducible representations of the infinite crystal space group G. In particular for the main region of a crystal consisting of $N = N_1N_2N_3$ unit cells, eqn. (42) is satisfied by $N_1N_2N_3$ $\mathbf{k}$ vectors:

$$\mathbf{k} = \frac{s_1}{N_1}\mathbf{b}_1 + \frac{s_2}{N_2}\mathbf{b}_2 + \frac{s_3}{N_3}\mathbf{b}_3 \qquad s_i = 0, ..., N_i - 1 \tag{43}$$

as the transformation matrix l is diagonal in this case.

For the cyclic cluster symmetry group $G^{\mathbf{A}}$ the orthogonality relations of irreducible representations may be written in the form

$$\begin{aligned} \frac{1}{L}\sum_{\mathbf{q}\epsilon Q} e^{-i\mathbf{q}\mathbf{a}_n^0} &= \delta_{\mathbf{a}_n^0,\mathbf{A}} \\ \frac{1}{L}\sum_{\mathbf{a}_n^0\epsilon T^{\mathbf{A}}} e^{-i\mathbf{q}\mathbf{a}_n^0} &= \delta_{\mathbf{q},0} \end{aligned} \tag{44}$$

In eqn. (44) the summation over $\mathbf{q}$ is over those L points of the BZ that satisfy eqn. (42). The summation over $\mathbf{a}_n^0$ is over those translation vectors of the original direct lattice that belong to translation subgroup $T^{\mathbf{A}}$ of the cyclic cluster symmetry group $G^{\mathbf{A}}$.

Relations (44) are well known for the particular case when the main region of a crystal is considered as a cyclic cluster. They were also used for $N_i = s_i, (i = 1, 2, 3)$ [39] where the s_i are called shrinking factors, defining $L = s_1s_2s_3$ sampling points $\mathbf{q}\epsilon Q$ in the BZ in the self-consistent DM calculations.

This consideration shows that the case of a diagonal matrix for l in eqn. (15) may be considered as a particular case of the cyclic cluster chosen to satisfy eqn. (44). In general relations (44) are satisfied also for cyclic clusters defined by non-diagonal matrices in eqn. (15) as these relations simply mean the orthonormality relations for the irreducible representations of $G^{\mathbf{A}}$.

This result is important for the implementation of the CCM in HF LCAO methods. It means that the Fourier series (eqns. (39) and (40)) for the DM in direct and reciprocal space may be written for any cyclic cluster if the summation in (39) is made over L $\mathbf{k}$ points satisfying (42) and the summation in (40) is over direct lattice translations $\mathbf{a}_n^0$ belonging to the cyclic cluster translation group $T^{\mathbf{A}}$.

The CCM ensures the balance in the direct and reciprocal lattice summation in DM. It has been used previously to overcome the non-physical divergence of direct lattice sums in the non-local HF exchange for crystals [35]. This divergence appears when the cubature formula is used to calculate integral (38) over the BZ

$$\rho_{\mathbf{r}\mathbf{r}'} = \frac{1}{V_B}\int P_{\mathbf{r}\mathbf{r}'}(\mathbf{k})d^3\mathbf{k} \approx \frac{1}{N_0}\sum_{j=1}^{N_0} P_{\mathbf{r}\mathbf{r}'}(\mathbf{k}_j) \tag{45}$$

Using (37) and the orthonormality of BF's $\psi_{n\mathbf{k}}(\mathbf{r})$ one obtains

$$\rho^2_{\mathbf{r}\mathbf{r}'} = \int_{NV_a} \rho_{\mathbf{r}\mathbf{r}''}\rho_{\mathbf{r}''\mathbf{r}'}d^3\mathbf{r}'' = \frac{2N}{N_0}\rho_{\mathbf{r}\mathbf{r}'} \tag{46}$$

The exact DM $\rho_{\mathbf{r}\mathbf{r}'}$ within the one-determinant approximation of the HF method satisfies the idempotency relation

$$\int_{NV_a} \rho_{\mathbf{r}\mathbf{r}''}\rho_{\mathbf{r}''\mathbf{r}'}d^3\mathbf{r}'' = 2\rho_{\mathbf{r}\mathbf{r}'} \tag{47}$$

as a consequence of the orthonormality of BF's in the main region of a crystal. Approximation (46) ensures relation (47) only for $N = N_0$. Since N_0 is usually supposed to be fixed by chosing some sampling set Q of N_0 $\mathbf{k}$ points, the lattice summation up to the limit of the infinite crystal gives

$$(\hat{\rho})^2_{\mathbf{rr}} = \int_{NV_a} |\rho_{\mathbf{rr}'}|^2 d^3\mathbf{r}' \to \infty \quad \text{when} \quad N \to \infty \tag{48}$$

The divergence of $\int |\rho_{\mathbf{rr}'}|^2 d^3\mathbf{r}'$ causes the divergence of the exchange energy when $N \to \infty$ as the non-local exchange term in eqn. (34) is connected with the calculation of the non-diagonal elements $\rho_{\mathbf{rr}'}$ of DM.

This problem also arises when the LCAO approximation is used giving the divergent lattice sums in the exchange part of the Fock matrix elements. As semiempirical versions of the LCAO HF scheme like CNDO, INDO, and NDDO take into account the non-local nature of the exchange term, the problem has also to be considered in all band structure calculations based on these approaches [35].

To overcome this divergence the idea of a cyclic cluster, generated by a LUC transformation in direct lattice, was considered for DM calculations both in the semiempirical [35] and the ab initio [37] realizations of the HF LCAO method.

Let us consider this in more detail following the outlines in the literature [39]. We restrict the present discussion to insulators with no partially filled energy bands. The set Q of $L = s_1 s_2 s_3$ sampling points in the BZ is used as input to solve the one-electron equations. The shrinking factors s_1, s_2, s_3 define in fact the diagonal matrix in (15) and the set of $\mathbf{k}$ points that satisfy eqn. (42). The LUC translation vectors $A_j = s_j \boldsymbol{a}_j (j = 1, 2, 3)$ define the superlattice and a corresponding Wigner-Seitz super-cell. The non-zero Fourier coefficients in (40) are taken only for those direct lattice vectors $\mathbf{a}^0_n$ which are inside or at the surface of this supercell. For the lattice vectors at the surface it is taken into account that they may be equivalent to each other through a superlattice vector $\mathbf{A}$, i.e. PBC for the supercell are introduced in the same way as in the CCM.

The orthonormality conditions (44) for cyclic clusters are used to define an approximate DM in $\mathbf{k}$ space in such a way that it coincides with the exact one at all sampling points and satisfies the periodicity condition in reciprocal lattice

$$\begin{aligned} \tilde{P}_{\mathbf{rr}'}(\mathbf{k}) &= \sum_{\mathbf{a}^0_n} \tilde{\rho}_{\mathbf{rr}'}(\mathbf{a}^0_n) e^{i\mathbf{k}\mathbf{a}^0_n} \\ \tilde{\rho}_{\mathbf{rr}'}(\mathbf{a}^0_n) &= \frac{1}{L} \sum_{\mathbf{q} \epsilon Q} \tilde{P}_{\mathbf{rr}'}(\mathbf{q}) e^{-i\mathbf{q}\mathbf{a}^0_n} \end{aligned} \tag{49}$$

Use of orthonormality conditions (44) allows to perform the integration over BZ in order to obtain the approximate DM $\tilde{\rho}_{\mathbf{rr}'}$ and to receive the non-zero Fourier coefficients in (17). For insulators these coefficients coincide with those corresponding to the CCM, for conductors they have a more complicated form as the DM calculation must be performed up to the Fermi level for each $\mathbf{q}$ point.

The DM approximation (49) may be realized in CRYSTAL [32, 40, 41] using the cyclic cluster calculation of the DM at each step of the SCF procedure [36]. The *cyclicity region* is defined in this case by the choice of the Q set of sampling points if the primitive unit cell is taken in direct lattice. However, if from the very beginning a LUC (super cell) is introduced, the cyclicity region of the DM will be defined by the matrix $l = l_A\, l_Q$ where the matrix l_A defines the LUC and l_Q is defined by the Q set choice. Of course, in the usual calculations of perfect crystals the primitive cell is always the best choice, but for the implementation of the CCM in calculations of point defects the cyclicity region of the DM is defined also by the choice of l_A.

In principle, an infinite number of two-electron integrals over atomic basis functions has to be calculated for the generation of the Fock matrix in direct lattice. In CRYSTAL this calculation is limited by the introduction of truncation parameters. The idempotency condition of the DM is dependent from the exchange series calculation regulated by two pseudo-overlap parameters $S^{\mathbf{g}}$ and $S^{\mathbf{n}}$ ($P^{\mathbf{g}} = -\log_{10}S^{\mathbf{g}}, P^{\mathbf{n}} = -\log_{10}S^{\mathbf{n}}$) so that the exchange integrals $(\chi_{\mu 0}\chi_{\nu 0}|\chi_{\lambda \mathbf{g}}\chi_{\sigma \mathbf{n}})$ are disregarded if either the overlap between χ_{μ} and χ_{λ} is less than $S^{\mathbf{g}}$ or if the overlap between χ_{ν} and χ_{σ} is less than $S^{\mathbf{n}}$. Here $\mu, \nu, \lambda, \sigma$ denote atomic orbitals in the central unit cell and **g**, **n** are translation vectors in direct lattice. It was demonstrated [39, 42] that the thresholds $P^{\mathbf{g}}$ and $P^{\mathbf{n}}$ must be different. The recommended difference $P^{\mathbf{n}} - P^{\mathbf{g}}$ ranges from 2 to 8. Only such a choice of these parameters ensures that long-range density matrix elements $\rho_{\nu\sigma}(\mathbf{g})$ are taken into account that can provide appreciable contributions to Fock matrix elements $F_{\mu\lambda}(\mathbf{n})$. In the limit of very strict tolerances $P^{\mathbf{n}}$, $P^{\mathbf{g}}$ this problem disappears.

Thus in CRYSTAL calculations the input tolerances $P^{\mathbf{n}}$, $P^{\mathbf{g}}$ define the *summation field* for exchange integrals and must be chosen in balance with the Q set of sampling points for the implementation of the CCM.

The other three integral tolerances used in CRYSTAL are connected with the truncation of one-electron, Coulomb, and exchange (direct-overlap truncation) and are usually set equal to parameter $P^{\mathbf{g}}$. Therefore the most critical influence on the direct lattice summation field size appears to be the choice of $P^{\mathbf{n}}$ (which has been named in the code as TOL5 [39]). The default value of $P^{\mathbf{n}}$ is 12 in CRYSTAL98 [41] which means that exchange integrals are disregarded if the overlap between atomic functions is smaller than 10^{-12}. The default values for the other tolerances are set to 6.

The use of the CCM in the direct lattice part of CRYSTAL allows to circumvent the lattice sum divergence of the non-local exchange part of the Fock matrix as the calculation of two-electron integrals is performed in such a way that only the *cyclicity region* is taken into account. However, this balance in the direct and reciprocal lattice summations is not achieved if the *summation field* is larger than the *cyclicity field*. Since CRYSTAL calculations are usually performed in such a way that the convergence of the results with respect to both the shrinking factors and direct lattice summation tolerances is achieved, the aforementioned balance is ensured automatically. However, when a cyclic cluster calculation is made, the balance has to be checked. This can be done by checking the idempotency property of the DM or by using simple relations based on local properties of the electronic structure. Such relations are discussed in the following section.

8 Density matrix idempotency in the HF-LCAO method for periodic systems

In LCAO approximation BF's $\psi_{i\mathbf{k}}(\mathbf{r})$ are represented as linear combinations of Bloch sums $\phi_{\mu\mathbf{k}}(\mathbf{r})$ of the atomic functions

$$\begin{aligned}\psi_{i\mathbf{k}}(\mathbf{r}) &= \sum_{\mu} C_{i\mu}(\mathbf{k})\phi_{\mu\mathbf{k}}(\mathbf{r}) \\ \phi_{\mu\mathbf{k}}(\mathbf{r}) &= \sum_{\mathbf{a}_n} e^{i\mathbf{k}\mathbf{a}_n}\chi_{\mu}(\mathbf{r}-\mathbf{R}_A-\mathbf{a}_n)\end{aligned} \tag{50}$$

The atomic orbital (AO) $\chi_{\mu}^{An} = \chi_{\mu}(\mathbf{r}-\mathbf{R}_A-\mathbf{a}_n)$ is centered on the atom A in the unit cell with translation vector $\mathbf{a}_n$. As the orbitals χ_{μ}^{An} are non-orthogonal, the overlap integrals of Bloch sums (50) are lattice sums of AO overlap integrals.

$$S_{\mu\nu}(\mathbf{k}) = \int \phi^{*}_{\mu\mathbf{k}}(\mathbf{r})\phi_{\nu\mathbf{k}}(\mathbf{r})d^3\mathbf{r} = \sum_{\mathbf{a}_n} e^{-i\mathbf{k}\mathbf{a}_n}\int \chi_{\mu}(\mathbf{r}-\mathbf{R}_A)\chi_{\nu}(\mathbf{r}-\mathbf{R}_B-\mathbf{a}_n)d^3\mathbf{r} \tag{51}$$

The elements of the DM $P(\mathbf{k})$ in the LCAO approximation have the following form

$$P_{\mu\nu}(\mathbf{k}) = 2\sum_{i}^{occ} C^{*}_{i\mu}(\mathbf{k})C_{i\nu}(\mathbf{k}) \tag{52}$$

where indices μ, ν denote AO's in the primitive unit cell and i denotes occupied energy bands.

In the LCAO approximation the idempotency relation (47) may be written in both reciprocal and direct space:

$$P(\mathbf{k})S(\mathbf{k})P(\mathbf{k}) = 2P(\mathbf{k}) \tag{53}$$

$$\sum_{\mathbf{a}_m\mathbf{a}'_m} \boldsymbol{\rho}(\mathbf{a}_m)\boldsymbol{S}(\mathbf{a}'_m-\mathbf{a}_m)\boldsymbol{\rho}(\mathbf{a}_n-\mathbf{a}'_m) = 2\boldsymbol{\rho}(\mathbf{a}_n) \tag{54}$$

The overlap integrals $S_{\mu\nu}(\mathbf{a}'_m-\mathbf{a}_m) = \int \chi_{\mu}(\mathbf{r}-\mathbf{R}_A-\mathbf{a}'_m)\chi_{\nu}(\mathbf{r}-\mathbf{R}_B-\mathbf{a}_m)d^3\mathbf{r}$ form the overlap matrix $\boldsymbol{S}(\mathbf{a}'_m-\mathbf{a}_m)$. Eqns. (53) and (54) are received from definition (35) of the DM for periodic systems, from relations (39) and (40) for the DM in direct and reciprocal space and from the LCAO approximation given by relations (50).

The idempotency relations (53) and (54) are satisfied both for the main region of a crystal and any cyclic cluster of smaller size, consisting of L primitive unit cells. The order of the DM in (54) is LM if M is the number of AO's in the primitive unit cell. Eqn. (53) is fulfilled for any of L **k** vectors in the BZ connected with the cyclic cluster choice, i.e. satisfying (42). DM **P(k)** and overlap matrix **S(k)** are $M \times M$ matrices. As mentioned in section 2, the idempotency of DM is connected with the balance between the choice of **k** vectors in (53) and direct lattice summation in (54).

If the Bloch sums (50) are transformed by a Löwdin orthogonalization

$$\phi^{\lambda}_{\mu\mathbf{k}}(\mathbf{r}) = \sum_{\nu} \mathbf{S}^{-1/2}{}_{\mu\nu}(\mathbf{k})\phi_{\nu\mathbf{k}}(\mathbf{r}) \tag{55}$$

the idempotency relation for the DM in the orthogonal basis $\boldsymbol{\rho}^{\lambda}(\mathbf{a}_n)$ is

$$\sum_{\mathbf{a}_m} \boldsymbol{\rho}^{\lambda}(\mathbf{a}_m)\boldsymbol{\rho}^{\lambda}(\mathbf{a}_n-\mathbf{a}_m) = 2\boldsymbol{\rho}^{\lambda}(\mathbf{a}_n) \tag{56}$$

As for the DM, the following evident relations are valid:

$$\begin{aligned} \rho_{\mathbf{rr}'}(\mathbf{a}_n) &= \rho^*_{\mathbf{r}'\mathbf{r}}(-\mathbf{a}_n) \\ P(\mathbf{k}) &= P^*(-\mathbf{k}) \end{aligned} \tag{57}$$

Eqn. (56) has the following form for $\mathbf{a}_n = \mathbf{0}$

$$\sum_{\mathbf{a}_m}\sum_{\nu} \rho^{\lambda}_{\mu\nu}(\mathbf{a}_m)\rho^{\lambda}_{\mu'\nu}(\mathbf{a}_m) = 2\rho^{\lambda}_{\mu\mu'}(\mathbf{0}) \tag{58}$$

Eqn. (58) allows to introduce a simple relation for checking the idempotency of the DM in HF-LCAO calculations. This is made by introducing the local properties of the electronic structure extended to periodic systems [43].

For an orthogonalized AO basis the bond orders $W_{AB}(\mathbf{a}_n)$ [43] of atom A in the central primitive unit cell ($\mathbf{a}_n = \mathbf{0}$) with atom B in a unit cell with translation vector $\mathbf{a}_n$ is expressed through DM $\boldsymbol{\rho}(\mathbf{a}_n)$ as

$$W_{AB}(\mathbf{a}_n) = \sum_{\mu\epsilon A,\nu\epsilon B} |\rho^{\lambda}_{\mu\nu}(\mathbf{a}_n)|^2 \tag{59}$$

From (58) one receives, using definition (59) and summing over $\mu\epsilon A$

$$\sum_{B}\sum_{\mathbf{a}_n} W_{AB}(\mathbf{a}_n) = 2\boldsymbol{p}^{\lambda}_A \tag{60}$$

where

$$\boldsymbol{p}^{\lambda}_A = \sum_{\mu\epsilon A} \rho^{\lambda}_{\mu\mu}(\mathbf{0}) \tag{61}$$

is the electron charge (sum of AO population for Löwdin population analysis) on atom A.

Defining an atomic covalence C_A as the sum of bond orders of atom A with all other atoms B of a crystal one obtains

$$C_A = \sum_{B\neq A} W_{AB}(\mathbf{0}) + \sum_{\mathbf{a}_n\neq\mathbf{0}}\sum_{B} W_{AB}(\mathbf{a}_n) = 2\boldsymbol{p}^{\lambda}_A - W_{AA}(\mathbf{0}) \tag{62}$$

From (62) it follows that the atomic covalence C_A in a periodic system may be calculated both as a lattice sum of bond orders $C^{(2)}_A$ and as an expression $C^{(1)}_A$, containing only those DM elements which refer to atom A. Such a possibility is a result of the DM idempotency in cyclic clusters modeling the periodic system.

By summing of C_A (62) over all atoms A in a primitive unit cell and by taking into account the DM normalization (36), one obtains

$$\sum_{A} C^{(2)}_A = \sum_{A}\sum_{B\neq A}\sum_{\mathbf{a}_n} W_{AB}(\mathbf{a}_n) + \sum_{A} W_{AA}(\mathbf{0}) = 2N_e \tag{63}$$

where N_e is the number of electrons in the primitive unit cell.

Relation (63) provides a simple way to check the idempotency property of the DM. Therefore it can be used to implement the CCM in HF LCAO calculations.

This implementation requires simple modifications in the property part of the CRYSTAL program. A Löwdin population analysis is introduced for self-consistent DM and the bond orders $C^{(2)}_A$ are calculated for atoms of the crystal. The lattice summation in (63) is made over the same part of the lattice that has been used in the integrals calculation for the self-consistent procedure. As it follows from the discussion in section 2, the lattice summation field is defined by the most severe tolerance used in the two-electron exchange integrals calculation.

In order to receive results for a cyclic cluster, relation (63) is checked for different choices of tolerances $P^{\mathbf{n}}$, $P^{\mathbf{g}}$ and Q sets of wave vectors in the BZ. The implementation of a cyclic cluster for perfect crystals is realized in most cases by using the primitive unit cell and taking into account that the *cyclicity field* of the DM is defined by the choice of Q. The implementation of the CCM in HF LCAO calculations with CRYSTAL95 [40] will be demonstrated in the next sections for the rutile structure.

9 Cyclic cluster representation of the rutile structure

Our choice of rutile TiO_2 to illustrate the cyclic cluster approach is explained by the following reasons. The system itself is an oxide of considerable technological interest, so that many theoretical studies of its electronic and geometrical structure and properties have been performed [44, 45, 46, 47, 48, 49, 50], including HF-LCAO pseudo-potential calculations [49, 50]. Also non-stoichiometric TiA_xO_{2-x} (A=F,Cl) and metal-doped rutile $M_xTi_{1-x}O_2$ (M=Ru,Fe,V,Li) has been studied with molecular clusters and periodic supercell models [5, 51, 52, 53, 54].

The cyclic cluster modeling of the perfect bulk rutile in the present study is useful for future applications of this model to defective rutile structures. In addition, also surface properties of rutile are intensively studied [9, 55, 56] and it is possible to extend the HF-LCAO CCM also to this field.

The tetragonal rutile structure belongs to space group $P4_2/mnm$ (D_{4h}^{14}) and contains two TiO_2 units per cell. The two Ti atoms are located at the Wyckoff $2a$ sites (0, 0, 0) and ($\frac{1}{2}$, $\frac{1}{2}$, $\frac{1}{2}$) with D_{2h} site symmetry, while the four O atoms are located at the $4f$ sites $\pm(u, u, 0)$ and $\pm(u + \frac{1}{2}, \frac{1}{2} - u, \frac{1}{2})$ with site symmetry C_{2v}. Each Ti atom is surrounded by a slightly distorted octahedron of O atoms with two different Ti-O distances. The inclusion of all nearest neighbors for a Ti atom requires at least a cyclic cluster with translation vectors $\mathbf{a}_1(0\ 0\ a)$, $\mathbf{a}_2(0\ a\ 0)$, $2\mathbf{a}_3(0\ 0\ 2c)$. The smallest cyclic cluster that contains the nearest neighbors to an O atom is formed by translation vectors $2\mathbf{a}_1(2a\ 0\ 0)$, $2\mathbf{a}_2(0\ 2a\ 0)$, $2\mathbf{a}_3(0\ 0\ 2c)$.

For the general choice of cyclic clusters, it is useful to perform a preliminary analysis of the connection between the distribution of neighbor atoms in spheres around a given central atom, and their distribution in primitive unit cells. The first distribution is due to the point symmetry of the crystal, the latter is a consequence of translation symmetry. In the CCM both distributions are important as this model allows to maintain the point symmetry of atoms in a crystal by introducing PBC's for translations.

It was shown [30] that for simple tetragonal lattices with translation vectors $\mathbf{a}_1(a\ 0\ 0)$, $\mathbf{a}_2(0\ a\ 0)$, $\mathbf{a}_3(0\ 0\ c)$ the symmetric LUC transformation (15) can be performed in four different ways maintaining tetragonal symmetry:

1. By simple extension of translation vectors $(n_1\mathbf{a}_1, n_1\mathbf{a}_2, n_2\mathbf{a}_3)$

   The corresponding transformation matrix is given in Table 5 at the bottom of the first column.

2. By rotation of vectors $\mathbf{a}_1$ and $\mathbf{a}_2$ by 45° in their plane and extension of the $\mathbf{a}_3$ vector.

   The transformation matrix is given in the second column of Table 5.

Table 5: Cyclic clusters used for the representation of the rutile bulk; special point (SP)-set and its accuracy J

		CL[a].1			CL[a].2			CL[a].3			CL[a].4		
n_1	n_2	cluster	SP-set	J	cluster	SP-set	J	cluster	SP-set	J	cluster	SP-set	J
1	1				C2.2	Γ, Z	1	C2.3	Γ, A	3	C4.4	$\Gamma, M, 2R$	4
1	2	C2.1	Γ, M	2	C4.2	Γ, M, A, Z	4	C4.3	$\Gamma, Z, 2V$	5	C8.4	$\Gamma, M, Z, A, 4W(p=\frac{1}{4})$	8
2	1	C4.1	$\Gamma, M, 2X$	1	C8.2	$\Gamma, M, 2X, 4\Sigma(p=\frac{1}{4})$	1	C8.3	$\Gamma, M, 2X, 4S(p=\frac{1}{4})$	4	C16.4	$\Gamma, M, 2X, 4\Sigma, 4U, 4T(p=\frac{1}{4})$	4
2	2	C8.1	$\Gamma, M, Z, A, 2X, 2R$	4	C16.2	$\Gamma, M, Z, A, 2X, 2R, 4\Sigma, 4S(p=\frac{1}{4})$	4	C16.3	$\Gamma, M, 2X, Z, A, 2R, 8\varepsilon(p=\frac{1}{4})$	15			
2	3	C12.1	$\Gamma, M, 2X, 2\Lambda, 2V, 2W(p=\frac{1}{3})$	9	C24.2	$\Gamma, M, 2X, 2V, 2\Lambda, 4W, 4\Sigma(p=\frac{1}{4}), 8(\frac{1}{4}, \frac{1}{4}, \frac{1}{3})$	9	C24.3	$\Gamma, M, 2X, 2V, 2\Lambda, 4W, 4S(p=\frac{1}{4}), 8(\frac{1}{4}, \frac{1}{4}, \frac{1}{6})$	19			
2	4	C16.1	$\Gamma, M, 2X, Z, A, 2R, 2\Lambda, 2V, 4W(p=\frac{1}{4})$	10									
			$L = n_1^2 n_2 = \begin{vmatrix} n_1 & 0 & 0 \\ 0 & n_1 & 0 \\ 0 & 0 & n_2 \end{vmatrix}$			$L = 2n_1^2 n_2 = \begin{vmatrix} n_1 & -n_1 & 0 \\ n_1 & n_1 & 0 \\ 0 & 0 & n_2 \end{vmatrix}$			$L = 2n_1^2 n_2 = \begin{vmatrix} 0 & n_1 & n_2 \\ n_1 & 0 & n_2 \\ n_1 & n_1 & 0 \end{vmatrix}$			$L = 4n_1^2 n_2 = \begin{vmatrix} -n_1 & n_1 & n_2 \\ n_1 & -n_1 & n_2 \\ n_1 & n_1 & -n_2 \end{vmatrix}$	

[a] L is the number of primitive unit cells in the large unit cell defining the cyclic cluster. The number m in CL.m defines the choice of the corresponding transformation matrix given at the bottom of each column.

3. By changing from simple to body-centered tetragonal lattice.

   This can be achieved in two different ways. The corresponding transformation matrices are given in columns 3 and 4 of Table 5.

In Table 5 cyclic clusters representing the rutile bulk structure are given that are generated from the primitive bulk unit cell by applying the abovementioned four types of LUC transformations. The cluster size was restricted to $L = 24$ (144 atoms) which is the largest cyclic cluster treatable with CRYSTAL95 within the available computational resources.

The CCM allows to establish a correspondence between the symmetry of electronic states of the finite model and the symmetry of states of an infinite crystal using the theory of Special Points (SP) of the Brillouin zone [57]. Generating the cyclic cluster LUC transformation (15) with matrix l is accompanied by the corresponding transformation of the reciprocal lattice generating a reduced Brillouin zone (RBZ) [28]. The L -**k**-points of the BZ, satisfying relation (10), become equivalent to the $\mathbf{k} = \mathbf{0}$ point of RBZ and form a SP-set, used in the DM calculation of the cyclic cluster. The accuracy of this set is defined by the number J of sphere of direct lattice translation vectors generated by LUC translation vectors $\mathbf{A}_j$ from eqn. (8) with smallest nonzero length. These vectors are connected by the crystal point group transformations. In [58] the radius of this sphere is called 'length cutoff' and used as a measure of the quality of completness of the corresponding **k**-mesh.

SP-sets may be split over the **k**-vector stars, defining different irreducible representations of space groups [28] (each star consists of symmetry-related non-equivalent **k**-vectors).

In Table 5 SP sets and their accuracy J are given for different cyclic clusters. Multiples before the **k**-vector labels mean the number of vectors in the star. The labels of the symmetry points and symmetry lines of the BZ for simple tetragonal lattice are taken from [59].

10 Pseudopotential Hartree-Fock calculations of cyclic clusters representing the rutile structure

The calculations were performed in the valence electron approximation with effective core potentials of Durand and Barthelat that are known to give results comparable to all-electron basis sets [60]. The lattice parameters were taken from the literature [49] where the optimized structure of rutile was found in good agreement with experiment (calculated values: a=4.555 Å, c=2.998 Å, experimental values: a=4.594 Å, c=2.959 Å).

In the present study the crystalline-orbital program CRYSTAL95 [40] was used with the modifications mentioned in previous sections allowing to check the DM idempotency [43].

The first step of the rutile cyclic cluster modeling was to perform bulk calculations corresponding to good and very good quality sets of the integral thresholds and shrinking factors to clarify their influence on the results for bulk properties. As it is seen from Table 6, the total energy per unit cell (E_{tot}) and the one-electron energy at the top of the valence band (ϵ_{top}) for fixed shrinking factors IS=6 converge when the DM becomes idempotent.

For integral thresholds 8 8 8 8 16 the twice number of electrons per unit cell $2N_e$ is closest to the exact number 64. In accordance with the discussion in section 2, it is found that for converged results the *cyclicity field* (in this case given by a six-fold increase of the primitive translation vectors) and the lattice summation field (defined by the threshold 10^{-16} for exchange integrals) are balanced. As it is seen from Table 6 the decrease of the threshold up to 10^{-17} or the increase of the shrinking factors up to 8 only gives rise to small changes in

Table 6: Effect of **k** set choice and integral thresholds on calculated properties of bulk rutile; total energy E_{tot} (a.u.), top of valence band ϵ_{top} (a.u.), Löwdin charge $q_{\rm Ti}$ (a.u.) on Ti, total number of electrons N_e obtained with eqn. (63)

Shrinking factor IS [a]	6	6	6	6	8
Integral thresholds N_1, N_2 [b]	6, 12	7, 14	8, 16	9, 17	8, 16
E_{tot}	-69.7758	-69.7819	-69.7840	-69.7845	-69.7840
ϵ_{top}	-0.3028	-0.3104	-0.3102	-0.3096	-0.3102
$q_{\rm Ti}$	+1.73	+1.72	+1.72	+1.72	+1.72
$2N_e$	63.79	63.94	63.97	63.96	63.97

[a] Monkhorst-Pack set [57]
[b] Accuracy is set to 10^{-N_1} and 10^{-N_2} for exchange integrals based on the overlap between atomic orbitals

the results for E_{tot} and ϵ_{top}. The charge on Ti atoms, given for Löwdin population analysis, is practically the same for all considered calculation parameters. This simply means that the diagonal DM elements are less sensitive to the lattice summation than the non-diagonal elements that are crucial for the DM idempotency. The calculated bond orders (see eqn. (59)) for Ti and O atoms to their neighbors distributed over spheres of equivalent atoms are given in Tables 7 and 9. It is seen that 1.) bond orders are stable when the defined exchange integral truncation thresholds are increased from standard values (6, 12) to severe values (8, 16). 2.) The bond orders are close to zero for atoms outside the sixth sphere of neighbors around Ti atoms and outside the tenth sphere around O atoms.

Table 7: Effect of integral thresholds on calculated properties of bulk rutile; bond orders to nearest neighbors for titanium[a] ($s_1 = s_2 = s_3 = 6$)

Distance (Å)	neighbor atom (total number)	integral thresholds		
		(6,12)	(7,14)	(8,16)
1.951	O (4)	0.557	0.557	0.557
1.972	O (2)	0.498	0.500	0.501
2.998	Ti (2)	0.030	0.030	0.030
3.454	O (4)	0.002	0.002	0.002
3.553	Ti (8)	0.013	0.014	0.014
3.588	O (4)	0.003	0.002	0.002
4.063	O (8)	0.001	0.001	0.001
4.470	O (2)	0.001	0.003	0.003
4.555	Ti (4)	0.000	0.000	0.000

[a] calculated using Löwdin basis with eqn. (59)

The calculation of local properties of the electronic structure (bond orders, atomic covalences) appears to be useful for the choice of cyclic clusters representing the bulk structure. This is seen from the results of cyclic cluster calculations for the rutile structure in Table 8. The results are given with respect to a band structure calculation of rutile with IS=6 corresponding to a cyclic cluster composed of 6×6×6 = 216 primitive unit cells, and the same integral thresholds, 6 6 6 6 12, that were used for all cyclic clusters in consideration.

The cyclic cluster calculations were performed using the SUPERCELL option of CRYSTAL95 and setting the shrinking factors IS to 1 (corresponding to the $\mathbf{k} = \mathbf{0}$ point in the BZ),

Table 8: Results of HF LCAO calculations of cyclic clusters for bulk rutile. Quality J, difference of the total energy ΔE and the energy of the valence band top $\Delta\epsilon$ (a.u.) with respect to the largest cluster C216.1, twice the total number of electrons per unit cell calculated using eqn. (63) within the corresponding cyclic cluster ($2N_e^C$) and within the interaction radius defined by the integral thresholds ($2N_e^S$), weighting coefficients of neighboring atoms in spheres i around Ti ($\omega_{\mathrm{Ti}}(i)$) and O ($\omega_{\mathrm{O}}(i)$), total number of atoms $N(i)$ in sphere i, Löwdin charge q_{Ti} on Ti

					spheres i	1 2 3 4 5 6	1 2 3 4 5 6 7 8 9 10	
					$N(i)$	4 2 2 4 8 4	2 1 1 8 2 2 2 2 2 1	
Cluster	J	ΔE	$\Delta\epsilon$	$2N_e^C$	$2N_e^S$	$\omega_{\mathrm{Ti}}(i)$	$\omega_{\mathrm{O}}(i)$	q_{Ti}
CL4.1	1	-4.4041	-0.2856	63.91	245.80	$\frac{1}{2}$ 1 0 1 $\frac{1}{2}$ 0	$\frac{1}{2}$ 1 1 $\frac{1}{2}$ 0 1 1 0 0 1	1.97
CL4.2	4	-0.0302	-0.0136	63.85	199.57	1 1 $\frac{1}{2}$ $\frac{1}{2}$ $\frac{1}{2}$ $\frac{1}{2}$	1 1 1 1 $\frac{1}{2}$ $\frac{1}{2}$ $\frac{1}{2}$ $\frac{1}{2}$ $\frac{1}{2}$ 0	1.60
CL4.3	5	-0.0138	-0.0134	63.89	185.90	1 1 1 $\frac{1}{2}$ $\frac{1}{2}$ 1	1 1 1 1 1 $\frac{1}{2}$ $\frac{1}{2}$ 1 1 0	1.60
CL4.4	4	-0.0248	-0.0170	63.73	248.79	1 1 $\frac{1}{2}$ 1 $\frac{1}{2}$ $\frac{1}{2}$	1 1 1 1 $\frac{1}{2}$ 1 1 1 $\frac{1}{2}$ $\frac{1}{2}$	1.68
CL8.1*	4	-0.0167	-0.0127	63.77	111.32	1 1 1 1 1 $\frac{1}{2}$	1 1 1 1 $\frac{1}{2}$ 1 1 $\frac{1}{2}$ $\frac{1}{2}$ 1	1.69
CL8.2	1	-4.4048	-0.2854	63.92	241.76	$\frac{1}{2}$ 1 0 1 $\frac{1}{2}$ 0	$\frac{1}{2}$ 1 1 $\frac{1}{2}$ 0 $\frac{1}{2}$ $\frac{1}{2}$ 0 0 1	1.97
CL8.3	4	-0.0183	-0.0124	63.74	108.95	1 1 $\frac{1}{2}$ 1 1 $\frac{1}{2}$	1 1 1 1 $\frac{1}{2}$ 1 1 $\frac{1}{2}$ $\frac{1}{2}$ 1	1.69
CL8.4	8	-0.0022	-0.0040	63.81	74.66	1 1 1 1 1 1	1 1 1 1 1 1 1 1 1 1	1.72
CL12.1	9	+0.0014	-0.0001	63.77	70.17	1 1 1 1 1 1	1 1 1 1 1 1 1 1 1 1	1.73
CL16.1	10	+0.0010	-0.0003	63.79	67.04	1 1 1 1 1 1	1 1 1 1 1 1 1 1 1 1	1.73
CL16.2	4	-0.0178	-0.0125	63.76	108.14	1 1 $\frac{1}{2}$ 1 1 $\frac{1}{2}$	1 1 1 1 $\frac{1}{2}$ 1 1 $\frac{1}{2}$ $\frac{1}{2}$ 1	1.69
CL16.3	15	+0.0018	-0.0002	63.83	63.86	1 1 1 1 1 1	1 1 1 1 1 1 1 1 1 1	1.73
CL16.4	4	-0.0178	-0.0125	63.75	108.15	1 1 $\frac{1}{2}$ 1 1 $\frac{1}{2}$	1 1 1 1 $\frac{1}{2}$ 1 1 $\frac{1}{2}$ $\frac{1}{2}$ 1	1.69
CL24.2	9	+0.0014	-0.0001	63.78	66.94	1 1 1 1 1 1	1 1 1 1 1 1 1 1 1 1	1.73
CL24.3	19	+0.0005	-0.0006	63.80	63.80	1 1 1 1 1 1	1 1 1 1 1 1 1 1 1 1	1.73
CL27.1*	9	+0.0008	+0.0005	63.79	66.93	1 1 1 1 1 1	1 1 1 1 1 1 1 1 1 1	1.73
CL64.1*	18	+0.0002	-0.0002	63.80	63.80	1 1 1 1 1 1	1 1 1 1 1 1 1 1 1 1	1.73
CL216.1*	45	0.0000	0.0000	63.79	63.79	1 1 1 1 1 1	1 1 1 1 1 1 1 1 1 1	1.73

*Band calculations (shrinking factors 2, 3, 4, 6 for CL8.1, CL27.1, CL64.1, and CL216.1, respectively)

defining the DM *cyclicity region* as the cyclic cluster region itself. For each cyclic cluster there are certain weighting coefficients of neighbor atoms in spheres i around Ti ($\omega_{\mathrm{Ti}}(i)$) and O ($\omega_{\mathrm{O}}(i)$). For inner atoms of a cyclic cluster $\omega(i) = 1$, while for border atoms $\omega(i) < 1$. For the calculation of the total number of electrons per unit cell within a cyclic cluster N_e^C these weighting coefficients have to be incorporated into eqn. (63).

$$\sum_A C_A^{(2)} = \sum_A \sum_{B \neq A} \sum_{\mathbf{a}_n} \omega_B(\mathbf{a}_n) W_{AB}(\mathbf{a}_n) + \sum_A W_{AA}(\mathbf{0}) = 2N_e \qquad (63a)$$

The abovementioned thresholds define the interaction radius for the exchange integrals truncation. For the given structural parameters and basis sets this radius is 8.16 Å around a Ti atom and 7.84 Å around an O atom, respectively. The total number of electrons per unit cell N_e^S within the interaction radius was calculated by summing bond orders over spheres i, containing a total number $N(i)$ of atoms as given in the second line of Table 8.

An analysis of Table 8 allows to choose a cyclic cluster of relatively small size in such a way that it still reproduces bulk properties with reasonable accuracy. Let us discuss this in more detail.

For the small size clusters ($L = 4, 8$) the unbalance of the *cyclicity region* and the interaction radius causes a huge difference between N_e^C and N_e^S as the sum of bond orders inside the interaction radius includes the neighboring cyclic clusters around the central cyclic cluster.

Table 9: Effect of integral thresholds on calculated properties of bulk rutile; bond orders to nearest neighbors for oxygen[a] ($s_1 = s_2 = s_3 = 6$)

Distance (Å)	neighbor atom (total number)	integral thresholds		
		(6,12)	(7,14)	(8,16)
1.951	Ti (2)	0.557	0.557	0.557
1.972	Ti (1)	0.498	0.500	0.501
2.498	O (1)	0.033	0.033	0.033
2.774	O (8)	0.022	0.023	0.023
2.998	O (2)	0.019	0.019	0.020
3.301	O (2)	0.001	0.001	0.001
3.454	Ti (2)	0.002	0.002	0.002
3.588	Ti (2)	0.003	0.002	0.002
3.902	O (2)	0.037	0.050	0.050
3.944	O (1)	0.034	0.044	0.044
4.063	Ti (4)	0.001	0.001	0.001

[a] calculated using Löwdin basis with eqn. (59)

The PBC's forces atoms outside the central cyclic cluster to be equivalent to the corresponding atoms inside the cluster. At the same time the summation of bond orders inside a *cyclicity region* gives N_e^C values that are largely model size independent, as the DM is calculated for the cyclic cluster in consideration.

From Table 8 one concludes also that even for the same size L of a cyclic cluster the difference of the calculated properties E_{tot}, ϵ_{top}, and atomic charges Q strongly depend on the *compactness* of a cluster. The compactness is defined by the number of spheres i of neighboring atoms with $\omega(i) = 1$ inside the cyclic cluster. This property of cyclic clusters is important as, for example, the more compact cluster CL4.3 gives results closer to the band calculation limit that the larger but less compact cluster CL8.2. In order to construct more compact clusters, the anisotropy of the rutile structure ($a \approx 1.5c$) has to be taken into account.

From Table 8 it follows that although for CL16.3 the integral thresholds ensure the DM idempotency ($2N_e^S - 2N_e^C = 0.03$) the larger cluster CL24.3 better approximates the infinite bulk limit of the calculated properties. For the cyclic cluster CL12.1 the calculated properties are close to those for CL16.3, but the difference $2N_e^S - 2N_e^C$ is much larger, 6.4, i.e. the DM is non-idempotent. It is well known that small changes of only the most severe integral threshold does not essentially affect the results [39]. Therefore, its decrease may restore the DM idempotency while the calculated properties are only slightly changed. It is demonstrated in Table 8 that by using TOL5=10 instead of 12, the difference $2N_e^S - 2N_e^C$ decreases drastically to 0.20 for the same cluster, CL12.1. For CL16.1 the influence of TOL5 is also small.

For a cyclic cluster that has been chosen properly, an increase of the cyclicity region has to give only small changes in the calculated results. For CL12.1 and CL16.3 it is seen (Table 8) that $2 \times 2 \times 2$ supercell calculations using these cyclic clusters as basis show practically the same results as the clusters themselves.

In CCM calculations of defective crystals such an increase of the cyclicity region allows to investigate the convergence of the results with cluster size.

As it is seen from Table 8, the SP-sets accuracy J is important to reproduce the total energy. The cyclic cluster increase without change of J (clusters CL4.1 and CL8.2 with $J = 1$, clusters CL8.1, CL16.1 and CL16.4 with $J = 4$) gives only small changes in total energy.

The cyclic clusters CL12.1 and CL16.3 generate highly accurate SP-sets ($J = 9$ and $J = 15$, respectively) and therefore the corresponding total energies are close to that of the infinite crystal. The same refers to the one-electron energy at the top of the valence band.

11 Cyclic cluster calculations based on the semiempirical method MSINDO

For comparison, we also performed cyclic cluster calculations for the perfect rutile bulk using the semiempirical SCF MO method MSINDO [61], a modified version of SINDO1 [62, 63, 64, 65, 66] that has been successfully applied to calculations of bulk and surface properties of oxides [67, 68]. The method has been designed for the quantum chemical treatment of molecular systems, but has recently been extended to calculations of cyclic clusters [25].

The underlying strategy of the CCM implementation in MSINDO was to start from a molecular system and to introduce PBC in such a way that both local point symmetry and stoichiometry of the system are conserved. This has been realized by an averaging of two-center integrals (overlap, nuclear attraction, electron repulsion) between the central atom of the interaction region and atoms at the boundaries [25]. In this approach, the *cyclicity* and *summation* regions are the same and not dependent on integral thresholds as for periodic methods.

In MSINDO only two-center two-electron integrals of the form $(\mu_A\mu_A|\nu_B\nu_B)$ are taken into account. Atomic orbitals μ and ν are taken as s-type functions. Therefore, these integrals have no angular dependence and an averaging over integrals between different atoms on the same radius around the considered atom causes no problems. It has to be mentioned that such a simple procedure is not possible for ab initio Hartree-Fock methods due to the presence of three- and four-center integrals.

In MSINDO also two-center one-electron integrals are replaced by empirical formulas that do not contain any angular dependence, so that the two-center elements of the Fock matrix have the following form [62]:

$$\begin{aligned} F_{\mu\nu} &= H_{\mu\nu} + P_{\mu\nu}(\mu\mu|\nu\nu) \\ &= \Delta H_{\mu\nu} + L_{\mu\nu} + P_{\mu\nu}(\mu\mu|\nu\nu) \qquad (\mu\epsilon A, \nu\epsilon B) \end{aligned} \tag{64}$$

with the two parametrized empirical functions

$$\begin{aligned} \Delta H_{\mu\nu} &= \frac{1}{4}(K_A + K_B)S_{\mu\nu}(f_A h_{\mu\mu} + f_B h_{\nu\nu}) \\ f_A &= 1 - e^{-\alpha_A R_{AB}} \\ L_{\mu\nu} &= -\frac{1}{2}(\zeta_\mu^2 + \zeta_\nu^2)\frac{S_{\mu\nu}(1 - S_{\mu\nu})}{1 + \frac{1}{2}(\zeta_\mu + \zeta_\nu)R_{AB}} \qquad (\zeta_\mu, \zeta_\nu \quad \text{orbital exponents}) \end{aligned} \tag{65}$$

where K_A, K_B, and α_A are empirical parameters optimized for a given set of reference molecules [62, 61]. In CCM calculations these matrix elements remain unchanged when atom B is in the inner part of the interaction region of atom A. In the case that B is on the border of the interaction region around A, the contributions from all $n_{AB'}$ atoms B' that are translationally equivalent to B and appear at the same radius are summed up into one matrix element with weighting coefficient $\omega_{AB'} = \frac{1}{n_{AB'}+1}$ [25]

$$\begin{aligned} H_{\mu\nu} &= \sum_{B'}^{\text{equiv}} \omega_{AB'} H_{\mu\nu'} \qquad \text{with } \nu' \text{ on } B' \\ G_{\mu\nu} &= \sum_{B'}^{\text{equiv}} \omega_{AB'} (\mu\mu|\nu'\nu') \end{aligned} \tag{66}$$

In this way the idempotency of the DM (which is always in direct space since the $\mathbf{k} = 0$ approximation is applied) is guaranteed for all cases. This was tested for the cyclic cluster CL12.1. The DM idempotency was tested in two ways:

1. by calculating the difference $C_A^{(1)} - C_A^{(2)}$ obtained with relation (62)
2. by directly calculating $\mathbf{P}^2 - 2\mathbf{P}$ and summing the absolute values of the differences over all matrix elements, $\Delta = \sum_{\mu\nu} |(\mathbf{P}^2)_{\mu\nu} - 2\mathbf{P}_{\mu\nu}|$

By both ways of calculation it was clearly demonstrated that the MSINDO CCM density matrix is exactly idempotent. The differences of bond orders $C_A^{(1)}$ and $C_A^{(2)}$ as well as Δ are in the same order of magnitude as the SCF threshold, 10^{-8}. At the same time the symmetry of the crystalline system was correctly reproduced as checked by counting the degenerate one-electron levels in the valence band region and comparison with the results of a symmetry analysis. The symmetry of the cyclic cluster is different from the molecular cluster due to the presence of additional translation symmetry elements. Therefore, highly degenerate one-electron levels appear that are not present in any molecular model of rutile. The same number of degenerate levels as for CRYSTAL for the cyclic cluster CL12.1 was obtained. The calculated binding energies per unit cell E_{tot}, HOMO orbital energies ϵ_{top}, and atomic Löwdin charges q for a number of cyclic clusters, CL8.1, CL12.1, CL24.2, and CL32.2, are given in Table 10 [69]. The geometry of the four clusters has been optimized in terms of the three structural parameters c, R_{TiO} and R'_{TiO}. At low temperatures, the experimental references are [70]: c=2.954 Å, R_{TiO} = 1.946 Å, and R'_{TiO} = 1.976 Å. For the largest cyclic cluster, CL32.3, the corresponding MSINDO values are 2.96 Å, 1.90 Å, and 1.91 Å, which is in reasonable agreement with experiment. From Table 10 it can be seen that the convergence of E_{tot} is much slower than for the periodic HF calculations, see Table 8. On the other hand, ϵ_{top} and q_{Ti} show a similar convergence behavior. The MSINDO binding energies per TiO_2 unit E_B vary between 1722 kJ/mol and 2096 kJ/mol and are close to the experimental value of 1900 kJ/mol [71]. The calculated HOMO orbital energies for MSINDO (Table 10) and HF (Table 8) are similar. The MSINDO ϵ_{top} is only 0.6 eV less negative than the HF result.

Table 10: MSINDO CCM results for bulk rutile; total energy per unit cell E_{tot} (a.u.), binding energy per unit E_B (kJ/mol), HOMO energy ϵ_{top} (a.u.), atomic Löwdin charge on Ti q_{Ti}.

Cluster	E_{tot}	E_B	ϵ_{top}	q_{Ti}
CL8.1	−70.2164	1880	−0.2947	+1.74
CL12.1	−70.0959	1722	−0.2910	+1.77
CL24.2	−70.3774	2092	−0.2898	+1.77
CL32.2	−70.3428	2046	−0.2885	+1.77

12 Conclusions

We considered in this paper SCM of point defects in MgO crystall and cyclic cluster models of rutile structure.

The density matrix idempotency relations described here may be easily extended to the unrestricted Hartree-Fock (UHF) method when the orbitals for α and β spins are treated independently.

In the most recent versions of CRYSTAL [41] the density functional theory (DFT) has been implemented so that Kohn-Sham equations are solved [72]. The local Gaussian basis allows to realize also hybrid DFT-HF calculations or HF plus *a posteriori* DFT correlation. Different local and non-local exchange and correlation functionals are available. For the non-local functionals it is expected that similar problems with the balance between direct lattice and BZ summation, as discussed in section 2, occur as for the HF method.

As it is seen from the HF calculations of bulk rutile structure the CCM technique allows the good representation of the perfect crystal, which is essential for point defect studies.

13 Acknowledgement

I highly appreciate the coorperation with Prof. K. Jug, Prof. V.P. Smirnov, Dr. T. Bredow, Dr. A.V. Leko, Dr. I.I. Tupytsyn, Dr. V.A. Veryazov — My coauthors in former publications. I am grateful to F. Janetzko for unpublished results of rutile cyclic cluster MSINDO calculations and his help in preparing this manuscript.

References

[1] P. Deák, phys. stat. sol. (B) **217** (2000) 9-21

[2] R.P. Messmer, G.D. Watkins, Inst. Phys. Conf. Ser. 16 (IOP, London, 1973) p. 255-271

[3] S.G. Louie, M. Schluter, J.P. Chelikowsky, and M.L. Cohen, Phys. Rev. B **13** (1976) 1654-1663

[4] A.M. Dobrotvorskii and R.A. Evarestov, phys. stat. sol. (B) **66** (1974) 83 -91

[5] R.A. Evarestov and V.P. Smirnov, phys. stat. sol. (B) **215** (1999) 949-956

[6] G. Makov, R. Shah, and M.C. Payne, Phys. Rev. B **53** (1996) 15513-15517

[7] R. Dovesi and R. Orlando, Phase Transitions **52** (1994) 151-164

[8] Y.F. Zhukovskii, E.A. Kotomin, P.W.M. Jacobs, and A.M. Stoneham, Phys. Rev. Lett. **84** (2000) 1256-1259

[9] T. Bredow, E. Aprà, M. Catti, and G. Pacchioni, Surf. Sci. **418** (1998) 150-165

[10] M. Hakala, M.J. Puska, and R.M. Nieminen, Phys. Rev. B **61** (2000) 8155-8161

[11] V. M. Bermudez, Solid Stat. Commun. **118** (2001) 569-574

[12] G. Mallia, R. Orlando, C. Roetti, P. Ugliengo, R. Dovesi, Phys. Rev. B **63** (2001) 2351021-2351027

[13] B. Aradi, A. Gali, P. Deak, J.E. Lowther, N.T. Son, E. Janzen, W.J. Choyke, Phys. Rev. B **63** (2001) 2452021-24520219

[14] Y.X. Wang, W.L. Zhong, C.L. Wang P.L. Zhang, Phys. Lett. A **285** (2001) 390-394

[15] C. Pisani, Phase Transitions **52** (1994) 123-154

[16] W.C. Mackrodt, Ber. Bunsenges. Phys. Chem. **101** (1997) 169-177

[17] A. Lichanot, C. Larrieu, C. Zicovich-Wilson, C. Roetti, R. Orlando, and R. Dovesi, J. Phys. Chem. Sol. **59** (1998) 7-12

[18] R.A. Evarestov and V.A. Lovchikov, phys. stat. sol. (B) **79** (1977) 743-751

[19] P.V. Smith, J.E. Szymanski, and J.A.D. Matthews, J. Phys. C **18** (1985) 3157-3174

[20] J. Noga, P. Baňacký, S. Biskupič, R. Boča, P. Pelikán, M. Svrček, A. Zajac, J. Comput. Chem. **20** (1999) 253-261

[21] P. Deák and L. Snyder, Phys. Rev. B **36** (1987) 9619-9627

[22] A.Shluger and E.Stefanovich,Phys. Rev. B **160** (1990) 9664-9672

[23] P. Persson, A. Stashans, R. Bergström, and S. Lunell, Int. J. Quantum Chem. **70** (1998) 1055-1066

[24] A. Poredda, V.A. Lovchikov, and K. Jug, *Cluster Models for Surface and Bulk Phenomena*, NATO ASI Series B, Vol. 283, 641-650, Eds. G. Pacchioni, P.S. Bagus, F. Parmigiani, Plenum, New York 1992

[25] T. Bredow, G. Geudtner, and K. Jug, J. Comput. Chem. **22** (2001) 89-101

[26] J. Miró, P. Deák, C.P. Ewels, and R. Jones, J. Phys. C: Condens. Matter **9** (1997) 9555-9562

[27] T.Bredow,R.A.Evarestov and K.Jug, phys.stat.sol(b) **222** (2000) 495-516

[28] R.A. Evarestov and V.P. Smirnov, *Site Symmetry in Crystals: Theory and Applications* (2nd ed.), Springer Series in Solid State Sciences, vol. 108, Springer, Berlin 1997

[29] R.A.Evarestov,A.V.Leko and V.P.Smirnov, phys. stat. sol(b) **128** (1985) 275-285

[30] R.A.Evarestov and V.P.Smirnov, J. Phys. C: Condens. Matter **9** (1997) 3023-3031

[31] R.A.Evarestov and V.P.Smirnov, phys. stat. sol(b) **201** (1997) 75-87

[32] R. Dovesi, V.R. Saunders, and C. Roetti, *Crystal92 User's Manual*, Università di Torino, Torino 1993

[33] T.Hahn, International Tables for Crystallography,vol.A,Reidel,Dordrecht,1983

[34] S.Lunnel,A.Stashans,L.Ojamae,H.Lindstrom and A.Hagfeldt,J.Amer.Chem.Soc. **119** (1997) 7374-7380

[35] R.A. Evarestov, V.A. Lovchikov, and I.I. Tupitsyn, phys. stat. sol. (b) **117** (1983) 417-427

[36] R.A. Evarestov, I.I. Tupytsyn, Russ. J. Phys. Chem. **74** Suppl.2 (2000) 363-375

[37] C. Pisani, E. Aprà, and M. Causà, Int. J. Quantum. Chem. **38** (1990) 395-417

[38] R.A. Evarestov, M.I. Petrashen, and E.M. Ledovskaya, phys. stat. sol. (b) **68** (1975) 453-462

[39] C. Pisani, R. Dovesi, and C. Roetti, *Hartree-Fock ab initio Treatment of Crystalline Systems*, Lecture Notes in Chemistry, vol. 48, Springer-Verlag, Berlin 1988

[40] R. Dovesi, V.R. Saunders, C. Roetti, N.M. Harrison, R. Orlando, and E. Aprà, *Crystal95 User's Manual*, Università di Torino, Torino 1996

[41] V.R. Saunders, R. Dovesi, C. Roetti, M. Causà, N.M. Harrison, R. Orlando, C.M. Zicovich-Wilson,, *Crystal98 User's Manual*, Università di Torino, Torino 1998
http://www.dl.ac.uk/TCS/Software/CRYSTAL
http://www.ch.unito.it/ifm/teorica/crystal.html

[42] C. Pisani, R. Dovesi, and R. Orlando, Int. J. Quantum Chem. **42** (1992) 5-34

[43] R.A. Evarestov, A.V. Leko, and V.A. Veryazov, Phys. Sol. State **41** (1999) 1286-1290

[44] P.K. Schelling, N. Yu, and J.W. Halley, Phys. Rev. B **58** (1998) 1279-1293

[45] D.R. Hamann, Phys. Rev. B **56** (1997) 14979-14984

[46] P. Reinhardt, B.A. Heß, and M. Causà, Int. J. Quantum Chem. **58** (1996) 297-306

[47] S.-D. Mo and W.Y. Ching, Phys. Rev. B **51** (1995) 13023-13032

[48] K.M. Glassford and J.R. Chelikovsky, Phys. Rev. B **46** (1992) 1284-1298

[49] B. Silvi, N. Fourati, R. Nada, and C.R.A. Catlow, J. Phys. Chem. Solids **52** (1991) 1005-1012

[50] A. Fahmi, C. Minot, B. Silvi, and M. Causà, Phys. Rev. B **47** (1993) 11717-11724

[51] S.-D. Mo, L.B. Lin, and D.L. Lin, J. Phys. Chem. Solids **55** (1994) 1309-1316

[52] K.M. Glassford and J.R. Chelikovsky, Phys. Rev. B **47** (1993) 12550-12553

[53] A. Stashans, S. Lunell, R. Bergström, A. Hagfeldt, and S.-E. Lindquist, Phys. Rev. B **53** (1996) 159-170

[54] A. Stashans, S. Lunell, and R.W. Grimes, J. Phys. Chem. Solids **57** (1996) 1293-1302

[55] T. Bredow and K. Jug, Surf. Sci. **327** (1995) 398-408

[56] J. Muscat and N.M. Harrison, Surf. Sci. **446** (2000) 119-127

[57] H.J. Monkhorst and J.D. Pack, Phys. Rev. **13** (1976) 5188-5192

[58] M. Moreno, A. Soler, Phys. Rev. B **45** (1992) 13891-13898

[59] C.J. Bradley, A.P. Cracknell, *The Mathematical Theory of Symmetry in Solids*, Clarendon Press, Oxford 1972

[60] B. Silvi, M. Causà, R. Dovesi, and C. Roetti, Mol. Phys. **67** (1989) 891-899

[61] B. Ahlswede and K. Jug, J. Comput. Chem. **20** (1999) 563-571

[62] D.N. Nanda and K. Jug, Theoret. Chim. Acta **57** (1980) 95-106

[63] K. Jug, R. Iffert, and J. Schulz, Int. J. Quantum Chem. **32** (1987) 265-277

[64] J. Li, P. Correa de Mello, and K. Jug, J. Comput. Chem. **13** (1992) 85-92

[65] K. Jug and T. Bredow, *Methods and Techniques in Computational Chemistry, METECC95*, eds. E. Clementi and G. Corongiu, STEF, Cagliari, 1995, p.89-139

[66] K. Jug, T. Bredow, SINDO1: Parametrization and Application, *Encyclopedia of Computational Chemistry*, Vol. 4, Eds. P. v. Ragué Schleyer, N.L. Allinger, T. Clark, J. Gasteiger, P.A. Kollman, H.F. Schaefer III, P.R. Schreiner, Wiley, New York 1998, 2599-2609

[67] B. Ahlswede, T. Homann, and K. Jug, Surf. Sci. **445** (2000) 49-59

[68] A.R. Gerson, and T. Bredow, Phys. Chem. Chem. Phys. **1** (1999) 4889-4895

[69] F. Janetzko, unpublished results

[70] J.K. Burdett, T. Hughbanks, G.J. Miller, J.W. Richardson, Jr., and J.V. Smith, J. Am. Chem. Soc. **109** (1987) 3639-3645

[71] D.R. Lide, *CRC Handbook of Chemistry and Physics*, 79th Ed., CRC Press, Boca Raton 1998/1999

[72] M.D. Towler, A. Zupan, and M. Causà, Comput. Phys. Commun. **98** (1996) 181-204

Computational Materials Science
C.R.A. Catlow and E.A. Kotomin (Eds.)
IOS Press, 2003

Vibrations in solids and small particles from first-principles calculations

A. V. Postnikov
Theoretical Low-Temperature Physics,
Gerhard Mercator University, D-47048 Duisburg, Germany

Abstract. The possibility to treat dynamical properties of solids and small particles on the basis of conventional *ab initio* schemes for evaluating total energy and/or forces is discussed. The emphasis is given on frozen-phonon approach and classical molecular dynamics. The frozen phonon results are derived from total energy hypersurface, mapped over a number of calculations for distorted structures, making use of full-potential linearized augmented plane waves method. The examples considered are layered titanium dichalcogenides ($TiSe_2$ and $TiTe_2$) and lithium niobate and tantalate, for which the frozen phonon calculation gives zone-center TO modes, straightforwardly accessible (but in part not known from experiment) in dichalcogenides, or quite complicated and in part controversial in $LiNbO_3$. The comparison with $LiTaO_3$ shows that the softening of all but one A_1-TO modes in tantalate, as compared to niobate, has little to do with higher Ta mass, but is instead due to an interplay of less rigid lattice in more covalent $LiTaO_3$ and different degree of anharmonicity of certain modes. A molecular dynamics treatment is tried on bulk and small clusters of TiC, using a numerical local orbitals method SIESTA. The results indicate the presence of a hard surface-related mode, dominating in small particles. Advantages and disadvantages of both approaches to the treatment of vibrations are discussed.

1 Introduction

Lattice vibrations in crystals can be in most cases satisfactorily described in a classical approach, making use of the Born–Oppenheimer approximation. Exceptions are some relatively rare cases where essentially quantum behaviour plays a role, like e.g. quantum paraelectrics, or vibronic systems. The Born–Oppenheimer approximation assumes that the electronic system is equilibrated at each moment, by solving (in principle) stationary Schrödinger equation, or (typically in practice) Kohn–Sham equations of time-independent density functional theory. The point charges of nuclei obey classical equations of motion in a potential field influenced by a momentary charge distribution. This is not the only scheme known. Historically, the Ehrenfest approach is known that allows a propagation of the electron wavefunction via the time-dependent Schrödinger equation. Since 1985, Car–Parrinello approach gained popularity, as it allows "fast"dynamics where the equilibration of the electronic system at each step is avoided, and propagation of nuclear and electronic degrees of freedom is done simultaneously, via solving a system of coupled equations. This is done at a price of deviation from (or, more precisely, small fluctuations about) the Born–Oppenheimer surface. An extensive literature exists on this subject; a good recent review is that by Marx and Hutter[1].

All the following discussion refers to Born–Oppenheimer dynamics. The corresponding total energy hypersurface is either probed in a sequence of precision calculations for a number of representative displacements from the equilibrium, that makes the essence of the frozen phonon method. Or, it is studied "on the fly", for a sequence of ion positions generated by Newtonian equation of motion, with forces coming from a precision electronic structure calculation. This second approach is *ab initio* molecular dynamics (MD). Both methods are briefly outlined below and some examples of their use given.

2 Dynamical matrix and symmetry considerations

Let us start with the frozen phonon approach. We denote in the following the deviations of individual atoms from their equilibrium positions by u_i, and total energy corresponding to each displaced configuration by $E(\{u_i\})$. The index i runs from 1 to $3N$ over all atoms in the unit cell (or in the molecule) and Cartesian coordinates of displacements. Then the matrix of force constants is

$$F_{ij} = \frac{\partial^2 E}{\partial u_i \partial u_j}\,. \tag{1}$$

The equation of motion

$$m_i \ddot{u}_i + \sum_{j=1}^{3N} F_{ij} u_j = 0\,, \tag{2}$$

when the solutions of harmonic form $u_i \sim A_i \cos(\omega t + \varepsilon)$ are searched for, reduces to the matrix equation:

$$\sum_{j=1}^{3N} F_{ij}\, A_j - \omega^2 m_i A_i = 0\,. \tag{3}$$

The values m_i are masses, equal for i values running through three Cartesian components of a given atom. Dividing each line by corresponding mass allows to reduce the problem to a conventional eigenvalue problem:

$$\sum_j \frac{F_{ij}}{m_i} u_j - \omega^2 u_i = 0\,. \tag{4}$$

In this formulation, each eigenvector is proportional to the displacement patterns in the corresponding mode. Attributing masses differently results in two alternative definitions of the eigenvector:

$$\sum_j \frac{F_{ij}}{m_j}(m_j u_j) - \omega^2 (m_i u_i) \;=\; 0\,; \tag{5}$$

$$\sum_j \frac{F_{ij}}{\sqrt{m_i m_j}}(\sqrt{m_j} u_j) - \omega^2 (\sqrt{m_i} u_i) \;=\; 0\,. \tag{6}$$

The last formulation has the advantage that the matrix to be diagonalized is symmetric, so corresponding diagonalization routines can be used. For two other cases, diagonalization must be done for the real matrix of general form, yielding ω^2 as complex eigenvalues (that must be of course real from physical reasons). Now, consider the situation when "primary" displacements are not purely Cartesian, but linear combinations of the latter. A typical motivation for such choice is the use of symmetry properties in order to block-diagonalize the dynamical matrix and reduce the size of a problem. We introduce a transformation to symmetry-adapted coordinates S_t via a (general, so far) transformation matrix:

$$S_t = \sum_j B_{tj} u_j \, . \tag{7}$$

The classical kinetic energy is then

$$\mathcal{T} = \sum_i \frac{m_i \dot{u}_i^2}{2} = \sum_{tt'i} (B^{-1})_{it} \, (B^{-1})_{it'} \frac{m_i}{2} \dot{S}_t \dot{S}_{t'} \, , \tag{8}$$

and the potential energy can be fitted to the second order in symmetry-adapted displacements from the results of calculation:

$$U = \frac{1}{2} \sum_{tt'} F_{tt'} S_t S_{t'} \, . \tag{9}$$

The Lagrangian equation of motion

$$\sum_{t'} \left[\sum_i (B^{-1})_{it} \, m_i \, (B^{-1})_{it'} \ddot{S}_{t'} + (F_{tt'} + F_{t't}) S_{t'} \right] = 0 \tag{10}$$

in the harmonic approximation, with the solution of the form $S_t \sim A_t \cos(\omega t + \varepsilon)$ leads to the matrix equation:

$$\left[\mathbf{F} - \omega^2 \mathbf{G} \right] \mathbf{A} = 0 \, , \tag{11}$$

where $\mathbf{F}$ is the force constant matrix, and $G_{tt'} = \sum_i B_{it} m_i^{-1} B_{it'}$ is the kinetic energy matrix. Similarly to Eqs. (5, 6), different rearrangement of masses leads to a conventional eigenvalue problem

$$(\mathbf{FG} - \omega^2)\mathbf{A} = 0 \, , \quad \text{or} \quad (\mathbf{GF} - \omega^2)\mathbf{A} = 0 \, , \tag{12}$$

with identical resulting frequencies but different meaning of eigenvectors, or in the symmetrized form:

$$\left(\mathbf{G}^{1/2} \mathbf{F} \mathbf{G}^{1/2} - \omega^2 \right) \mathbf{A} = 0 \, . \tag{13}$$

Thus selected symmetry-adapted coordinates guarantee that frozen phonons corresponding to each irreducible representation of the space group in question can be calculated independently on the others. In first-principles calculations where the dominating computational load is related with calculation of force constants, splitting F_{ij} into smaller blocks may reduce the amount of calculations quite considerably. Projecting out three translational modes per unit cell (for solids), or three translational and three rotational modes (for molecules or finite clusters) may further reduce the size of the dynamical problem.

3 Frozen phonon calculations for some crystals

In the following, some recent results are presented of frozen phonon calculations on moderately complex systems. The analysis of vibrations follows from the study of the total energy as function of displacements from equilibrium geometry, therefore, a precision evaluation of total energies in low-symmetry structures was essential. In the examples below, the necessary calculations have been done using the full-potential linear augmented plane-wave method, described e.g. in Ref. [2], as implemented in the WIEN97 code[3]. Apart from total energy, the forces on atoms were also available and used in some calculations.

Table 1: Calculated and measured Γ phonons in $TiSe_2$ and $TiTe_2$. Infrared and Raman active modes are indicated by IR and R, correspondingly

	$TiSe_2$		$TiTe_2$	
	Calc.	Exp.	Calc.	Exp.
A_{1g} (Γ_1^+; IR)	193	195* 194†	133	127[2]
A_{2u} (Γ_2^-; R)	293	250*	255	
E_g (Γ_3^+; R)	134	134* 134†	162	144†
E_u (Γ_3^-; IR)	143	142*	169	

*Ref. [6]
†A. N. Titov and Y. S. Ponosov (unpublished)

3.1 Titanium dichalcogenides

Titanium dichalcogenides $TiCh_2$ (Ch = S, Se, Te) are semimetals with hexagonal crystal structure (space group D_{3d}^1, CdI_2 type). The $TiCh_2$ layers are separated by van der Waals-type gaps. Inside each layer, Ti is surrounded by 6 chalcogen atoms in almost perfect octahedral configuration. The van der Waals gap can host different other atoms, in particular transition metals, giving rise to ordered or disordered intercalated phases. The electronic structure of pure and intercalated Ti dichalcogenides has been studies in a number of publications – see Ref. [4] and references therein. The dynamical properties, apart probably from those of TiS_2, have not yet been subject to *ab initio* treatment.

The symmetry analysis of zone-center (transversal) phonons shows that they are split into A_{1g}, A_{2u}, E_g and E_u blocks, each containing just one vibrational mode. Therefore, the vibration patterns (eigenvectors) follow immediately from symmetry and are schematically shown in Fig. 1. Two-dimensional representations E_g and E_u include two partners, which represent in-plane displacements (of either only chalcogen in the E_g, or chalcogen vs. Ti in the E_u) along two orthogonal directions in the plane.

With the lattice vectors as in conventional hexagonal setting, $(a,\ 0,\ 0)$ $(-a/2,\ a\sqrt{3}/2,\ 0)$ $(0,\ 0,\ c)$, and positions of atoms in the unit cell Ti $(0,\ 0,\ 0)$, Ch $(0,\ a\sqrt{3}/3,\ \pm\Delta c)$ (Ch=Se, Te), the symmetry coordinates introduced by Eq. (7) (after projecting out the uniform dis-

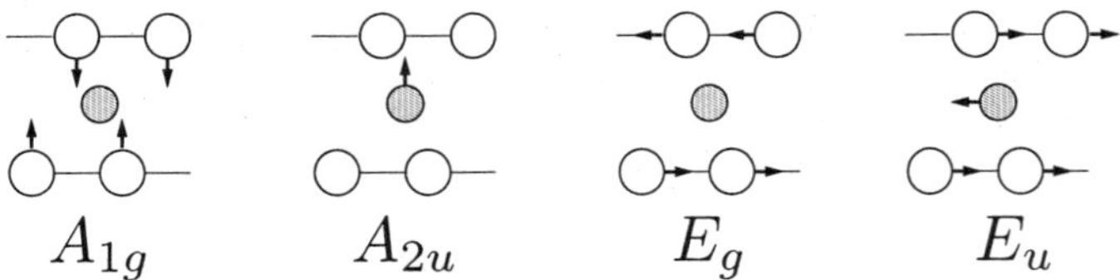

Figure 1: Four vibration modes in Ti dichalcogenides

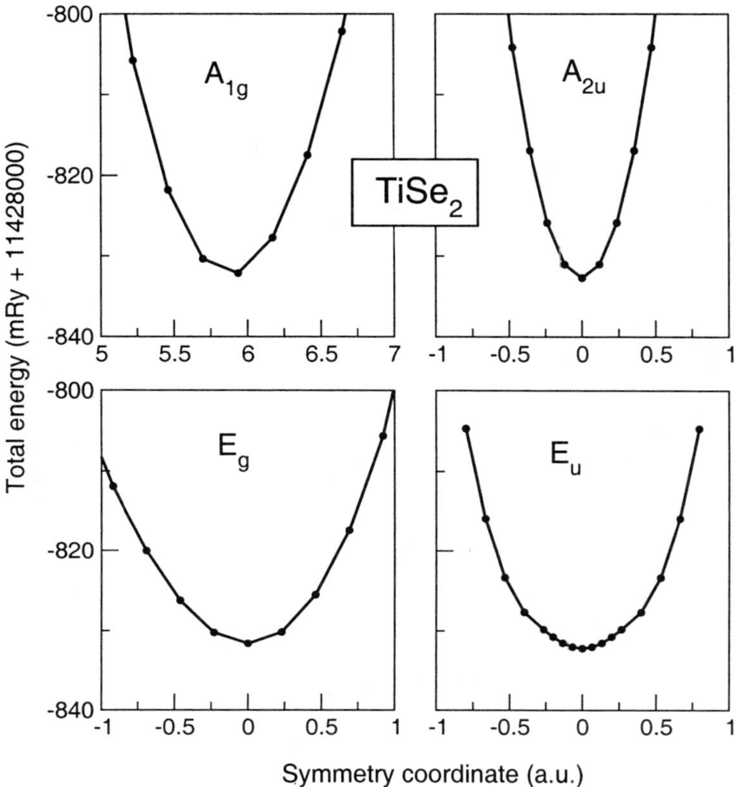

Figure 2: Total energy profiles for Γ phonons in $TiSe_2$.

placement mode) read:

$$\begin{aligned} S_{A_{1g}} &= Z_{\mathrm{Ch1}} - Z_{\mathrm{Ch2}}\,; \\ S_{A_{2u}} &= Z_{\mathrm{Ti}} - \frac{1}{2}(Z_{\mathrm{Ch1}} + Z_{\mathrm{Ch2}})\,; \\ S_{E_g} &= X_{\mathrm{Ch1}} - X_{\mathrm{Ch2}}\,: \\ S_{E_u} &= X_{\mathrm{Ti}} - \frac{1}{2}(X_{\mathrm{Ch1}} + X_{\mathrm{Ch2}})\,. \end{aligned}$$

There is an ambiguity in choosing symmetry coordinates for two partners of two-dimensional irreducible representations E_g and E_u, e.g., Y displacements may be taken instead of X, or any two normal directions in the (XY)-plane. Lattice parameters a, c, Δ equal 3.519 Å, 6.280 Å and 0.247 (for $TiSe_2$) and 3.752 Å, 6.932 Å and 0.248 (for $TiTe_2$), as optimized in a total energy calculation[5]. The total energy as depending on corresponding symmetry coordinate is plotted in Figs. 2 and 3. $\partial^2 E/\partial S^2$ at the minimum yielded the force constants and allowed to calculate corresponding frequencies, straightforwardly in this case. However, the full profile of the total energy away from equilibrium allows to estimate an importance of anharmonic effects for different modes. One should note that for two different possible

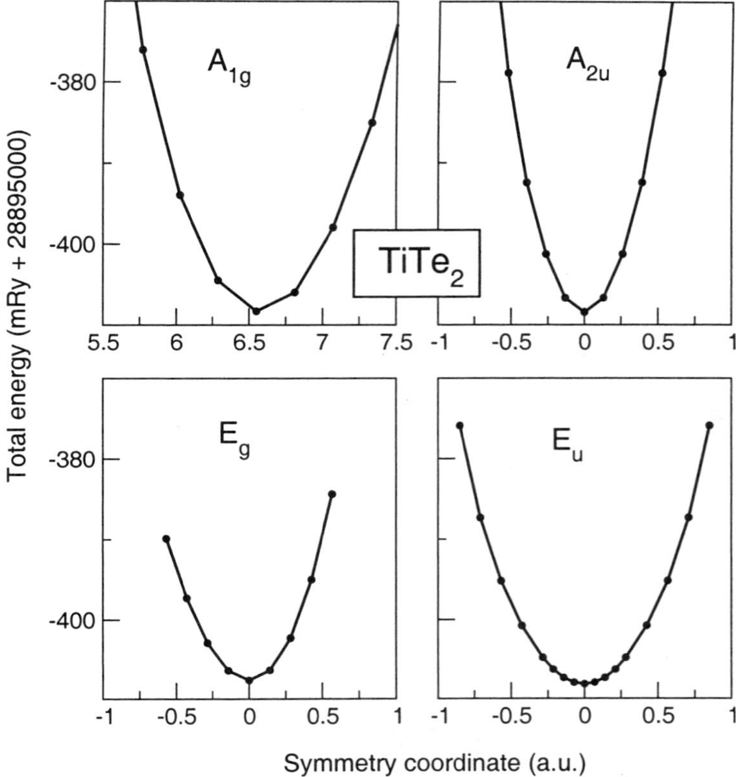

Figure 3: Total energy profiles for Γ phonons in $TiTe_2$.

choices of symmetry coordinates for two-dimensional representations E_g and E_u the total energy profile has different shape (it is easy to see that it is symmetric for displacements in the X-direction and asymmetric for Y-displacements), but this doesn't affect the curvature at the minimum that yields the same frequency in the harmonic approximation.

One can see that the A_{1g} energy profile is steeper in $TiSe_2$, as consistent with slightly lower covalency, compared to $TiTe_2$. The A_{2u} profiles are almost identical, and the difference in corresponding frequencies comes from a higher mass of Te. The most pronounced differences are in E_g and in E_u modes: the first one has much steeper energy profile, resulting in much higher frequency in $TiTe_2$; for the E_u mode the overall shape of the potential well (over large region of displacement magnitudes) is similar in both systems, but $TiSe_2$ exhibits a considerable anharmonicity of this mode, with the harmonic estimate of corresponding frequency much lower than in $TiTe_2$. One can expect a considerable temperature dependency of the E_u line in $TiSe_2$. Comparing calculated and measured values of Γ-TO frequencies in Table 3.1, one notes good agreement, with the exception of the A_{2u} mode. Such drawback in the interpretation of a seemingly simple vibration mode invites for additional investigations by both experiment and theory. Much higher frequency value in the calculation lets to assume that other channels of energy transfer in the lattice are coupling to this mode. As regards the experimental situation for $TiTe_2$, the unpublished data listed in Table 3.1 seem to be the only ones available at the moment. The authors admit instability of samples under laser irradiation and possible contributions from elemental tellurium. Again, additional studies would be helpful in elucidating the situation.

3.2 *Lithium tantalate, as compared to lithium niobate*

$LiNbO_3$ and $LiTaO_3$ are ferroelectrics with relatively high transition temperatures (1480 and 950 K, correspondingly) both possessing just one ferroelectric phase with almost identical lattice parameters. Differently from many other materials of similar composition, they have a crystal structure not immediately related to cubic perovskite, but a rhombohedral one, with space group $R3c$ (C_{3v}^6) and two formula units per unit cell. Low symmetry complicates a precision total-energy analysis of structural trends in $LiNbO_3$ and $LiTaO_3$. Nevertheless, the first system has been studied in Ref. [7] (some previous works are cited therein), yielding the information about frequencies and eigenvectors of all zone-center TO phonons. A new attribution of some experimentally measured Raman lines belonging to a complicated (9 modes) E-block has been proposed, resolving certain controversies. Recently, a similar analysis has been done for $LiTaO_3$, albeit for the A_1 block only, with the aim to explain a counter-intuitive relation between phonon frequencies in two compounds: whereas three modes get softened when going from niobate to tantalate (TO_1: 252→206, TO_2: 275→253, TO_4: 632→597 cm^{-1}), the TO_3 mode is noticeably hardening (332→356 cm^{-1}). This is the more so strange that both compounds have very similar electronic band structures. Since Ta ion is much heavier than Nb, the above trend could be only understood by tiny differences in chemical bonding, probably combined with different participation of Ta and Nb in corresponding modes of both compounds. This has been discussed and explained in Ref. [8].

30 degrees of freedom in combined displacements of 10 atoms per unit cell are split in case of $Li(Nb/Ta)O_3$, after projecting out three translational modes, into $4A_1$, $5A_2$ and 9 (doubly degenerate) E TO-modes. Simplifying, A_1 modes are primarily vibrations of Li and transition metal atom against oxygen sublattice along the trigonal axis, and A_2 modes – vibrations of each atom against its counterpart (in the doubled formula unit), again along the trigonal axis. But the displacement of oxygen atoms leads not only to compression but also to twist and torsion of O_6 octahedra, so that corresponding "pure oxygen" modes are present in both A_1 and A_2 blocks. The analysis of all eigenvectors and the predictions of isotope shift of frequencies in $LiNbO_3$ is given in Ref. [7].

The peculiarity of $LiTaO_3$ in this context is that in a (slightly) more expanded lattice the potential energy profile associated with the Li z-displacement becomes more flat (see Fig. 4). As a consequence, the Li contribution that in $LiNbO_3$ was present predominantly (of four A_1 TO modes) in the TO_2 mode, is in $LiTaO_3$ shared between TO_1 and TO_2.

The eigenvectors calculated for A_1 modes in $LiTaO_3$ are shown in Table 3.2, as reproduced from Ref. [8]. The entries in the table are in the hexagonal setting, where the (relaxed) positions of atoms are $(0\ 0\ 0)$ and $(0\ 0\ 0.5)$ for Ta, $(0\ 0\ 0.282)$ and $(0\ 0\ 0.782)$ for Li, and $(0.049\ 0.343\ 0.071)$ for the "prototype" oxygen atom, whose five colleagues in the rhombohedral unit cell can be generated as $(\bar{y}, x{-}y, z)$, $(y{-}x, \bar{x}, z)$, $(y{-}x, y, z)$, $(\bar{y}, \bar{x}, z)$, $(x, x{-}y, z)$. The transformation to the Cartesian projections X, Y, Z (for both atom positions and eigen-

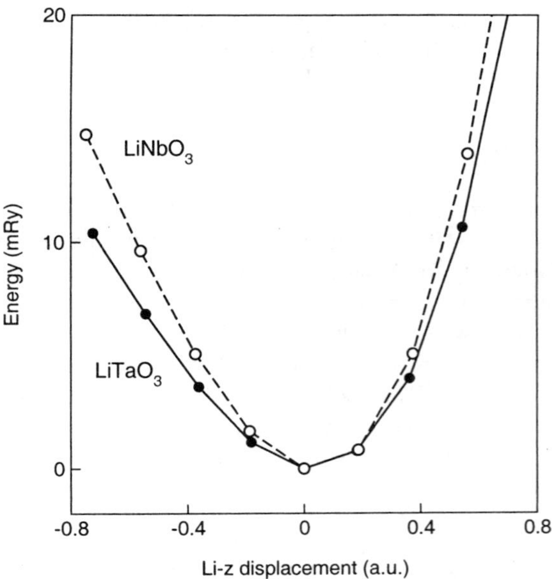

Figure 4: Potential energy profile associated with Li z-displacements in A_1-TO_3 mode of $LiNbO_3$ (LNO) and $LiTaO_3$ (LTO). Positive movements shorten the distance between Li and O ions, while the negative movements shorten that between Li and Nb(Ta) ions.

vector components) proceeds by

$$\begin{pmatrix} X \\ Y \\ Z \end{pmatrix} = \begin{pmatrix} 0 & \sqrt{3}/2 & 0 \\ -1 & 1/2 & 0 \\ 0 & 0 & c/a \end{pmatrix} \begin{pmatrix} x \\ y \\ z \end{pmatrix} .$$

Contrary to the Li modes that are softer in tantalate, the TO_3 and TO_4 modes that correspond, correspondingly, to almost rigid rotation about the z axis and to the torsion (rotation of three top atoms against the three bottom ones) of O_6 octahedra, are more hard in $LiTaO_3$. This is not related to the difference in Nb and Ta masses and can be traced to a slightly higher degree of covalency in tantalate. Higher covalency enhances interaction between lattice-dynamical degrees of freedom, softening the softest modes and hardening the

Table 2: Eigenvectors of TO phonons in $LiNbO_3$ and $LiTaO_3$ in the hexagonal setting (from Ref. [8])

	$LiTaO_3$			$LiNbO_3$		
	Ta	Li	O	Nb	Li	O
TO_1	0. 060	0. 168	(0. 096 0. 132 −0. 104)	0. 143	0. 035	(0. 036 0. 085 −0. 123)
TO_2	0. 109	−0. 191	(−0. 047 0. 038 −0. 080)	0. 068	−0. 254	(0. 015 −0. 014 0. 001)
TO_3	−0. 007	0. 051	(−0. 441 −0. 101 −0. 003)	0. 007	−0. 001	(−0. 463 −0. 156 −0. 006)
TO_4	0. 029	0. 022	(−0. 127 −0. 440 −0. 037)	0. 023	0. 015	(−0. 082 −0. 437 −0. 022)

hardest. In $LiNbO_3$, the TO_4 mode exhibits strong anharmonicity (as was deduced in Ref. [7] based on the shape of corresponding total energy profile) that shifts its experimentally measured frequency beyond that of TO_4 in $LiTaO_3$ (where anharmonicity is almost negligible). Taken together, this explains an unexpected relation between A_1-TO frequencies in both systems.

4 Methods accessing phonon dispersion

If calculation of phonon dispersion or of phonon density of states is a priority, the frozen phonons method has quite limited applicability. The choice of symmetry-adapted coordinates can be done for an arbitrary $\mathbf{q}$ value, and the force constant matrix splits into blocks corresponding to the symmetry (if any) of a particular $\mathbf{q}$ point. However, even if the size of such blocks remains moderate (for high-symmetric $\mathbf{q}$ points), the evaluation of total energy must be done for particular displacement patterns in a supercell *commensurate* with the wave of given $\mathbf{q}$. This limits the number of $\mathbf{q}$ that can be studied and makes a calculation technically demanding, as correspondingly multipled supercells must be introduced. In practice, direct evaluation of not-Γ phonons by frozen-phonon method is limited to zone-boundary phonons, which calculations typically demand to double the unit cell in an appropriate way.

Systematic ways to treat arbitrary wavevector in a calculation are provided by linear response theory and within a so-called 'direct method'. The first approach considers the reaction of electronic system to a small periodic perturbation, that is, in this context, a wave of atomic displacements with given (arbitrary) $\mathbf{q}$ in a crystal. Calculations, done independently for each $\mathbf{q}$, include the solution of the Sternheimer equation self-consistently and simultaneously with the Kohn-Sham equation(s). The practical implementation of this scheme is described in the recent review by Baroni *et al.*[9], concentrating primarily on pseudopotential techniques and plane-wave bais set. The implementation in all-electron schemes is more difficult due to the necessity to take into account the displacement of basis functions (centered on atoms) as a by-effect of perturbation. This has been done by Savrason within the linear muffin-tin orbitals (LMTO) formalism[10] and by Yu and Krakauer[11] in the full-potential linearized augmented plane wave (LAPW) method. Applications of both methods are numerous by now; since the calculations are quite demanding, they are typically done for not very complicated structures, in most cases with less than 10 atoms per unit cell.

The "direct method" relies on evaluation of forces in a sequence of supercells, with subsequent Fourier transformation to $\mathbf{q}$-dependent force constants [12, 13, 14]. It is related to frozen-phonon formalism in a sense that for selected high-symmetry $\mathbf{q}$ points, the force constants extracted from a calculation for the corresponding supercell yield "exact" results. Calculation for general (i.e. non commensurate with any supercell) $\mathbf{q}$ values are only exact if the real-space decay of force constants is confined in the largest supercell considered, that is in practice seldom the case. A number of direct method calculations of phonon dispersion for different systems[15, 16, 14, 17, 18, 19] has been performed by K. Parlinski who created a software for recovering phonon dispersions and other related information from the forces available from any accurate electronic-structure code.

5 Molecular dynamics (MD) simulations

A yet another approach to *ab initio* dynamics is via real-time simulations, using a sufficiently long sequence of (equal) time steps and a subsequent analysis of trajectories. This can be done using straightforward Newtonian dynamics (e.g., the Verlet algorithm), that propagates the ion of mass M at the moment $t + \delta t$ as

$$\mathbf{r}(t+\delta t) = 2\mathbf{r}(t) - \mathbf{r}(t-\delta t) + (\delta t)^2 \frac{\mathbf{F}(t)}{M}\,, \tag{14}$$

depending on its positions at two precedent steps t and $t - \delta t$, and the force acting on the ion at the last step $\mathbf{F}(t)$. Once some sufficiently long simulation history is accumulated as set of trajectories of all particles $i = 1, \ldots, N$ in coordinate or/and velocity space, $\{\mathbf{v}_i(t)\}$, a useful information can be extracted e.g. from different correlation functions, first of all, velocity autocorrelation function:

$$C_v(\tau) = \frac{1}{N}\sum_{i=1}^{N} \frac{1}{t_{\max}} \sum_{t_0=1}^{t_{\max}} \left[\mathbf{v}_i(t_0)\cdot\mathbf{v}_i(t_0+\tau)\right]\,. \tag{15}$$

Here, an averaging out over initial time instants formally removes the dependence of the result on starting conditions. In addition to this, it is considered essential in MD to skip some initial stage of simulations, several hundred steps or so, in order to achieve "equilibration", a stable behaviour relatively independent on the start configuration. A Fourier transformation of $C_v(\tau)$

$$G(\omega) = \int_{-\infty}^{\infty} d\tau C_v(\tau) e^{-i\omega\tau}$$

or, more precisely, the spectral density of it $I(\omega) = |G(\omega)|^2$ yields the vibration density of states.

A good introduction in dynamical simulations on a real-time scale can be found in a book by Allen and Tildesley[20]; specific aspects of modern *ab initio* molecular dynamics algorithms are discussed in a recent review by Marx and Hutter[1].

The advantage of this approach is that all information about dynamics in a system can be collected from a single run. The temperature can be imposed as a parameter, and the vibrations are not constrained to be harmonic. So, in principle the development of vibration properties with temperature may be addressed in a simulation.

The disadvantage of the MD simulation is that a considerable number of time steps is necessary for obtaining a reasonable and well resolved Fourier spectrum. The length of a time step is limited (from above) by the fact that it determines the highest frequency in the resulting Fourier spectrum (see e.g. Ref. [21], Chapter 12 for a discussion) that must be well beyond all physically relevant frequencies in the system. The number of simulation steps must be large enough to ensure the independence of results on starting conditions and for reducing statistical errors. Moreover, the total length of simulation run t_{run} determines the frequency resolution, $\Delta\nu \leq t_{\text{run}}^{-1}$. The error in the autocorrelation function (see, e.g., Ref. [20], p. 197) develops as

$$\sigma\langle\mathbf{v}(t)\mathbf{v}\rangle \sim \left(\frac{\tau_{\mathbf{v}}}{N t_{\text{run}}}\right)^{1/2}, \tag{16}$$

where τ_v is a representative correlation time (related to dominating frequency), N is the number of atoms in a simulation (i.e. per supercell) and t_{run} – full time of simulation.

The way the temperature is introduced in a simulation is either – in the simplest case – by appropriate scaling the velocities of particles to keep the predetermined value of the kinetic energy, or – more sophisticated – to introduce additional degrees of freedom in a system (e.g. a Nosé thermostat, see Ref. [22] for a review) for enforcing canonical distribution in the ensemble in question. The disadvantage of the first approach is that it spoils the deterministic dynamics provided e.g. by the Verlet algorithm; the Nosé thermostat, on the other hand, brings in an artificial parameter of effective mass that slightly affect the overall dynamic properties.

5.1 Force constants and molecular dynamics in titanium carbide

As an example of trying both approaches on the same system, we consider bulk TiC and several stoichiometric TiC clusters. With its crystal structure of NaCl type, this material has only one Γ-TO mode of T_{1u} symmetry, corresponding to (triply degenerate) relative displacement of Ti and C sublattices. Electronic structure of stoichiometric TiC has been subject to a number of calculations – see, e.g., Ahuja *et al.*[23] and references therein. I calculated equilibrium lattice constant a and bulk modulus B with the WIEN97 implementation of the LAPW method[3], using generalized gradient approximation; the results are a=8.07 a.u. and B=2.96 MBar, to be compared with experimental values 8.16 a.u. and 2.4 MBar, correspondingly. These numbers are also close to those given in Ref. [23]. The analysis of dynamical properties has been done based on the results of calculation with first-principles tight-binding package SIESTA[24, 25] that uses numerical and strictly spatially constrained basis functions and norm-conserving pseudopotentials. Apart from many advantages as e.g. allowing the order-N scaling of computational effort with system size (that was not used in the present study), SIESTA for any given geometry works as a quite sensitive total-energy tool, where the forces on atoms also available with good accuracy. The SIESTA estimate of bulk characteristics of TiC, based on the energy-volume curve, is a=8.18 a.u. and B=2.53 MBar. The pseudopotential used for titanium was generated for the electron configuration $3p^6\,3d^{2.5}\,4s^1$. The phonon frequency following from the fit of the total energy vs. Ti–C relative displacement is 14.2 THz, or 475 cm^{-1}. A calculation by Jochym *et al.*[26] using the "direct method" and Hellmann-Feynman forces calculated with the CASTEP (pseudopotential planewave) package from individual atom displacements in the 2×2×2 supercell (64 atoms in total) yields around 17 THz (experimental value 16.1 THz). A somehow underestimated value of calculated frequency (as in our calculation using SIESTA) is common and usually reflects the effect of anharmonicity neglected in the calculation. An opposite trend in the calculation of Ref. [26] may be due to an unsufficient convergency of real-space force constants within the supercell (that, via truncated lattice summation, must influence the results even for q=0.[1] Another possible sources of discrepancy are the attribution of Ti3p states to the core in the generation of pseudopotential in Ref. [26]. One should note however that apart from this quite small numerical discrepancy, full phonon dispersion and density of vibrational states are available in Ref. [26] with overall good agreement with experiment.

[1]It was mentioned above that the force constants at selected high-symmetry q values, including q=0, can be estimated "exactly" from a supercell calculation. However, the procedure of *single atom* displacements from equilibrium positions and the analysis of resulting forces, as done in Ref. [26], does not probe a purely Γ-frozen phonon and allows to access the latter only via Fourier transformation of real-space force constants.

As a counterpart to this frozen-phonon treatment, we switch now to the analysis of real-time simulation. In the simplest case, the MD run can be done on the smallest unit cell, containing just one atom of each type. One would expect just one dominating frequency in the vibrational spectrum, corresponding to the Γ-TO mode. Indeed, the velocity autocorrelation function exhibits periodic behaviour and no dissipation – a consequence that a single mode cannot exchange energy with other degrees of freedom and remains correlated with itself over long time intervals. The spectral function exhibits a single peak near ν=16 THz, slightly higher than the (zero-temperature) frozen phonon result and in good agreement with experiment. The frequency shift reflects the fact that a finite temperature (900 K in the above example) and correspondingly elevated amplitude of velocities allow to go beyond the harmonic approximation.

The above example is artificial in a sense that in a solid, there is no strict periodicity of vibration patterns, and a realistic simulation must be done on a larger unit cell. This would allow different wavelengths of vibrations, hence the parts of the Brillouin zone apart from Γ will be sampled and contribute to the vibrational density of states. An additional advantage follows from statistics, represented by Eq. (16) above. As different modes begin to exchange energy, correlations in the displacements of ions (either in neighboring primitive cells, or at the same place but over longer time intervals) get gradually lost, the time-dependent velocity autocorrelation function falls down faster and becomes less "noisy". The price paid for such refinement of results is the necessity to perform simulations on a large supercell over many time steps.

Dynamical simulations have been done for two clusters, a stoichiometric Ti_4C_4 representing the smallest cubic fragment of the crystal, and $Ti_{14}C_{13}$, that is a larger cube with carbon as central atom. The latter cluster seems to be quite common in Ti–C cluster chemistry (see a review by Rohmer *at al.*[27]). Prior to dynamical simulation, the static relaxation has been performed that showed a slight contraction in both clusters. In Ti_4C_4, the distances from the center of mass are 1.672 Å for Ti and 1.668 Å for C, that must be compared with $a\sqrt{3}/8$=1.847 Å in the bulk. In $Ti_{14}C_{13}$, the distances from the central (C) atom are 2.042, 2.957 and 3.403 Å for first (Ti), second (C) and third (Ti) neighbours, correspondingly, whereas in the bulk the corresponding distances are 2.133; 3.017 and 3.695 Å. The velocity autocorrelation function for the smaller cluster is shown in Fig. 5, left panel, its corresponding spectral density – in the right panel. A considerable hardening of the main peak compared to the bulk is seen, related to reduced number of neighbors of the atoms in cluster. The dominant mode associated with this peak is "breathing" (in antiphase) of two penetrating tetrahedra, Ti_4 and C_4.

Preliminary studies of vibrations in a larger $Ti_{14}C_{13}$ cluster shows that the dominant (surface) peak is still present at $\approx$ 22–25 THz, whereas the spectral density gets enhanced in the "bulk" region of 8–16 THz. A more detailed description of corresponding vibrational modes will be given elsewhere.

Summarizing, the analysis of vibrations on the basis of conventional calculation schemes, that give access to accurate total energy and forces, may prove efficient even for moderately large systems (with several tens of atoms), with a big advantage that no, or only small, modification of existing electronic-structure codes is needed. Such approaches can hardly compete with schemes based on response theories in what regards the analysis of phonon dispersions; however, they may be quite sufficient for the interpretation of Raman spectra and for studying vibrational density of states and displacement patterns induced at different frequencies.

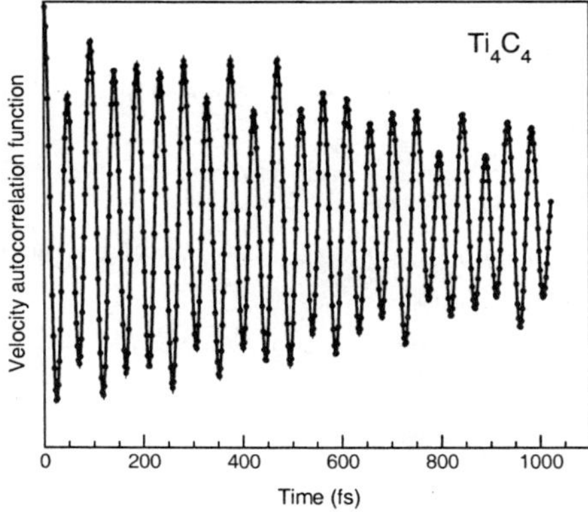

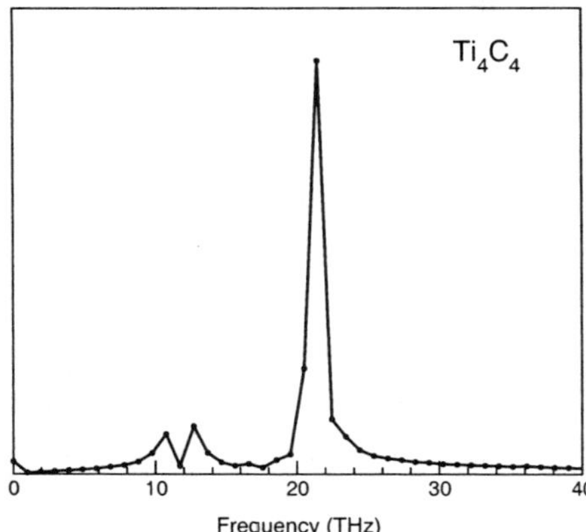

Figure 5: Velocity autocorrelation function (left panel) and its corresponding spectral power function (right panel) for the Ti_4C_4 cluster.

Acknowledgments

Part of the results presented here have been obtained within the *Sonderforschungsbereich 445* supported by the German Research Society. The calculations for the TiC clusters used the SIESTA program; many useful discussions with P. Ordejón are appreaciated in this relation. I am grateful to P. Entel for permanent support of the work, to A. L. Ivanovskii for the introduction into the TiC problematics, and to A. I. Titov for many discussions about Ti dichalcogenides and for providing me with his unpublished data.

References

[1] D. Marx and J. Hutter, *Ab initio molecular dynamics: Theory and Implementation*, In: Modern Methods and Algorithms of Quantum Chemistry: Winterschool, 21.–25. Februar 2000, Forschungszentrum Jülich, Proceedings, Band 1, pp. 301–449, http://www.fz-juelich.de/nic-series/Volume1/Volume1.html.

[2] D. J. Singh, *Planewaves, pseudopotentials and the LAPW method*. ISBN 0-7923-9421-7, Kluwer Academic Publishers, Boston, 1994.

[3] P. Blaha, K. Schwarz and J. Luitz, WIEN97, Vienna University of Technology (1997). Improved and updated Unix version of the original copyrighted WIEN-code, which was published by P. Blaha, K. Schwarz, P. Sorantin, and S. B. Trickey, in Comput. Phys. Commun. 59, (1990) 339.

[4] A. N. Titov, A. V. Kuranov, V. G. Pleschev, Y. M. Yarmoshenko, M. V. Yablonskikh, A. V. Postnikov, S. Plogmann, M. Neumann, A. V. Ezhov and E. Z. Kurmaev, Electronic structure of Co_xTiSe_2 and Cr_xTiSe_2. *Phys. Rev. B* **63** (2001) 035106–1–035106–8.

[5] A. V. Postnikov, M. Neumann, S. Plogmann, Y. M. Yarmoshenko, A. N. Titov and A. V. Kuranov, Magnetic properties of $3d$-doped $TiSe_2$ and $TiTe_2$ intercalate compounds. *Comput. Mater. Sci.* **17** (2000) 450–454.

[6] J. A. Wilson, Modelling the contrasting semimetallic characters of TiS_2 and $TiSe_2$. *Physica Status Solidi (b)* **1** (1978) 11–36.

[7] V. Caciuc, A. V. Postnikov and G. Borstel, *Ab initio* structure and zone-center phonons in $LiNbO_3$. *Phys. Rev. B* **61** (2000) 8806–8813.

[8] V. Caciuc and A. V. Postnikov, *Ab initio* zone-center phonons in $LiTaO_3$: comparison to $LiNbO_3$. (submitted to Phys. Rev. B).

[9] S. Baroni, S. de Gironcoli, A. Dal Corso and P. Giannozzi, Phonons and related crystal properties from density-functional perturbation theory. *Reviews of Modern Physics* **73** (2001) 515–562.

[10] S. Y. Savrasov, Linear response calculations of lattice dynamics using muffin-tin basis sets. *Phys. Rev. Lett.* **69** (1992) 2819–2822.

[11] R. Yu and H. Krakauer, Linear-response calculations within the linearized augmented plane-wave method. *Phys. Rev. B* **49** (1994) 4467–4477.

[12] K. Kunc and P. Gomes Dacosta, Real-space convergence of the force series in the lattice dynamics of germanium. *Phys. Rev. B* **32** (1985) 2010–2021.

[13] S. Wei and M. Y. Chou, Ab initio calculation of force constants and full phonon dispersions. *Phys. Rev. Lett.* **69** (1992) 2799–2802.

[14] K. Parlinski, Z. Q. Li and Y. Kawazoe, First-principles determination of the soft mode in cubic ZrO_2. *Phys. Rev. Lett.* **78** (1997) 4063–4066.

[15] K. Parlinski and Y. Kawazoe, *Ab initio* study of phonons in hexagonal GaN. *Phys. Rev. B* **60** (1999) 15511–15514.

[16] K. Parlinski, Z. Q. Li and Y. Kawazoe, *Ab initio* calculations of phonons in $LiNbO_3$. *Phys. Rev. B* **61** (2000) 272–278.

[17] K. Parlinski and Y. Kawazoe, Ab initio study of phonons in the rutile structure of SnO_2 under pressure. *Eur. Phys. J. B* **13** (2000) 679–683.

[18] P. T. Jochym and K. Parlinski, Ab initio lattice dynamics and elastic constants of ZrC. *Eur. Phys. J. B* **15** (2000) 265–268.

[19] K. Parlinski and Y. Kawazoe, Ab initio study of phonons and structural stabilities of the perovskite-type $MgSiO_3$. *Eur. Phys. J. B* **16** (2000) 49–58.

[20] M. P. Allen and D. J. Tildesley, *Computer Simulation of Liquids*. ISBN 0-19-855645-4. Oxford University Press, Oxford, 1987.

[21] W. H. Press, B. P. Flannery, S. A. Teukolsky and W. T. Vetterling, *Numerical Recipes. The Art of Scientific Computing*. ISBN 0-521-38330-7. Cambridge University Press, Cambridge, New York, Port Chester, Melbourne, Sydney, 1989.

[22] S. Nosé, Constant temperature molecular dynamics methods. *Progress of Theoretical Physics Supplement* **103** (1991) 1–46.

[23] R. Ahuja, O. Eriksson, J. M. Wills and B. Johansson, Structural, elastic, and high-pressure properties of cubic TiC, TiN, and TiO. *Phys. Rev. B* **53** (1996), 3072–3079.

[24] E. Artacho, D. Sánchez-Portal, P. Ordejón, A. García and J. M. Soler, Linear-scaling ab-initio calculations for large and complex systems. *Physica Status Solidi (b)* **215** (1999) 809–817.

[25] D. Sánchez-Portal, P. Ordejón, E. Artacho and J. M. Soler, Density-functional method for very large systems with LCAO basis sets. *Int. J. Quant. Chem.* **65** (1997) 453–461.

[26] P. T. Jochym, K. Parlinski and M. Sternik, TiC lattice dynamics from ab initio calculations. *Eur. Phys. J. B* **10** (1999) 9–13.

[27] M.-M. Rohmer, M. Bènard and J.-M. Poblet, Structure, reactivity, and growth pathways of metallocarbohedrenes M_8C_{12} and transition metal/carbon clusters and nanocrystals: A challenge to computational chemistry. *Chemical Reviews* **100** (2000) 495–542.

Computational Materials Science
C.R.A. Catlow and E.A. Kotomin (Eds.)
IOS Press, 2003

Quantum Chemical Approach to Excited States in Material Science

Coen de Graaf[1], Carmen Sousa[2] and Francesc Illas[2]

1) Departament de Química Física i Inorgànica and Institut d'Estudis Avançats, Universitat Rovira i Virgili, Plaça Imperial Tàrraco 1, 43000 – Tarragona, Spain

2) Departament de Química Física and Centre de Recerca en Química Teòrica, Universitat de Barcelona, C/ Martí i Franquès 1, 08028 – Barcelona, Spain

Abstract. It is shown that the combination of the embedded cluster approach and powerful ab initio quantum chemical methods provides useful information in the study of excited states in solid state compounds. After reviewing the cluster model approach and discussing two multiconfigurational wave function based methods, five illustrative examples of the predictive and interpretative character are given. The examples are chosen such that they cover the energy interval that covers a wide range of energies, from the meV for magnetic interaction, to 100 eV for transitions involving core electrons including the 1 eV range for valence transitions.

1. Introduction

Many interesting phenomena in solid state physics and chemistry can be approached by considering the electronic ground state only. However, a substantial number of fascinating solid state phenomena are inherent to excited electronic states. A first and obvious example is the color of a crystal which is the consequence of the absorption of light of certain wave lengths either by the pure crystal or induced by the presence of point defects. Other examples include the electric and magnetic properties of crystals. The conductivity (or the absence of it) of a given material is a consequence of its electronic structure and in particular of the nature of the ground state and the low-lying electronic states. This distinguishes metals form semiconductors and these from insulators. Magnetic properties are also intimately related to excited states. The interplay between magnetic and electrical properties seems to be at the heart of the mechanism of high-T_c superconductivity.

A crystal is a macroscopic object that can be understood as built from a small unit cell that is repeated in all three dimensions. For this reason, a very natural approach is the introduction of translational symmetry by imposing periodic boundary conditions on the unit cell, leading to the band theory of electrons moving in a periodic potential. Density functional theory, based on the Hohenberg-Kohn theorem [1] and the Kohn-Sham equations [2], provides a rigorous way to determine the ground state energy of the N-

electron system. The combination of DFT and band structure theory has been (and still is) widely applied in material science and is certainly very important to understand and interpret many experimental observations. An important advantage of DFT is that many body effects (electron correlation) are incorporated without losing the physically appealing one-electron model. However, DFT also has several well-established shortcomings. Beside the lack of a well-defined hierarchy of the different exchange-correlation functionals (LDA, GGA, hybrid functionals, etc.), there is another fundamental problem with DFT which concerns the description of excited states. In principle, it is only possible to obtain detailed information about the ground state. Recently, Time Dependent DFT opened the possibility to obtain information about the relative energies and oscillator strengths of the different excited states [3-5]. However, even in this case it is not possible to obtain the excited state density. This feature constrains the study of excited state phenomena to the energetics only and the detailed analysis of the character of the excited state is not accessible. In addition, the fact that Kohn-Sham theory maps the density onto a single determinant wave function makes it, in general, not possible to construct proper spin eigenfunctions or to correctly describe the electronic states having substantial multiconfigurational character. This last point is especially important in the study of excited states, where as opposite to the ground state several different electronic configurations contribute to the corresponding electronic wave function.

The well-established computational methods of quantum chemistry, based on a *N*-electron wave function description of the electronic states of interest offers an interesting alternative to DFT to study excited states in material science. These schemes provide a sound hierarchy of methods of increasing accuracy and permit to rigorously define spin eigenfunctions for any spin multiplicity. Furthermore, the complex electronic structure often found for excited states can be assessed by means of a proper multiconfigurational computational scheme. However, these methods are not normally used because of the difficulty to implement periodic symmetry in multiconfigurational wave function based methods. Recently, a wave function based method to incorporate many-body effects in band structure calculations [6] has been reported although this approach is limited to ground state properties only.

Therefore, we leave behind the periodic description of the electronic structure of a crystal and turn our attention to a local approach to describe excited states in solid state compounds. The approach consists in modeling the crystal by a small number of atoms at positions as they occur in the real crystal and to embed this collection of atoms in a potential that accounts for the part of the crystal not explicitly treated. This way of representing the crystal is generally known as the embedded cluster approach and allows us to apply any of the quantum chemical wave function based methods developed for the study of molecules. For the electronic ground state of ionic materials built from closed-shell ions, e.g. NaF or CuCl, it is known that the cluster model and periodic approaches lead to identical results at the Hartree-Fock and DFT levels of theory. The wave functions obtained from a sufficiently large (embedded) cluster calculation are equivalent to so-called Wannier functions, i.e. crystal wave functions that are peaked at the lattice positions. These Wannier functions can be transformed into delocalized Bloch functions by a unitary transformation, hence connecting the embedded cluster approach and the band theory [7]. The situation changes for open-shell systems, both for the electronic ground state, and for excited states.

The dissimilarity between band theory and the cluster approach can be illustrated for the ionization from a Cu-3*d* level in CuCl [7]. As argued above for the neutral, unionized, system the band and cluster approaches are equivalent in the limit of sufficiently large clusters. In the delocalized band description of the ionization process an electron is removed from a Bloch state at the top of the filled band with mainly Cu-3*d* character. Because this state is delocalized over all copper sites, the change in the electronic density at each site is infinitely small and no relaxation effects occur. On the other hand, in the local description an electron is removed from any of the Cu-3*d* Wannier states, which are strongly peaked around the Cu sites. In this localized description a large change in the local electronic density occurs with a concomitant response in the resulting electronic wave function. The electronic relaxation is rather similar to that encountered when an isolated $Cu^{+}(3d^{10})$ ion is ionized, i.e. a large decrease in energy of about 6 eV. Electron delocalization governs the band width; a wide band indicating large delocalization effects and a narrow band highly localized electrons. For systems with large local relaxation effects, compared to those arising from delocalization, the lowest energy is obtained by first accounting for the local relaxation and after that restoring the translational symmetry. This is particularly the case in the study of properties connected with the 3*d* states of the Transition Metal (TM) atoms in TM materials, hence the local approach is the more natural basis for describing these properties.

In this chapter we will discuss the suitability of the embedded cluster model approach to study excited states in material science. The energy range that can be studied is very wide and runs from the meV range for magnetic interactions, via excitation energies of a few eV for optical transitions, to the study of core excitations with typical excitation energies of the order of 100 eV or more. In the remainder of this chapter we will first give a more detailed description of the embedded cluster model approach. A general procedure will be given to cut the cluster from the crystal and how to incorporate the effect of the rest of the crystal. Next, we will discuss various computational wave function based schemes that have proven their ability to accurately describe excited state phenomena. In Section 4, we will discuss the energy range of excited states that can be covered with the methods mentioned in Sections 2 and 3 and review five representative examples of the study of excited states in this energy range. These examples have been chosen in order to illustrate that the combination of cluster model approach and state-of-the-art multiconfigurational wave function based methods provides a computational scheme that does not only accurately reproduce available experimental data but also exhibits predictive and, very important, interpretative power.

2. Material model – the embedded cluster model approach

This section describes the strategy to construct an accurate representation of an ionic crystal from a local point of view. We will illustrate each step with two examples: the modeling of a neutral silver impurity in KCl [8, 9] and the construction of a cluster model to investigate the magnetic coupling between two spin moments localized on the nickel ions in NiO [10, 11].

2.1. Quantum cluster region

The first step in the construction of an adequate embedded cluster model is to divide the physical system into a local and an outer region. The local region is to be treated accurately whereas the outer region ought to provide an appropriate representation of the rest of the crystal. Next, one needs to identify which are the atoms that have to be included in the local region. This region necessarily includes the atoms that play a fundamental role in the property under study. These are then described with high precision quantum chemical computational schemes. The choice of the local region is crucial and will always be a balance between precision and feasibility. Too many atoms in the local region will lead to unnecessarily large calculations, while a too conservative choice will result in an unrealistic description of the physics under study. In the case of an impurity in a host lattice, the smallest local region comprises the silver impurity and its nearest neighbors, which in the first of the examples mentioned above leads to an $AgCl_6$ cluster with octahedral symmetry. For the magnetic coupling problem in our second example at least two magnetic centers (Ni^{2+} ions) and the bridging ligand in between them, i.e. a Ni_2O cluster, have to be included. However, previous cluster model studies have shown that the complete oxygen coordination shell of each nickel must be added to the local region, which gives us a Ni_2O_{11} cluster with D_{2h} symmetry. As a general rule, the local quantum region contains the atoms directly involved in the property under study extended with the first shell of atoms around those centers. However, note that there are many cases (e.g. geometry optimizations) for which such a local region is not sufficient and a larger number of atoms has to be considered explicitly in this region.

2.2 Static short-range interactions

The embedding of the local region of the cluster model must account for three different interactions. In the first place, we consider the short-range electrostatic repulsion between the cluster atoms and those surrounding them. The quantum mechanical nature of this interaction makes that one cannot rely on a simple embedding based on classical models and more elaborate schemes have to be invoked. Several different approaches have been developed over the years, but we will only mention here the Total Ion Potential (TIP) embedding [12, 13] and the more rigorous *Ab Initio* Embedding Model Potential (AIEMP) scheme [14, 15]. Both embedding schemes absorb the interaction of the cluster electrons with other electrons into an effective 1-electron Hamiltonian acting in the electrons of the atoms in the local region only. In other words, the static short range repulsion of the cluster with it surroundings can be accounted for by adding specific 1-electron matrix elements to the standard energy expression of the total cluster energy. Hence both methods provide a computational scheme that accounts for the short range repulsion at a low computational cost.

A very pragmatic way to obtain these matrix elements is to approximate the charge distribution of the ions external to the cluster with effective core potentials, so called Total Ion Potentials (TIP's). This approach has first been applied by Winter, Pitzer and Temple in their study of a Cu^+ impurity in NaF [12]. The sodium ions around the CuF_6 cluster region were modeled by the large core Hay and Wadt pseudopotentials [16]. This potential includes all electrons up to the $3p$ shell, which corresponds exactly to the electronic configuration of the Na^+ ion found in NaF. Although the charge distribution Na $[1s^2...3p^6]$ is of course not exactly the same for the isolated atom as for the Na^+ ion found in ionic lattices, the pseudopotential gives a satisfactorily description of the short-range repulsion between cluster and surrounding. The method is, however, restricted to cations only, since sufficiently large core pseudopotentials cannot be constructed for anions, to say the charge distribution of the O^{2-} ion does not have any resemblance with the charge distribution represented by the Hay and Wadt (or any other) effective core potential for oxygen. In our example of constructing a cluster model for NiO, the coordination of each oxygen atom in the Ni_2O_{11} quantum cluster region is completed with Ni^{2+} TIP's to avoid the artificial delocalization of the oxygens in the cluster. We now have a $Ni_2O_{11}Ni_{28}$ embedded cluster, where the latter 28 Ni^{2+} ions are represented by TIP's [17]. No further extension of the second region is possible since no TIP's can be constructed for O^{2-}.

The *Ab Initio* Embedding Model Potential (AIEMP) approach [14, 15], an extension of the model potential approach developed by Huzinaga and co-workers [18, 19], provides a more complete and more rigorous embedding method. The potentials are derived by optimizing the wave function of each ion in the lattice separately in the field of all other ions [20]. The wave function for each ion is then converted into a model potential by writing the Coulomb interaction with other ions in the lattice as an analytical function and the non-local exchange operator by its spectral representation in the function space used for the ions. These new model potentials are then used in the next iteration to get a better representation of the environment of each ion and in this way better wave functions for each ion. This iterative cycle leads to an optimal model potential representation of the charge distribution of any ion in an ionic lattice. The AIEMP's are material specific and need to be constructed for each crystal separately. Therefore, to extend the cluster of our first example, Ag(0) in KCl, we first need to derive AIEMP's for K^+ and Cl^- in the KCl lattice. Once this has been done, the $AgCl_6$ cluster can be embedded in AIEMP's to account for the short-range repulsion. Since in this case there is no restriction to cations only, as is the case for TIP's, as many layers as desired can be included in the outer region. However, representing all atoms (except those in the quantum cluster region) in a cube of length $2a$ or $4a$ (a being the lattice parameter) centered at the central cluster atom by AIEMP's normally gives converged results as far as the short-range repulsion is concerned.

2.3 Long-range electrostatic interactions

In the present models examples this interaction arises from the electric field generated by all ions in the crystal and is easily incorporated in the material model by calculating the Madelung field of the ions not included in the cluster and adding this potential to the cluster Hamiltonian. To this purpose first the exact Madelung field is calculated by an Ewald summation [21] with formal point charges at the lattice positions. From this total potential the contribution of the cluster ions is subtracted. The use of formal charges is consistent with the assignment of an integer number of electrons to the cluster, while the use of fractional charges — e.g. Mulliken charges, to calculate the Madelung field — introduces some severe conceptual problems. For example, the violation of the overall charge neutrality, taking NiO as an example q_{Ni} is in general not equal to $-q_O$, where q_X is the charge calculated by some suitable population analysis. Furthermore, the number of electrons to be assigned to the cluster is no longer unique.

There are essentially three methods to include the effects of the resulting Madelung field. The first one is the Evjen method [22], which consists in setting the value of the point charges at the lattice positions in the cluster edge and far from the local region according to the restriction of overall charge neutrality. This restriction often results in a fast convergence of the Madelung potential with the number of point charges included in the outer region. The overall charge neutrality is achieved by setting formal charges at all lattice positions except for the charges in the outermost shell. For a cubic lattice half of the formal charge is used for the faces; a quarter for the edges and one eight for the vertices. The method is however not applicable for any type of crystal and does not give accurate representations of the Madelung field except for the simplest lattices like fcc [23]. A more general scheme is fitting a small set of point charges at lattice positions within a certain radius around the cluster to the exact value of the potential at a large number of points in a grid around the cluster local region. Finally, we mention a third approach in which all charges within a sphere of radius r around the center of the cluster are taken [24]. The radius is, however, taken such that the overall charge of all centers within the sphere is zero in order to ensure a good convergence of the resulting potential.

For the two examples discussed in the previous two subsections, all three methods to account for the long-range electrostatic interaction are equally valid, since both systems correspond to fcc lattices.

2.4 Long-range polarization

The response of the crystal to changes in the electronic structure of the quantum cluster region is more difficult to account for than the static interactions discussed in the previous subsection. A real *ab initio* scheme to include this effect has been described by Barandiarán and Seijo [25], but the method demands computational resources that exceeds the present computational facilities.

The polarization E_{pol} induced by a charge on the surrounding lattice can be estimated by means of the classical Born formula [26]:

$$E_{pol} = -(1 - 1/\varepsilon)q^2/2R, \tag{2.1}$$

where ε is the dielectric constant of the material, q is the absolute value of the charge and R is the radius of the spherical cavity in which the charge is distributed. Since a certain degree of ambiguity remains in the definition of R, this correction is only qualitative.

A more refined approach is the so-called shell model [27, 28], which has been used for the study of the ground state of oxygen vacancies (F centers) in MgO [29] and implemented in the ICECAP code [30]. Instead of point charges, the cluster is embedded in polarizable ions represented by a point charge core and a shell connected to it by a spring to simulate its dipole polarizability. In this way the polarization response of the host can be self-consistently taken into account up to infinite distance. However, the shell model is still not sufficiently well developed so as to be widely applicable and at present its use is restricted to a few materials only for which accurate shell model parameters exists. An even more rigorous treatment to account for the long-range polarization has been developed by Pisani and co-workers [31, 32]. This is the perturbed cluster method based on the EMBED computer program [33]. The method relies on the knowledge of the one-electron Green function G^f for the unperturbed host crystal, which is obtained by means of the periodic program CRYSTAL [34, 35]. A cluster C containing the adsorbate is defined with respect to the rest of the host H. The molecular solution for the cluster C in the field of H is corrected self-consistently by exploiting the information contained in G^f in order to allow a proper coupling of the local wave function to that of the outer region.

The long range polarization is, however, only important when charges are created in the quantum cluster region. A classical example is provided by NiO, for an NiO_6 cluster modeling this oxide, Janssen and Nieuwpoort [36] estimated the correction to the ionization energy to be 3–4 eV. However, the charge of the cluster is maintained in all the examples that will be discussed in Section 4 and therefore no action has been undertaken to account for the polarization of the lattice.

2.5 The embedded cluster model for covalent materials

The discussion above is only valid for ionic materials. The embedded cluster model approach can, however, also be applied to covalent materials such as GaAs or SiO_2 although completely different strategies need to be adopted. In Si and SiO_2 the cluster dangling bonds are usually saturated by H atoms [37-41], the geometry of the cluster is usually optimized, especially for cluster models representing point defects, but with the embedding hydrogen atoms fixed in space to provide a simple yet efficient representation of mechanical embedding. The saturation of the cluster broken bonds is an important aspect of the embedding but not the only one. In fact, in this way the crystalline Madelung field is almost completely neglected except for the electrostatic potential built in the quantum mechanical region. While the Madelung field is certainly of less importance in the more covalent materials like bulk silicon or silica, it is crucial in the description of surfaces of solid materials with a more ionic character such as the II-V or III-IV semiconductor families.

3. Methods

Electron correlation effects, defined by reference to a Hartree-Fock description, play an important role in the accurate description of electronic excited states in almost all compounds including solid state materials. Various attempts have been made to divide the electron correlation energy into different contributions. We will follow the scheme which divides the correlation energy in dynamical (mainly atomic) electron correlation, and non-dynamical electron correlation. Dynamical correlation effects arise from the instantaneous electron-electron repulsion. Electrons avoid each other more effectively than predicted by a mean-field approach. Non-dynamical electron effects are caused by the inability to describe the N-electron wave function with a single configuration. Often, non-dynamical correlation effects are related to the inadequate balance of neutral and ionic valence bond components in the Hartree-Fock wave function.

The most straightforward way to improve the mean-field approximation inherent to the Hartree-Fock approach is to extent the N-electron wave function with more determinants and variationally optimize the expansion coefficients. For systems where the Hartree-Fock wave function is a good starting point, the inclusion of determinants connected with single and double excitations usually recovers a large part of the electron correlation effects. However, for more complicated systems a multiconfigurational reference wave function is required and this implies a much faster increase of the size of the computational problem than for a single reference wave function. A simple way to reduce the computational cost is to leave out the configurations that either have a small contribution to the correlation energy or to the N-electron wave function. It is however extremely difficult to find a selection threshold (for the energy contribution or coefficient in the wave function, respectively) that gives a balanced description of the fundamental state and the excited state of interest. In principle, the excitation energy should be stable against a consecutive decrease of the threshold for selection. Nevertheless there are many cases in which this is not the case until very small thresholds are applied, and hence, the computational cost is hardly reduced.

Hence, alternative schemes need to be applied that combine accuracy and efficiency in the description of excited states. The Difference Dedicated CI (DDCI) scheme [42-44] is based on the understanding that many external determinants contribute equally to the correlation energy of the electronic states involved in the process under study. Therefore, a selection is made and only those determinants are included that contribute to the energy difference between the states. First a Complete Active Space Configuration Interaction (CASCI) wave function is constructed that has included for the essential physics of the problem, i.e. accounts for the largest part of the non-dynamical correlation. From this CASCI wave function all single and double replacements are generated to include dynamical correlation effects in the wave function. It has, however, been proven that up to second-order perturbation theory the determinants connected to double replacements of electrons from the inactive orbitals to the virtual space do not contribute to the energy difference between two states and can be left out of the calculation. This class of determinants is the most numerous and hence this selection criterion considerably reduces the computational cost. Note that the selection of determinants is made beforehand and is not based on the energy contribution or the coefficient in the CI expansion. Therefore, the DDCI scheme does not suffer from the instability observed for the selected CI methods. Another advantage with respect to conventional multireference CI methods is that the DDCI approach is (almost) size consistent.

A second computational scheme to account for the electron correlation effects in systems with a prominent multiconfigurational character is the Complete Active Space Second-order Perturbation Theory (CASPT2) [45, 46]. Instead of shortening the CI expansion, this methods includes the complete list of single and double excitations from a CAS Self Consistent Field (CASSCF) reference wave function, but treats the effect of these determinants up to second-order only. Notice that the difference between CASCI and CASSCF lies in the orbital optimization procedure which in the first case are obtained from a previous Hartree-Fock calculation whereas in the second case they are optimized simultaneously to the configuration mixing coefficients. In both cases, CASCI and CASSCF, the non-dynamical electron correlation is basically treated at the CAS level. The CASPT2 procedure efficiently accounts for most of the remaining, mostly dynamical, correlation,.

4. The hierarchy of excited states in Materials Science

It has already been commented that excited states play an important role in many properties of materials and therefore its knowledge and prediction is of crucial importance. Depending on the energy of the radiation used to induce the excitation from the ground state, very different types of excited states are found. In the meV region, the radiation is so weak that only alterations of the spin ordering are induced. There is a large number of materials that show a characteristic magnetic order at low temperatures due to the alignment of local spin moments, which result from the open-shell ions in an ionic crystal. NiO is a prototype of these materials [47]. Increasing the energy of the radiation up to the 1 eV range, may induce transitions in the optical region of the spectra. Taking again NiO as example, the rich optical spectrum is consistent with an interpretation in terms of d-d transitions localized on the Ni^{2+} ions in the lattice. In the same region, but often slightly above, rather well defined complex spectra are found that are interpreted as originated by impurities or defects in the crystal. The neutral Ag impurity in KCl and the spectra of color centers in MgO will be used to illustrate this domain. At even higher up in energy, ~100 eV, interesting spectroscopic features appear. These involve necessarily internal (or core) electrons and can be either connected to core-valence transitions or core-level ionizations. A particularly interesting case is the electronic transition from the nearly atomic metal-2*p* state in MgO, Al_2O_3 and SiO_2 to the lowest unoccupied levels which results in a clear excitonic feature. The hole-particle interaction results in a localized excited state, often called a Frenkel exciton, which appears in the optical band gap. These examples will be discussed in the forthcoming subsections. The CASPT2 calculations discussed in the present chapter have all been performed with the MOLCAS 4 program package [48] whereas the DDCI calculations have been carried out with the CASDI code [49]. In spite of the apparent complexity of these calculations a large number of results have been obtained using work stations, use of amore powerful supercomputer being limited to a few cases.

4.1 The meV range; magnetic coupling

The existence of unpaired electrons in the partly filled $3d$ shells in TM compounds gives rise to magnetic moments which are highly localized on the metal sites. The interaction of these magnetic moments, although being rather small, cannot be explained by a pure dipolar interaction only and additional mechanisms need to be invoked. The work of Anderson and Nesbet in the late fifties elucidated an exchange mechanism where the ligand bridging two metal centers plays a fundamental role in defining the so-called superexchange interaction [50-52].

The experimental determination of the magnetic coupling interaction strength is usually done by fitting magnetic susceptibility data to an analytical expression which contains the parameter J. This analytical expression is based on the phenomenological Heisenberg Hamiltonian that describes spin-spin coupling. For a two-electron system it is easy to show that the interaction can indeed be described by means of such an effective spin Hamiltonian. For this purpose let us recall that in a two electron system one has a singlet and a triplet spin eigenstates. The barycentric average energy of these two spin state is given by $E_{av} = 1/4\ \{E(S{=}0) + 3\ E(S{=}1)\}$ because the singlet and triplet have one and three components respectively. Now, let us define the magnetic coupling parameter J as the energy difference between the triplet and singlet spin states, i.e. $E(S{=}1) - E(S{=}0)$. With these definitions the energy of each of the two spin states can be expressed as

$$E_k = 1/4\ \{E(S{=}0) + 3\ E(S{=}1)\} - 1/2\ k\ \{E(S{=}1) - E(S{=}0)\} = E_{av} - 1/2\ k\,J \qquad (4.1)$$

where k is a parameter such that for $k = 3/2$ the expression gives the energy of the singlet and $k = -1/2$ the energy of the triplet, both with respect to the average energy of the configuration.. Now, with the total spin operator of the two electron system S defined as $S_1 + S_2$, we can write down the double of the product of the individual spin operators as:

$$2S_1{\cdot}S_2 = S^2 - S_1^{\,2} - S_2^{\,2} = S^2 - 3/2. \qquad (4.2)$$

From this equation it can be readily seen that the eigenvalues of $2S_1{\cdot}S_2$ are -3/2 for the singlet and 1/2 for the triplet. These numbers coincide with the negative values of k for which the equation 4.1 results in energies for the singlet and triplet, respectively. Therefore, the energies of the spin states of the two electron case are given by the operator:

$$\mathrm{H} = E_{av} - J\,S_1{\cdot}S_2. \qquad (4.3)$$

Since we are only interested in the relative energies of the different spin states, the first term in the Hamiltonian can be left out. In this way, J parameterizes the magnitude of the spin-spin interaction with the usual understanding that $J > 0$ leads to ferromagnetism (triplet ground state) and $J < 0$ to antiferromagnetism (singlet ground state). This Hamiltonian can be generalized to many-electron systems and to more than two centers yielding the well-known Heisenberg Hamiltonian for the description of spin-spin coupling in molecules and material science:

$$\mathrm{H} = -\sum_{i>j} J_{ij}\, S_i \cdot S_j \qquad (4.4)$$

For a cluster model with just two magnetic centers this Heisenberg Hamiltonian reduces to that of Equation 4.3 and under the assumption of a common orbital part for the different possible spin states that can be constructed, the magnetic coupling parameter J is directly related to differences in the energy values of the full cluster Hamiltonian. In the case of two S=1/2 particles –two hydrogen atoms at large distances or two Cu^{2+} ions– J is obtained as the difference of the total energy of the singlet and the triplet. The mapping of the energy eigenvalues onto the Heisenberg Hamiltonian for two S=1 particles (e.g. two Ni^{2+} ions) leads to $J = E(S) - E(T)$ and $3J = E(S) - E(Q)$, where $E(S)$, $E(T)$ and $E(Q)$ are the energies of the singlet, triplet and quintet respectively. Moreover, this particular case permits to investigate whether the system under consideration behaves according to the Heisenberg Hamiltonian because J can be extracted from two different mappings.

As stated in the Introduction, it is generally not possible to construct proper spin eigenfunctions for the density functional theory based methods and the mapping onto the Heisenberg Hamiltonian can no longer be performed. The same is true for unrestricted Hartree-Fock (UHF) and other spin-polarized computational schemes. The broken symmetry approach of Noodleman and Davidson [53, 54] provides an alternative way to derive magnetic coupling parameters. In this approach one can relates the eigenvalues of the Ising Hamiltonian:

$$\mathrm{H} = -\sum_{i>j} J_{ij}\, \mathrm{S}_{zi} \cdot \mathrm{S}_{zj} \tag{4.5}$$

to the energy eigenvalues of spin symmetry broken solutions and from this mapping extract the magnetic coupling parameter of interest. The Ising Hamiltonian does only consider the z-component of the total spin, and hence only m_s is a good quantum number. For the case of two particles with S=1/2, we now have to relate the energy of the high-spin (HS) state (m_s=±1) and the low-spin (LS) state (m_s=0) to the magnetic coupling parameter J. The eigenvalue of the Ising Hamiltonian of the HS state is $-J$ and that of the low-spin state equals J. Hence, the magnetic coupling parameter J is defined as $1/2\{E(\mathrm{LS}){-}E(\mathrm{HS})\}$. Nevertheless, there seems to be no consensus about the exact details of the mapping of the energy eigenvalues of the Ising Hamiltonian and the energy difference of the spin symmetry broken states. A large amount of articles has been published on this subject and there exist at least three slightly different ways to perform the mapping [55-59]. The reader is referred to these articles for a more profound discussion of the broken symmetry approach.

Since the *ab initio* determination of the magnetic coupling parameter involves very small energy differences, it is important to first establish the validity of the cluster model approach and the computational method for two rather simple, well-defined systems, namely $KNiF_3$ and K_2NiF_4. Both compounds have perovskite-like structure and exhibit antiferromagnetic order with magnetic coupling constants of -95±7 K (-8.2±0.6 meV) and -100±4 K (-8.6±0.3 meV), respectively [60, 61]. The validity of the cluster model will be checked by comparing to the periodic calculations of Ricart *et al.* [62] carried out at the same level of theory. From the UHF energies of the ferromagnetic and antiferromagnetic double unit cell, these authors derived values of –29.8 K for $KNiF_3$ and –31.3 for K_2NiF_4. Embedded cluster model calculations (Ni_2F_{11} embedded in TIP's and point charges for both compounds) give magnetic coupling parameters of –31.4 K and –33.6 K for the two nickel perovskites applying the same approximation for the N-electron wave function as in the periodic calculations and the mapping procedure described above [63]. Although these values only recover about 30% of the experimental figure, it is clear that the cluster results coincide with the values derived from the periodic calculations. The difference is less than 5% and of the order of a few Kelvin only. The close resemblance between periodic and cluster calculations is not unique for the two cases mentioned here, but has also been observed for La_2CuO_4, NiO, Sr_2CuO_3, Ca_2CuO_3, $KCuF_3$, K_2CuF_4, and CuF_2, among others [24, 64-69]. Hence, it can be concluded that the embedded cluster model gives a valid representation of the material to study magnetic coupling mechanisms in ionic solids. The limitation of the quantum cluster region to two magnetic moments carrying metal centers, and the representation of the nearest neighbors of the central Ni_2F_{11} unit by rather simple pseudopotentials do not introduce serious artifacts in the description of the magnetic coupling.

Next, we study the performance of the two computational schemes mentioned in Section 3, CASPT2 and DDCI. The reference wave function for both methods is constructed as a Complete Active Space Configuration Interaction (CASCI) expansion using a common set of Molecular Orbitals (MO) in the case of DDCI or individually optimized MO's for CASPT2. This CASCI contains all Slater determinants that can be built by distributing the 4 unpaired electrons over the active orbitals in all possible ways. The active orbitals are those corresponding to the partially filled $3d$ shell in each Ni^{2+} ion, i.e. the $3d(x^2\text{-}y^2)$ and $3d(z^2)$ orbitals defining the e_g manifold. The CASCI description, which essentially contains the terms included in the so-called Anderson model of superexchange, predicts the compounds to be antiferromagnetic (see Table 1, numbers refer to results obtained using the Restricted Open-shell Hartree-Fock (ROHF) triplet MO's), although the absolute value is a factor of ~4 too small. In spite of the disagreement with experiment it is remarkable that the ratio $J(KNiF_3)$ / $J(K_2NiF_4)$ is very similar to the experimental one. CASCI gives too small a J-value because of the lack of dynamical electron correlation external to the CAS in the N-electron wave function. The inclusion of these effects, either by DDCI or CASPT2, largely improves the calculated magnetic coupling constant. For both compounds, we now obtain results that are almost within the experimental error bars. Notice that all levels of descriptions give a slightly larger J for K_2NiF_4 than for $KNiF_3$ in agreement with the experimental observations.

In summary, the computation of the magnetic coupling constants of the prototype materials $KNiF_3$ and K_2NiF_4 applying a combination of the embedded cluster model approach and modern quantum chemical schemes (DDCI or CASPT2) serves as an illustration that such a theoretical approach is able to reproduce very accurately the experimental data. Hence, it is expected that the same approach applied to materials for experimental data is either scarce, contradictory or absent can provide interesting predictions as will be illustrated in the next paragraphs.

The discovery in the early 1990s of the $Sr_{n-1}Cu_nO_{2n-1}$ (with $n \geq 2$) series by Hiroi *et al.* [70] opened a new direction in the copper oxide chemistry and physics. The compounds in this series are built from n one-dimensional chains in which the copper ions in adjacent chains are coupled by oxygen centers to form so-called n-leg spin ladders. The magnetic interaction between the ladders is rather small since ladders are connected to each other by a Cu - O - Cu bond of ~90°. In principle the spin ladders interpolate between the one-dimensional and the two-dimensional case: two interacting chains for n=2, while for very

Table 1. Magnetic coupling constants (in Kelvin) for $KNiF_3$ and K_2NiF_4 obtained at different levels of theory using an embedded Ni_2F_{11} cluster model.

Method	$KNiF_3$	K_2NiF_4
CASCI	-20.5	-22.8
DDCI	-86	-94
CASPT2	-78	-84
Experimental value	-95±7	-100±4

large n-values a two-dimensional CuO_2 plane appears. However, intensive theoretical and experimental investigations showed that the progression from one- to two-dimensions is far from smooth [71]. Even-leg ladders show a spin gap, neither observed in the copper oxide chains nor in two-dimensional antiferromagnets. On the other hand the odd-leg ladders do not possess a spin gap and behave as effective one-dimensional chains. The appearance of a spin-gap in the even-leg ladders gives rise to a finite spin-spin correlation length as $T \to 0$, whereas the spin-spin correlation function of the odd-leg ladder is similar to the one of the single chain. The spin ladders also attracted much attention because of the appearance of superconductivity upon doping the ladders with holes. Comprehensive reviews of all interesting phenomena of the spin ladders are given by Dagotto and Rice [71], Rice [72], and Maekawa [73].

In principle, there are now two different magnetic coupling parameters, one for the interaction along the legs, J_{leg}, and one along the rungs, J_{rung}, of the ladder. Although both legs and rungs are built from similar linear Cu – O – Cu bonds, the possibility cannot be dismissed that J_{leg} differs from J_{rung}. Given the similarity in Cu–O bond distances along legs and rungs, one would certainly expect a nearly isotropic situation ($J_{rung}/J_{leg} = 1$). However, both neutron scattering experiments [74] and magnetic susceptibility measurements [75] have been interpreted with a strongly anisotropic coupling and suggested $J_{rung}/J_{leg} = 0.5$. This observation has been confirmed by ^{63}Cu and ^{17}O NMR studies [76]. A possible origin for this anisotropy might be the different Cu – Cu distances in $SrCu_2O_3$ (3.934 Å for the leg and 3.858 Å for the rung). Note, however, that the shorter distance, and hence presumably larger magnetic interaction, is along the rung. This contradicts the experimental determination $J_{rung}/J_{leg} \approx 0.5$. Another possibility for the origin of this anisotropy lies in the differences in the Madelung potential for the oxygen centers mediating the superexchange interaction. In $SrCu_2O_3$, there is a difference of nearly 1 eV, the Madelung potential is 22.44 eV for the leg and 21.35 eV for the rung, assuming formal charges of +2, +2 and –2 for Sr, Cu and O, respectively. However, the magnitudes of the oxygen Madelung potentials imply that the O-$2p$ $\to$ Cu-$3d$ charge transfer energy is smaller along the rung. This suggest a larger J along the rung and $J_{rung}/J_{leg} > 1$, again in contradiction with the interpretation of experiment.

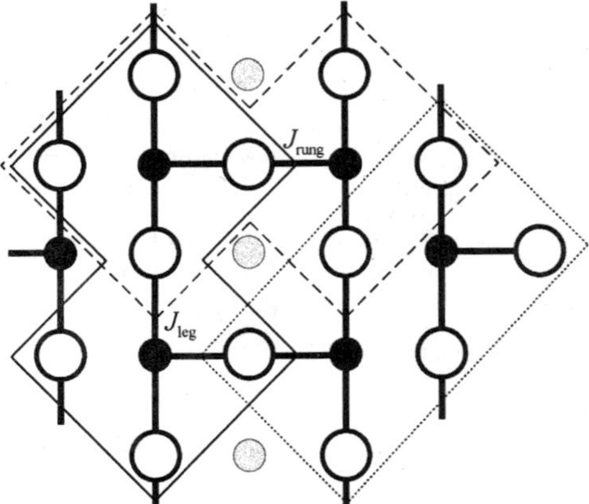

Figure 1. Schematic view of the structure of the ladder compound $SrCu_2O_3$. Black circles represent Cu; open circles O; and gray circles Sr, which are situated above and below the Cu_2O_3 plane. The centers inside the area enclosed by the thin solid line form the Cu_2O_7 quantum cluster region for the leg and those inside the area enclosed by the thin dashed line the cluster for the rung. The thin dotted line encloses the quantum cluster area for the inter ladder interaction. The thick lines represent the strongly antiferromagnetic Cu–O–Cu bonds, from which emerges the ladder structure.

Thus two unexpected results are found in for these ladders. The first is the large anisotropy, and the second is that the ratio is opposite to what we would expect. Here, we present the results of an *ab initio* quantum chemical study to intend to clarify these points. The spin ladder $SrCu_2O_3$ is modeled by two different clusters, as illustrated in Figure 1. One Cu_2O_7 cluster with local D_{2h} symmetry is used to extract a J-value for the rung and another Cu_2O_7 cluster with local C_{2v} symmetry models the interaction along the leg. We also use a Cu_2O_6 cluster to calculate the inter-ladder magnetic interaction. All these clusters are embedded in an electrostatic background represented by optimized point charges that reproduce the Madelung potential in the whole cluster region with an accuracy better than 1 meV. To avoid the artificial polarization of the electrons of the cluster towards the point charges, we also include the short-range electrostatic repulsion between the cluster atoms and their near neighbors by representing the ions in the direct environment of the cluster with Total Ion Potentials (TIP's) as explained in more detail in Section 2.

Table 2 gives the results of the CASCI, DDCI2 and DDCI calculations for leg and rung. DDCI2 considers a subset of the determinants of the full DDCI list, it includes only those determinants that up to second-order perturbation theory contribute to the singlet-triplet difference [77]. It is significantly cheaper than DDCI, but usually does not give as good results as DDCI does. The numbers in Table 2 clearly show that the ratio between J_{rung} and J_{leg} does not show a very strong dependence of the computational scheme applied; all of them indicate a slightly smaller magnetic interaction for the rung than for the leg. However, as known from previous studies, the absolute values of the calculated J's differ strongly between one method and the other. We observe a relative small antiferromagnetic interaction in the CASCI calculations, which is greatly enhanced by the inclusion of external electron correlation (DDCI2 or DDCI). The fact that DDCI2 cause an enlargement of J by a factor of three has been observed before in many other compounds, e.g. La_2CuO_4 and NiO [10, 78], but the doubling of J by adding the determinants connected to a relaxation of the charge transfer excitations to the wave function (DDCI) is significantly larger than has been observed for La_2CuO_4 [79], for which an increase of about 40% has been observed. This indicates that covalent interactions are relatively more important in $SrCu_2O_3$, which is also expressed in the values of J obtained with DDCI, which are about 25% larger than in La_2CuO_4. Our final and most reliable *ab initio* estimates for J_{rung} and J_{leg} are 139.3 meV and 154.8 meV, and hence, the ratio of both parameters equals 0.9.

Finally, we consider a different type of magnetic interaction in $SrCu_2O_3$ arising from the interaction of the spin moments on Cu-ions located on the legs of different ladders. This interaction is assumed to be very weak because it involves an interaction in which the two O-2*p* orbitals which participate are strongly orthogonal to each other. The explicit calculation of the magnitude of this interaction can be carried out by means of a Cu_2O_6 cluster and indeed yields a very small and ferromagnetic interaction of 0.95 meV at the CASCI level. However, the inclusion of the external electron correlation effects through

Table 2. *Ab initio* estimates of the magnetic coupling parameter (in meV) for leg, rung and inter ladder in $SrCu_2O_3$.

Method	J_{rung} / J_{leg}	J_{rung}	J_{leg}	J_{inter}
CASCI	0.91	-23.9	-21.7	1.0
DDCI2	0.91	-83.8	-76.1	
DDCI	0.89	-155.8	-139.3	12.5

DDCI increases the calculated magnitude of J_{inter} up to 12.5 meV (see Table 2). The inter ladder interaction is now ~10% of the magnetic interactions along the legs and rungs. It may need to be considered in the fitting of experimental data and the assumption of isolated ladders in $SrCu_2O_3$ reconsidered. Recent publications [80-82], posterior to the *ab initio* study mentioned here [83], indicate that the ratio between the coupling constants of rung and leg is indeed closer to 1 than to 0.5.

4.2 The eV range; d-d excitations in Transition Metal Oxides

According to the simplest textbook definition, insulators and semiconductors are characterized by a region of forbidden energy separating the occupied and unoccupied states [47]. This energy gap is evident in the optical spectra of these materials and can be rationalized in terms of the band theory of solids. In some cases, however, the optical spectra of these materials is not so simple and well-defined spectroscopic features appear in the forbidden energy region. These features are evidenced by the onset of optical absorption at energies below the interband continuum threshold. The origin of this particular spectroscopic feature is interpreted as due to the formation of Frenkel excitons. The partly filled 3*d*-shell in TM oxides is highly localized on the metal ions and the different $3d^n$ states appear as well-defined peaks in the optical absorption spectra in the band gap. In this section, we first discuss the performance of the embedded cluster model approach for the prototype material NiO.

To establish the validity of the cluster model approach and to access the different contributions to the excitation energies of the d-d transitions, a Ni^{2+} ion in NiO bulk is embedded in different environments. The first material model is a Ni^{2+} ion in vacuum. In the second step, the atomic energy levels are perturbed by point charges that represent the Madelung field of bulk NiO. Next, the material model is improved by replacing the point charges directly surrounding the central Ni^{2+} ion with frozen oxygen charge distributions. The frozen charge distributions are obtained from a Hartree-Fock calculation on an $[O_6]^{12-}$ cluster embedded in the point charges plus a formal +2 charge replacing the Ni^{2+} ion. Next, we consider a more common quantum cluster region, NiO_6 embedded in point charges. Finally, a material model is constructed that accounts for the short-range repulsion as explained in Section 2. The NiO_6 unit is embedded in Ni^{2+} and O^{2-} AIEMP's optimized for NiO [84]. In the first three models, the reference wave function is a CASSCF wave function constructed in an active space that contains 10 electrons and 10 orbitals, namely the Ni-3*d* orbitals and a correlating set of virtuals, the so-called $3d'$ orbitals. For the remaining models the active space is extended with two orbitals with strong O-2*p* character to allow for an accurate treatment of the ligand to metal charge transfer contributions. CASPT2 estimates the remaining correlation effects of the Ni-3*s*, 3*p* and 3*d* electrons, and, when appropriate, also of the O-2*s* and 2*p* electrons.

The atomic energy levels for a Ni^{2+} ion in vacuum agree satisfactorily with the experimental values [85]. A small improvement can be obtained by adding the Ni-3*s* and 3*p* orbitals to the active space [86], but this is beyond our scope. As can be seen in Figure 2, these atomic energy levels are split in the next model, which is a Ni^{2+} ion in the field of point charges. In the figure the highest level, ^1S or b^1A_{1g}, is not included, because of the high excitation energy. The lowest excitation energy, which corresponds to $10D_q$ in the ligand field parameterization scheme, is only 0.40 eV. Obviously a point charge or crystal field description of the surroundings yields an inadequate picture of the energy splittings in the real material.

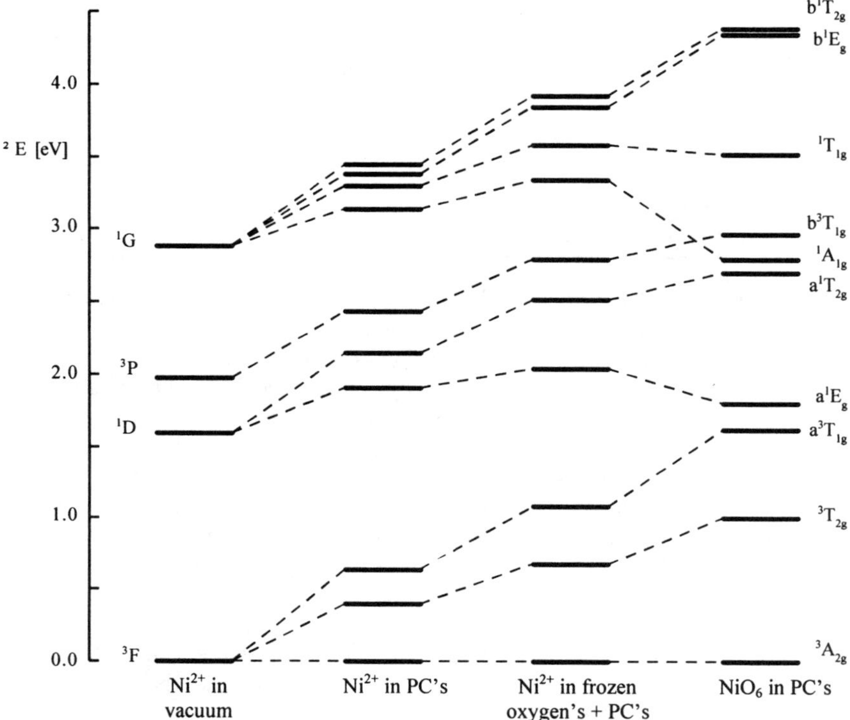

Figure 2. Excitation energies of the d-d transitions in bulk NiO obtained with CASPT2 within different representations of the real material

The improved material model —a Ni^{2+} ion embedded in frozen oxygen charge densities— causes a further splitting of the atomic levels, without changing the relative order of the states. There is a clear separation between the levels that arise from different atomic states, which indicates that the frozen oxygen charge distributions in the present material model supply a weak field on the Ni^{2+} center. Although $10D_q$ has increased to 0.67 eV, still a rather large difference remains with the experimental value of ~ 1.1 eV [87]. The main part of this difference must be ascribed to the lack of covalency effects in the wave function, because most of the electron correlation on nickel is included and the effect of correlating the oxygen electrons is expected to be much smaller than the remaining discrepancy. Figure 2 also includes a comparison with the result obtained from a NiO_6 quantum cluster region, which compared to the previous Ni^{2+} ion embedded in frozen oxygen charge distributions and point charges introduces the possibility for covalent interactions between Ni and O. Obviously the influence of these covalency effects is quite large; for almost all states the excitation energy changes significantly and the separation between energies of states arising from different free-ion states disappears.

It is improbable that the remaining, small discrepancy of $10D_q$ with experiment is due to an incorrect or incomplete treatment of the electron correlation effects, since other computational schemes applied to the NiO_6 cluster embedded in point charges yield very similar results. We mention here the First Order CI calculations of Janssen and Nieuwpoort [88], the DDCI calculation of Lorda, Illas and Bagus [89], and the MC-CEPA calculations of Staemmler and co-workers [90]. Therefore, it is an interesting exercise to compute $10D_q$ going beyond the simple point charge description of the quantum cluster region surroundings.

Table 3. 10*Dq* (in eV) in bulk NiO within different materials models

Material model	CASSCF	CASPT2
NiO_6	0.91	1.01
+ 18 Ni^{MP}	1.16	1.28
+ 18 Ni^{MP} (1*s*,1*p*)	1.06	1.17
+ 18 frozen Mg^{2+}	1.04	1.15
+ 18 Ni^{MP} (1*s*,1*p*) + 8 O^{MP} (1*s*,1*p*)	1.06	1.17

Table 3 shows how the ligand field splitting depends on details of the embedding in NiO. The basic NiO_6 cluster embedded in point charges leads to a value for $10D_q$ of 1.0 eV. However, the inclusion of the Pauli repulsion of the cluster atoms with the next shell of ions raises $10D_q$ by a small, although not negligible, amount. The representation of the 18 cations around the basic cluster by AIEMP's (1*s*, 1*p*) or by a frozen charge distribution both result in a ligand field splitting of ~1.15 eV. We observe that the bare potentials overestimate the effect of the Pauli repulsion and that $10D_q$ is insensitive to the precise details of the representation of the next shell of oxygen ions. Furthermore, Table 3 indicates that the embedding only affects the CASSCF value. Independent of the embedding scheme, CASPT2 raises $10D_q$ by 0.1 eV. The present theoretical estimate of the ligand field splitting for NiO is in good agreement with the value of 1.13 eV deduced from experiments.

The increase of $10D_q$ is accompanied by an increase of the covalent character of the Ni–O bond. In terms of natural orbitals this increase does not occur through an increase of the coefficients of the configurations connected with charge transfer excitations in the CASSCF wave function, but through an increase of the Ni character in the bonding e_g orbital with mainly O-2*p* character. A semiquantitative measure for this effect is given by the Mulliken gross population of the active orbitals. The total $d(e_g)$-population —defined as the Mulliken Ni-3*d* gross population times the natural occupation number of the orbital, summed over the e_g orbitals— for the ground state increase from 2.09 for the basic cluster embedded in point charges to 2.17 for the embedding by AIEMP's (1*s*,1*p*) or by a frozen charge distribution. For the bare AIEMP's, the $d(e_g)$-population was determined as 2.27. Note that the absolute values of these *d*-populations cannot be related to total charges on the Ni-ion, since the Mulliken populations are sensitive to the details of the basis sets [91]. Nevertheless, the values confirm the trend of an increasing importance of the covalent interactions when the description of the cluster surroundings is improved.

4.3 The eV range; Spectroscopy of impurities and defects – KCl:Ag^0

The embedded cluster model approach is an ideal tool to study the wide variety of defects and impurities that can be formed inside insulating lattices like alkali halides or earth alkali oxides. One important class of impurities are originated by the addition of metal halides as AgX, CuX, TlX to an alkali halide. The positively charged metals (Ag^+, Cu^+, etc.) can occupy alkali cation positions with a typical concentration of one metal cation over 200-10000 alkali ions. Thereafter, other defects can be created by X-irradiation which causes the formation of free positive and negative charge carriers. The irradiated electrons can be trapped either in halide vacancies to form F and F' centers, or creating neutral metal centers in cationic positions. On the other hand, the holes can be trapped by the anions originating the so called V_K centers, and also by the cations giving paramagnetic centers (Ag^{2+}, Cu^{2+}) [92].

The formation of neutral atoms like Ag^0, Hg^0, Cu^0, Tl^0 as impurity occupying cationic positions in ionic lattices is less common, but has been proved to exist through Electronic Paramagnetic Resonance (EPR) and Electron-Nuclear Double Resonance (ENDOR) [93]. In the case of Ag^0 in KCl, the hyperfine constant is only about 5% smaller than the corresponding free silver atom value, showing that the electronic configuration of silver is very close to $4d^{10}5s^1$. Nevertheless, the system cannot be described in terms of a free Ag atom. For instance the optical absorption spectrum of KCl:Ag^0 shows the presence of an intense band at 6.3 eV while it appears at 5.7 eV for KBr:Ag^0 [94]. Based on this shift, this band was tentatively assigned to a $Cl^- \rightarrow Ag^0$ charge transfer (CT) transition [95]. It must however be stated that the appearance of a CT transition at such low energies might be surprising because the Ag impurity changes from a neutral state to a negatively charged state, whereas a positively charged oxidation state is more common for silver. Moreover, the excitation energy is in contradiction with the empirical law for CT transition energies for divalent impurities of Simonetti and McClure [96], which states E_{CT} (eV) = $C - I(M^{n-1})$, where the empirical parameter C = 22.4 eV and $I(M^{n-1})$ is the ionization energy of the free M^{n-1} ion. Assuming that this law can also be applied for $n \neq 2$, the CT energy for Ag^0 impurities in KCl would be expected to appear around 20 eV, since $I(Ag^-)$ = 1.2 eV.

For an accurate description of the spectroscopy of this neutral silver impurity, a first requirement is the determination of the geometry distortion around the impurity. However, the equilibrium distance (R_e) between Ag^0 and its nearest neighbors (Cl^- ions in the present case) is quite difficult to measure. Basically, two reasons can be given for this difficulty. In the first place, the concentration of the impurities is usually quite low and in the detection limit of the Extended X-Ray Absorption Fine Structure (EXAFS) technique [97]. Moreover, the coexistence of various oxidation states of the silver atom –usually the impurity is accompanied by Ag^+ impurities– complicates the interpretation of such spectrum, since the spectra of the two species would appear around the same X-ray absorption energy.

Therefore, theoretical methods, and more specifically, the embedded cluster model approach can give valuable information about the geometry distortion in this type of impurities. The general considerations given in Section 2 to construct an accurate representation of the material lead us to an $AgCl_6$ cluster embedded in AIEMP's (all remaining atoms in a cube running from $-a$ to a, being a the lattice parameter of the pure KCl crystal). Following the conclusions of Ref. [98] the six AIEMP's representing the K^+ ions at ($1a$, 0, 0) are provided with a ($1s$, $1p$) basis set to maintain the strong-orthogonality between cluster and surroundings. In addition, the long-range electrostatic interaction is incorporated by an array of Evjen point charges to complete a cube of $4a$ around the central atom. The result of the geometry optimization with an ROHF wave function indicates an elongation of the bond length of approximately 9% in comparison to the pure crystal. The incorporation of the electron correlation effects by second-order perturbation theory gives a slightly smaller elongation, but the difference with ROHF is small enough to continue the study of the geometry relaxation at the Hartree-Fock level only.

For such a large relaxation it is expected that the simple $AgCl_6$ cluster is not large enough and that the presence of the impurity is also noted by other ions. Therefore, we extend the quantum cluster region with the 6 K^+ ions at ($1a$, 0, 0), for which the largest effect can be expected. The number of ions represented by AIEMP's is also augmented in order to provide a correct representation of the surroundings for the potassium ions added to the quantum cluster region. The ROHF geometry optimization of this larger $AgCl_6K_6$ embedded cluster results in an outward relaxation of R_e as large as 18%.

Although this large outward relaxation cannot be confirmed directly with experimental data, there is some indirect information that give support to the ROHF geometry

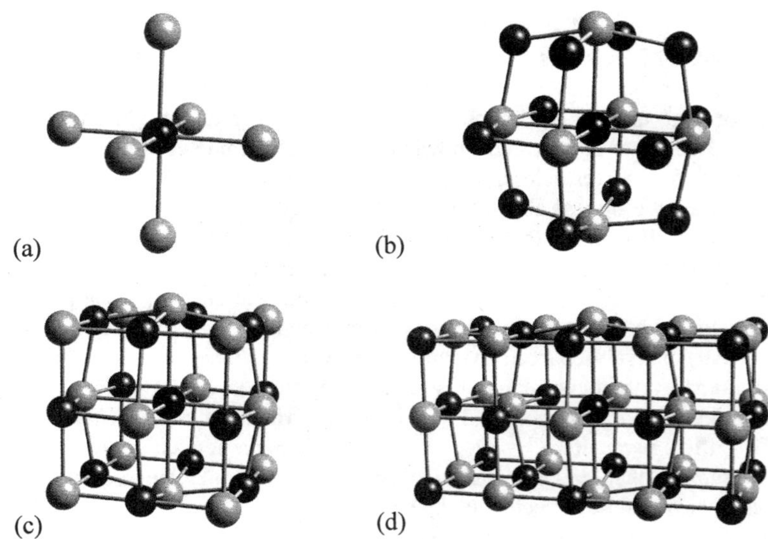

Figure 3. Quantum cluster regions of Ag^0 impurity in KCl at the optimized geometry. The black sphere represent the Ag0 impurity, the dark gray spheres refer to K ions, and light gray spheres to Cl ions. From a to d the following clusters are shown: (a) $AgCl_6$, (b) $AgCl_6K_{12}$, (c) $AgCl_6K_{12}Cl_8$, (d) $AgCl_6K_{12}Cl_8K_2Cl_8K_8$

optimization result. In the first place, the unpaired spin density on the silver atom (f_{Ag}), which is directly related to the hyperfine constant A, is about 0.85 and has its maximum at R_e. This small reduction of f_{Ag} compared to the free silver atom is in agreement with the experimental A(KCl:Ag^0) [93], which is rather similar to A(Ag). Another indirect confirmation of the large relaxation is given by the vibrational frequency of the A_{1g} mode, where only the displacement of the six Cl^- ligands is considered. This frequency is calculated as 130 cm^{-1}. This number is supported by the temperature dependence of A, which implies a phonon frequency of ~110 cm^{-1} [93].

The spectroscopy of the KCl:Ag^0 system is studied in a series of four cluster models for which the quantum cluster region is progressively increased. The following quantum cluster regions are considered [see Figures 3(a)-3(d)]: $AgCl_6$, $AgCl_6K_{12}$, $AgCl_6K_{12}Cl_8$, and $AgCl_6K_{12}Cl_8K_2Cl_8K_8$. These quantum cluster regions are embedded in AIEMP's. The atoms contained in a cube of length $2a$ centered in the impurity plus the atoms located at the (3/2,0,0), (3/2,1/2,0), and (3/2,1/2,1/2) unique positions are described with AIEMP's optimized for the pure KCl crystal. The long-range electrostatic interaction is included in the model by an array of Evjen charges to complete a cube of $8a$ around the Ag^0 impurity.

Table 4 summarizes the excitation energies of the different bands obtained with the four clusters mentioned before. For the largest cluster only CASSCF energies are listed because the CASPT2 is computationally too demanding. In order to analyze the experimental spectrum of the Ag^0 impurity in KCl, we discuss each transition separately. In the first place, we observe that the excitation energy of the $^2T_{1u}$ continuously decreases with increasing cluster size. Moreover the final state of the $^2A_{1g} \rightarrow {}^2T_{1u}$ transition is as delocalized as the cluster allows, and hence, to have a correct description of this state much larger cluster models are needed. Nevertheless, our calculations confirm the delocalized character of this state with mainly K-4s and 4p character, i.e. a transition into the conduction band, and cannot be considered as an atomic-like $5s \rightarrow 5p$ transition. The second and third band in the spectrum arise at 4.11 and 4.73 eV, respectively and have been assigned to a localized transition on the Ag^0 impurity, involving the Ag-4d and Ag-5s

Table 4. CASSCF and CASPT2 excitation energies (in eV) relative to the $^2A_{1g}$ ground state for different quantum cluster regions

State	$AgCl_6$	$AgCl_6K_{12}$	$AgCl_6K_{12}Cl_8$	$AgCl_6K_{12}Cl_8K_2Cl_8K_8$	Exp.
$^2T_{1u}$	3.77 / 3.82	3.61 / 3.64	3.52 / 3.54	3.44 / -	2.92
^2E_g	4.72 / 4.22	4.75 / 4.18	4.74 / 4.15	4.74 / -	4.11
$^2T_{2g}$	4.78 / 4.29	4.80 / 4.25	4.79 / 4.20	4.80 / -	4.73
b^2A_{1g}	13.47 / 13.48	7.58 / 7.52	5.95 / 5.75	5.59 / -	5.35
$^2T_{1u}$ (CT)	12.37 / 10.71	8.13 / 6.46	8.12 / 6.38	8.13 / -	6.29

orbitals [95]. The numbers in Table 4 show that the calculated transition energies of the ^2E_g and $^2T_{2g}$ states are completely independent of the cluster size and hence corroborate the localized character of the transitions. Mulliken Population Analysis (MPA) shows that the excited state open shell orbitals has a very strong Ag-4*d* character. Since spin-orbit effects are not considered in the calculations, the splitting between the ^2E_g and $^2T_{2g}$ states is much smaller than the experimentally observed splitting of 0.62 eV. The calculated transition energy of the $a^2A_{1g} \rightarrow b^2A_{1g}$ transition, which in an isolated silver atom corresponds to a 5*s* → 6*s* transition, rapidly decreases as the size of the cluster increases and seems to converge to the experimental transition energy of 5.35 eV of band IV in the spectrum. MPA shows that the open shell orbital in the excited state extends all over the cluster, and hence, belongs to the conduction band of the crystal. Therefore, it is not appropriate to relate this orbital to a 6*s*-like atomic orbital.

Finally, we discuss in some detail the last peak of the spectrum, assigned to a local charge transfer (CT) transition from the Cl-3*p* orbital to an a_{1g} orbital with strong Ag-5*s* character [95, 99]. In all cluster models, the open shell of the $^2T_{1u}$ excited state has a strong Cl-3*p* character and the $^2A_{1g} \rightarrow {}^2T_{1u}$ transition can indeed be ascribed to a local charge transfer transition. The excitation energy for the smallest cluster is very high, larger than 10 eV. Obviously, such a small cluster cannot account for an excitation in which a hole is formed on the edge of the cluster. The ions in the direct cluster environment are described with model potentials which do not allow these ions to respond to the changes in the charge distribution between Ag and Cl. This artifact seems to be largely removed in the next cluster which includes the twelve K ions surrounding the Cl ions, the excitation energy experiences an important lowering of about 4 eV. The fact that the excitation energy does not change anymore for the larger clusters considered shows that the polarization of the lattice in response to the CT transition is not as long ranged as for ionization processes. Two reasons can be given to rationalize this observation. In the first place, the transfer of an electron from the ligand-*p* shell to the Ag-5*s* orbital is accompanied by a redistribution of the density in the closed shell orbitals. This causes the hole created on the ligands to be effectively screened and therefore the net transfer of charge is smaller than one. For the transition studied here, the net transfer estimated by MPA is ~0.5 electrons. The second reason for the rather short ranged character of the lattice relaxation can be found in the fact that electron reorganization occurs within a local region. This makes that the atoms located further away from the impurity do not experience large changes in the Coulomb potential due to a local CT transition.

Note that the effect of CASPT2 is much larger than for any of the other states considered. The large stabilization of the CT state by CASPT2 has been observed before by Geleijns *et al.* [100] in the study of the local charge transfer processes in NiO. Two effects contribute to the differential electron correlation. The first one is caused by the different Ag-*s* and Cl-*p* occupations in the ground and excited state, which tends to stabilize the state with higher *s* occupation, as has been shown by Sousa *et al.* in the study of the copper halide molecules [101]. The second effect is the recuperation of the orbital relaxation effects [36, 102, 103]. In the CASSCF wave function of the $^2T_{1u}$(CT) state, an electron is removed from a molecular orbital that is a linear combination of two Cl atomic orbitals. This implies that only part of the orbital relaxation due to the removal of an electron is included in this wave function. CASPT2 has the complete orbital relaxation included up to first order and this causes a further stabilization of the charge transfer state.

The calculated excitation energy for the CT state is converged with respect to the cluster size. We observe that this transition is genuinely local and can be accurately described with a rather small cluster. The excitation energy is computed to be 6.4 eV which compares very well with the experimental transition energy of band VI, 6.3 eV. Spin-orbit effects are relatively unimportant as the open shell orbital in the excited state is highly localized on the Cl atom. The atomic spin-orbit splitting for Cl is only 0.1 eV, and hence, it is not expected that the $^2T_{1u}$(CT) state is significantly affected by spin-orbit coupling effects.

The dependence of the CT energy on the Ag–Cl distance gives us an additional way to approximately estimate the relaxation of the geometry around the impurity. Therefore, this transition has been calculated at four different distances in the $AgCl_6K_{12}Cl_8$ cluster. Beside R_e (3.707 Å), we also apply 3.146 Å, which corresponds to the K–Cl distance in the pure crystal, and 3.427 Å, which is the minimum of the potential-energy surface varying only the Ag–Cl distance. Furthermore, we calculate the CT energy at 3.800 Å, being a slightly larger distance than R_e. From short to larger distances the resulting CASPT2 excitation energies are 7.40, 7.25, 6.38, and 6.10 eV. This diminution of the transition energy with increasing distance has been rationalized by a more pronounced destabilization of the one-electron levels associated with the metal than those with the ligand when the metal-ligand distance decreases [104, 105]. The results show that the CT energy only coincides with the experimental value when the outward relaxation of the Cl ions surrounding the silver impurity is of the order of 20%, in excellent agreement with the ROHF geometry optimization.

4.4 The eV range; Spectroscopy of impurities and defects – F centers in MgO

The oxygen vacancies in solid MgO give rise to interesting features in the optical spectra. The so-called F centers are characterized by the trapped electrons in the cavity left by removing the anion. The removal of an O^- ion or a neutral O atom results in F^+ or F centers with one or two electrons trapped, respectively. It is widely accepted that these electrons are localized at the center of the vacancy [106] and therefore, the combination of cluster model approach and highly accurate *ab initio* methodologies is a reliable technique to study the spectroscopy of these vacancies. It is well-known that both F and F^+ centers give rise to an intense absorption band around 5 eV, which is composed of two bands, one at 4.96 eV due to F^+ centers and one due to neutral F centers at 5.0 eV [107, 108]. In a previous work, Illas and Pacchioni studied the F centers in bulk and surface MgO by means of cluster models and MRCI wave functions [106]. The computed excitation energies for bulk F and F^+ centers were 6.0 and 5.8 eV respectively, with an error of 0.8-1 eV with respect to the experiment, due to limitations in the level of calculation, mainly the basis sets size. Applying the same error estimation, these authors could assess the surface optical transitions at 2.2-2.5 eV for F and F^+ defects. The interpretation of the optical measurements of MgO surface is much less clear. The bands assigned in the literature to surface F centers vary from 2-2.3 eV to 1.15 eV while other bands remain unassigned (see Ref. [109] and references therein).

Here, we present the results of the optical transitions associated to F centers formed at low-coordinated sites of the MgO (100) surface. As these excitations will appear at lower energy than those in the regular surface we will check whether they can be responsible of some of the unassigned features of the spectrum.

To model a regular surface, a step and a corner, we use the following cluster models: a Vac $(Mg^{2+})_5$ $(O^{2-})_{12}$ $(Mg^{2+})_{12}$ $(TIP)_{12}$ cluster embedded in an array of 634 point charges for the surface, a Vac $(Mg^{2+})_4$ $(O^{2-})_{10}$ $(Mg^{2+})_{20}$ cluster embedded in an array of 563 point charges for the step, and a Vac $(Mg^{2+})_3$ $(O^{2-})_6$ $(Mg^{2+})_{10}$ cluster embedded in 323 point charges for the corner, where Vac refers to the vacancy left by an O anion and the TIP's model the 12 outermost Mg^{2+} cations. All atoms are treated as effective core potentials (ECP) that include the 1*s*, 2*s* and 2*p* electrons for the Mg cations and the 1*s* core orbital for the O anions. The basis sets used are the following: an optimized (3*s*, 2*p*, 1*d*) uncontracted basis set for the vacancy, a contracted [4*s*, 1*p*] / (2*s*, 1*p*) basis for the next Mg^{2+} neighbors, a [4*s*, 4*p*] / (2*s*, 2*p*) for O^{2-} and a [4*s*] / (1*s*) contraction for the outermost Mg^{2+} ions (for more details concerning basis set and cluster models see Ref. [109]). Since the creation of a vacancy can induce non-negligible geometrical relaxation, the positions of the Mg nearest neighbors and the O next nearest neighbors of the vacancy are optimized at the Hartree-Fock level for the ground state of the F and F^+ centers for each cluster. The rest of the atoms and point charges are fixed at the ideal crystal position. DDCI calculations have been performed for the lowest excited states for both F and F^+ centers in the three clusters. The model space is a CAS which involves two orbitals and either two or one electron for F and F^+ centers, respectively. To construct the CAS a set of molecular orbitals has to be chosen, however excitation energies computed by DDCI are not very dependent on the molecular orbitals used [109].

In Table 5 we summarize the results of the DDCI calculations on the lowest excitation energies for terrace, step and corner F and F^+ centers. As expected, for a given transition the excitation energy decreases as the coordination of the defect decreases. For F centers, the allowed singlet to singlet lowest transition occurs at 3.4 eV for the surface, 2.9 eV for the step and 2.6 eV for the corner. The same trend is found for the F^+ centers, where the first doublet to doublet transition, goes from 3.6 eV for the surface, to 2.6 eV for the step

Table 5. Summary of excitation energies (in eV) of F and F+ centers on the MgO surface, step and corner sites (a)

	Surface		Step		Corner	
Center	Transition	DDCI	Transition	DDCI	Transition	DDCI
F	$^1A_1 \to {}^3A_1$ $(1s \to 2p_z)$	1.98	$^1A' \to {}^3A''$ $(1s \to 2p_z)$	1.63	$^1A \to {}^3E$ $(1s \to 2p_{x,y})$	1.39
	$^1A_1 \to {}^1A_1$ $(1s \to 2p_z)$	3.39	$^1A' \to {}^1A''$ $(1s \to 2p_z)$	2.92	$^1A \to {}^1E$ $(1s \to 2p_{x,y})$	2.60
	$^1A_1 \to {}^1E$ $(1s \to 2p_{x,y})$	4.72	$^1A' \to {}^1A'$ $(1s \to 2p_{x,y})$	3.82		
F^+	$^2A_1 \to {}^2A_1$ $(1s \to 2p_z)$	3.56	$^2A' \to {}^2A''$ $(1s \to 2p_z)$	2.60	$^2A \to {}^2E$ $(1s \to 2p_{x,y})$	2.43

(a) DDCI calculations for F centers have been performed using the orbitals of the $^3A_1$ state for the surface, and those of the $^3A'$ and ^3A states for the step and corner, respectively. For F^+ centers the orbitals of the $^2A_1$ state have been used for the surface and $^2A'$ and ^2A for the step and corner.

and 2.4 eV for the corner. As observed previously for the bulk calculations, excitations for the F^+ centers appear around 0.2 eV below the excitations due to F centers. The excitations at the step and corner sites are shifted about 0.5-1.2 eV compared to the surface transitions indicating that, if these defects are present on the surface, two different optical bands should be detected in the spectrum.

Comparison of our results with experiment is difficult because, except for the band appearing at 5 eV, the rest of the bands are not unambiguously assigned. For bulk MgO previous calculations showed an overestimation of 0.8-1 eV of the excitation energy, that is, an error of ~15%. Applying a similar correction to our present calculations, the singlet to singlet allowed transitions are predicted to occur at ~2.9 and ~3.0 eV for F and F^+ surface centers, at ~2.5 and ~2.2 eV for F and F^+ step centers and at ~2.2 and ~2.1 eV for F and F^+ corner centers. These values allow us to tentatively assign the bands observed around 2-2.3 eV to low-coordinated (step and corner) F centers while the transitions due to surface vacancies should be seen around 3 eV. EELS measurements on MgO ultrathin films grown on Ag(100) have shown bands at 3.3 eV and 2.1 eV, which have been assigned to surface F centers [110]. These new experimental data give additional support to our theoretical assignments. The assignment we propose implies that the oxygen vacancies detected by optical spectroscopy are localized both at regular surface and at low-coordinated sites.

A final point concerns the comparison of the DDCI method and the CASPT2 approach. We have performed some test calculations for the step cluster using all electron basis sets of moderate size for all cluster atoms. These calculations not only permit a direct comparison of the two approaches for the inclusion of electron correlation but also provide a test of the ECP and basis used in the previously commented calculations. Compared to the results listed in Table 5 for the surface, the DDCI excitation energies obtained with all electron basis sets change less than 0.1 eV for all states considered. Moreover, CASPT2 excitation energies are also very similar to the DDCI energies obtained with the all electron basis sets. The perturbative scheme gives energies which are about 0.2 eV lower. Hence, we conclude that the present choice of ECP and basis set are sufficient to compute excitation energies of F centers in the MgO (100) surface.

4.5 The 100 eV range; Core excitons

Core level spectroscopy is often applied to obtain information about the valence electronic structure of solid state materials. Widely used techniques are X-ray Photoelectron Spectroscopy (XPS), based on the ionization of an electron from a core level, and X-ray Absorption Spectroscopy (XAS), in which a core electron is excited into an unoccupied valence orbital. When the hole-particle interaction is strong enough a localized state arises in the latter case, a so-called core exciton. Previous theoretical work on MgO [111] showed that there are no excitonic states when the hole is made in one of the oxygen orbitals, neither in the valence O-2*p* orbital, nor in the deep core O-1*s* level. The singly occupied molecular orbital (SOMO) arising from such excitations extends all over the cluster and the expectation value of the r^2 operator is at least as large as the size of the cluster. However, theoretical evidence was given for the existence of a localized excited state when the core hole was made in the Mg-2*p* orbital. The Hartree-Fock excitation energy was calculated as 54.6 eV, which is only 1 eV in difference with the experimental value of O'Brien and co-workers, who reported a peak around 0.3 eV below the MgO conduction band and give an exciton energy of 53.4 eV [112]. The difference of 1 eV can either be ascribed to the lack of electron correlation or to a cluster size effect. Hence, to illustrate the possibilities of the embedded cluster model approach, this section discusses the extension of the previous study on MgO with a method to account for the electron correlation effects, we compare three cluster model of increasing size, and also apply the methodology to the more complex oxides Al_2O_3 and SiO_2.

For MgO we consider MgO_6, $Mg_{13}O_{14}$ and $Mg_{19}O_{14}$ cluster, whereas for Al_2O_3 the quantum cluster regions considered are AlO_6 and Al_8O_9. The low point group symmetry of the corundum cluster does not allow further extension of the size of the quantum cluster region as in the case of MgO. These clusters are embedded in point charges that reproduce the Madelung potential, although we investigate the effect of including the short-range repulsion by representing the atoms in the first shells around the $Mg_{13}O_{14}$ unit by AIEMP's. The covalent nature of SiO_2 makes the embedding with point charges (either formal or effective charges) less realistic. Therefore, the method shortly discussed in Section 2.6 is applied. The quantum cluster region is completed by saturating the dangling bonds with hydrogen atoms. Here, we consider a SiO_4H_4 and a $SiO_4Si_4O_{12}H_{12}$ cluster model. The atoms in the cluster are described with Atomic Natural Orbitals (ANO), specially designed to accurately describe electron correlation effects, being yet as compact as possible [113, 114]. The central metal atom in all clusters is described with a (5*s*,4*p*,1*d*) contracted basis set, while this basis is reduced to (4*s*,3*p*) for the next shell of metal centers. The oxygens directly coordinating the central metal ion have a (4*s*,3*p*,1*d*) basis set, while all other oxygens are assigned a (3*s*,2*p*) basis. A (2*s*) basis is used for the embedding hydrogens.

In the first place we discuss the degree of localization of the $M(2p^6) \rightarrow$ SOMO excitation. For this purpose, we first calculate the expectation value of the r^2 operator of the SOMO, and compare $<r^2>^{1/2}$ with the distance of the central metal ion to the atoms at the edge of the cluster. An alternative measure is given by the difference in the Mulliken net charges of the cluster atoms between the ground state and the excitonic state with a M-2*p* electron promoted into the lowest unoccupied molecular orbital (LUMO). The spatial extent of the SOMO for the largest clusters considered are 3.72, 5.29 and 3.98 Å for MgO, Al_2O_3 and SiO_2, respectively. The largest distance in the respective clusters are 3.64, 3.50, and 5.15 Å. Hence, it can be concluded that the SOMO is reasonable localized within the cluster region for MgO and SiO_2, but that the $<r^2>^{1/2}$ value for Al_2O_3 is perhaps too large to support the localized character of this exciton. Obviously a larger cluster would be very helpful to settle this point, but the low symmetry of the cluster makes such calculation prohibitively large. Nevertheless, the changes in the charge distribution obtained from MPA shows that the creation of a core hole and a particle in the LUMO essentially only affects the central metal ion and the first shell of ligands (see Table 6). We stress again that

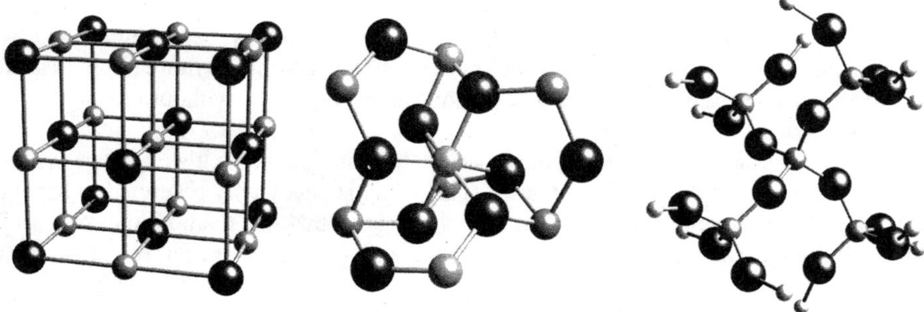

Figure 4. Schematic representation of the quantum cluster region to model MgO, Al_2O_3 and SiO_2. Dark spheres represent oxygens and light spheres metals. Depicted are from left to right: the $Mg_{13}O_{14}$ cluster; the Al_8O_9 cluster; and the $SiO_4Si_4O_{12}H_{12}$ cluster.

Table 6. Net Mulliken charges of the different shells of atoms in the ground state (GS) and the excited state (ES) and the difference of both quantities.

	Shell of atoms				
MgO	Mg	O_6	Mg_{12}	O_8	
GS	2.11	-2.20	2.10	-2.01	
ES	1.30	-2.10	2.12	-2.00	
Diff.	0.81	-0.10	-0.02	-0.01	
Al_2O_3	Al	O_6	Al_7	O_3	
GS	2.72	-2.02	3.06	-2.02	
ES	2.61	-1.84	2.91	-1.99	
Diff.	0.11	-0.18	0.15	-0.03	
SiO_2	Si	O_4	Si_4	O_{12}	H_{12}
GS	1.38	-0.92	2.48	-1.16	0.53
ES	0.46	-0.61	2.45	-1.14	0.49
Diff.	0.92	-0.31	0.03	-0.02	0.04

the absolute values have to be regarded with caution because of the limitations of the MPA, all the more in the present case, since different basis sets are used for the different shells of atoms. However, the similarities of the net charges of the more external shells of atoms in ground and excited state is an indication of the local character of the excitation. Notice, that the differences for Al_2O_3 are in agreement with the slightly more delocalized character found with the comparison of cluster size and $\langle r^2 \rangle^{1/2}$.

Having established the capacity of the cluster models used here to describe the excitonic core hole state, the next step is to analyze the energetics involved in these excitons. In the first place, we comment on the dependence of the excitation energy at the Hartree-Fock level on the cluster size. Table 7 shows that this dependency is rather weak, with exception of the MgO_6 cluster, which apparently is too small to give a good description of the core exciton. For MgO, the ROHF values are converged within approximately 0.4 eV with respect to the cluster size. Such a statement cannot be made for the other two materials because no large cluster can be constructed, although the results in Table 6 indicate that no large changes are expected for larger quantum cluster regions. The inclusion of electron correlation effects by CASPT2 (the M-2*p* and the O-2*s*,2*p* electrons are included in the perturbational treatment) lowers the excitation energy by approximately 1 eV for all three compounds. The results are in general in better agreement with experiment, but do not change the overall picture obtained from the more simple Hartree-Fock treatment, hence experimental observation can be explained without loosing the one-electron or molecular orbital model.

On the other hand, the effect of the inclusion the short-range repulsion between cluster and surroundings is still to be investigated. This study is necessarily restricted to MgO and SiO_2 only, because the above commented results suggest that the quantum cluster region is too small in the case of corundum. For MgO, we consider a cluster of 125 quantum mechanically treated atoms. This model, $Mg_{63}O_{62}$, consists of a cube of 2*a* (7.28 Å), and is further embedded in point charges. The quantum cluster region is $Mg_{13}O_{14}$, the other ions are described with model potentials: TIP's for the cations and AIEMP's for the anions. The thirty Mg^{2+} TIP's nearest to the quantum cluster region are provided with a (1*s*,1*p*) basis set in order to permit the SOMO to further delocalize. However, the inclusion of the Pauli

Table 7. Hartree-Fock and CASPT2 excitation energies (in eV) for MgO, Al_2O_3 and SiO_2. Small, medium and big refers to MgO_6, $Mg_{13}O_{14}$ and $Mg_{19}O_{14}$ for MgO. For Al_2O_3, small and medium stand for AlO_6 and Al_8O_9, whereas they refer to SiO_4H_4 and $SiO_4Si_4O_{12}H_{12}$ for SiO_2.

	MgO		Al_2O_3		SiO_2	
Cluster	ROHF	CASPT2	ROHF	CASPT2	ROHF	CASPT2
Small	52.1	50.7	78.7	77.2	108.8	107.0
Medium	55.0	53.9	78.3	77.5	109.5	108.0
Big	54.6	53.8				
Experiment	53.4		78.6		106.1	

repulsion leads to a spatial extent of the SOMO of 3.49 Å, which is somewhat shorter than the 3.72 Å estimated from the $Mg_{13}O_{14}$ cluster in point charges only. This higher degree of localization leads to a slightly higher excitation energy: 56.1 and 55.3 eV, for ROHF and CASPT2, respectively. The embedding with hydrogens of the clusters that represent SiO_2 is in principle sufficient to avoid discontinuities in the material model. Nevertheless, it is not completely clear what distance should be taken for the O–H bond. The most common choice is to take a distance of 0.98 Å, corresponding to the average equilibrium O–H distance in several chemical compounds. The stability of the excitation energy with respect to this parameter is checked by calculating the exciton energy in SiO_4H_4 and $SiO_4Si_4O_{12}H_{12}$ varying the O–H distance between 0.9 and 1.2 Å. For the small cluster model the energy decreases by 2.5 eV, going from 107.7 eV for an O–H distance of 0.9 Å to 105.2 for 1.2 Å. The effect is much less pronounced in the larger cluster. In the same interval the energy only changes 0.9 eV: from 108.1 eV at 0.9 Å to 107.2 eV for 1.2 Å. Hence, it can be concluded that the embedding with hydrogens is rather insensitive to the position at which the hydrogen are placed, given that the quantum cluster region is reasonably large.

A final comparison of the theoretical value of the exciton energies with the experimental ones shows that the cluster model approach is able to reasonably reproduce the energetics of these phenomena. This allows for a detailed analysis of the character of the final state, which has been shown to be rather localized but cannot be considered as a M-3*s* orbital as has been suggested. The remaining difference with experiment must be ascribed to a combination of small effects due the limited quantum cluster region, the approximate representation of the short-range repulsion and the finite one-electron basis sets used in the calculations. For a better agreement with experiment it is expected that all three effects need to be improved.

5. Conclusions and summary

The examples presented in this chapter show that the combination of the cluster model approach and multiconfigurational wave function based methods provides a powerful tool to study excited states in material science in a very wide energy range. Magnetic coupling parameters can be calculated in a straightforward manner within experimental accuracy for prototype magnetic insulators like $KNiF_3$ and K_2NiF_4. This allows us to predict the magnetic coupling parameters for other materials for which less information exist. Moreover, quantum chemistry can provide essential information about the mechanisms that determine the magnitude of the magnetic coupling. A careful analysis of the material model shows that the localized d-d excitations in NiO are significantly affected by the covalent interactions between the metal centers and the oxygen ligands coordinating it. A small, although not negligible effect has been observed from the next shell of cationic centers. Quantum chemical cluster model studies are also ideal to obtain detailed information about the geometry distortions around an impurity. The method predicts a rather large outward relaxation of the first shell of Cl ions around the neutral silver impurity in the ionic KCl crystal host. The subsequent analysis of the electronic transitions due to this impurity confirmed the existence of a charge transfer transition at a rather low energy. An example of the study of defects has also been given, namely the spectroscopy of the F-centers in MgO. The careful comparison of different low coordinated sites at the MgO surface helped to interpret recent experimental data about this system. Finally, we have applied the methodology to core-level transitions. Analyzing the energetics and the character of the wave function of the final states, the local character of the M-2$p \rightarrow$ LUMO transition could be confirmed, although the interpretation as a purely atomic transition has been shown not to be correct.

Acknowledgments

We thank Ibério de P. R. Moreira, Ria Broer, Miguel Moreno, Antonio Aramburu, M[a] Teresa Barriuso, and Richard Martin for the collaborations and stimulating discussions. This work has been financed by the Spanish "Ministerio de Educación y Ciencia" CICyT Project No. PB98-1216-CO2-01 and, in part, by the Generalitat de Catalunya SGR1999-SGR0040. C. de G. acknowledges financial support through the TMR activity "Marie Curie research training grants", Grant No. FMBICT983279 established by the European Community. Part of the computer time was provided by CESCA/CEPBA through research grants from the University of Barcelona.

References

[1] P. Hohenberg and W. Kohn, Phys. Rev. **136** (1964) B864.

[2] W. Kohn and L.J. Sham, Phys. Rev. **140** (1965) A1133.

[3] E.K.U. Gross, J.F. Dobson and M. Petersilka, *Density Functional Theory*, Springer-Verlag, Heidelberg, 1996.

[4] M.E. Casida, C. Jamorski, K.C. Casida and D.R. Salahub, J. Chem. Phys. **108** (1998) 4439.

[5] S. Hirata, M. Head-Gordon and R.J. Bartlett, J. Chem. Phys. **111** (1999) 10774.

[6] A. Shukla, M. Dolg, P. Fulde and H. Stoll, Phys. Rev. B **60** (1999) 5211.

[7] W.C. Nieuwpoort and R. Broer, in: *Cluster models for Surface and Bulk Phenomena*, edited by G. Pacchioni, P.S. Bagus and F. Parmigiani, p. 505 (Plenum Press, New York, 1992).

[8] J.A. Aramburu, M. Moreno, I. Cabria, M.T. Barriuso, C. Sousa, C. de Graaf and F. Illas, Phys. Rev. B **62** (2000) 13356.

[9] C. Sousa, C. de Graaf, F. Illas, M.T. Barriuso, J.A. Aramburu and M. Moreno, Phys. Rev. B **62** (2000) 13366.

[10] C. de Graaf, F. Illas, R. Broer and W.C. Nieuwpoort, J. Chem. Phys. **106** (1997) 3287.

[11] C. de Graaf, R. Broer and W.C. Nieuwpoort, Chem. Phys. Lett. **271** (1997) 372.

[12] N.W. Winter, R.M. Pitzer and D.K. Temple, J. Chem. Phys. **86** (1987) 3549.

[13] N.W. Winter and R.M. Pitzer, J. Chem. Phys. **89** (1988) 446.

[14] Z. Barandiarán and L. Seijo, J. Chem. Phys. **89** (1988) 5739.

[15] L. Seijo and Z. Barandiarán, *Computational Chemistry: Reviews of Current Trends*, World Scientific, Singapore, 1999.

[16] P.J. Hay and W.R. Wadt, J. Chem. Phys. **82** (1985) 270.
[17] R.L. Martin and P.J. Hay, unpublished.
[18] V. Bonifacic and S. Huzinaga, J. Chem. Phys. **60** (1974) 2779.
[19] S. Huzinaga, L. Seijo, Z. Barandiarán and M. Klobukowski, J. Chem. Phys. **86** (1987) 2132.
[20] L. Seijo and Z. Barandiarán, J. Chem. Phys. **94** (1991) 8158.
[21] P.P. Ewald, Ann. Physik **64** (1921) 253.
[22] H.M. Evjen, Phys. Rev. **39** (1932) 675.
[23] C. Sousa, J. Casanovas, J. Rubio and F. Illas, J. Comp. Chem. **14** (1993) 680.
[24] P. Reinhardt, M.P. Habas, R. Dovesi, I. de P.R. Moreira and F. Illas, Phys. Rev. B **59** (1999) 1016.
[25] Z. Barandiarán and L. Seijo, in: *Cluster models for Surface and Bulk Phenomena*, edited by G. Pacchioni, P.S. Bagus and F. Parmigiani, p. 565 (Plenum Press, New York, 1992).
[26] M. Born, Z. Physik **1** (1920) 45.
[27] B.G. Dick and A.W. Overhauser, Phys. Rev. **112** (1958) 90.
[28] C.R.A. Catlow, M. Dixon and W.C. Mackrodt, in: *Computer Simulation of Solids*, edited by C.R.A. Catlow, p. 130 (Springer, Berlin, 1982).
[29] R. Pandey and J.M. Vail, J. Phys.: Condens. Matter **1** (1989) 280.
[30] J.H. Harding, A.H. Parker, P.B. Keegstra, R.Pandey, J.M. Vail and C. Woodward, Physica **131B** (1985) 151.
[31] C. Pisani and U. Birkenheuer, Comp. Phys. Comm. **96** (1996) 152.
[32] C. Pisani, F. Corà, R. Nada and R. Orlando, Comp. Phys. Comm. **82** (1994) 139.
[33] C. Pisani, U. Birkenheuer, F. Corà, R. Nada and S. Casassa, EMBED96 User's Manual, University of Torino, Torino, 1996.
[34] C. Pisani, R. Dovesi and C. Roetti, *Hartree-Fock ab-initio treatment of crystalline systems*, Lecture Notes in Chemistry Vol. 48. Springer Verlag, Heidelberg, 1988.
[35] V.R. Saunders, R. Dovesi, C. Roetti, M. Causà, N.M. Harrison, R. Orlando and C.M. Zicovich-Wilson, CRYSTAL98 User's Manual, University of Torino, Torino, 1998.
[36] G.J.M. Janssen and W.C. Nieuwpoort, Phys. Rev. B **38** (1988) 3449.
[37] G. Pacchioni, P.S. Bagus and F. Parmigiani, *Cluster Models for Surface and Bulk Phenomena*, NATO ASI Series B Vol. 283. Plenum Press, New York, 1992.
[38] J. Sauer, P. Ugliengo, E. Garrone and V.R. Saunders, Chem. Rev. **94** (1994) 2095.
[39] G. Pacchioni, Heter. Chem. Rev. **2** (1995) 213.
[40] P.S. Bagus and F. Illas, in: *The Surface Chemical Bond in Encyclopedia of Computational Chemistry*, edited by P.V. Schleyer, N.L. Allinger, T. Clark, J. Gasteiger, P.A. Kollman, H.F. Schaefer III and P.R. Schreiner, p. 2870 (John Wiley & Sons, Chichester, UK, 1998).
[41] F. Illas, L. Roset, J.M. Ricart and J. Rubio, J. Comp. Chem. **14** (1993) 1534.
[42] J. Miralles, J.P. Daudey and R. Caballol, Chem. Phys. Lett. **198** (1992) 555.
[43] J. Miralles, O. Castell, R. Caballol and J.-P. Malrieu, Chem. Phys. **172** (1993) 33.
[44] V.M. García, O. Castell, R. Caballol and J.-P. Malrieu, Chem. Phys. Lett. **238** (1995) 222.
[45] K. Andersson, P.-Å. Malmqvist, B.O. Roos, A.J. Sadlej and K. Wolinski, J. Phys. Chem. **94** (1990) 5483.
[46] K. Andersson, P.-Å. Malmqvist and B.O. Roos, J. Chem. Phys. **96** (1992) 1218.
[47] N.W. Ashcroft and N.D. Mermin, *Solid State Physics*, Holt, Reinhart and Winston, New York, 1976.
[48] K. Andersson, M.R.A. Blomberg, M.P. Fülscher, G. Karlström, R. Lindh, P.-Å. Malmqvist, P. Neogrády, J. Olsen, B.O. Roos, A.J. Sadlej, M. Schütz, L. Seijo, L. Serrano-Andrés, P.E.M. Siegbahn and P.-O. Widmark, MOLCAS version 4, University of Lund, Sweden, 1997.
[49] D. Maynau and N. Ben Amor, CASDI suite of programs, Toulouse, 1997
[50] P.W. Anderson, Phys. Rev. **115** (1959) 2.
[51] R.K. Nesbet, Ann. Phys. **4** (1958) 87.
[52] R.K. Nesbet, Phys. Rev. **119** (1960) 658.
[53] L. Noodleman and J.G. Norman Jr., J. Chem. Phys. **70** (1979) 4903.

[54] L. Noodleman and E.R. Davidson, Chem. Phys. **109** (1986) 131.

[55] R. Caballol, O. Castell, F. Illas, I. de P.R. Moreira and J.-P. Malrieu, J. Phys. Chem. A **101** (1997) 7860.

[56] J. Cano, P. Alemany, S. Alvarez, M. Verdaguer and E. Ruiz, Chem. Eur. J. **4** (1998) 476.

[57] F. Illas, I. de P.R. Moreira, C. de Graaf and V. Barone, Theor. Chem. Acc. **104** (2000) 265.

[58] C. Blanchet-Boiteux and J.-M. Mouesca, J. Phys. Chem. A **104** (2000) 2091.

[59] T. Soda, Y. Kitagawa, T. Onishi, Y. Takano, Y. Shigeta, H. Nagao, Y. Yoshioka and K. Yamaguchi, Chem. Phys. Lett. **319** (2000) 223.

[60] M.E. Lines, Phys. Rev. **164** (1967) 736.

[61] L.J. de Jongh and R. Miedema, Adv. Phys. **23** (1974) 1.

[62] J.M. Ricart, R. Dovesi, C. Roetti and V.R. Saunders, Phys. Rev. B **52** (1995) 2381.

[63] I. de P.R. Moreira and F. Illas, Phys. Rev. B **55** (1997) 4129.

[64] Y-S Su, T.A. Kaplan, S.D. Mahanti and J.F. Harrison, Phys. Rev. B **59** (1999) 10521.

[65] M.D. Towler, N.L. Allan, N.M. Harrison, V.R. Saunders, W.C. Mackrodt and E. Aprà, Phys. Rev. B **50** (1994) 5041.

[66] C. de Graaf and F. Illas, Phys. Rev. B **63** (2001) 014404.

[67] M.D. Towler, R. Dovesi and V.R. Saunders, Phys. Rev. B **52** (1995) 10150.

[68] I. de P.R. Moreira and F. Illas, Phys. Rev. B **60** (1999) 5179.

[69] P. Reinhardt, I. de P.R. Moreira, C. de Graaf, R. Dovesi and F. Illas, Chem. Phys. Lett. **319** (2000) 625.

[70] Z. Hiroi, M. Azuma, M. Takano and Y. Bando, J. Solid State Chem. **95** (1991) 230.

[71] E. Dagotto and T.M. Rice, Science **271** (1996) 618.

[72] T.M. Rice, Z. Phys. B **103** (1997) 165.

[73] S. Maekawa, Science **273** (1996) 1515.

[74] R.S. Eccleston, M. Uehara, J. Akimitsu, H. Eisaki, N. Motoyama and S. Uchida, Phys. Rev. Lett. **81** (1998) 1702.

[75] D.C.Johnston, Phys. Rev. B **54** (1996) 13009.

[76] T. Imai, K.R. Thurber, K.M. Shen, A.W. Hunt and F.C. Chou, Phys. Rev. Lett. **81** (1998) 220.

[77] J.-P. Malrieu, J. Chem. Phys. **47** (1967) 4555.

[78] J. Casanovas, J. Rubio and F. Illas, Phys. Rev. B **53** (1996) 945.

[79] I. de P.R. Moreira, F. Illas, C.J. Calzado, J.F. Sanz, J.-P. Malrieu, N. Ben Amor and D. Maynau, Phys. Rev. B **59** (1999) 6593.

[80] M. Matsuda, K. Katsumata, R.S. Eccleston, S. Brehmer and H.-J. Mikeska, J. Appl. Phys. **87** (2000) 6271.

[81] F. Naef and X. Wang, Phys. Rev. Lett. **84** (2000) 1320.

[82] S. Sugai, T. Shinoda, N. Kobayashi, Z. Hiroi and M. Takano, Phys. Rev. B **60** (1999) 6969.

[83] C. de Graaf, I. de P.R. Moreira, F. Illas and R.L. Martin, Phys. Rev. B **60** (1999) 3457.

[84] L. Seijo, unpublished (1997)

[85] C. Corliss and J. Sugar, J. Phys. Chem. Ref. Data **10** (1981) 197.

[86] K. Pierloot, E. Tsokos and B.O. Roos, Chem. Phys. Lett. **214** (1993) 583.

[87] R. Newman and R.M. Chrenko, Phys. Rev. **114** (1959) 1507.

[88] G.J.M. Janssen and W.C. Nieuwpoort, Int. J. Quantum Chem. Symp. **22** (1988) 679.

[89] E. Lorda, F. Illas and P.S. Bagus, Chem. Phys. Lett. **256** (1996) 377.

[90] A. Freitag, V. Staemmler, D. Cappus, C.A. Ventrice Jr., K.Al Shamerey, H. Kuhlenbeck and H. J. Freund, Chem. Phys. Lett. **210** (1993) 10.

[91] P.S. Bagus, F. Illas, C. Sousa and G. Pacchioni, in: *Electronic properties of Solids using Cluster Methods*, edited by T.A. Kaplan and S. D. Mahanti, p. 93 (Plenum Press, New York, 1995).

[92] A.M. Stoneham, *Theory of Defects in Solids; Electronic structure of defects in insulators and semiconductors*, Oxford University Press, Oxford, 1996.

[93] S.V. Nistor, D. Schoemaker and I. Ursu, Phys. Stat. Sol. B **185** (1994) 9.

[94] M. Saidoh and N. Itoh, J. Phys. Soc. Jap. **35** (1973) 1122.

[95] M. Moreno, J. Phys. C **13** (1980) 6641.
[96] J. Simonetti and D.S. McClure, J. Chem. Phys. **71** (1979) 793.
[97] J.H. Barkyoumb and A.N. Mansour, Phys. Rev. B **46** (1992) 8768.
[98] J.L. Pascual, L. Seijo and Z. Barandiarán, J. Chem. Phys. **98** (1993) 9715.
[99] I. Cabria, M.T. Barriuso, J.A. Aramburu and M. Moreno, Int. J. Quantum Chem. **61** (1997) 627.
[100] M. Geleijns, C. de Graaf, R. Broer and W.C. Nieuwpoort, Surf. Sci. **421** (1999) 106.
[101] C. Sousa, W. A. de Jong, R. Broer and W.C. Nieuwpoort, Mol. Phys. **92** (1997) 677.
[102] H. Ågren, P.S. Bagus and B.O. Roos, Chem. Phys. Lett. **82** (1981) 505.
[103] R. Broer and W.C. Nieuwpoort, Theor. Chim. Acta **73** (1988) 405.
[104] J.A. Aramburu, M. Moreno and M.T. Barriuso, J. Phys.: Condens. Matter **8** (1996) 6901.
[105] K. Wissing, M.T. Barriuso, J.A. Aramburu and M. Moreno, J. Chem. Phys. **111** (1999) 10217.
[106] F. Illas and G. Pacchioni, J. Chem. Phys. **108** (1998) 7835.
[107] Y. Chen, R.T. Williams and W.A. Sibley, Phys. Rev. **182** (1969) 960.
[108] J.H. Crawford Jr., Semiconductors and Insulators **5** (1983) 599.
[109] C. Sousa, F. Illas and G. Pacchioni, Surf. Sci. **429** (1999) 217.
[110] D. Peterka, C. Tegenkamp, K.-M. Schröder, W. Ernst and H. Pfnür, Surf. Sci. **431** (1999) 146.
[111] P.S. Bagus, F. Illas and C. Sousa, J. Chem. Phys. **100** (1993) 2943.
[112] W.R. O'Brien, J. Jia, Q.-Y.Dong, T.A. CallCott, J.-E. Rubenson, D.L. Mueller and D.L. Ederer, Phys. Rev. B **44** (1991) 1013.
[113] P.-O. Widmark, P.-Å. Malmqvist and B.O. Roos, Theor. Chim. Acta **77** (1990) 291.
[114] P.-O. Widmark, B. J. Persson and B.O. Roos, Theor. Chim. Acta **79** (1991) 419.

Computational Materials Science
C.R.A. Catlow and E.A. Kotomin (Eds.)
IOS Press, 2003

Computer simulations of polar oxide surfaces

Fabio Finocchi, François Bottin and Claudine Noguera
Laboratoire de Physique des Solides, Bâtiment 510, Université Paris-Sud, F-91405 Orsay, France

Abstract. In this lecture we resume the basic principles that are needed to understand the physical and chemical behavior of the polar oxide surfaces. As a prototype, we consider the (111) orientation of magnesium oxide, for which a few computer simulations exist. We point out the advantages and the drawbacks of the periodic slab model for the simulations of surfaces with a net dipole. We discuss a way to overcome the numerical difficulties and mimic virtually isolated slabs by using periodic boundary conditions.

1 Introduction

Oxides are very common in nature. They are always present, although in a sometime uncontrolled way, whenever a material is in contact with the ambient atmosphere. They play a fundamental role in corrosion, friction, lubrication processes. The understanding of the properties of oxide surfaces is thus a key issue in many disciplines, such as geology, electrochemistry, and catalysis. As far as the latter one is concerned, large terraces obtained by cleavage have a very low reactivity, while powders that show various orientations may be very active catalysts. On the other hand, nano-structured surfaces begin to be used nowadays, as substrates to grow artificial structures with specific morphology. The presence of inequivalent sites on reconstructed surfaces with large unit cells influences the atomic diffusion and the adsorption thermodynamics, and makes it possible to drive specific growth modes, favoring for example the formation of size-controlled clusters.

In such a context, polar surfaces of compound materials are of prominent interest. By cutting the crystal along a polar orientation, each repeat unit in the direction perpendicular to the surface bears a non-zero dipole moment. An electrostatic instability results from the presence of that macroscopic dipole, which can only be healed by the introduction of compensating charges in the outer planes. This can be achieved either by a deep modification of the surface electronic structure, with an anomalous filling of the surface states, or by non-stoichiometric reconstructions that imply a different concentration of the atomic species with respect to the bulk. A third and very effective mechanism to achieve the stabilization of polar orientations, which is not discussed here, is the absorption of charged species. Indeed, polar orientations are generally much more reactive than cleavage planes.

In this lecture, we summarize the basic theoretical principles needed to understand the properties of polar orientations, and address some technical points regarding their computer modeling. Some examples of simulations are also given as an illustration.

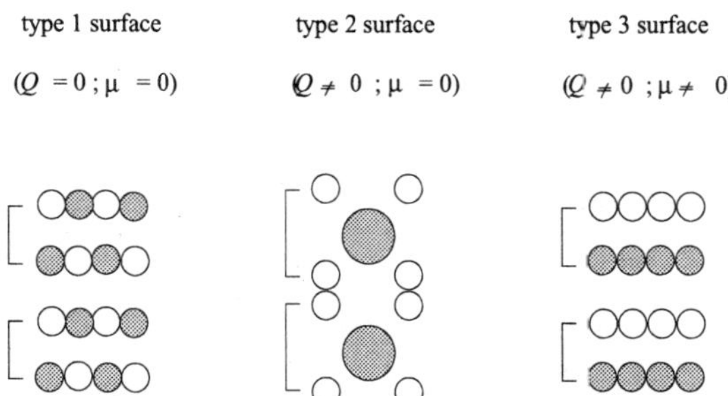

Figure 1: Classification of insulating surfaces according to Tasker [1]. Q and $\vec{p}$ are the layer charge density and the dipole moment in the repeat unit perpendicular to the surface (indicated by a bracket), respectively.

2 Polar surfaces: basic concepts

According to classical electrostatic criteria, the stability of a compound surface depends on the characteristics of the charge distribution in the structural unit which repeats itself in the direction perpendicular to the surface, as sketched in Figure 2 [1]. Type 1 or 2 surfaces – which differ by the charge Q borne by their layers – have a zero dipole moment $\vec{p}$ in their repeat unit and are thus potentially stable. At variance, polar type 3 surfaces have a diverging electrostatic surface energy [2] due to the presence of a non-zero dipole moment not only on the outer layers (which would not distinguish them from non-polar rumpled or reconstructed surfaces), but on *all* the repeat units throughout the material.

The simplest representation of a crystalline compound cut along a polar direction is given in Figure 2a. Two inequivalent layers of opposite charge densities equal to $\pm\,\sigma$ alternate along the normal to the surface, with inter-layer spacings R_1 and R_2. Each repeat unit bears a dipole moment density equal to $p = \sigma R_1$, and, as a result, the electrostatic potential increases monotonically across the system by an amount $\delta V = 4\pi\sigma R_1$ per double layer. δV is large, typically of the order of several tens eV in an ionic material like MgO. The total dipole moment $P = N\sigma R_1$ of N bilayers is proportional to the slab thickness, and the electrostatic energy amounts to $E = 2\pi N R_1 \sigma^2$. It is very large, even for thin films. In the limit $N \to \infty$, the electrostatic contribution to the surface energy per unit area diverges. This is the origin of the surface instability.

The classification of surfaces relies on the characteristics of the polarization in the bulk unit cell, on the surface orientation $\hat{n}$, and on the nature of the crystal termination. Recently, a generalized definition of the bulk electric polarization $\vec{P}$ of insulating crystals has been given in terms of the centers of charge of the Wannier functions of the occupied bands. Considering a surface of orientation $\hat{n}$, the bound charge density Q_b which accumulates at a surface is given by $Q_b = \vec{P} \cdot \hat{n}$. It is defined modulo e/A, with A the surface cell area. If $Q_b = 0$ (modulo e/A), the surface is non-polar; otherwise it is polar [3]. $\vec{P}$ can be estimated from the knowledge of the ground state electronic distribution in the bulk unit cell, which is quite accurately provided either by high resolution X ray diffraction experiments, as recently shown e.g. in alumino-silicate compounds [4], or by first principles methods, such as those which are based on the density functional theory (DFT) [5].

Actually, in most cases, more simplistic models for the electronic structure may easily tell whether a surface is polar or not. In binary compounds, for example, the difference in

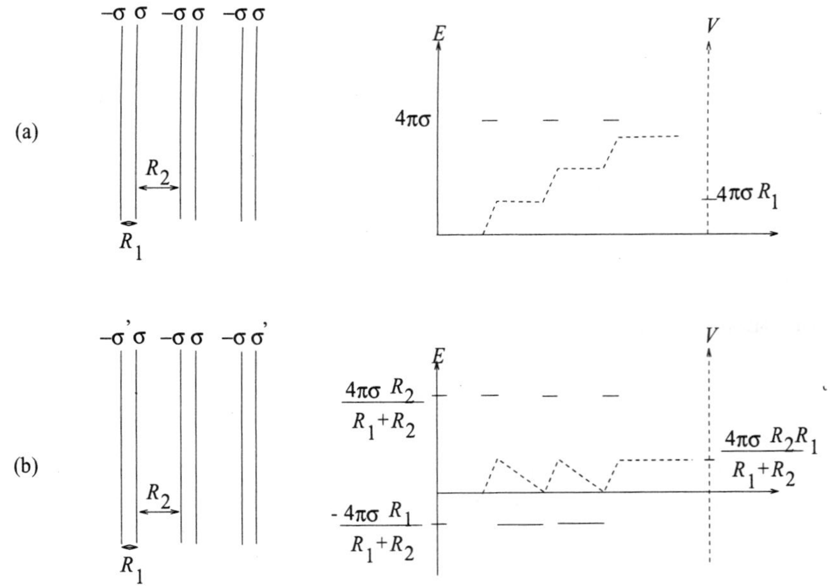

Figure 2: Spatial variations of the electrostatic field E and potential V in a slab cut along a polar direction.

electro-negativity of the constituents readily points out the sign of the charge transfer between the ions. Thus, for simple crystal structures and for orientations such that layers containing cations only and anions only alternate – (100) and (111) zinc-blende surfaces; (111) rock-salt surfaces (see Section 4), etc. –, the presence of a dipole moment in the repeat unit is unquestionable, whatever the charge values – provided they are non-zero. These surfaces are unambiguously polar. The same is also true for some surfaces of ternary compounds, such as the (110) and (111) faces of ABO_3 perovskites. In $SrTiO_3$, for example, the (110) and (111) repeat units contain alternating SrTiO and O_2 layers in the first case, and alternating SrO_3 and Ti layers in the second one. In both cases, one of the layers in the repeat unit contains a single constituent, which in no way can be considered as neutral.

There are less obvious cases, among which is the (100) perovskite surface (see Section 3.6.1). In $SrTiO_3$, it presents alternating layers of SrO and TiO_2 composition. If formal charges are assigned to the ions (Sr^{2+}, Ti^{4+} and O^{2-}), then each layer is charge neutral, the repeat unit bears no dipole moment, and the orientation is considered as non-polar. This is the statement most often encountered in the literature. However, $SrTiO_3$ is not fully ionic. Its gap width, equal to 3eV, places it on the border line between semiconductors and insulators and the Ti-O bond presents a non-negligible part of covalent character. The actual charges are thus likely not equal to the formal ones and there is little chance that $Q_{Sr}+Q_O$ and $Q_{Ti}+2Q_O$ vanish. $SrTiO_3$(100) should thus be considered as a polar surface and this exemplifies how careful one should be in the classification of surfaces.

In addition, the orientation $\hat{n}$ is not always sufficient to characterize a surface, especially when various terminations may be produced. In the rutile structure, for example, in which some transition metal oxides MO_2 crystallize, the bulk repeat unit in the (110) direction is made of three layers either of O or $(MO)_2$ composition, and there exist three chemically inequivalent terminations, which expose a single oxygen layer ($O/(MO)_2/O$ sequence), two oxygen layers ($O/O/(MO)_2$ sequence) or one mixed cation-oxygen layer ($(MO)_2/O/O$ sequence). Only in the first case, the repeat unit bears no dipole moment. Similarly, on the basal (0001) face of the corundum structure, met in some M_2O_3 sesquioxides, three chemically distinct terminations may be produced, with a single cation layer, two cation layers or one oxygen layer in contact with vacuum. Only the first one is non-polar and is met under standard preparation conditions. However, when experimental conditions are varied, for example, in the process of fabrication of ultra-thin films, or under bombardment, reducing or oxidizing conditions, some variations of stoichiometry in the surface layers may take place and polarity arguments have to be re-examined.

According to classical electrostatic, ideal polar surfaces are thus unstable. However, specific modifications of the charge density in the outer layers may cancel out the macroscopic component of the dipole moment and heal the polarity. Within the geometry displayed in Figure 2b, this can be achieved, for example, by assigning a value $\sigma' = \sigma R_2/(R_1 + R_2)$ to the charge density on the outer layers of the slab, and it results in a total dipole moment $P = \sigma R_1 R_2/(R_1 + R_2)$ which is no longer proportional to the slab thickness. The monotonic increase of the electrostatic potential is also suppressed. More generally, when m outer layers are modified ($|\sigma_j| \neq \sigma$ for $1 \leq j \leq m$ and $|\sigma_{m+1}| = \sigma$), the condition for the cancellation of the macroscopic dipole moment reads [6]:

$$\sum_{j=1}^{m} \sigma_j = -\frac{\sigma_{m+1}}{2} \left[(-1)^m - \frac{R_2 - R_1}{R_2 + R_1} \right] \tag{1}$$

A polar surface can thus be stabilized provided that the charge compensation dictated by Eq.(1) is fulfilled. It implies that either the charges or the stoichiometry in the surface layers are modified with respect to the bulk, and thus, several scenarios are conceivable in order to heal polarity:

- one or several surface layers have their composition which differs from the bulk stoichiometry. This may lead to a phenomenon of reconstruction or terrace formation depending upon how the vacancies or ad-atoms order. However, if no order takes place, surface diffraction patterns exhibit a (1×1) symmetry and, unless quantitative analysis is performed, give no information on the surface stoichiometry.

- foreign atoms or ions, coming from the residual atmosphere in the experimental set-up, provide the charge compensation.

- on stoichiometric surfaces, charge compensation may result from an electron redistribution in response to the polar electrostatic field. This mechanism is well exemplified in *self-consistent* electronic structure calculations.

Which process actually takes place firstly depends upon energetic considerations. As it will be explained in the following, polar surfaces that are found in nature never show a macroscopic dipole moment, since its magnitude would not be compatible at all with the surface stability. In a real material, the electronic degrees of freedom can always be at work to reach charge compensation through the third mechanism. However, in most cases, the resulting surface energy is high and other processes may be more effective and better explain the experimental situation. If experiments are performed in thermodynamic equilibrium conditions, the observed surface configuration is that with the lowest relevant thermodynamic potential (e.g. the grand potential if the surface is in contact with a reservoir – an oxygen rich atmosphere, for example). If thermodynamic equilibrium is not reached, the observed surface configuration is that of lowest energy, which is *kinetically* accessible [7]. Another important point to stress is that all previous considerations apply only for the case of polar terminations of bulk crystals. If, on the other hand, few atomic planes oriented along a polar direction are deposited onto a substrate, they can be stable even if the condition for polarity cancellation (Eq.1) is not fulfilled, since the average electrostatic potential drop across the film is proportional to the layer thickness, and may thus be small. Also in this case, the stability of the deposited film depends on the detailed electron distribution, and on thermodynamic and kinetic considerations, as for other non-polar over-layers.

3 The slab model and plane waves

On the theoretical side, the simulation of polar surfaces requires special attention. First, as already mentioned, since it is necessary to describe accurately the charge distribution in the outer layers, only electronic structure calculations that account for the self-consistent relation between the electron density and the electrostatic potential can yield reliable results. In the following, we will essentially discuss the class of *fully* self-consistent calculations which can describe the electron redistribution at the surface in a quantum mechanical framework and thus treat the various mechanisms that are able to compensate the macroscopic dipole on the same footing.

As far as the *model* to represent the surface is concerned, we will mainly focus on the slab model and show that it is suited to simulate well ordered surfaces, provided that some caution is taken. The cluster models that are sometimes employed for the representation of a polar surface may be easily biased by finite-size effects, especially when the part of the cluster that is treated quantum mechanically is small. Embedding the cluster in an array of charges or pseudo-potentials surely alleviates the bias given by finite-size effects [8]. Nevertheless, because of the long-range Coulomb interaction, and the self-consistent relation between the charge and the electrostatic potential, any anomalous filling of surface states may considerably alter the embedding potential with respect to standard situations, which may thus imply a delicate numerical procedure. This explains why most of the recent numerical simulations make use of slabs, which seem to be more robust than cluster models when treating polar orientations.

The use of a plane wave basis set in conjunction with slab models has become very popular. Here, we give only the basic ideas relevant to the simulation of polar surfaces. The surface is simulated by a film consisting of some atomic layers, plus a vacuum region surrounding it. The slabs have to be thick enough to display bulk characteristics in their central layers. Then, periodic boundary conditions (PBC) are applied in the three spatial directions. The model system thus consists of a regular array of indefinitely wide (in the parallel dimensions) films arranged in a super-lattice, alternating atomic layers and a void region along the surface normal (see Figure 3). The use of the discrete translational symmetry permits the use of the Bloch theorem and all the concepts that have been introduced for the simulation of perfect crystals. In particular, the repeat unit is called the *supercell*, which is analogous to the unit cell of the crystal [8]. Provided that the interaction between the repeated atomic layers can be neglected (and we will discuss how that interaction can be minimized in the following), the slab model gives a reliable picture of large terraces.

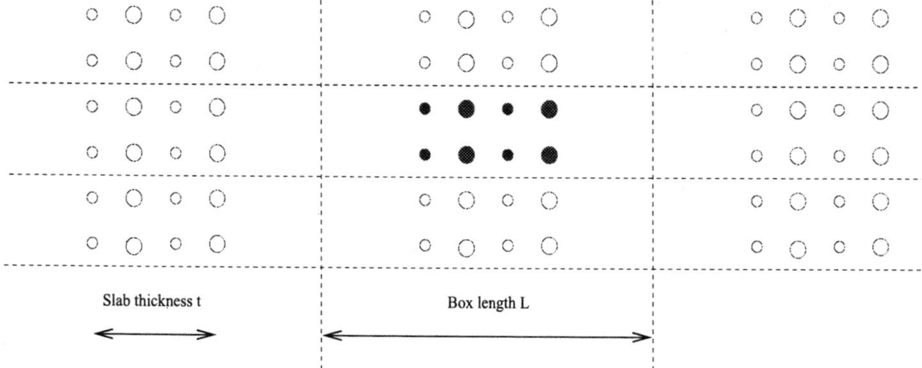

Figure 3: slab and periodisation

As far as the choice of the basis set to expand the Kohn-Sham orbitals is concerned, there are two big families, consisting of *localized* or *delocalized* functions, among which the plane waves (PW) are especially popular and widely used. While localized functions can be periodically repeated or not, PW's are generally used in conjunction with the supercell scheme, within the framework of the DFT [11]. The Kohn-Sham orbitals needed to represent the electron distribution of the valence electrons are expanded in plane waves:

$$\psi_{j,\mathbf{k}}(\mathbf{r}) = \sum_{\mathbf{G}} c_j(\mathbf{k}+\mathbf{G}) e^{i\,\mathbf{r}\cdot(\mathbf{k}+\mathbf{G})} \tag{2}$$

where j is the band index that enumerates the Kohn-Sham orbitals, $\mathbf{k}$ is a vector in the Brillouin Zone and $\mathbf{G}$ a reciprocal lattice vector. PW's have thus the full three-dimensional periodicity of the supercell, and one must thus worry about spurious interactions between the repeat slabs along the surface normal, an issue that is addressed in the following. The expansion in eq.(2) includes all $\mathbf{k}+\mathbf{G}$ vectors such that $\hbar^2|\mathbf{k}+\mathbf{G}|^2/2m \leq E_{\text{cut}}$. The larger the cutoff energy E_{cut}, the finer the corresponding real-space grid, and the better the accuracy of the representation of the wave-functions, the electron density and the effective potential. Therefore, the quality of the plane-wave basis set may be routinely improved by increasing the unique parameter E_{cut}. Usually, its value is chosen in such a way that the numerical values of the relevant physical quantities to be computed (such as the total energy, the atomic forces, the band structure, etc.) are little affected by raising E_{cut} further. Such a computational scheme has several advantages: firstly, the plane waves basis set corresponds to a uniform real-space grid. This implies that the derivatives of the total energy with respect to the atomic displacements, i.e. the atomic forces, can be easily calculated through the Hellmann-Feynman theorem [9]. At odds with the case of localized basis sets – often consisting of finite-range functions centered on the atomic sites – the quality and the accuracy of the description of the electron density is completely independent of the atomic positions. Therefore, when calculating the atomic forces using plane waves, one has not to evaluate Pulay forces [10], which are due to the variation of the total energy with respect to the sole variation of the centers of the basis functions (that is, both atomic positions and orbitals should be frozen). That makes plane waves an optimal and easy-to-use basis set whenever the stable atomic configurations are not a-priori known, i.e. when geometry optimization or molecular dynamics runs have to be carried out. Secondly, the availability of Fast Fourier Transforms (FFT) permits to perform each specific operation either in real space or in reciprocal (**G**) space, according to the numerical convenience, thus avoiding any convolution product. Thirdly, the implementation of analytical formulas in plane waves is simple and transparent, so that reading and improving plane-wave based computer codes is generally an easier task than when using a localized basis set.

On the other hand, plane waves imply some drawbacks: firstly, their number is generally very large (usually there are of the order of 10^2 or even 10^3 PW's per valence orbital). As a result, the computational effort is quite big and the memory requirements for one-hundred-atom systems may easily approach one GByte. The output files are huge and some physical quantities are hindered in such a big amount of information. In order to reduce the computational effort, PW are usually employed to describe valence electrons only. Indeed, in most cases, core electrons (whose orbitals are localized and rapidly varying, needing many PW's for their expansion) can be considered as spectator, since they do not participate to chemical bonds. The effect of core electrons on the valence orbitals are reproduced by means of pseudo-potentials. The interested reader can refer to the existing literature [11, 12, 13, 14] for more details. Secondly, a lot of useless accuracy is employed in region of space where no electron density is prèsent. Thirdly, the use of PW generally implies PBC in the three spatial directions, which does not correspond to the physical reality in the case of isolated systems such as molecules or clusters or to semi-infinite systems such as surfaces. It is the purpose of the following sections to explain how the periodic slab can describe intrinsically non periodic phenomena such as dipolar fields.

When a plane wave basis set is used for the expansion of the electronic states, the slab is thus repeated in the direction perpendicular to the surface, adding a fictitious periodicity that might bias the numerical results. In order to get rid of spurious interactions between the images, the vacuum region has to be thick enough and the slabs should be symmetric with a zero total dipole moment. However, this is not always possible, for example along the (0001) orientation of wurtzite, since the stacking sequence is of the ABAB type, with two different inter-plane distances. Several schemes have been proposed in such cases. In one of them, two equivalent slabs, with one the mirror image of the other, are put in contact and fractional charges are introduced in the central layers in order to prevent charge transfer and bonding between alike-atoms [15]. Another solution consists in saturating the surfaces with non-integer charged hydrogen-like atoms, with a suitably fixed chemical potential [16]. In the case of III-V compounds, one has also tried to introduce a central layer of an element belonging to the IV-column of the periodic table [17].

More satisfactorily, other authors subtract the dipole-dipole interaction from the electrostatic energy by using cubic boxes [18] or simulation cells of arbitrary shape [19]. This method, which has been originally applied to molecular systems, has a severe limitation since it cannot be implemented self-consistently. More recently, Neugebauer and Scheffler [20], and Bengtsson [21], introduced a pseudo-system which is truly periodic and no longer presents a dipole moment in the cell, which makes possible to apply the standard supercell formalism and to restore the full self-consistent relation between the effective potential and the electron density.

In the following, we consider the (111) surface of MgO as a prototype, summarizing the theoretical results that have been obtained. Then, we discuss the technique to compensate the slab in some detail. We give a physical interpretation of this particular pseudo-system and carry out some test calculations in the framework of the DFT, which complement and correct some unclear points in previous works. We highlight the presence of an artificial field when PBC are used with a net dipole in the cell, then we show how to eliminate this effect and finally we give a numerical example.

4 Theoretical results on the MgO(111) surface

The literature about polar oxide surfaces has known a fast increase in the last years. From the experimental point of view, the preparation of polar orientations, either as single crystal terminations or as thin films grown on metal supports, has impressively progressed and nowadays it is not rare that polar surfaces of quality compared to cleavage faces can be produced. On the theoretical side, the peculiar properties of polar surfaces has stimulated many investigations, most of them based on computer simulations, by using ab-initio theories such as Hartree-Fock or DFT, or empirical inter-atomic potentials. A non exhaustive list of low-index polar oxide surfaces that have been the object of theoretical calculations includes:

- The (111) orientations of rock-salt oxides (MgO) [22, 23, 24, 25, 26];
- Polar (111) and (100) surfaces of inverse spinel compounds (Fe_3O_4) [27] ;
- The corundum (0001) surface (Fe_2O_3 [28], Al_2O_3 [29, 30]) ;
- The wurtzite (0001) and (000$\bar{1}$) surfaces (ZnO) [2, 31];
- The (110) and (111) termination of perovskite compounds ($SrTiO_3$, $BaTiO_3$) [32].

Some of the calculations carried out by using a localized basis set [27, 31, 32], while other ab-initio simulations were performed adopting the slab model with PBC in the three spatial directions [28, 25, 30]. In the latter cases, the dipole problem is avoided by choosing symmetric slabs. One exception is given by the study of alumina nucleation on TiC(111) [33], in which the PW-based scheme, with PBC, has been used in conjunction with the dipole correction following Ref.[21]. For the sake of conciseness, we consider here the (111) termination of magnesium oxide as a prototype for the computer simulation of polar oxide surfaces. The interested reader can refer to recent review articles [34] for a more comprehensive outline of the recent experimental and theoretical advances on other polar oxide surfaces.

The rock-salt structure consists of two interpenetrating face-centered cubic lattices of anions and cations. This structure is one of the most stable for highly ionic solids [35]. It is met in alkaline earth oxides (MgO, CaO, SrO, BaO) and in some transition metal oxides such as NiO, CoO, FeO, TiO, VO, etc with cations in a +2 oxidation state. Some of the latter ones are however very often non-stoichiometric in the cation and/or anion sub-lattices. The polar orientation of lowest indices is (111). Its two-dimensional unit cell is hexagonal and the surface atoms are three-fold coordinated. A crystal cut along (111) presents alternating layers of metal and oxygen composition, which are equidistant ($R_2 = R_1$). In an ionic picture, the two-dimensional unit cell bears a charge ± 2, so that a reduction of charge by a factor of 2 is required in the outer layers, according to Equation 2.1. When charge compensation is provided by changes in stoichiometry, simple electrostatic arguments suggest two stable surface configurations. One is obtained by removing every other atom in the outermost layer, which yields a missing row surface structure with a (2×1) reconstructed unit cell. This configuration may be thought of as a stacking of non-dipolar M/2O/M repeat units, with zero inter-unit distance. A second stable surface configuration is obtained by removing 75% of the atoms in the outermost layer and 25% in the layer beneath, in a way which produces a (2×2) surface unit cell. Several arrangements of atoms consistent with this surface stoichiometry are conceivable.

Single crystal (111) surfaces as well as thin films grown on various substrates have been produced for NiO, CoO and MgO. Most studies report reconstructed $p(2 \times 2)$ surface cells. On air-annealed NiO(111), the structural measurements indicates that the $p(2 \times 2)$ surface is of the octopolar type [36], with $\{100\}$ nanofacets and is terminated by Ni atoms, as predicted by molecular dynamics simulations of the surface [37]. After in-situ annealing and oxidation, however, the $p(2 \times 2)$ single crystal surface is better described as that of an hypothetical Ni_3O_4 inverse spinel structure [38]. In the case of MgO(111), the low temperature phase is analyzed as a mixture of octopolar and spinel reconstructions [39].

Several quantum mechanical studies of the unreconstructed MgO(111) surfaces have been performed, using a DV-Xα method (Discrete Variational) on a cluster model [22], a semi-empirical Hartree-Fock method [23], or a DFT approach [25], both in a slab geometry . All of them evidence an open-shell character of the outer layers, arising from important shifts and overlap of the surface valence and conduction bands. This produces a large reduction of charge in the surface layers, which fulfills the electrostatic criterion. At variance with ionic models, the calculated surface energy is large ($E_s \geq 5$ J/m^2), but not diverging. This however suggests that the unreconstructed surface is not the lowest energy configuration.

The stability of unreconstructed polar (111) films, on the other hand, seems possible when the thickness does not exceed a few monolayer. The theoretical study of VO ultra-thin films [40] shows that the O/V/O (111) tri-layer unit has a formation energy comparable to the O/VO/VO/O (100) or (110) thin films. It should be stressed that these bi- or tri-layers do not pertain to the semi-infinite polar surface family. The concept of cancellation of the *macroscopic* dipole moment does not apply to it, due to the presence of so few layers.

Simulations of reconstructed surface structures have been performed either using classical pair potential methods [24, 26]. They have considered the (2×1), the octopolar (2×2), the micro-facetted $(\sqrt{3} \times \sqrt{3})$R30° and $(2\sqrt{3} \times 2\sqrt{3})$R30° surfaces. The (2×2) surface energy is found to be lower than the (2×1) one. However, due to the lack of grand potential calculation, one cannot tell in which range of oxygen partial pressure, each surface termination is thermodynamically stable.

We have recently carried out extensive simulations of the (2×2) reconstructions of the MgO(111) surface. The surface is modeled by a symmetric slab consisting of nine atomic layers. Both Mg and O terminations are considered, for various surface stoichiometry. The plane-wave pseudo-potential scheme is employed. The exchange and correlation effects are described through the generalized gradient correction to the LDA [41]. For each configuration, we start from an educated guess, and optimize the geometry. All those simulations were completed by molecular dynamics runs on the ps scale and at temperatures as high as $\simeq 1200$ K, in order to explore the stability of the structures that have been found by local minimizations, and the possible dynamical paths between them. In the following, we summarize our numerical results that are listed according to the surface stoichiometry of the configurations that have been simulated. A more extensive report will be published [42]. The concentration and the chemical nature of the atoms in the two outermost planes identify the stoichiometry without ambiguity: for instance, (1/4 O, 3/4 Mg) means that the surface plane has one O out of four, and that three Mg out of four are present in the sub-surface plane. All the configurations described in the following are compensated, and correspond to a lower surface energy than the unreconstructed termination, computed within the same scheme.

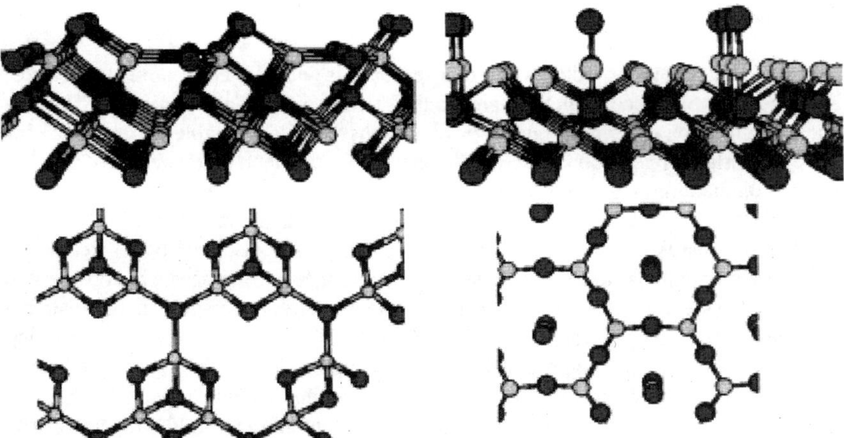

Figure 4: Left panel: side (above) and top (below) views of the O-terminated 2 × 2 octopole reconstruction of MgO(111). Oxygen and Magnesium atoms are drawn in dark and light grey, respectively. Right panel: same as before, for the 2 × 2 spinel reconstruction of MgO(111). The simulations were carried out for a nine-layer slab, for which any laterally-averaged quantity is symmetric under the inversion of the coordinates along the surface normal. The O atoms in the central layer are fixed at their bulk positions. In the side views, five atomic layers are drawn, the outermost plane being the one on the top, while the bulk layer is drawn below. In the views from the top, only the three outermost atomic planes are drawn for the sake of simplicity.

- *(1/4 O, 3/4 Mg)* Two configurations are found: the octopolar and the spinel-like reconstructions, in agreement with the structural determination performed by GIXS [39]. Their structural characteristics are sketched in Figure 4. Note that the atomic relaxations are large, even on the octopolar reconstruction, which strongly differs from the rigid structure proposed by Wolf on the basis of electrostatic arguments [36]. The spinel reconstruction is less stable than the octopolar one [44], and it turned into the latter one after a simulated annealing run at a temperature $\simeq$ 800 K as high, on the ps time scale. The bigger stability of the octopolar configuration can be easily explained by the larger coordination number of most surface atoms with respect to the spinel termination.

- *(2/4 O, 4/4 Mg)* This surface termination may be reached by removing two out of four surface oxygens from the unreconstructed face. Two different configurations are found by molecular dynamics runs, either corresponding to a (2×2) or a (2×1) reconstruction, as drawn in Figure 4. The surface energies of the two configurations are very close, consistently with the similar first-neighbor environments for both them.

- *(1/4 Mg, 3/4 O)* The octopolar and spinel configurations found for such a surface concentration are almost identical to those obtained for the case *(1/4 O, 3/4 Mg)*, by inverting the chemical nature of the atoms. Nevertheless, within the DFT-GGA, the Mg-termination of the spinel reconstruction is found to be metallic whereas the octopolar configuration is closed-shell and much more stable. This is in agreement with the experimental findings which suggest that the Mg-terminated octopole (that is, O-rich) is the stable reconstruction when raising the O partial pressure.

5 The dipole problem in one dimension

As already said in the previous sections, the slab approach is widely used to simulate surfaces, especially when complex reconstructions are studied. However, the slab approach, when coupled to three-dimensional periodic boundary conditions (PBC), is unable to reproduce intrinsically non-periodic phenomena, such as the electrostatic potential generated by a dipole distribution. It is the aim of the present section to introduce the problem in one dimension, along the surface normal. In the Poisson's equation, the electrostatic potential is uniquely determined by the charge density if and only if the actual BC are specified: two different BC give two different potentials (Fig. 5). In order to illustrate these differences, we consider the one-dimensional system that can be obtained by using a laterally averaged charge distribution along the direction perpendicular to the surface (z axis), which is a first necessary step to address the problem of modeling a slab with a net dipole moment along z. The charge density $\rho(\vec{r})$ is assumed to be periodic. It can generate different potentials (periodic or aperiodic) depending on whether the PBC are applied to the potential. In the following we restrict ourselves to neutral systems ($\int \rho(\vec{r})d^3\vec{r} = 0$). The generalization to charged surfaces is in progress [43]. The charge density is split in two parts. Firstly, we define a laterally averaged charge density as

$$\rho_{av}(z) = \frac{1}{S_{cell}} \iint \rho(\vec{r})dxdy \tag{3}$$

where the integration on the xy plane is restricted to the two-dimensional unit cell of area S_{cell}. The residual part of the density $\rho^{'}(\vec{r})$, which does not contribute to the dipole moment, is defined simply as:

$$\rho(\vec{r}) = \rho_{av}(z) + \rho^{'}(\vec{r}) \qquad \text{with} \iint \rho^{'}(\vec{r})dxdy = 0 \quad \forall z. \tag{4}$$

Using this decomposition and according to the superposition principle, the Poisson's equation (in three dimensions) is split in two parts:

$$\frac{\partial^2}{\partial z^2}V_{av}(z) = -4\pi\rho_{av}(z) \quad \text{and} \quad \Delta V^{'}(\vec{r}) = -4\pi\rho^{'}(\vec{r}). \tag{5}$$

The first one of equations (5) can be easily solved to give the laterally averaged potential $V_{av}(z)$ corresponding to $\rho_{av}(z)$:

$$V_{av}(z) = -2\pi \int_{-\infty}^{+\infty} \rho_{av}(z^{'})|z - z^{'}|dz^{'} \tag{6}$$

As displayed in Fig. 5, the natural boundary conditions for the potential of an isolated slab are $V_{av}(\pm\infty) = \pm 2\pi p_z$ (+ sign on the left and - sign on the right of the slab) where p_z is the z-projected dipole moment per slab and unit surface, such as:

$$p_z = \int_{-L/2}^{+L/2} \rho_{av}(z^{'})z^{'}dz^{'} = \frac{1}{S_{cell}} \iiint_{cell} z\rho(\vec{r})d^3\vec{r} \tag{7}$$

L is the dimension of the cell in the z direction. The solution of the second equation (5) gives the corrugated potential arising from the residual density $\rho^{'}(\vec{r})$. For this purpose, one can use the Fourier expansion. The generic reciprocal lattice vector is decomposed as $\vec{G} = \vec{G}_z + \vec{G}_{||}$, in two components that are normal and parallel to the surface plane, respectively:

$$V'(z,\vec{G}_{||}) = \begin{cases} \frac{2\pi}{|\vec{G}_{||}|}\int_{-\infty}^{+\infty} \rho'(\vec{G}_{||},z')\exp(-|z-z'||\vec{G}_{||}|)dz' & \text{if } \vec{G}_{||} \neq \vec{0} \\ 0 & \text{if } \vec{G}_{||} = \vec{0} \end{cases} \tag{8}$$

$V'(z,\vec{G}_{||})$ decays exponentially with z. We therefore focus on the laterally averaged potential and neglect the corrugated potential in the following.

We point out that all the previous equations can be used for a periodic as well as for an isolated system. For the latter one, let us keep the notation $V_{av}(z)$ for the averaged electrostatic potential. For a periodically repeated slab in the z direction, within the framework of the PBC, the potential that we note $V_{av}^{PBC}(z)$ is forced to be periodic and the 1D-Poisson's equation (5) in the reciprocal space reads:

$$\hat{V}_{av}^{PBC}(G_z) = \frac{4\pi\hat{\rho}_{av}(G_z)}{G_z^2} \tag{9}$$

In real space, by using inverse Fourier transform, $V_{av}^{PBC}(z)$ is equal to:

$$V_{av}^{PBC}(z) = \sum_{G_z\neq 0}\frac{4\pi}{G_z^2}\left[\frac{1}{L}\int_{-L/2}^{+L/2}\rho_{av}(z')\exp(-iG_z z' dz')\right]\exp(iG_z z). \tag{10}$$

In this expression the term $G_z = 0$ vanishes for neutral systems. For a fixed charge density $\rho(z)$, we emphasize that $V_{av}^{PBC}(z)$ differs from $V_{av}(z)$, and that the electrostatic energies are different whenever a dipole moment is present in the cell. For this purpose, we introduce the electrostatic energies per slab with ($\mathcal{E}_{es}^{PBC}$) or without PBC – that is, for an isolated slab – ($\mathcal{E}_{es}$) as:

$$\mathcal{E}_{es}^{PBC} = \frac{S_{cell}}{2}\int_{-L/2}^{+L/2}\rho_{av}(z)V_{av}^{PBC}(z)dz \qquad \mathcal{E}_{es} = \frac{S_{cell}}{2}\int_{-L/2}^{+L/2}\rho_{av}(z)V_{av}(z)dz. \tag{11}$$

The potential corresponding to the uncompensated system that is subject to PBC can be shown to be (see Appendix A):

$$V_{av}^{PBC}(z) = \frac{\partial\mathcal{E}_{es}^{PBC}}{\partial\rho_{av}(z)} = V_{av}(z) - 4\pi p_z(\frac{z}{L}-\frac{1}{2}) \qquad \text{with } -L/2 < z \leq +L/2. \tag{12}$$

The constant term in equation (12) is set in such a way that PBC are satisfied: $V_{av}^{PBC}(-L/2) = V_{av}^{PBC}(+L/2)$. With equations (11) and (12), the electrostatic energy per slab associated with the averaged potentials for the isolated and periodic systems, respectively, reads:

$$\mathcal{E}_{es}^{PBC} = \mathcal{E}_{es} - 2\pi S_{cell}\frac{p_z^2}{L} \tag{13}$$

As an illustration, we plot in Fig. 5 the very simple case of a charge distribution consisting of two delta peaks per cell with a finite surface charge density σ and such as $p_z = d\sigma$:

$$\rho_{av}^{(\uparrow\downarrow)}(z) = \sum_{n=-\infty}^{n=+\infty}\sigma\left[\delta(z-d/2+nL) - \delta(z+d/2+nL)\right] \tag{14}$$

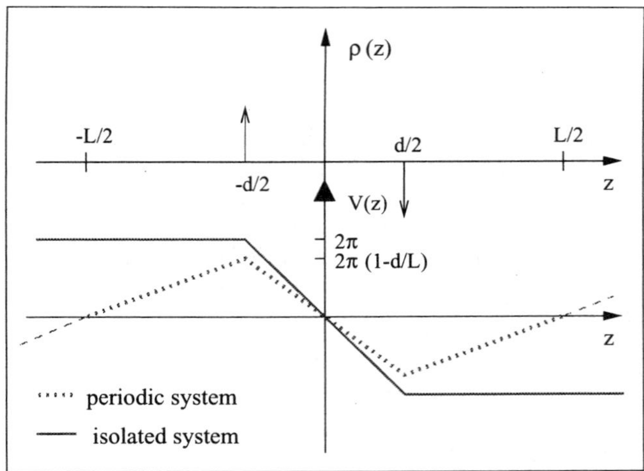

Figure 5: The laterally averaged potentials of the system with PBC ($V_{av}^{PBC}(z)$) or without ($V_{av}(z)$) for a model of charge density. The potentials are given in units of p_z.

As displayed in Fig. 5 and equations (12) and (13), the two potentials $V_{av}^{PBC}(z)$ and $V_{av}(z)$ and the associated energies are different. The system with PBC does not describe the isolated system but rather a slab in its depolarizing electric field. For a repeated system with a net dipole moment, the actual potential is aperiodic, which cannot be accounted for by PBC. In other terms, both PBC and discrete Fourier series are not suited to systems with a net dipole moment in the unit cell. Moreover, the additional term in potential and energy is long range and decays as $1/L$ with the dimension of the cell. It is impossible to make it negligible, unless unreasonably big cells are used, which implies a waste of computational resources.

6 The slab with the compensating dipole

A way to overcome the dipole problem was recently suggested by Schultz [45]. A model charge distribution is constructed to match the low multipole moments of the actual charge density, and the long-range part of the electrostatic potential is obtained from the model charge as if the system were isolated. The potential corresponding to the remainder charge density, which is short-range, is then evaluated within PBC. Such a method is not restricted to surfaces that are represented as a slab. However, it implies a fitting procedure on the charge density that may add some numerical complications. In the case of surfaces, we adopt the point of view of keeping the PBC and we introduce an ansatz to cancel out the net dipole moment (or equivalently the aperiodic component of the effective potential), while preserving the properties of the isolated slab such as charge density and electrostatic potential in the periodic system. For this purpose, we add a compensating charge distribution $\rho_{dip}(z)$ with a dipole moment per surface area $-p_z$ equal and opposite to that of the slab, such that:

$$\int_{-L/2}^{+L/2} \rho_{dip}(z)\, z\, dz = -p_z \qquad (15)$$

This compensating dipole must be located in the vacuum region, in such a way that it does not interact with the real system, as it is discussed in the following. The fictitious system is now formed by the charge distributions of the physical system, $\rho(\vec{r})$, and by the compensating one, $\rho_{dip}(z)$, under PBC. The *total* averaged electrostatic potential is now $V_{tot}^{PBC}(z) = V_{dip}^{PBC}(z) + V_{av}^{PBC}(z)$. *In the space outside the dipole distribution* $\rho_{dip}(z)$, the compensating potential $V_{dip}(z)$ is given by:

$$V_{dip}^{PBC}(z) = 4\pi p_z(\frac{z}{L} - \frac{1}{2}). \quad (16)$$

Since there is no dipole moment left in the cell, both density and total potential are actually periodic, that is, $V_{tot}^{PBC}(z) = V_{tot}(z)$. We can safely use the PBC and the Poisson's equation in the Fourier space can be written as :

$$\hat{V}_{tot}(G_z) = \hat{V}_{tot}^{PBC}(G_z) = -\frac{4\pi\hat{\rho}_{av}(G_z)}{G_z^2} - \frac{4\pi\hat{\rho}_{dip}(G_z)}{G_z^2} = \hat{V}_{av}^{PBC}(G_z) + \hat{V}_{dip}^{PBC}(G_z) \quad (17)$$

The first term in equation (17) is exactly the potential of the system with PBC (Eq. 9) and the second one $\hat{V}_{dip}^{PBC}(G_z)$ is the periodic component of the potential that is created by the compensating dipole in each cell, in reciprocal (Fourier) space. The central issue is to determine under which conditions $V_{tot}^{PBC}(z)$ is a good approximation to the real potential $V_{av}(z)$ in the region where the charge density of the physical system is appreciably non zero. A condition on electrostatic energy thus results, which allows us to restrict the class of compensating charge distributions to be used in the numerical scheme. Here we give an heuristic derivation, while a more rigorous one can be found in Appendix C.

In presence of the compensating potential $V_{dip}^{PBC}(z)$, the electrostatic energy of the charge distribution $\rho_{av}(z)$ of the physical system is given by:

$$\mathcal{E}_{comp}^{PBC} = \frac{S_{cell}}{2} \int_{-L/2}^{+L/2} \rho_{av}(z) \left[V_{dip}^{PBC}(z) + V_{av}^{PBC}(z)\right] dz \quad (18)$$

which differs from the electrostatic energy of the real system $\mathcal{E}_{es}$ (equation 11). In order to compute the latter one from periodic quantities, we make the hypothesis that:

$$\int_{-L/2}^{+L/2} \rho_{av}(z) \left[V_{dip}^{PBC}(z) + V_{av}^{PBC}(z)\right] dz \simeq \int_{-L/2}^{+L/2} \rho_{av}(z) V_{av}(z) dz \quad (19)$$

and that $\rho_{av}(z)$ and $\rho_{dip}(z)$ have zero overlap. A sufficient condition for equation (19) to hold is that in the region where $\rho_{av}(z) \neq 0$, $\left[V_{dip}^{PBC}(z) + V_{av}^{PBC}(z)\right] = V_{av}(z)$. An example for the one-dimensional case is given in Fig.6.

More generally, the source of the compensating potential V_{dip}^{PBC} could consist of two delta peaks but, as displayed in Fig. 6, such a correction introduces a discontinuity in the potential or at least in its derivatives. For practical reasons, we choose the compensating dipole density distribution as a sum of two gaussian functions of opposite charges. The distance between the centers of the two gaussians and their width are chosen in such a way as to avoid a slow decay of the Fourier components of the electrostatic potential and the overlap with $\rho_{av}(z)$. Under the previous assumptions, the energy $\mathcal{E}_{comp}^{PBC}$ defined in equation(18) is a good approximation to $\mathcal{E}_{es}$. To summarize, from a practical point of view, in a PW-based code, one has to: (1) evaluate the surface dipole moment p_z of the charge density; (2) add a compensating dipole

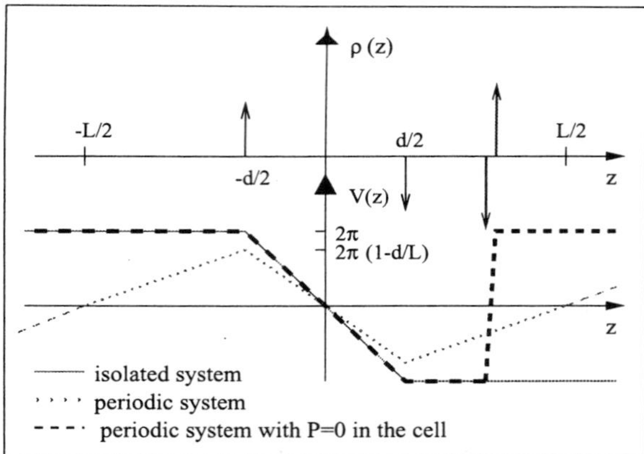

Figure 6: The case of a planar charge distribution given by two delta functions as in Fig.5 is presented, with a compensating dipole which is taken as two delta peaks.

distribution that satisfies the requirements previously stated; (3) compute the total electrostatic potential as usual, by using Fourier transforms (and thus apply PBC, as in equation 17); (4) evaluate $\mathcal{E}^{PBC}_{comp}$ in two steps: the first term, $\int dz\rho_{av}(z)V^{PBC}_{av}(z)$, by using Fourier transforms; the second term, $\int dz\rho_{av}(z)V^{PBC}_{dip}(z)$, by working out the direct space expression of $V^{PBC}_{dip}(z)$ and performing the integration.

7 Application to a water monolayer

Up to that point, the charge density was considered as fixed, independently of the presence of the compensating dipole. Now, we take into account the self-consistent relation between the electron density and the effective potential as in the Kohn-Sham equations. Therefore, the correction on the potential that takes into account the correct BC, implies a correction on the charge density. The charge density, the dipole moment and the corrected potential are thus calculated at each step of the minimization of the Hohenberg-Kohn total-energy functional:

$$\mathcal{E}_{tot} = \mathcal{T} + \mathcal{E}^{PBC}_{comp} + \mathcal{E}_{xc} \tag{20}$$

with $\mathcal{T}$ the kinetic energy $\mathcal{E}_{xc}$ the exchange and correlation energy that we treat within the generalized gradient correction to the local density approximation [41]. The system that was used by Bengtsson [21] to demonstrate the precision of the method is a water monolayer. Although it does not represent a polar oxide surface it has a large intrinsic dipole moment and can be used as a benchmark. The distance between the molecules of water in xy plane is fixed to $3\AA$ and the relaxation of all other structural degrees of freedom is allowed. The convergence of the energy and dipole moment with respect to the length of the cell is carefully tested. By using a PW cutoff energy of 50Ha and a (4,4,2) Monkhorst-Pack [46] k-point mesh for the Brillouin-zone integration, the calculated total energy is converged within a precision much better than that corresponding to the magnitude of the dipole correction. The monolayer is considered to be fully relaxed when the forces on the ions are less than 0.02 eV/Å.

The constant straight line displayed in the left panel of Figure 7 represents the reference

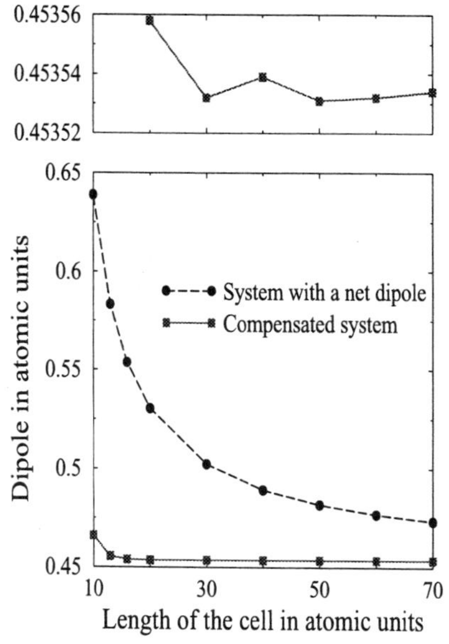

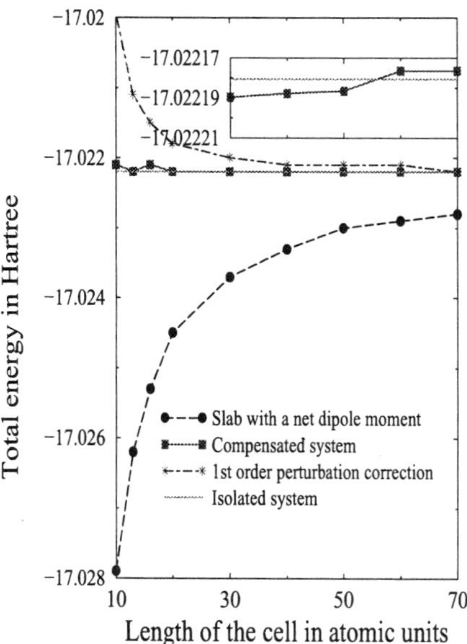

Figure 7: Convergence of the energy and the dipole moment of a water monolayer with respect to the length of the cell. Periodic boundary conditions are used throughout. The solid line in the left panel at $\simeq -17.02218$ Hartree represents the reference energy of an isolated system (see text for explanations).

energy of the isolated system. It is calculated by using two opposite slabs separated by 35 a.u. in a wide supercell (L=70 a.u.). By using Eq.(18) we compute the total electrostatic energy in three cases: (1) a slab with a net dipole (no compensation), within PBC; (2) a system with a compensating dipole, with a frozen charge density equal to that of case (1), which can be computed through perturbation theory by evaluating simply V_{dip}^{PBC} and $\mathcal{E}_{comp}^{PBC}$ (Equation 18); (3) the energy of the compensated system for which the self-consistent relation between the electron density and the effective potential is restored. Although the total energy computed via perturbation theory – case (2) – shows a fast convergence with respect to the cell dimension, the compensated system (3) tends more rapidly to the reference energy, apart for cell lengths smaller than 20 a.u. For cells of such a size, the energy of the compensated system is not equal to the reference energy because the potential of the real system does not come to a plateau in the region of the compensating dipole. This is an example of the *overlap* between the compensating dipole and the electron density of the physical system, which is discussed in detail in Appendix C. For instance, the cell length used by Bengtsson [21] (L= 6Å) to demonstrate the precision of the method, gives a non negligible overlap that implies a limited precision (of about 20 meV) on the electrostatic energy.

A more noticeable correction (of about 30%) appears on the dipole moment of the slab (see the right panel of Figure 7), when the system with a net dipole is compensated through the introduction of $\rho_{dip}(z)$. The cancellation of the artificial depolarizing electric field due to PBC induces a shift of the charge density which reduces the dipole moment. The correction is thus relevant whenever a surface dipole moment (or equivalently the work function) is evaluated in the slab geometry by using PBC.

Moreover, we point out that although the energy correction for usual cell dimensions,

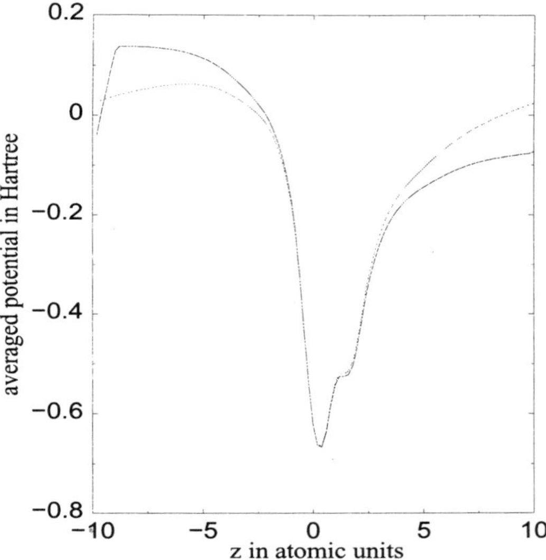

Figure 8: The laterally averaged potentials of the not compensated (dotted line) and compensated (solid line) slabs for the water monolayer, plotted along the surface normal z.

may be weak – of the order of a few mHa – the choice of the reference energy (i.e. the energy of the physical system without PBC) may be delicate due to the slow convergence of $\mathcal{E}_{tot}^{PBC}$ towards $\mathcal{E}_{es}$ (equation 19) with respect to the cell size. By using the correct reference, a cell $\sim$ 20 a.u. wide, a compensating dipole distribution consisting of two gaussian functions of opposite charges [47] , we note that a very good precision on both energy and dipole moment can be safely reached. As one can see in the inset of figure 7, a precision as large as 1 meV and 1 mDebye are reached on the total energy and the dipole moment of the virtual isolated system, respectively.

As a last point, the difference between the potentials of the not compensated and the compensated systems in Figure 8 is noteworthy. The two potentials are practically identical in the region where the electron density is appreciable, although in the void region near to the surface they sensitively differ. Some care must thus be paid whenever one is interested in physical phenomena happening at the vicinity of the surface. For instance, when calculating the adhesion energy of layers with a net dipole moment as a function of the inter-layer distance, the reliability of the model based on the dipole correction must be accurately checked.

Appendix A: The depolarizing electric field within the periodic boundary conditions

Here we give a detailed derivation of Equation (12), and show that the additional depolarizing field is an artifact due to PBC when a dipole field is present in the unit cell. Furthermore, some results of this calculation are used in Appendix C.

$$\begin{aligned} & \frac{\delta}{\delta\rho_{av}(z)}\left[\frac{1}{2}\int\rho_{av}(z^{'})V_{av}^{PBC}(z^{'})dz^{'}\right] \\ = \; & \frac{\delta}{\delta\rho_{av}(z)}\left[\frac{1}{2}\int\rho_{av}(z^{'})\sum_{G_z\neq 0}\frac{4\pi}{G_z^2}\left[\frac{1}{L}\int_{-L/2}^{+L/2}\rho_{av}(z^{''})\mathrm{e}^{-iG_zz^{''}}dz^{''}\right]\mathrm{e}^{iG_zz^{'}}dz^{'}\right] \end{aligned} \tag{21}$$

The functional derivative of this expression gives two terms that are identical:

$$\frac{\delta\mathcal{E}_{es}^{PBC}}{\delta\rho_{av}(z)} = \sum_{G_z\neq 0}\frac{4\pi}{G_z^2}\left[\frac{1}{L}\int_{-L/2}^{+L/2}\rho_{av}(z^{''})\,\mathrm{e}^{-iG_zz^{''}}dz^{''}\right]\mathrm{e}^{iG_zz} = V_{av}^{PBC}(z) \tag{22}$$

We perform the summation ($G_z = 2\pi n_z/L$) before the integration and obtain [48]:

$$\begin{aligned} \frac{\delta\mathcal{E}_{es}^{PBC}}{\delta\rho_{av}(z)} &= \int_{-L/2}^{+L/2}\rho_{av}(z^{''})\left[\sum_{G_z\neq 0}\frac{4\pi}{G_z^2}\frac{1}{L}\mathrm{e}^{iG_z(z-z^{''})}\right]dz^{''} \qquad (23) \\ &= \int_{-L/2}^{+L/2}\rho_{av}(z^{''})\left[\frac{2L}{\pi}\sum_{n_z>1}\frac{\cos\left[\frac{2\pi n_z}{L}(z^{''}-z)\right]}{n_z^2}\right]dz^{''} \\ &= 2L\pi\int_{-L/2}^{+L/2}\rho_{av}(z^{''})\left[\frac{|\,z^{''}-z\,|^2}{L^2}-\frac{|\,z^{''}-z\,|}{L}+\frac{1}{6}\right]dz^{''} \quad \text{with} \quad 0\leq z^{''}-z\leq L \end{aligned}$$

For a neutral system, some terms vanish and :

$$\frac{\delta\mathcal{E}_{es}^{PBC}}{\delta\rho_{av}(z)} = -2\pi\int_{-L/2}^{+L/2}\rho_{av}(z^{''})\,|\,z^{''}-z\,|\,dz^{''}+\frac{2\pi}{L}\int_{-L/2}^{+L/2}\rho_{av}(z^{''})z^{''2}dz^{''}-\frac{4\pi z}{L}\int_{-L/2}^{+L/2}\rho_{av}(z^{''})z^{''}dz^{''} \tag{24}$$

The first term in the above equation is exactly $V_{av}(z)$, the potential without PBC, the second one does not depend on the z variable and the third one gives the z-projected dipole moment per unit cell p_z:

$$\frac{\delta\mathcal{E}_{es}^{PBC}}{\delta\rho_{av}(z)} = V_{av}(z)-\frac{4\pi p_z}{L}z+constant = V_{av}^{PBC}(z) \tag{25}$$

The constant is determined by the actual boundary conditions as explained in the main text. By using equations (22) and (24) we obtain Eq.(12).

Appendix B: The constraint on the compensating dipole distribution

The total dipole moment must be equal to zero in the cell. As a consequence, a compensating distribution $\rho_{dip}(z)$ is introduced such that:

$$\int_{-L/2}^{+L/2} z^{'}\left[\rho_{av}+\rho_{dip}\right](z^{'})dz^{'} = 0 \tag{26}$$

The previous equation provides a functional dependence of $\rho_{dip}(z)$ on ρ_{av}; therefore

$$\frac{\delta}{\delta\rho_{av}}\{\int_{-L/2}^{+L/2} z' \left[\rho_{av}+\rho_{dip}\right](z')dz'\} = 0 \tag{27}$$

or, equivalently:

$$\int_{-L/2}^{+L/2} z' \frac{\delta\rho_{dip}(z')}{\delta\rho_{av}(z)} dz' = -z \tag{28}$$

a result that is used in Appendix C.

Appendix C: Derivation of the compensating potential in one dimension

In this appendix, we give an alternative derivation for the expression of $\mathcal{E}^{PBC}_{comp}$. Let us assume that we can only calculate quantities within PBC. We introduce in our simulation box a compensating charge distribution $\rho_{dip}(z)$, whose corresponding potential, within PBC, is $V^{PBC}_{dip}(z)$. Both $\rho_{dip}(z)$ and $V^{PBC}_{dip}(z)$ have to be determined, within the condition that the electrostatic total potential of the compensated system $V^{PBC}_{comp}(z)$ be representative of the electrostatic potential $V_{av}(z)$ of the virtually isolated system. By considering the electrostatic energy, we require that $\mathcal{E}^{PBC}_{comp}$ as calculated in the computer code approaches $\mathcal{E}_{es}$ given in equation (19):

$$\mathcal{E}^{PBC}_{comp} = \frac{S_{cell}}{2}\int_{-L/2}^{+L/2} \rho_{av}(z)\left[V^{PBC}_{dip}(z)+V^{PBC}_{av}(z)\right]dz \simeq \mathcal{E}_{es} = \frac{S_{cell}}{2}\int_{-L/2}^{+L/2} \rho_{av}(z)V_{av}(z)dz \tag{29}$$

We can deduce the potential in the real space V^{PBC}_{comp} which is consistent with $\mathcal{E}^{PBC}_{comp}$, under the condition (15) (see Appendix B). We take into account the fact that the charge density of the compensating dipole depends on the charge density of the real system $\rho_{av}(z)$. The functional derivative of the electrostatic energy with respect to the charge density $\rho_{av}(z)$ reads:

$$\begin{aligned}
\frac{1}{S_{cell}}\frac{\delta\mathcal{E}^{PBC}_{comp}}{\delta\rho_{av}(z)} &= \frac{1}{2}\int dz'\frac{\delta\rho_{av}(z')}{\delta\rho(z)}\left[V^{PBC}_{av}(z')+V^{PBC}_{dip}(z')\right] \\
&+\frac{1}{2}\int dz'\rho_{av}(z')\frac{\delta V^{PBC}_{av}(z')}{\delta\rho(z)}+\frac{1}{2}\int dz'\rho_{av}(z')\frac{\delta V^{PBC}_{dip}(z')}{\delta\rho(z)} \\
&= V^{PBC}_{av}(z)+\frac{1}{2}V^{PBC}_{dip}(z)+\frac{1}{2}\int dz'\int dz''\rho_{av}(z')\frac{\delta V^{PBC}_{dip}(z')}{\delta\rho_{dip}(z'')}\frac{\delta\rho_{dip}(z'')}{\delta\rho_{av}(z)}
\end{aligned} \tag{32}$$

where we have used the chain rule for derivation. Now, we consider that $\rho_{dip}(z')$ is the source of the compensating potential $V^{PBC}_{dip}(z')$. Therefore, by using the results of Appendix A, and replacing the subscripts av with dip, one obtains :

$$\frac{\delta V^{PBC}_{dip}(z')}{\delta\rho_{dip}(z'')} = 2\pi L\left[\frac{|z''-z'|^2}{L^2}-\frac{|z''-z'|}{L}+\frac{1}{6}\right] \tag{33}$$

By integrating over z', and using equation(6), one obtains :

$$\int dz^{'} \rho_{av}(z^{'}) \frac{\delta V_{dip}^{PBC}(z^{'})}{\delta \rho_{dip}(z^{''})} = \left[V_{av}(z^{'}) - \frac{4\pi p_z z^{''}}{L} + constant \right] \quad (34)$$

Inserting the previous result in equation (32), and using the constraint on $\rho_{dip}(z)$ (equation 28), one obtains :

$$V_{comp}^{PBC}(z) = V_{av}(z) + \frac{1}{2}\left[V_{dip}^{PBC}(z) - \frac{4\pi p_z}{L} z + constant \right] + \frac{1}{2}\int dz' V_{av}(z^{'}) \frac{\delta \rho_{dip}(z')}{\delta \rho_{av}(z)} \quad (35)$$

Apart from the last term in the previous equation, we see that if we construct, $V_{dip}^{PBC}(z) = \frac{4\pi p_z}{L} z$ in the region of space where $\rho_{av}(z) \neq 0$, the electrostatic potential $V_{es}^{PBC}(z)$ is equal to the laterally averaged potential $V_{av}(z)$ of the real system (unless for a constant term), plus the last term on the right side of equation (35), that we note $\Delta V_{es}(z)$. To our knowledge, the existence of $\Delta V_{es}(z)$ has never been reported so far, in any of the paper dealing with the dipole correction.

In order to understand the physical meaning of $\Delta V_{es}(z)$, we consider the simple yet representative case $\rho_{dip}(z) = p_z f(z)$, with $f(z)$ an arbitrary smooth function independent of $\rho_{av}(z)$. Therefore, by considering that $p_z = \int dz\, z\, \rho_{av}(z)$ and $\delta\rho_{dip}(z')/\delta\rho_{av}(z) = z f(z') = z \rho_{dip}(z')/p_z$:

$$\Delta V_{es}(z) = \frac{z}{2p_z} \int dz' V_{av}(z^{'}) \rho_{dip}(z') \quad (36)$$

which can sensitively be different from zero and thus contribute to $V_{comp}^{PBC}(z)$ whenever the electrostatic potential of the virtual isolated system varies in the region where the compensating dipole distribution is non zero. Such a correction on the electrostatic potential implies that the term

$$\Delta \mathcal{E}_{comp} = \frac{S_{cell}}{2} \int dz' V_{av}(z^{'}) \rho_{dip}(z^{'}) \quad (37)$$

should accordingly corrects the energy. Therefore, two conditions on the compensating distribution $\rho_{dip}(z)$ must be fulfilled in order that $V_{comp}^{PBC}(z)$ which was defined in equation (35) be equal to $V_{es}(z)$ in the region where $\rho_{av}(z) \neq 0$: (1) the dipole of the compensating distribution must be equal and opposite to that of the average charge density of the real system (equations 28 and 35); (2) the dipole distribution must be placed in a region where the electrostatic potential of the virtually isolated system is constant, i.e. $\rho_{dip}(z)$ and $\rho_{av}(z)$ must not overlap.

References

[1] P.W. Tasker, J. Phys. C: Solid State Phys. **12** 4977 (1979).

[2] R. W. Nosker, P. Mark and J.D. Levine, Surf. Sci **19**, 291 (1970).

[3] D. Vanderbilt and R. D. King-Smith Phys. Rev. B **48**, 4442 (1993).

[4] S. Kuntzinger, N. E. Ghermani, Y. Dusausoy, C. Lecomte, Acta Cryst. B **54**, 819 (1998).

[5] see, e.g., R. G. Parr and W. Yang, *Density Functional theory of atoms and molecules* (Oxford University Press, Oxford, 1989)

[6] C. Noguera, A. Pojani, F. Finocchi, J. Goniakowski, in *Chemisorption and reactivity on Supported Clusters and Thin Films: towards an understanding of microscopic processes in catalysis* NATO ASI Series E: Applied Sciences, Vol 331. Ed. R. M. Lambert, G. Pacchioni (Kluwer Academic Publishers, Dordrecht, 1997).

[7] C. B. Duke, Chem. Rev. **96**, 1237 (1996).

[8] P. Deàk, Phys. Stat. Sol. (b) **217**, 9 (2000).

[9] R. P. Feynman, Phys. Rev. **56**, 340 (1939).

[10] P. Pulay, Mol. Phys. **17**, 197 (1969).

[11] M.C. Payne, M.P. Teter, D.C. Allan, T.A. Arias, and J.D. Joannopoulos, Rev. Mod. Phys. **64**, 1045 (1992).

[12] W.E. Pickett, Comp. Phys. Rep. **9**, 116 (1989).

[13] N. Troullier and J.L. Martins, Phys. Rev. B **43**, 1993 (1991).

[14] D. Porezag, M.R. Pederson and A.Y. Liu, Phys. Stat. Sol. (b) **217**, 219 (2000).

[15] K. Shiraishi, J. Phys. Soc. Japan **59**, 3455 (1990).

[16] J. Yamauchi, M. Tsukada, S. Watanabe and O. Sugino, Phys. Rev. B **54**, 5586 (1996).

[17] S. Mankefors, Phys. Rev. B **59** 13151 (1999) 13151

[18] G. Makov and M. C. Payne, Phys. Rev. B **51**, 4014 (1995).

[19] L. N. Kantorovich, Phys. Rev. B **60**, 15476 (1999); L. N. Kantorovich and I. I. Tupitsyn, J. Phys. Condens. Matter **11**, 6159 (1999)

[20] J. Neugebauer and M. Scheffler, Phys. Rev. B **46**, 16067 (1992).

[21] L. Bengtsson, Phys. Rev. B **59**, 12301 (1999).

[22] M. Tsukada and T. Hoshino, J. Phys. Soc. Jpn. **51** 2562 (1982).

[23] A. Pojani, F. Finocchi, J. Goniakowski, C. Noguera, Surf. Sci. **387**, 354 (1997).

[24] K. Hermansson, M. Baudin, B. Ensing, M. Alfredsson, M. Wojcik, J. Chem. Phys. **109** 7515 (1998).

[25] J. Goniakowski, C. Noguera, Phys. Rev. B **60**, 16120 (1999).

[26] S. C. Parker, N. H. de Leeuw, S. E. Redfern, Faraday Discuss. **114** 381 (1999).

[27] J. Ahdjoudi et al. Surf. Sci. **443**, 133 (1999).

[28] X.G. Wang et al, Phys. Rev. Lett. **81**, 1038 (1998).

[29] V.E. Puchin et al, Surf. Sci. **370**, 190 (1997).

[30] C. Verdozzi, D.R. Jennison, P.A. Schultz and M.P. Sears, Phys. Rev. Lett. **82**, 799 (1999).

[31] A. Wander et al., Phys. Rev. Lett. **86**, 3811 (2001).

[32] A. Pojani, F. Finocchi and C. Noguera, Surf. Sci. **442**, 179 (1999); F. Bottin, F. Finocchi and C. Noguera, to be published.

[33] B.I. Lundqvist et al., Surf. Sci. **493**, 253 (2001).

[34] C. Noguera, J. Phys. Cond. Matt. **12**, R367 (2000).

[35] J. C. Phillips, Rev. Mod. Phys. **42**, 317 (1970).

[36] D. Wolf, Phys. Rev. Lett. **68** 3315 (1992).

[37] P. M. Oliver, G. W. Watson, S. C. Parker, Phys. Rev. B **52** 5323 (1995).

[38] A. Barbier et al. Phys. Rev. Lett. **84**, 2897 (2000).

[39] A. Barbier and J. Jupille, private communication.

[40] G. Kresse, S. Surnev, M.G. Ramsey, and F.P. Netzer, Surf. Sci. **492**, 329 (2001).

[41] J.P. Perdew, K. Burke and M. Enzerhof, Phys. Rev. Lett. **77**, 3865 (1996).

[42] F. Finocchi, A. Barbier, J. Jupille and C. Noguera, to be published.

[43] F. Bottin and F. Finocchi, to be published.

[44] The computed surface energies of the O-terminated octopolar and spinel reconstructions are 2.1 and 4.4 J/m^2, respectively.

[45] P.A. Schultz, Phys. Rev. B **60**, 1551 (1999).

[46] H.J. Monkhorst and J.D. Pack, Phys. Rev. B **13**, 5188 (1976).

[47] The two gaussian functions are 0.05 a.u. wide and their centers are shifted by 1 a.u. with respect to each other. The amplitude of the gaussian functions is a dynamical variable, which means that it can varies during the simulation runs in order to compensate the dipole of the slab within the numerical accuracy. The mass center of the compensating dipole distribution is located at mid-distance between two periodic images of the slab, in order to minimize the overlap with the slab electron density without having to increase the thickness of the void region too much.

[48] I.S. Gradshteyn and I.M. Ryzhik, *Table of integrals, series and products* (Academic Press, London, 1980).

Computational Materials Science
C.R.A. Catlow and E.A. Kotomin (Eds.)
IOS Press, 2003

Computer Simulation of Surfaces of Metals and Metal-Oxide Materials

Nora H. DE LEEUW[1], Timothy G. COOPER[1], Christopher J. NELSON[1,2]
Donald MKHONTO[3] and Phuti E. NGOEPE[3]

[1] *Department of Chemistry, University of Reading, Whiteknights, Reading RG6 6AD, UK*

[2] *Davy Faraday Research Laboratory, Royal Institution of Great Britain, Albemarle Street, London W1X 4BS, UK*

[3] *Materials Modeling Center, University of the North, Sovenga 0272, South Africa*

1. Introduction

The aim of this chapter is to describe two approaches to modelling the structures and energetics of surfaces of metals and inorganic solids, using both classical atomistic simulations and electronic structure calculations. We will illustrate the application of these techniques by describing some of our recent work using electronic structure calculations to elucidate the structure and energies of both perfect and defective surfaces of silver and calcium oxide as well as atomistic simulations of adsorption of small molecules at the surfaces of the more complex oxide materials calcite and apatite. Understanding extended defects such as surfaces and interfaces is important as they control the physical and chemical properties of ceramics including electronic and ionic conduction, diffusion, creep, crystal growth, corrosion, reactivity and catalysis [1,2]. Experiments that probe the structure and stability at the atomic level are often difficult and hence computer simulation can been used as a complementary tool for understanding the structure and properties of such defects [1,3,4].

The classical atomistic simulation of the surfaces of ionic solids was pioneered by the work of Tasker [1,5] and Mackrodt and Stewart [6]. Much of the early work was confined to modelling planar surfaces of the cubic rocksalt oxides MgO, CaO and NiO [7-9], later extended to more complicated materials such as Cr_2O_3 [10], Al_2O_3 [11] and Fe_3O_4 [12]. Further complexity has been afforded by the development of models for oxy-anions such as SO_4^{2-}, CO_3^{2-} and PO_4^{3-} which has led to surface simulations of for example $CaCO_3$ [13,14] and $BaSO_4$ [15] and in this work $Ca_{10}(PO_4)_6F_2$. More recently, the surface-water interface is being addressed, as the growing interest in crystal growth and dissolution processes means that we can no longer meaningfully model surfaces in a vacuum but have to take into account the rôle of water in these processes [e.g. 16-18].

In addition, the increase in computer power and memory over the last few years has made it possible to model surfaces using electronic structure calculations, usually employing methods based on the Density Functional Theory. For example, Manassidis and Gillan [19] investigated surfaces of α-alumina while the work of Schroer *et al* [20] concentrated on ZnO and CdS surfaces. Since then, a wide range of materials has been studied, from metals such as iridium [21] and copper [22] to oxides like MgO [23], SnO_2 and TiO_2 [24,25] and even semiconductors such as $ErSi_{1.7}$ [26] and InAs [27]. Another aspect of surfaces which has been modelled extensively during the last few years is adsorption of various small molecules on metals, such as CO on palladium [28] and CO and CH_4 on nickel [29,30], and oxides such as CO and water on TiO_2 [31,32] and NH_3 on MgO [33]. Apart from these static calculations, electronic structure molecular dynamics simulations have also been performed by various groups, as for instance the dissociative chemisorption of Cl_2 on silicon [34,35]. However, these methods are, at present, still restricted in the number of atoms that can be treated and hence we find that the application of a combination of accurate electronic structure techniques, together with more approximate classical simulations for larger systems, is an efficient approach to model detailed surface structures and complex interfaces.

2. Simulating Surfaces

There are two approaches to the computer modelling of solid surfaces, employing either a two-dimensional or a three-dimensional simulation cell. In both cases, periodic boundary conditions are usually employed to limit the computer time required to model a system of realistic size. Periodic boundary conditions enable a finite number of particles to generate realistic behaviour by mimicking an infinite system in such a way that when a particle leaves the simulation box an image rejoins the box on the opposite side. The box of particles is thus surrounded by images of itself, producing an infinite repeat of the simulation cell in either two or three dimensions. Periodic boundary conditions are therefore very effective in modelling crystalline materials and it is also adequate for liquids or amorphous materials provided that any long-range properties are not considered.

2.1 Two-dimensional Periodicity

In the computer code METADISE [36], employed in this work for the simulation of the calcite and apatite systems, the crystal is considered to consist of a series of (charged) planes parallel to the surface and periodic in two dimensions. The crystal is then divided into two blocks, each of which is divided into two regions, region I and region II (figure 1a). The ions in region I are allowed to relax explicitly while those in region II are held fixed at their bulk equilibrium positions, although the two region IIs are allowed to move relative to each other. It is necessary to include region II to ensure that the potential of an ion at the bottom of region I is modelled correctly [5]. A surface is created when block II is removed with the top of region I as the free surface (figure 1b). Interfaces such as stacking faults and grain boundaries can be studied by fitting two surface blocks together in different orientations.

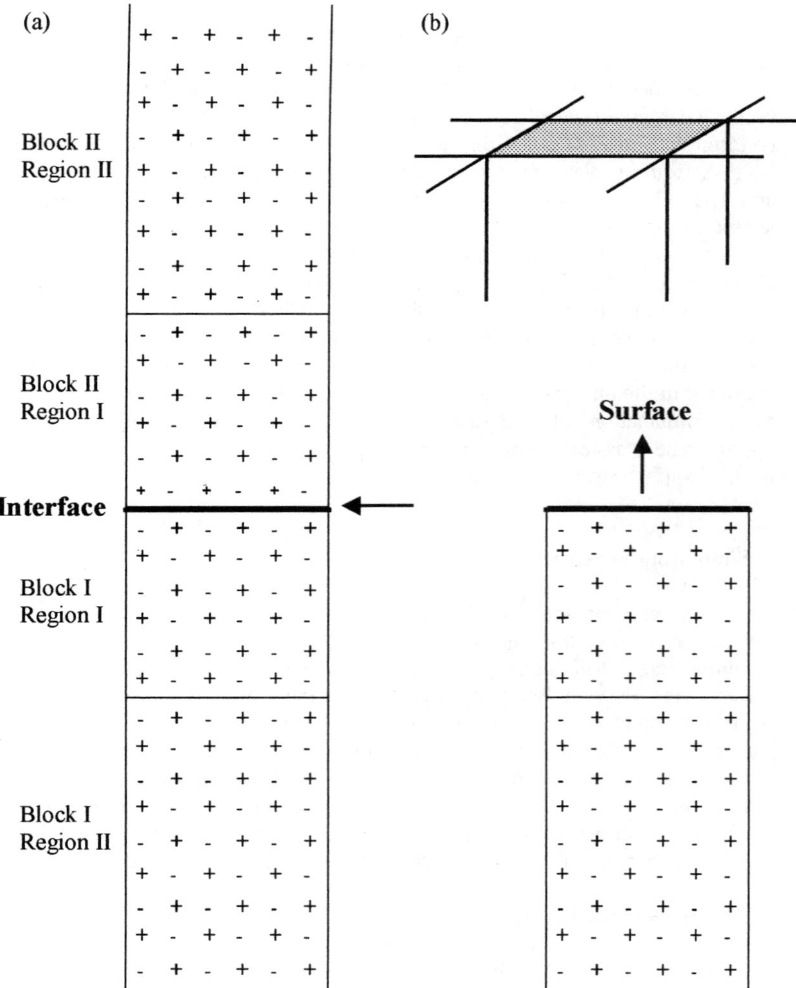

Figure 1. The two region approach used in METADISE, (a) the complete crystal and (b) half a crystal, exposing a surface, periodic in two dimensions.

The energy of the crystal is made up of two parts:

$$U_{latt} = U_1 + U_2 \tag{1}$$

where U_1 and U_2 are the energies of the combined ions in region I and region II respectively. The energy of region I is given as follows:

$$U_1 = \sum_{\substack{i \in I \\ j \in I}} \sum_{l} \Psi_{ij}\left(\left|r_{ij} - r_l\right|\right) + \frac{1}{2} \sum_{\substack{i \in I \\ j \in II}} \sum_{l} \Psi_{ij}\left(\left|r_{ij} - r_l\right|\right) \tag{2}$$

where Ψ_{ij} is some suitable pair potential. The first term includes interactions between the ions in region I only and the second term describes the interactions between the ions in region I and those in region II, which is referred to as the boundary interaction energy. The energy of region II consists only of this boundary interaction energy because the ions are kept fixed and hence the interaction between the ions in region II itself is unchanged and can therefore be ignored. The energy contribution is thus as follows:

$$U_2 = \frac{1}{2}\sum_{\substack{i \in I \\ j \in II}} \Psi_{ij}\left(\left|r_{ij} - r_1\right|\right) \tag{3}$$

2.2 Three-dimensional Periodicity

When we use three-dimensional periodic boundary conditions, as for example in our electronic structure calculations of silver and CaO, the approach for modelling a surface is first to simulate the bulk crystal, which has to be orientated in such a way that two of the three lattice vectors lie in the plane of the surface that we are investigating. A simulation of the bulk crystal is run to obtain the relaxed lattice constants and crystal structure. The surface is then created by expanding the lattice vector in the stacking direction (this is the third lattice vector which is the only one to have a component perpendicular to the surface), which creates a slab of material with two surfaces and a void between itself and its images in the stacking direction (figure 2). It is important that the void is big enough so that no interactions occur between one surface of the slab and its image across the void. Care must also be taken that the slab itself contains enough layers of material to behave as bulk crystal in the middle of the slab and to prevent ions at the surface experiencing interactions with the other face through the slab itself. Once created, the surface structures and energies can be calculated keeping the lattice vectors of this system of slab and void fixed at the bulk-relaxed values, to prevent the surfaces collapsing across the void. The atoms in the slab itself are of course allowed to relax during the calculation.

3. Surface Defects

Just as there are two approaches to the modelling of surfaces, either employing a two- or three-dimensional periodic simulation cell, similarly there are also two approaches to the modelling of surface defects, such as vacancies or adsorbates at the surface.

3.1 Supercell approach

In most computer simulation codes, such as METADISE and VASP used in this work, any defects in the bulk or on the surface are periodically repeated throughout the crystal. As this periodic repeat may give rise to unnaturally large concentrations in the crystal, large supercells made up of multiple simulation cells are used to model defects at low concentration. However, especially with computationally intensive calculations, the size of the supercells is limited by the available computational resources and often smaller than is required to give a realistic defect concentration. Some codes therefore employ an alternative approach, whereby the defect is truly isolated and embedded within a perfect crystal, hence simulating defects at infinite dilution.

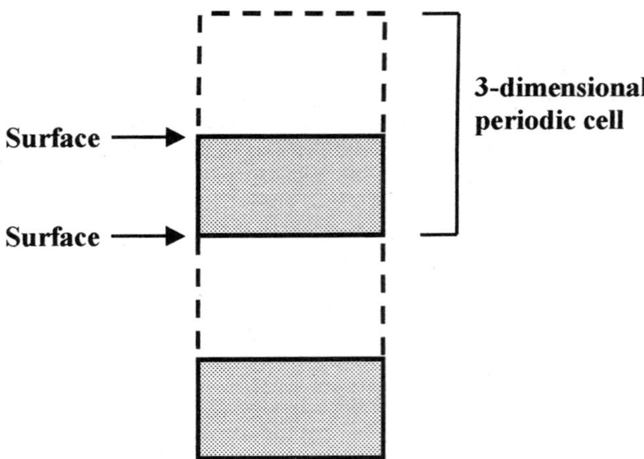

Figure 2. Schematic representation of the three-dimensional surface simulation cell.

3.2 Isolated defects

One code which allows for the modelling of isolated surface defects, such as charged vacancies or adsorbed molecules, is CHAOS, a surface simulation code developed by Duffy and Tasker [8] which calculates the structures and energies of point defects at the surface of crystals. In this code the crystal is again divided into a region I and a region II. Region II is further subdivided into two regions, IIa and IIb. In this case the regions are hemispheres centred upon the point defect on the surface (figure 3).

The interactions between ions in region I and region IIa are calculated explicitly just like the interactions between the ions in regions I and II described above. The interactions between region I and the rest of the crystal are approximated [37] assuming the rest of the crystal is a dielectric continuum. In addition, the defect at the surface, which is a dielectric discontinuity, creates a dipole which induces dipoles in region IIb. The interactions between region I and IIb are therefore modified to include these charge induced dipoles. These approximations assume that the effect of the defect on the local geometry of the crystal is only noticeable in the immediate surroundings of the defect but that the energy of the rest of the crystal is modified by the charged defect. The total energy of the system therefore consists of three terms:

$$U(\underline{x},\underline{y}) = U_1(\underline{x}) + U_2(\underline{x},\underline{y}) + U_3(\underline{y}) \tag{4}$$

where $U_1(\underline{x})$ is the energy of region I, $U_3(\underline{y})$ is the energy of region II and $U_2(\underline{x},\underline{y})$ is the energy of interaction between regions I and II. The vectors $\underline{x}$ and $\underline{y}$ represent the co-ordinates of the ions in region I, which are calculated explicitly, and the displacements in region II respectively. As $U_3(\underline{y})$ contains an infinite number of displacements, it cannot be solved exactly and we need to make assumptions. We assume that the distance between the defect and the innermost region of region II is large enough to consider the outer region to equal the perfect crystal. The charge induced dipoles due to the defect cause harmonic displacements of the ions only and hence $U_3(\underline{y})$ can be represented by a quadratic function of $\underline{y}$:

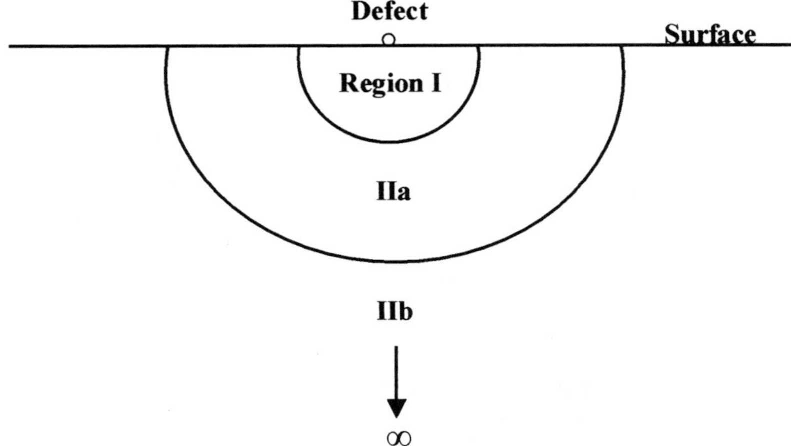

Figure 3. Isolated Defect

$$U_3(\underline{y}) = \frac{1}{2}\underline{y}\underline{\underline{A}}\underline{y} \tag{5}$$

where $\underline{\underline{A}}$ is the force constant of the harmonic displacement. At equilibrium, when $\underline{y} = \underline{\bar{y}}$, the total force which is the derivative of the total energy is zero and hence, when we substitute equation (5) into expression (4), we obtain:

$$\frac{\partial U(\underline{x},\underline{y})}{\partial \underline{y}} = \frac{\partial U_2(\underline{x},\underline{y})}{\partial \underline{y}} + \underline{\underline{A}}\underline{y} = 0 \tag{6}$$

If we multiply this equation by $\underline{y}/2$, this becomes:

$$-\left(\frac{\underline{\underline{y}}}{2}\right)\frac{\partial U(\underline{x},\underline{y})}{\partial \underline{y}} = \frac{1}{2}\underline{y}\underline{\underline{A}}\underline{\bar{y}} \tag{7}$$

and thus, when $\underline{y} = \underline{\bar{y}}$ and the outer displacements are at equilibrium with a given set of displacements of the inner region we obtain an expression for the total energy as a function of $U_1(\underline{x})$ and $U_2(\underline{x},\underline{y})$ only:

$$U(\underline{x},\underline{y}) = U_1(\underline{x}) + U_2(\underline{x},\underline{y}) - \frac{1}{2}\frac{\underline{\underline{y}}\partial U_2(\underline{x},\underline{y})}{\partial \underline{y}} \tag{8}$$

By iteratively minimising with respect to $\underline{x}$ until the forces on each ion in region I is zero, *i.e.*:

$$\frac{\partial U(\underline{x},\underline{y})}{\partial \underline{y}} = 0 \tag{9}$$

we obtain the total energy of the relaxed system including the defect. The energy is calculated in the same way as in equations (1-3) and the defect energy is then obtained from the difference between the energy of the perfect surface and that with point defect, obtained by CHAOS. One consequence of using CHAOS to calculate defects at surfaces rather than in the bulk is that the calculation of the ionic displacements and the energy of the continuum needs to be modified with respect to a bulk calculation. In a bulk calculation, the energy of the continuum is calculated by a r^{-4} summation (where r is the distance from the defect), which assumes that away from the defect in region II there is no structural deviation from the perfect crystal structure. This is an unrealistic assumption when surfaces are concerned and the energy is therefore calculated as a sum of planar integrals around the surface for the planes in region I and II defined in figure 1 and a volume integral over the rest of the crystal. These planar integrals take explicit account of the dilation of the crystal at the surface and hence structural relaxation is allowed.

$$E_{IIb} = -\frac{Q^2}{2}(E_{planar} + E_{volume}) \tag{10}$$

$$E_{planar} = \sum_{p \in I,IIa} \sum_{j} q_j M_j \int_{(R_{IIb}^2 - r_p^2)^{1/2}}^{\infty} \frac{1}{(r^2 - r_p^2)} 2\pi r dr \tag{11}$$

$$E_{volume} = \sum_{j} q_j M_j \int_{R_{IIb}}^{\infty} \frac{1}{r^4} 2\pi r^2 dr \tag{12}$$

where Q is the total charge of region I, q_j is the defect charge, M_j is the Mott-Littleton displacement factor, R_{IIb} is the cut-off radius, r_p is the perpendicular distance from the origin to plane p and r is the distance from the defect.

As noted above, an additional consequence of charged defects at surfaces is that a surface causes a discontinuity in the dielectric constant. As a result the charged point defect induces an image charge situated at half an inter-planar distance above the plane containing the defect. The field due to the image charge needs to be taken into account when calculating the ion displacements in region IIa and the polarisation energy of region IIb. This image charge interaction is given by:

$$q_i = q_{defect}\left(\frac{\varepsilon_1 - \varepsilon_2}{\varepsilon_1 + \varepsilon_2}\right) \tag{13}$$

where q_{defect} is the net charge of the defect, q_i is the charge of the image, ε_1 and ε_2 are the different dielectric constants of the adjoining materials and $\varepsilon_1 > \varepsilon_2$. When only surfaces in vacuum are considered, the value of ε_2 is usually set to 1, the value of free space.

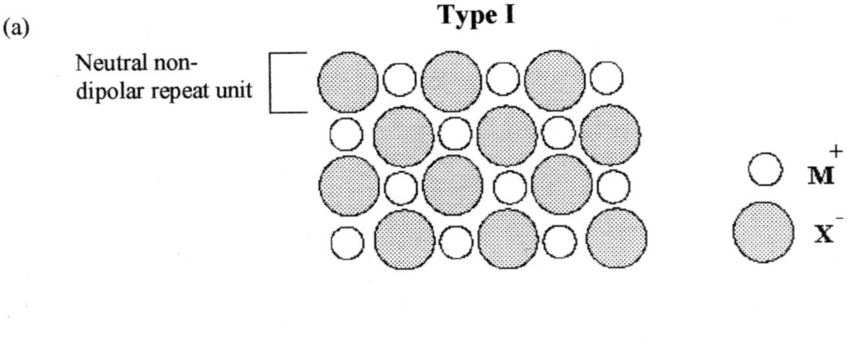

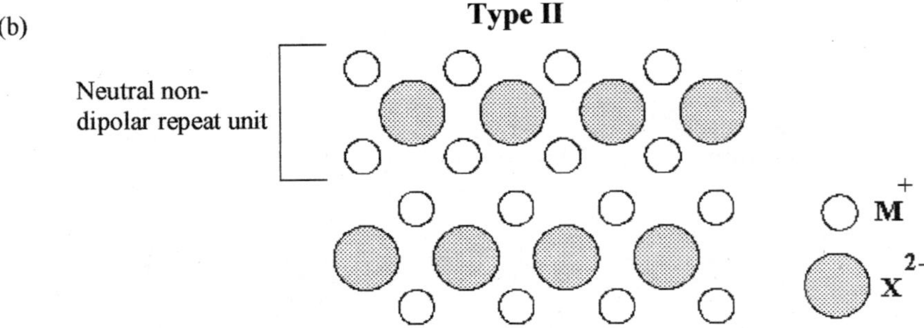

Figure 4. Stacking sequences showing type I surfaces consisting of charge neutral layers of positive and negative ions, and type II surfaces consisting of positively and negatively charged planes but with a charge neutral and non-dipolar repeat unit.

4. Types of Surfaces

As noted in section 2 above, the crystal surfaces can be considered as a stack of planes. Tasker [5] then identified three different types of surfaces: In a type I surface each plane has overall zero charge as it consists of both anions and cations in stoichiometric ratio (figure 4a). A type II surface has a stacking sequence of charged planes but the repeat unit consists of several planes in a symmetrical configuration and as a result there is no dipole moment perpendicular to the surface (figure 4b). As type I and type II surfaces have no dipole moment perpendicular to the surface the electrostatic energy converges.

Type III surfaces, however, are made up of a stack of alternately charged planes (figure 5a), producing a dipole moment perpendicular to the surface. Bertaut [38] showed that, when a dipole moment perpendicular to the surface is present in the unit cell, the surface energy diverges and is infinite. In order to study this type of surface we need to remove the dipole. One way of achieving this is to remove half the ions from the surface layer at the top of the repeat unit and transfer them to the bottom [e.g. 39] (figure 5b). As a result we obtain a surface which is partially vacant, either in cations or anions. These half-vacant planes are usually very unstable and are often particularly reactive towards impurity uptake or hydration [e.g. 40] or are found to reconstruct [e.g. 41]. Once a surface is generated, which has no dipole moment perpendicular to the surface, the surface energy can be calculated.

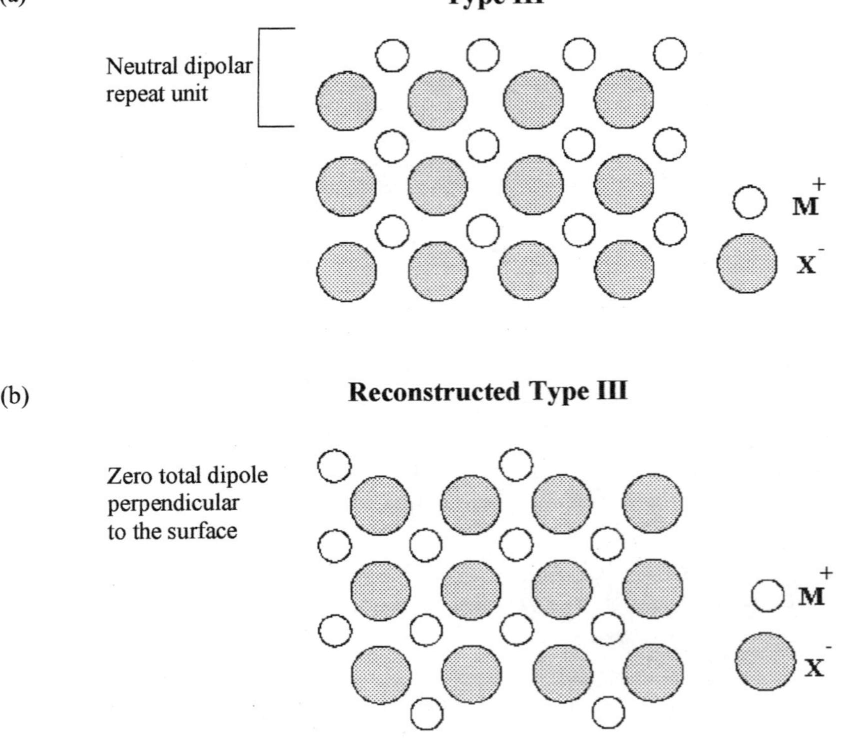

Figure 5. Stacking sequences showing a type III surface, consisting of alternating layers of positive and negative ions giving rise to a dipolar repeat unit, and a reconstructed Type III surface where half the surface ions have been shifted to the bottom of the unit cell, which has removed the dipole in the repeat unit.

5. Surface Stabilities

The free surface energy of a plane is a measure of the stability of that particular surface, where a small, positive value indicates a stable surface. In this work, static lattice energy minimisations have been used, employing both electronic structure calculations and classical atomistic simulation techniques, to obtain bulk, surface and adsorption energies. The vibrational properties of the crystal are not taken into account and in effect the simulation is of a crystal at zero Kelvin ignoring the zero point energy. Although this may seem like an unrealistic approximation, nevertheless these simulations give good agreement with experiment [e.g. 5, 42]. Furthermore, the vibrational and the zero point energy contributions to the surface free energies are small.

The excess energy for the surface has been defined by Stoneham [43] as:

$$E_S = [\textit{cohesive energy of finite crystal}] - [\textit{number of atoms}] \times [\textit{cohesion energy per atom in infinite crystal}] \qquad (14)$$

Throughout this work the surface energy γ is then defined per unit area as follows:

$$\gamma = \left(\frac{E_S}{\text{Area}} \right) \quad (15)$$

In liquids this expression is equivalent to the surface tension since a liquid surface cannot support a stress [44]. When stretching a liquid surface the number of atoms will increase to minimise the stress [5].

The energies for surface and bulk blocks of the two-dimensional periodic crystal are evaluated using the approach outlined in section 2. The energy of block I comprising region I and II is the energy of the finite crystal containing the surface, and the energy of blocks I and II together yield the bulk energy or energy per atom in an infinite crystal. Regions I and II of both blocks need to be sufficiently large to ensure convergence of the surface energies. In the three-dimensional periodic slab/void system, the surface energy is simply the difference between the bulk and surface unit cell divided by the total surface energy, and convergence of the surface energy with slab thickness needs to be evaluated for each surface.

5.1 Crystal Morphology

Gibbs [45] proposed that the equilibrium morphology of a crystal should possess a minimum (free) energy for a given volume. According to Wulff's Theorem [46], which is a variation of Gibbs theorem, the equilibrium shape of a crystal is determined by the surface energies of its various surfaces, in such a way that the equilibrium morphology is the shape of the crystal with minimum total surface free energy. Or, if the crystal is limited in space by n flat faces, then:

$$\frac{\sigma_1}{h_1} = \frac{\sigma_2}{h_2} = \ldots\ldots\ldots\ldots = \frac{\sigma_i}{h_i} = \frac{\sigma_n}{h_n} \quad (16)$$

where σ_i is the specific free energy of the ith face and h_i is the distance from the centre of the crystal to the plane of the ith face and is normal to the face [47]. Thus the height of a face is directly proportional to its specific free energy. Thus if two faces, for instance with directions [100] and [110], have the same specific free energy they will have the same height. Figure 6. gives a schematic representation of the two surfaces where the bold line indicates the resulting equilibrium morphology.

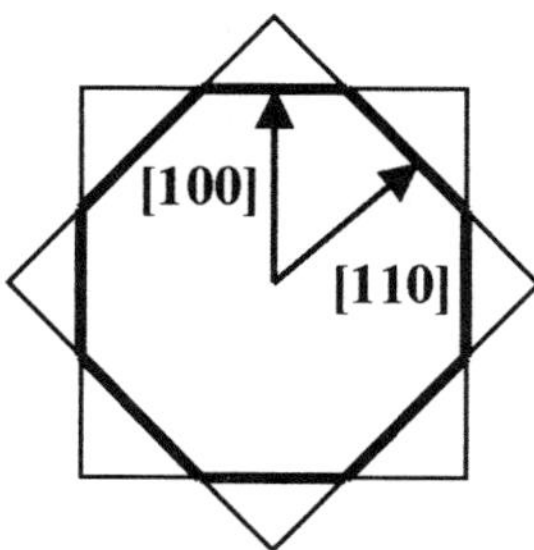

Figure 6. Schematic representation of two-dimensional equilibrium morphology due to two faces, (100) and (110), with equal specific free energy.

At zero Kelvin, the specific free energy is a close approximation of the surface energy as calculated by static lattice simulations. Masri and Tasker [48] found that the entropy term included in the surface free energy is small compared to the enthalpy term because the difference between the energies of the bulk and the surface is small. Thus the surface energies can be assumed to determine the equilibrium morphology of the crystal as follows:

$$E_S = \sum_i \gamma_i A_i = \textit{minimum for constant volume} \qquad (17)$$

where E_S is as defined in equation 14 [43] and γ_i and A_i are the surface energy and surface area of the ith crystallographic face. Again, if two surfaces have the same surface energy and the same area, the morphology will be as in figure 6 above. However, in the more likely case that the surface energies and areas are different, the surface with the highest surface energy will grow out fastest resulting in a small surface area (or a long distance from the crystal centre), while the face with the lower surface energy grows more slowly, resulting in a large surface area (or short distance from the crystal centre), and hence will be expressed in the equilibrium morphology. Figure 7 illustrates this process.

This method of predicting crystal morphologies on the basis of thermodynamic factors has been very successful when the results were compared with experimentally found crystal shapes [13]. However, rather than being determined thermodynamically, experimental morphologies often depend on kinetic factors [44]. Hartman and Bennema [49] have tried to overcome this problem by assuming attachment energies are proportional to the growth rate for each surface and hence determine the kinetic morphology. In this simple phenomenological model, a slice of material is added to an already existing face on the bulk and the energy released is the attachment energy, which is considered to be directly proportional to the growth rate [50]. Unfortunately, attachment energies assume bulk termination of the surface and do not take into account surface relaxation and reconstruction which was shown to be of utmost importance [e.g. 51]. Gay and Rohl [15] have attempted to address this problem by including the relaxation energy but it is unclear what this corresponds to within the attachment energy model. As such, although attachment energies are a useful and often successful concept, they are not as yet entirely satisfactory [44].

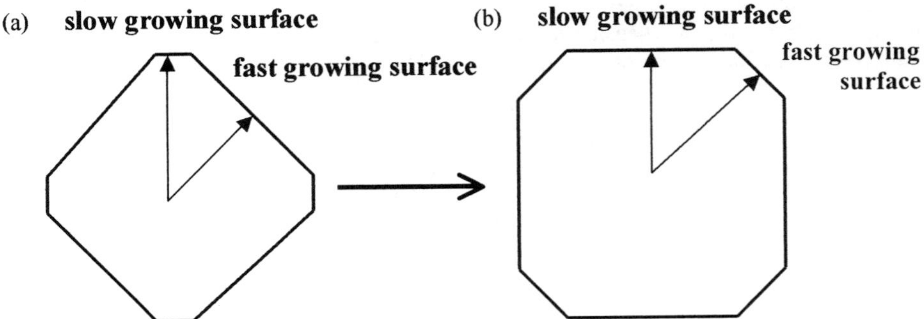

Figure 7. Schematic representation of the equilibrium morphology resulting from faces with different surface energies where (a) the initially large, fast growing surface in (b) has become small compared to the slow growing surface.

6. Applications

In the remainder of this chapter we will discuss various examples where we have modelled perfect and defective surfaces, including the effect of hydration on crystal morphology and surface adsorption of organic matter. Accurate electronic structure calculations were used to investigate and compare the geometry and electronic structure of two cubic material surfaces, namely the (111) surface of silver metal and the (100) and facetted (111) surfaces of the insulator calcium oxide. The surfaces of the more complex calcite $CaCO_3$ and fluorapatite $Ca_{10}(PO_4)_6F_2$ minerals were modelled using classical energy minimisation techniques, where large unit cells could be employed to investigate adsorption of small molecules at a host of surface features.

6.1. Silver and CaO Surfaces

The iso-structural silver metal and calcium oxide materials both have industrial importance. Silver is an important catalyst, for example in the selective partial oxidation of ethene to its epoxide which is an important pre-cursor for many chemical products, while CaO is an important component of cements and its reactivity towards water may lead to a destabilising expansion of the cement material. As such, the surface structures and reactivities of both materials need to be investigated if we are to understand these processes at the atomic level.

The total energy and structure of the Ag and CaO surfaces were determined using the Vienna Ab Initio Simulation Program (VASP) [52-55]. The basic concepts of density functional theory (DFT) and the principles of applying DFT to pseudopotential plane-wave calculations are reviewed in chapters ????? . Furthermore, this methodology is well established and has been successfully applied to the study of adsorbed atoms and molecules on the surface of ionic materials [56-59]. Within the pseudopotential approach only the valence electrons are treated explicitly and the pseudopotential represents the effective interaction of the valence electrons with the atomic cores. The valence orbitals are represented by a plane-wave basis set, in which the energy of the plane-waves is less than a given cutoff (E_{cut}). The magnitude of E_{cut} required to converge the total energy of the system has important implications for the size of the calculation when studying elements such as oxygen, which has tightly bound 2p electrons. The VASP program employs ultra-soft 'Vanderbilt'-like pseudo-potentials [60,61], which allows a smaller basis set for a given accuracy. In our calculations the core consisted of orbitals up to and including the 1s orbital for oxygen, the 3s orbital for Ca and the 4d orbital for Ag. The calculations were performed within the generalized-gradient approximation (GGA), using the exchange-correlation potential developed by Perdew and Wang [62], rather than the Local Density Approximation (LDA). Although this approach has been shown to give reliable results for a range of metals and ionic materials, [e.g. 63-65], we have nevertheless compared the two methods for silver metal, by calculating the cohesive energy and bulk modulus (equation 18).

$$B = \frac{-\Delta P}{(\Delta V / V)} \tag{18}$$

The experimental [66] and calculated values are shown in Table 1, from which it is clear that for these properties GGA gives better agreement with experiment than LDA.

Table 1. Calculated and Experimental Cohesive Energies and Bulk Moduli of Silver

Method	$E_{cohesive}$ (eV atom^{-1})	B (x10^{11} Nm^{-2})
LDA	3.707	1.305
GGA	2.818	1.069
Experiment	2.950	1.007

For surface calculations, where two energies are compared, it is important that the total energies are well converged. The degree of convergence depends on a number of factors, two of which are the plane-wave cutoff and the density of k-point sampling within the Brillouin zone. We have investigated these by undertaking a number of calculations for bulk silver and CaO with different values for these parameters. We have, by means of these test calculations, determined values for E_{cut} and the size of the Monkhorst-Pack [67] k-point mesh so that the total energy is converged to within 0.05 eV. Figure 8 shows a graph of the E_{cut} versus total energy for silver metal, from which we can see that an E_{cut} of 300 eV is sufficient for our silver calculations. We obtained an E_{cut} for CaO of 500 eV in the same way. In addition, we checked the convergence of the total energy for the slab calculations and found that the most suitable value of E_{cut} is the same as for the bulk materials. A k-point mesh of 3 x 3 x 3 for CaO and 10 x 10 x 10 for Ag gave the total energy convergence required.

The optimisation of the atomic coordinates (and unit cell size/shape for the bulk materials) was performed via a conjugate gradients technique which utilises the total energy and the Hellmann-Feynman forces on the atoms (and stresses on the unit cell). We used the usual approach for modelling the surfaces using three-dimensional periodic boundary conditions by considering slabs of Ag and CaO. In addition to the k-point density and E_{cut} discussed above, the convergence in surface calculations is also dependent on the thickness of the slab of material and the width of the vacuum layer between the slabs. Again we checked convergence by running a series of test calculations with different slab thicknesses and gap widths. For example, Table 2 shows the average energy per CaO unit of the CaO {100} surface as a function of these parameters. Following these test calculations the CaO bulk and {100} supercells used in the calculations contained four layers of CaO with 8 CaO units per simulation cell while the supercell for the {111} surface contained 10 CaO units.

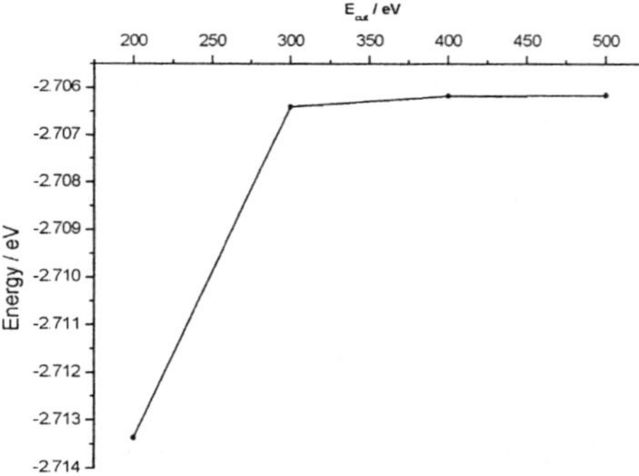

Figure 8. Graph of plane-wave cutoff energy versus total energy for silver metal.

Table 2. Average energy per CaO unit of the CaO {100} surface as a function of slab thickness and width of vacuum layer between the slabs.

E_{total} of CaO {100} surface / CaO unit (eV)				
Number of layers of crystal	Width of vacuum layer (Å)			
	10	15	20	25
3	-	-	-12.7304	-
4	-12.8068	-12.8073	-12.8070	-12.8074
5	-	-	-12.8508	-

Similarly, figure 9 shows a graph of the surface energy of the silver (111) surface versus slab thickness, from which it is clear that while a nine-layer slab will give the most accurate results the surface energy has converged to within 0.01 Jm^{-2} at five layers thickness. In each case the simulation cells were constructed so that there was a vacuum layer of at least 10 Å between the images of the surfaces either side of the void.

Both silver and CaO have face-centred cubic structure. In the bulk CaO crystal, each cation and oxygen ion is six-coordinate and the perfect crystal shows cubic morphology with one dominant surface, namely the (100), containing both calcium and oxygen ions where all ions are five-coordinate. In silver, the most stable surface is the (111) surface [e.g. 65], which is a hexagonal close packed plane of silver atoms. However, the CaO (111) surface is a highly unstable surface [40] as this surface is a dipolar Type III surface, as described in section 4 above. As a result of its low stability the CaO (111) surface reconstructs to form a facetted surface, which consists of three-sided pyramids with four-coordinated edges and either a three-coordinated calcium or an oxygen ion at the apex. The sides of the pyramids are five-coordinated (100) planes and the stability of the (111) surface was calculated to increase with increasing facetting of the originally dipolar (111) surface, leading to ever increasing areas of these (100) planes [41].

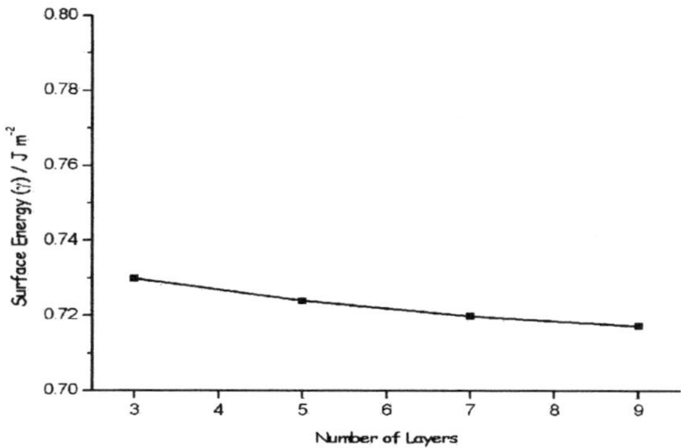

Figure 9. Graph of surface energy versus number of layers of silver (111) surface.

Table 3. Surface energies of CaO and Ag surfaces

Surface	γ (Jm^{-2})
CaO (100)	0.61
CaO (111)	1.49
Ag (111)	0.72

The calculated bulk lattice parameters of the two materials are in good agreement with experiment at a = b = c = 4.164 for Ag (exp. 4.101 [66]) and a = b = c = 4.752 Å for CaO (exp. 4.799 Å). The calculated surface energies for the three surfaces are collected in Table 3, from which it is clear that despite facetting of the CaO (111) surface, the CaO (100) surface is still the more stable plane.

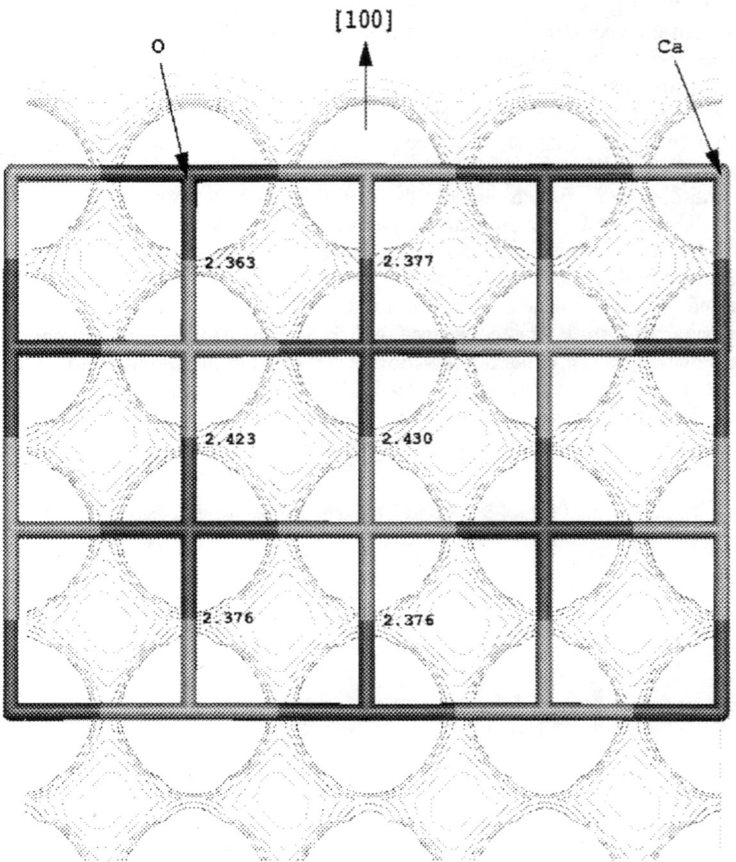

Figure 10. Sideview of the relaxed {100} surface showing electron density contour plots and interatomic distances. (Ca = mid grey, O = dark grey, contour levels are from 0.05 to 0.35 e/Å^3 at 0.05 e/Å^3 intervals, bondlenghts in Å)

Figure 10 shows the relaxed CaO (100) surface with a contour plot of the electron density through the CaO planes, from which it is clear that the crystal shows a strongly ionic character with the electron density centred on the calcium and oxygen ions. It can be seen that the distortion of the electron density round the surface ions is minimal, which leads to a negligible ionic relaxation of the surface region. The interatomic distances shown in figure 10. indicate that relaxation leads to only a small rumpling of the surface layer, due to a larger net increase in the Ca-Ca spacing (from a bulk spacing of 4.76 Å to 4.81 Å) than in the O-O spacing (from 4.76 Å to 4.79 Å), between the topmost ions and those one unit cell below. Hence, in the relaxed surface there is a difference in height between the surface calcium and oxygen ions of 0.02 Å with the calcium ions protruding above the oxygen ions.

Figure 11 shows the relaxed micro-facetted CaO (111) surface with electron density contour plots through two edges of the pyramids including the three-coordinated calcium ions at the apices. There is no discernible distortion of the electron densities on the four-coordinated edges, which are still centred on the individual ions but some of the electron density round the three-coordinated corner site has been re-distributed over the bonds between the calcium ion at the apex and the neighbouring four-coordinated oxygen ions below. As a result of this re-distribution of the electron density along the four-coordinated pyramid edges, relaxation of the micro-facetted (111) surface is more pronounced. The three-coordinated calcium ions at the pyramid apices move into the surface through contraction of the Ca-O bonds in the topmost two layers, from the bulk value of 2.38 Å to 2.16 Å. The bondlengths between these four-coordinated oxygen ions to the next layer of calcium ions also contract, but to a lesser extent (2.29 Å).

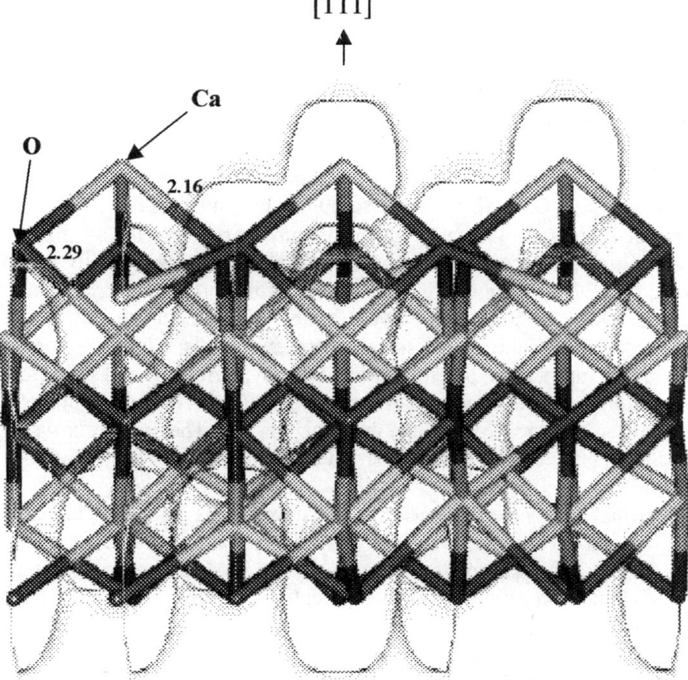

Figure 11. Sideview of the relaxed micro-facetted CaO {111} surface with electron density contour plots through apex calcium ions and two edges (Ca = grey, O = black, contour levels from 0.05 to 0.50 e/Å^3 at 0.05 e/Å^3 intervals, bondlenghts in Å).

The (111) surface of silver metal on the other hand, shown in figure 12, is a non-dipolar surface. Unlike the ionic CaO crystal, where the unreconstructed (111) surface consists of alternating planes of calcium and oxygen ions parallel to the surface leading to a dipole perpendicular to the surface, the silver planes parallel to the surface are charge-neutral and there is therefore no dipole perpendicular to the surface. As the (111) is also a close-packed plane of silver atoms it is the major silver surface with the lowest surface energy (Doll and Harrison), calculated in this work at 0.72 Jm^{-2}. Surface relaxation is negligible, with practically bulk termination of the two identical surfaces either end of the slab. The spacings between the first two surface layers have contracted by about 0.6-1.6% of the bulk interlayer spacing, while the interatomic Ag-Ag distances in the surface layers are approximately 0.4% shorter than those in the bulk material (figure 12). The electron density contour plots (through the silver atoms in the third, sixth and ninth layers from the top surface) show a large degree of localization of the electron density around the individual silver atoms in the bulk, although with some density distributed along Ag-Ag bonds. The electron density at the surface is far more delocalised over the surface silver atoms.

These DFT calculations of a metal and a purely ionic crystal have shown the effect of dipolarity on the relative stability of the CaO (111) surface compared to the otherwise equivalent Ag (111) surface, and the greater delocalisation of electron density along the metal surface compared to the surfaces of the ionic crystal.

6.2. The Effect of Surface Hydration on the Morphology of Fluor-Apatite

Apatites are important materials in both geological and biological situations. Hydroxy-apatite is the major constituent of mammalian bones and tooth enamel, while fluorapatite is the most abundant of the apatite group and the major phosphorus-bearing mineral in the environment. Furthermore, because of their affinity for impurity ions, apatites may well become important as environmental sinks for the uptake of heavy metals and radio-active contaminants, such as lead and uranium. Because of its importance as a natural bone material, apatite is a possible candidate in the manufacture of artificial bones, while fluoridation of hydroxy-apatite can suppress dental decay, for example by the production of fluorapatite ($Ca_{10}F_2(PO_4)_6$ which is less soluble than hydroxyapatite.

The surface structure and reactivity of apatites is relevant for all the above areas and hence we describe our recent work, modelling the structure and stability of both dehydrated and hydrated surfaces and the effect of surface hydration on the crystal morphology. The computer modelling technique employed for the work in this section is classical energy minisation, using two-dimensional surface simulation cells. Classical atomistic simulation methods employ interatomic potential models for the description of the forces between the component atoms and energy minimisation techniques such as conjugate gradients or Newton-Raphson methods to calculate the total energies. The background to various energy minimisation techniques and interatomic potential functions is fully described elsewhere in this volume.

The potential parameters describing the interactions between the atoms within the apatite crystal and between the water molecules and apatite surfaces were fitted empirically to various experimental parameters. The Ca-O and Ca-F parameters were taken from the calcium carbonate and calcium fluorite potentials developed by Pavese et al. [68] and Catlow et al. [69] respectively. The Ca-F and P-O interactions were fitted to the experimental crystal structure and elastic constants, bond dissociation energies and vibrational data of the phosphate group. All short-range interactions involving anions are between the shells apart from the Morse potential in the phosphate group, which acts between the phosphorus and oxygen cores.

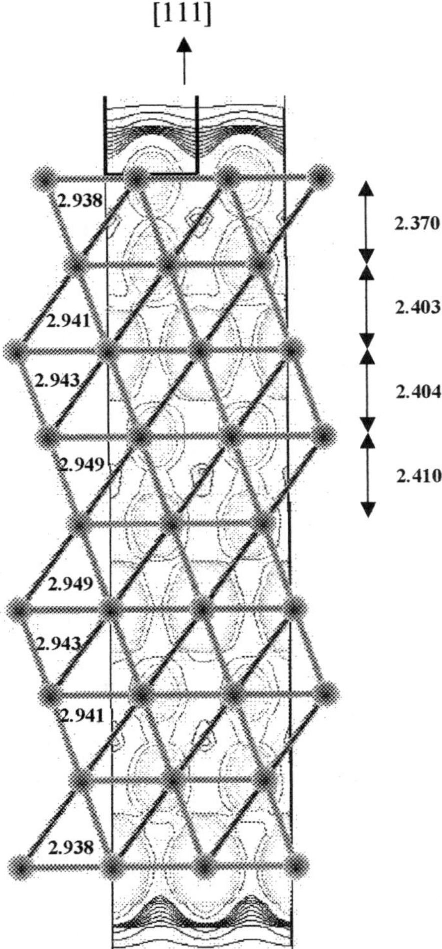

Figure 12. Sideview of the relaxed nine-layer Ag slab with (111) surfaces at either end, showing both interatomic and interlayer spacings (in Å) and electron density contour plots through the silver atoms of the 3rd, 6th and 9th layers (contour levels are from 0.03 to 0.50 e/Å^3 at 0.05 e/Å^3 intervals).

Fluorapatite has a hexagonal crystal structure with spacegroup $P6_3/m$ and a = b = 9.370 Å, c = 6.880 Å, where $\alpha = \beta = 90^0$, $\gamma = 120^0$ [70], which upon energy minimisation of the bulk crystal were calculated at a = b = 9.365 Å, c = 6.868 Å, $\alpha = \beta = 90^0$, $\gamma = 120^0$. We then created a range of surfaces with low Miller indices as these surfaces have the smallest interplanar spacings and as a result are often the most stable. The surfaces studied were the {0001}, $\{10\bar{1}0\}$, $\{10\bar{1}1\}$, $\{11\bar{2}0\}$, $\{1\bar{1}21\}$ and $\{10\bar{1}3\}$ planes and their symmetry-related surfaces. The (0001) and to a lesser extent $(10\bar{1}0)$ surfaces are experimental cleavage planes, while the $(11\bar{2}1)$ and $(10\bar{1}3)$ planes occur as twin planes in the natural crystals. Previous electronic structure calculations of fluorapatite have shown that the phosphate groups themselves have considerable covalent character but as a group act as a polyanion and we therefore kept the phosphate groups intact when creating the surfaces. We first calculated both the unrelaxed and relaxed surface energies of the dry surfaces (Table 4), from which we see that the (0001) surface is by far the most stable surface with the lowest energy, both before and after relaxation. This stability is rather remarkable as this surface is a dipolar type III surface and hence contains surface calcium vacancies, which were introduced to remove the dipole, but it agrees very well with the fact that the (0001) surface is the major cleavage plane of apatite [71].

Table 4. Surface and Hydration Energies of Fluorapatite Surfaces

Surface	$\gamma_{unrelaxed}$ (Jm^{-2})	$\gamma_{relaxed}$ (Jm^{-2})	γ_{hydr} (Jm^{-2})	E_{hydr} (kJmol^{-})
(0001)	1.19	0.77	0.45	-73
$(10\bar{1}0)$	2.84	1.32	0.75	-97
$(10\bar{1}1)$	1.81	1.00	0.61	-83
$(11\bar{2}0)$	2.28	1.12	0.72	-97
$(10\bar{1}3)$	2.98	1.94	1.67	-78
$(11\bar{2}1)$	2.62	1.12	0.88	-81

We next hydrated the surfaces up to full monolayer coverage, which is here defined as the maximum number of water molecules which could be accommodated on the surface before formation of a second water layer occurred. The surface energies of the hydrated surfaces are also collected in Table 4, and we see that hydration has a stabilising effect on all the surfaces considered, but the (0001) surface is still the dominant plane.

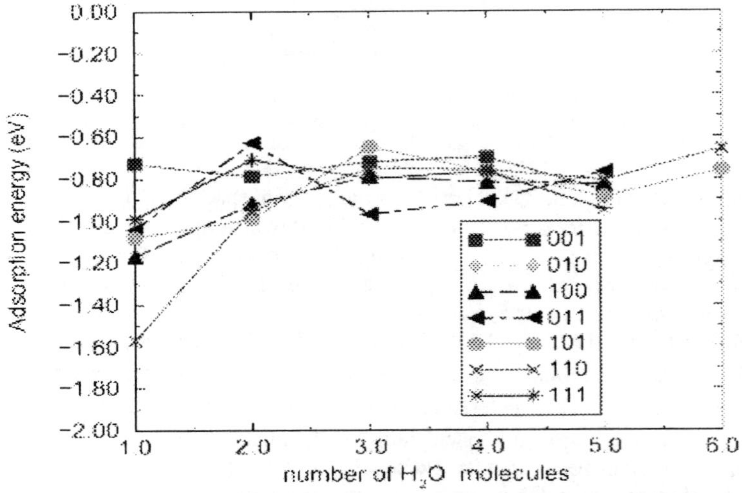

Figure 13. Sequential energies of adsorption of successive water molecules at the apatite surfaces.

The average hydration energies of between –73 and –104 kJmol^{-1}, listed in Table 4, indicate physisorption of the (molecular) water at the surfaces. The graph in figure 13. shows the sequential hydration energies for the different surfaces, where the energy plotted versus number of water molecules is the energy released at each addition of a successive water molecule at the surface in question. It is clear from the graph that most surfaces exhibit more or less Langmuir behaviour with the notable exception of the $(11\bar{2}1)$ surface, which on the whole shows a decrease in adsorption energy with the addition of successive water molecule.

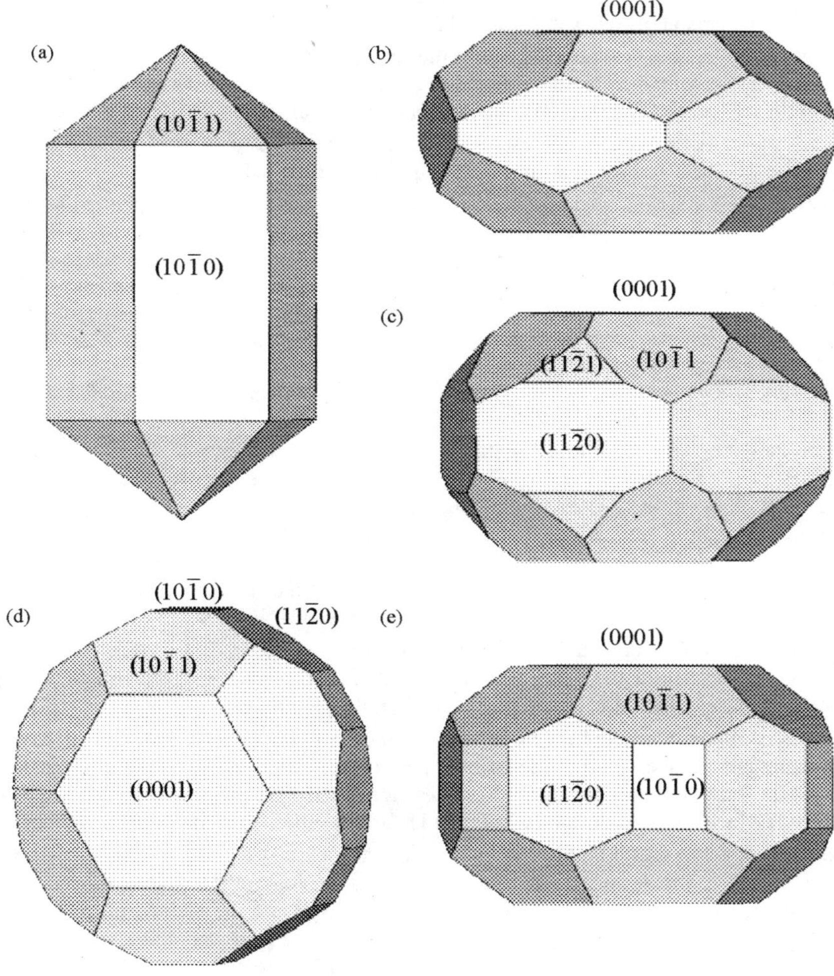

Figure 14. Morphologies of the fluoapatite crystal: (a) according to the Donnay-Harker method and calculated from (b) the unrelaxed, (c) the relaxed and (d,e) the hydrated surface energies.

As described in section 5.1 above, we can calculate equilibrium crystal morphologies from the surface energies of a particular material. Figure 14 shows the calculated morphologies of the fluorapatite crystal, indicating the surfaces expressed in the morphologies. Figure 14(a) is the morphology calculated using the Donnay-Harker method, which is based on the relative heights of the surfaces with respect to the centre. Only the $(10\bar{1}1)$ and $(10\bar{1}0)$ surfaces are expressed in this morphology, which is unrealistic as experimentally the (0001) surface is found to be the dominant surface and hence should be expressed in the crystal morphology. Figures 14(b) and (c) show the morphologies calculated respectively from the unrelaxed and relaxed surface energies in Table 4, where now the (0001) surface is expressed. When surface relaxation is taken into account, the $(11\bar{2}1)$ surface has become relatively more stable than the $(10\bar{1}1)$ and $(11\bar{2}0)$ surface and hence appears in the relaxed crystal morphology (figure 14c), in agreement with experimental crystal growth where the $(11\bar{2}1)$ surface is found to the twinning plane for the intergrowth of apatite crystals [71]. When the surfaces are hydrated, the morphology (figure 14e) changes again due to the dissimilar stabilisation of the different surfaces by the adsorption of water. The $(11\bar{2}1)$ surface has disappeared again, but now the $(10\bar{1}0)$ surface has become relatively more stable and is hence expressed in the hydrated morphology. The presence of both (0001) and $(10\bar{1}0)$ surfaces in the hydrated morphology agrees with experiment, where both these surfaces are found as cleavage planes [71] and as crystal growth features [72]. Figure 14(d) is a different view of the same hydrated morphology as in figure 14(e), looking down upon the (0001) surface. We can see that the hydrated crystals have a rounded hexagonal shape and are fairly flat platelets, in agreement with experimental apatite crystals in bone which are found to be nano-sized platelets [73,74].

6.3. Modelling the Adsorption of Organic Material at Calcite and Fluorite Surfaces

Investigating the adsorption of organic molecules at calcite and fluorite surfaces is important for a number of reasons. Firstly, organic surfactants are widely used as reagents in mineral flotation techniques, separating mixtures including calcite, fluorite and apatite; secondly, organic materials are found to inhibit crystal growth, while biological calcium minerals are found to form on an organic matrix. In this section, we have again employed classical energy minimisation techniques to investigate the interaction of methanoic acid (CHOOH) with stepped calcite ($CaCO_3$) and fluorite (CaF_2) surfaces.

Calcite has a rhombohedral crystal structure with space group $R\bar{3}c$, while fluorite has a cubic crystal structure with space group Pm3m. The dominant surface in calcite is the $\{10\bar{1}4\}$ surface, which is a type I surface with both calcium and carbonate oxygen ions in the surface plane, while the major surface of fluorite is the (111) plane, which is type II terminating in a layer of fluoride ions at the surface, although the calcium ions in the second layer are still accessible to adsorbates. We created six different steps on the calcite $\{10\bar{1}4\}$ surface and two on the fluorite (111) surface. On the calcite surface, we obtained two steps which were terminated by both calcium ions and carbonate groups alternating along the edges, which we will call type I steps in an analogy to the type I surfaces. One of the steps is acute, where the angle between the step wall and the underlying plane is 80^0, whereas the other step is obtuse with an angle between the step wall and underlying plane of 105^0. The other four steps are also either acute or obtuse steps, but these are type III steps, terminated by either calcium ions or carbonate groups. The surface energies of the planar and stepped $\{10\bar{1}4\}$ surfaces are collected in Table 5, from which it is clear that the presence of the steps destabilises the surface. However, the increase in surface energy from 0.59 Jm^{-2} to 0.71 to 0.82 Jm^{-2} for the steps indicates that the destabilisation is due as much to the mere presence of a step on the surface as to the detailed structure of the individual steps, which leads to a variation in surface energies of only 0.1 Jm^{-2}.

The energies of adsorption of methanoic acid at the planar and stepped surfaces are also collected in Table 5. It is clear that the adsorption sites at the steps are more reactive than on the plane, especially at the type I steps, which are experimentally found to occur widely. The adsorption energies at the type III steps are not as large, due to the fact that only one type of ion (cation or polyanion) is present at the step edge. Methanoic acid mainly adsorbs by its oxygen ion(s) to surface calcium ions, but also forms hydrogen-bonded interactions to surface oxygen ions. When both species are present on the edge, more interactions between surface ions and the methanoic acid molecules can be formed, and the adsorption energy is therefore often larger than at the edges, where only ion type can interact with the methanoic acid molecule.

Figure 15 shows the type I obtuse step edge, with adsorbed methanoic acid. The methanoic acid molecule coordinates to the surface by bridging from the step to the plane below. It bonds by its doubly bonded oxygen to a calcium ion on the step (Ca-O distance of 2.19 Å), while the hydrogen of the OH group is coordinated to a carbonate oxygen ion on the plane (H-O distance of 2.47 Å). The hydrogen atom bonded to the carbon is pointing away from the surface. Although there is some coordination between the carbon atom and the oxygen atoms on the step, this is not a strong interaction and the adsorption energy (-108.1 $kJmol^{-1}$) is thus not at large as on the acute type I step edge (-168 $kJmol^{-1}$), where the methanoic acid molecule forms a whole range of interactions with the surface, both to the step edge and to the plane below. It bonds by its doubly bonded oxygen to a calcium ino on the the terrace below the step (Ca-O 2.26 Å), while the hydrogen of the OH group forms hydrogen-bonded interactions to two oxygen atoms on the step edge, both from the same carbonate group (O-H 2.4-2.44 Å). The hydrogen of the carbon atom is also involved in the surface adsorption process, by bridging between one oxygen atom on the terrace (O-H 2.65 Å) and one on the step edge (O-H 2.55 Å).

Table 5. Surface and Adsorption Energies of Planar and Stepped Calcite $\{10\bar{1}4\}$ Surfaces

Surface	**γ (Jm^{-2})**	**$E_{adsorption}$ ($kJmol^{-1}$)**
Planar $\{10\bar{1}4\}$	0.59	-37.5
Acute type I step	0.82	-168.0
Obtuse type I step	0.71	-108.1
Acute type III step Ca	0.78	-46.1
Acute type III step CO_3	0.75	-70.9
Obtuse type III step Ca	0.74	-74.6
Obtuse type III step CO_3	0.71	-58.9

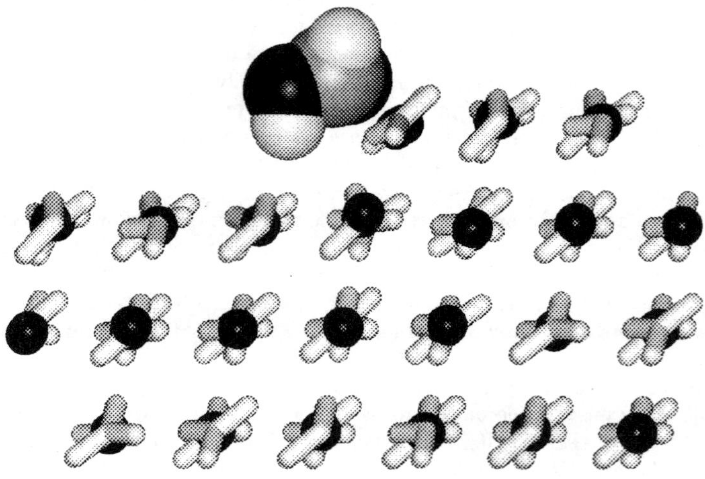

Figure 15. Calcite $\{10\bar{1}4\}$ surface with obtuse type I step and adsorbed methanoic acid molecule (Ca, O = black, CO_3 group and H = white, C = grey).

The adsorption energies of methanoic acid onto fluorite surfaces (Table 6) are larger than for the planar calcite and type III stepped surface, but they are much less than the acute type I step edge in calcite. The reason for this is the different pattern of adsorption of the methanoic acid molecule for fluorite and calcite. On the planar fluorite surface, the two oxygen ions of the methanoic acid molecule bridge across two surface calcium ions and both hydrogen atoms form hydrogen-bonded interactions to surface fluoride ions. On the planar calcite surface, however, the methanoic acid only bonds by its doubly bonded oxygen atom to once surface calcium and forms one hydrogen-bond to a surface oxygen ion. Hence, the adsorption of methanoic acid at the planar fluorite {111} surface releases more energy than at the planar calcite $\{10\bar{1}4\}$ surface. However, we saw above that adsorbing methanoic acid molecules at the type I steps of calcite form very close interactions with the steps, bridging between edge and terrace, leading to large adsorption energies. In contrast, interactions between methanoic acid molecules and the two steps on the fluorite {111} surface, which can also be classified as an acute and an obtuse step, are not as extensive and hence less energy is released upon adsorption. At the obtuse step, shown in figure 16, the methanoic acid molecule bridges between the terrace and the step, by coordinating to a terrace calcium ion by its doubly bonded oxygen and to edge calcium and fluoride ions by its hydroxyl oxygen and hydrogen ions. On the acute edge, the interactions between methanoic acid and surface are more extensive, coordinating by the doubly bonded oxygen atom to two terrace calcium ions, while the hydroxyl oxygen also bonds to two calcium ions, one on the terrace and one on the step edge. Finally, the hydroxyl hydrogen atom forms a hydrogen-bonded interaction to a fluoride ion on the edge. As a result of these interactions, the adsorption energy of methanoic acid at the acute fluorite step is larger than both the planar and the obtuse stepped surfaces, comparable to the obtuse step I surface of calcite.

Table 6. Surface and Adsorption Energies of Planar and Stepped Fluorite (111) Surfaces

Surface	γ_{pure} **(Jm^{-2})**	**E$_{adsorption}$ (kJmol^{-1})**
Planar {111}	0.52	-56.3
Acute	0.66	-90.8
Obtuse	0.54	-79.5

7. Conclusions

In this chapter we have described a variety of simulation methods to model a host of different surface features and properties. We have illustrated some of these methods by the application of two different surface modelling techniques to investigate a range of surface processes. Electronic structure calculations of two iso-structural materials, silver metal and CaO, showed how the stability of otherwise equivalent surfaces was affected by the dipolarity of the ionic surface. Classical energy minimisation simulations of the complex oxide material apatite have shown the power of these techniques to successfully predict macroscopic crystal morphologies from calculations of processes at the atomic level. We saw that the effect of surface relaxation and surface hydration on the crystal morphology needs to be taken into account if the predicted morphologies are to agree with experimentally found cleavage and crystal growth information. Finally, the complex organic-solid interface was addressed in a series of calculations of adsorption of methanoic acid at calcite and fluorite surfaces, including stepped features to better mimic “real” surfaces, where we compared the reactivity of the various surface sites between the two materials. The applications have shown how modern computer modelling techniques can now be applied routinely to investigate complex materials and processes, which are often difficult to study experimentally.

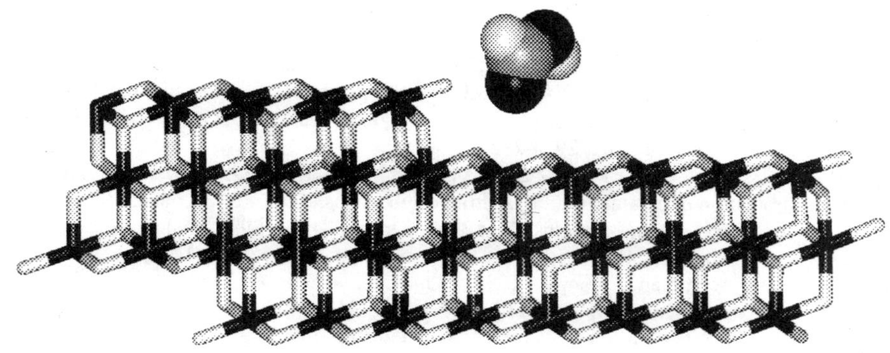

Figure 16. Fluorite {111} surface with obtuse step and adsorbed methanoic acid molecule (Ca, O = black, F,H = white, C = grey).

8. Acknowledgements

NHdL acknowledges The Wellcome Trust (Grant no. 065067), NERC (Grant no. NER/M/S/2001/00068) and the Royal Society (Grant no. 22292) for financial support. TGC acknowledges EPSRC (Grant No. GR/N65172/01) and the Chemistry Department of the University of Reading for a studentship. DM and PEN acknowledge the Royal Society and the NRF and CSIR, South Africa, for financial support. NHdL and CJN acknowledge the Materials Chemistry Consortium for provision of computer time on the Cray T3E high performance computing system.

9. References

1. Duffy D.M. (1986) J. Phys. C: Solid State Phys., **19**, 4394
2. Sakaguchi, I., Yurimoto, H. and Sueno, S. (1992) *Solid State Commun.*, **84**, 889.
3. Tasker, P.W. (1987) *U.K.A.E.A. Report AERE-R-9130*, United Kingdom Atomic Energy Authority, Harwell, UK.
4. Puls, M.P., Woo, C.H. and Norgett, M.J. (1977) *Phil. Mag.*, **36**, 1457.
5. Tasker, P.W. (1979) *J. Phys. C: Solid State Phys.*, **12**, 4977.
6. Mackrodt, W.C. and Stewart, R.F. (1979) *J. Phys. C: Solid State Phys.*, **12**, 431.
7. Colbourn, E.A., Mackrodt, W.C. and Tasker, P.W. (1983) *J. Mater. Sci.*, **18**, 1917
8. Tasker, P.W and Duffy, D.M. (1984) The structure and properties of the stepped surfaces of MgO and NiO, *Surf. Sci.* **137**, 91-102.
9. Tasker, P.W, Colbourn, E.A. and Mackrodt, W.C. (1985) Segregation of isovalent impurity cations at the surfaces of MgO and CaO, *J. Am. Ceram. Soc.* **68**, 74-80.
10. Lawrence, P.J., Parker, S.C. and Tasker, P.W. (1987) Computer simulation studies of perfect and defective surfaces of Cr_2O_3, *Adv. Ceram.*, **23**, 247.
11. Mackrodt, W.C. (1989) *J. Chem. Soc. Faraday Trans. 2*, **85**, 541.
12. Davies, M.J., Parker, S.C. and Watson, G.W. (1994) Atomistic simulation of the surface structure of spinel, *J. Mater. Chem.*, **4**, 813.
13. de Leeuw, N.H. and Parker, S.C. (1997) Atomistic simulation of the effect of molecular adsorption of water on the surface structure and energies of calcite surfaces, *J. Chem. Soc. Faraday Trans.*, **93**, 467.
14. Titiloye, J.O., de Leeuw, N.H. and Parker, S.C. (1998) Atomistic simulation of the differences between calcite and dolomite surfaces, *Geochim. Cosmochim. Acta*, **62**, 2637-2641.
15. Gay, D.H. and Rohl, A.L. (1995) *J. Chem. Soc. Faraday Trans.*, **91**, 925.
16. de Leeuw, N.H. and Parker, S.C. (1998) Surface structure and morphology of calcium carbonate polymorphs calcite, aragonite and vaterite: an atomistic approach, *J. Phys. Chem. B*, **102**, 2914-2922.
17. de Leeuw, N.H., Parker, S.C. and Harding, J.H. (1999) Molecular Dynamics simulation of crystal dissolution from calcite steps, *Phys. Rev. B*, **60**, 13792-13799.
18. de Leeuw, N.H., Higgins, F.M. and Parker, S.C. (1999) Modelling the surface structure and stability of α-quartz', *J. Phys. Chem. B*, **103**, 1270-1277.
19. Manassidis, I. and Gillan, J. (1994) *J. Am. Ceram. Soc.*, **77**, 335.
20. Schroer, P., Kruger, P. and Pollman, J. (1994) *Phys. Rev. B*, **49**, 17092.
21. Barrett, N.T., Guillot, C., Villette, B., Treglia, G. and Legrand, B. (1991) *Surf. Sci.*, **251**, 717.
22. Klepeis, J.E. and Terminello, L.J. (1996) *Phys. Rev. B*, **53**, 16035.
23. Kantorovich, L.N., Holender, J.M. and Gillan, M.J. (1995) *Surf. Sci.*, **343**, 221.
24. Goniakowski, J., Holender, J.M., Kantorovich, L.N. and Gillan, M.J. (1996) *Phys. Rev. B*, **53**, 957.
25. Purton, J., Bullett, D.W., Oliver, P.M. and Parker, S.C. (1995) *Surf. Sci.*, **336**, 166
26. Stauffer, L., Mharchi, A., Saintenoy, S., Pirri, C., Wetzel, P., Bolmont, D. and Gewinner, G. (1995) *Phys. Rev. B*, **52**, 11932.

27. Swanston, D.M., McLean, A.B., McIlroy, D.N., Heskett, D., Ludeke, R., Munekata, H., Prietsch, M. and DiNardo, N.J. (1994) *Surf. Sci.*, **312**, 361.
28. Mijoule, C. and Bouteiller, Y. (1991) *Surf. Sci.*, **253**, 375.
29. Gorling, A., Ackermann, L., Lauber, J., Knappe, P. and Rosch, N. (1993) *Surf. Sci.*, **286**, 26.
30. Burghgraef, H., Jansen, A.P.J. and van Santen, R.A. (1993) *Faraday Discuss.*, **96**, 337.
31. Pacchioni, G., Ferrari, A.M. and Bagus, P.S. (1996) *Surf. Sci.*, **350**, 159.
32. Goniakowski, J. and Gillan, M.J. (1996) *Surf. Sci.*, **350**, 145.
33. Pugh, S. and Gillan, M.J. (1994) *Surf. Sci.*, **320**, 331.
34. Payne, M.C., Stich, I., De Vita, A., Gillan, M.J. and Clarke, L.J. (1993) *Faraday Discuss.*, **96**, 151.
35. Stich, I., De Vita, A., Payne, M.C., Gillan, M.J. and Clarke, L.J. (1994) *Phys. Rev. B*, **49**, 8076.
36. Watson, G.W., Kelsey, E.T., de Leeuw, N.H., Harris, D.J. and Parker, S.C. (1996) Atomistic simulation of dislocations, surfaces and interfaces in MgO, *J. Chem. Soc. Faraday Trans.*, **92**, 433.
37. Mott, N.F. and Littleton, M.J. (1938) *Trans. Faraday Society*, **34**, 485.
38. Bertaut F. (1958) *Compt. Rendu.*, **246**, 3447.
39. Oliver, P.M., Parker, S.C. and Mackrodt, W.C. (1993) *Model. Mater. Sci. Eng.*, **1**, 755.
40. de Leeuw, N.H., Watson, G.W. and Parker, S.C. (1995) Atomistic simulation of the effect of dissociative adsorption of water on the surface structure and stability of calcium and magnesium oxide, *J. Phys. Chem.*, **99**, 17219.
41. de Leeuw, N.H., Watson, G.W. and Parker, S.C. (1996) Atomistic simulation of adsorption of water on three-, four- and five-coordinated surface sites of magnesium oxide, *J. Chem. Soc. Faraday Trans.*, **92**, 2081.
42. Parker, S.C. (1983) *Solid State Ionics*, **8**, 179.
43. Stoneham, A.M. (1976) *Defects and their Structure in Non-metallic Solids*, edited by B. Henderson and A.E. Hughes, Plenum, New York.
44. Harding, J.H. (1997) *Computer Modelling in Inorganic Crystallography*, ed. C.R.A. Catlow, Academic Press, London.
45. Gibbs, J.W. (1928) *Collected Works*, Longman, New York.
46. Wulff, G (1901) Zur Frage der Geschwindigkeit des Wachstums und der Auflösung der Kristallflachen, *Z. Kristallogr. Kristallgeom.*, **39**, 449.
47. Tauson, V.L., Abramovich, M.G., Akimov, V.V. and Scherbakov, V.A. (1993) *Geochim. Cosmochim. Acta*, **57**, 815.
48. Masri, P. and Tasker, P.W. (1985) Surface phonons and surface reconstruction and calcium-doped magnesium oxide, *Surf. Sci.* **149**, 209-225.
49. Hartman, P. and Bennema, P. (1980) *J. Cryst. Growth*, **49**, 145.
50. Woensdregt, C.F. (1992) *Phys. Chem. Miner.*, **19**, 52.
51. Mackrodt, W.C., Davey, R.J., Black, S.N. and Docherty, R. (1987) *J. Cryst. Growth*, **80**, 441.
52. Kresse, G. and Hafner, J. (1993) *Phys. Rev. B*, **47**, 5858.
53. Kresse, G. and Hafner, J. (1994) *Phys. Rev. B,* **49**, 14251.
54. Kresse, G. and Furthmüller, J. (1996) *Comput. Mater. Sci.*, **6**, 15.
55. Kresse, G. and Furthmüller, J. (1996) *Phys. Rev. B*, **54**, 11169.
56. Refson, K., Wogelius, R.A., Fraser, D.G., Payne, M.C., Lee, M.H. and Milman, V. (1995) *Phys. Rev. B*, **52**, 10823.
57. Lindan, P.J.D., Muscat, J., Bates, S., Harrison, N.M. and Gillan, M. (1997) *Faraday Discuss.* **106**, 135.
58. Lindan, P.J.D., Harrison, N.M. and Gillan, M.J. (1998) *Phys. Rev. Lett.*, **80**, 762.
59. Bates, S.P., Kresse, G. and Gillan, M.J. (1998) *Surf. Sci.*, **409**, 336.
60. Vanderbilt, D. (1990) *Phys. Rev. B*, **41**, 7892.
61. Kresse, G. and Hafner, J. (1994) *J. Phys.: Condens. Matter*, 6, 8245.
62. Perdew, J.P., Chevary, J.A., Vosko, S.H., Jackson, K.A., Pederson, M.R., Singh, D.J.

and Fiolhas, C. (1992) *Phys. Rev. B Condens. Matter,* **46**, 6671.
63. Goniakowski, J. and Gillan, M.J. (1996) Surf. Sci., **350**, 145.
64. de Leeuw, N.H., Purton, J.A., Parker, S.C., Watson, G.W. and Kresse, G. (2000) Density Functional Theory calculations of adsorption of water at calcium oxide and calcium fluoride surfaces, *Surf. Sci.*, **452**, 9.
65. Doll, K. and Harrison, N.M. (2001) Theoretical study of chlorine adsorption on the Ag(111) surface, *Phys. Rev. B*, **63**, 165410.
66. Kittel, C. (1996) *Introduction to Solid State Physics*, John Wiley & Sons Inc., New York.
67. Monkhorst, H.J. and Pack, J.D. (1976) *Phys. Rev. B*, **13**, 5188.
68. Pavese, A., Catti, M., Parker, S.C. and Wall, A. (1996) *Phys. Chem. Miner.*, **23**, 89.
69. Catlow, C.R.A., Norgett, M.J. and Ross, A. (1977) *J. Phys. C.* **10**, 1630.
70. Elliott, J.C. (1994) *Structure and chemistry of the apatites and other calcium orthophosphates*, Studies in inorganic chemistry, **18**, Elsevier.
71. Deer, W.A., Howie, R.A. and Zussman, J. (1992) *An introduction to the rock-forming minerals*, Longman, UK.
72. Rakovan, J. and Reeder, R.J (1994) Differential incorporation of trace elements and dissymmetrization in apatite: The role of surface structure during growth, *Am. Miner.* **79**, 892-903.
73. Weiner, S. and Wagner, H.D. (1998) The material bone: structure-mechanical function relations, *Ann. Rev. Mater. Sci.* **28**, 271-298.
74. Lee, W.T., Dove, M.T. and Salje, E.K.H. (2000) Surface relaxations in hydroxyapatite, *J. Phys.: Condens. Matter* **12**, 9829-9841.

Computational Materials Science
C.R.A. Catlow and E.A. Kotomin (Eds.)
IOS Press, 2003

Surface relaxation in solids and nanoparticles

A. V. Postnikov
Theoretical Low-Temperature Physics,
Gerhard Mercator University, D-47048 Duisburg, Germany

Abstract. While very accurate methods for electronic structure calculations within the density functional theory exist for ordered crystalline solids, in the study of nanoparticles, with symmetry properties largely lost, one often needs to bargain between treatable number of atoms and accuracy of calculated properties. The same problem is often encountered in the study of surfaces. The choice of appropriate simulation method is often difficult. In this contribution, several systems are considered where surface relaxation is non-negligible and associated with the variation of local characteristics, like magnetic moments. Calculations of electronic structure and (unconstrained) relaxation are done by an accurate tight-binding SIESTA method that uses numerical strictly localized orbitals as basis functions. Rutile-type SnO_2 crystal, traditionally used as a gas sensor, exhibits large and deeply going relaxation on its (110) surface. The morphology of the surface is not well known from experiment. The results of several available calculations are now refined and extended over larger supercells. The same method has been applied to the study of a number of Fe nanoparticles (consisting of approximately 30 – 60 atoms), chosen as spherical fragments of bcc or fcc bulk and subject to further relaxation. All particles exhibit large contraction within 2-3 outer layers, accompanied by an enhancement of local magnetic moments to approximately 3 μ_B.

In this contribution I review the results of first-principles simulations recently done for two different systems – bulk tin dioxide and Fe nanoparticles – that show considerable relaxation in the surface layers. From computational point of view, these studies weve demanding because high accuracy in the calculation of total energy and forces on atoms had to be achieved, while treating the systems of $\sim$60 atoms possessing no symmetry that could be exploited. Some earlier results were available, but their accuracy was contestable and needed to be checked. The computational approach applied was an *ab initio* scheme, using norm-conserving pseudopotentials and numerical localized basis functions (pseudoatomic orbitals) of finite extension range. The corresponding computer code SIESTA has been already reviewed in a number of publications [1, 2, 3], and successful applications to date include crystalline and molecular systems (see Ref. [4] for a review) with different degree of electron localization and character of chemical bonding.

Essential for the presently addressed applications is the ability of the code to include semicore states, like Sn$4d$ or Fe$3p$, in the valence band, that is essential for a precise energetics but may face certain problems in planewave pseudopotential schemes. Moreover, the Hamiltonian and overlap matrices are compact (due to efficiency of numerical basis) and sparse, that allows to profit from using an N-scaling algorithm in the treatment of large (over $\sim 10^2$ atoms) systems. The systems discussed below are not that large, so straightforward diagonalization was used for solving the Kohn–Sham equations. Accurate forces on atoms are available and have been used to drive structure relaxation by conjugated gradient method, starting from a fragment of bulk as a starting configuration.

The results reviewed below for slabs (SnO_2) and finite clusters (Fe) will be in full published elsewhere[5, 6]. Correspondingly, the data that serve exclusively test properties, like optimization of bulk crystalline structure and the accuracy of elastic properties obtained with SIESTA, the effect of different exchange-correlation schemes etc. will be skipped here. Apart from arriving at equilibrium geometry, one can gain additional information by simulating dynamical properties either in a frozen phonon scheme, or by running molecular dynamics. Some discussion to this subject can be found in Ref. [7].

1 Tin dioxide

Tin dioxide is nominally a wide-band insulator which however in normal conditions behaves like an n-type semiconductor, due to the presence of oxygen vacancies. A broad-range change of conductivity depending on ambient conditions allows the use of tin dioxide in gas sensors – see e.g. Ref. [8]. The performance of such sensors is essentially determined by the morphology of, and chemical reactions at, the surface. The properties of SnO_2 nanoparticles where the surface effects dominate are therefore of interest for both experimental and theoretical studies. Another field of possible applications of tin dioxide is related to a dependence of its optical band gap on the size of particles. Here again, an insight into electronic structure and its relation to structural properties by means of theory is highly interesting. The electronic structure of ideal bulk SnO_2 is well studied by *ab initio* calculations [9, 10], and the phonon dispersion has been recently determined from first principles by Parlinski and Kawazoe[11]. The study of surfaces is quite computationally demanding and advances rather slowly. Rantala *et al.*[12] studied the electronic structure of clean (unrelaxed) surface, also in the presence of adsorbed NO molecules, by a tight-binding cluster method; Manassidis *et al.*[13] calculated surface energy and structure relaxations near the (110) surface by a pseudopotential method. Later on, Rantala *et al.*[14, 15] studied relaxation of the (110) surface by two different techniques, a semiempirical tight-binding scheme and plane-wave pseudopotential calculations. Recently Oviedo and Gillan[16] performed a more detailed study of several SnO_2 surfaces, comparing their energies of formation, and started to look into mechanisms of the formation of oxygen vacancies, making use of the VASP code, that incorporates ultra-soft pseudopotentials and plane waves as basis functions.

A more detailed description of the results obtained with SIESTA can be found in Ref. [5]. Below, the relaxation pattern on (110) and (001) surfaces is addressed, compared with previous results (for small supercells) whenever available, and the conclusions about the width of the relaxed surface region are drawn.

The search for the equilibrium structure at the surface has been done in the slab geometry, with the smallest unit cell possible in the surface plane and the necessary number of atomic levels in the normal direction to sum up to the $Sn_{10}O_{20}$ composition in a supercell. The (110) and (001) surfaces, which in a sequence of recent plane-wave pseudopotential calculations [16] have been shown to be the most favorable and the next favorable surfaces, according to their formation energies, were addressed in the present simulations by SIESTA. The (110) surface seems to be the most extensively studied one in different simulations to date. Fig. 1 depicts two largest supercells used in our simulation for both surfaces in question. The crystal structure parameters have been taken as optimized for the bulk material (see previous section), the length of the supercells was fixed at 27.8 Å for the (110) and at 23.1 Å for the (001) surfaces, just in order to avoid direct overlap of the compact basis functions across the vacuum

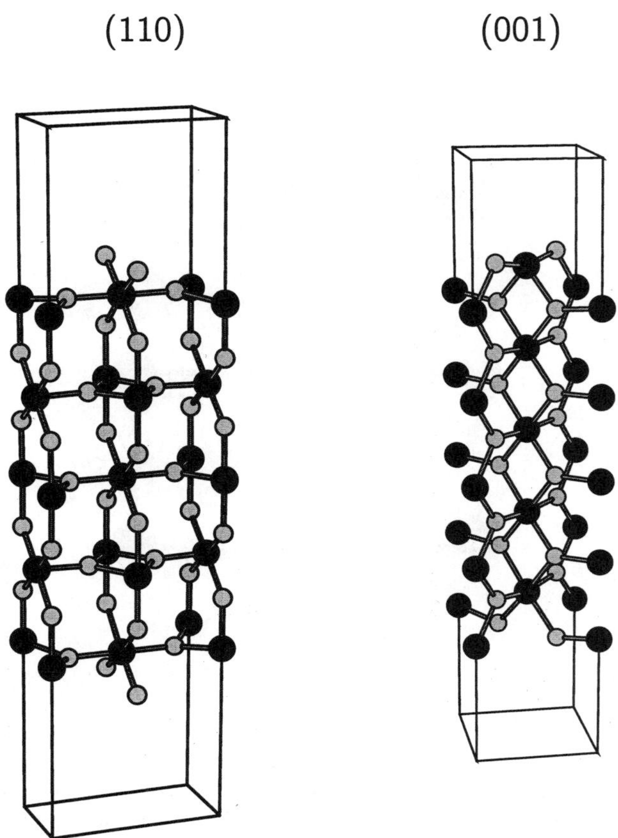

Figure 1: Stoichiometric $Sn_{10}O_{20}$ supercells used in the slab calculations for the (110) and (001) surfaces.

with those in a translated supercell. The coordinates of all atoms were free to relax. The (001) supercell shown in Fig. 1 represents the final convergent structure.

For the sake of more detailed discussion, in Fig. 2 the smallest supercell is shown that has been used for the simulation of the (110) surface, the stoichiometric Sn_6O_{12}. The removal of the topmost (bridge) oxygen atom on each side leaves the reduced Sn_6O_{10} supercell. The choice of these small supercells was motivated by the fact that such geometry was earlier addressed by Rantala *et al.*[14] in a first-principles plane-wave calculation, so that one could compare the results obtained by different methods. The values of structural parameters as introduced in Fig. 2 are listed in Table 1 according to our present calculation and two earlier ones.

In all cases, the present calculations essentially confirm the plane wave pseudopotential results of Ref. [14], even when they differ from tight-binding calculation results by Godin and LaFemina[17]. In particular, the tin atom between two bridging oxygen atoms (in the middle of the unit cell's upper face) relaxes slightly outwards, and the outward relaxation of the upper in-plane oxygen is larger in the relaxed supercell than in the stoichiometric one. However, the difference in the magnitude of the corresponding parameter $\Delta_{1\perp2}$ between

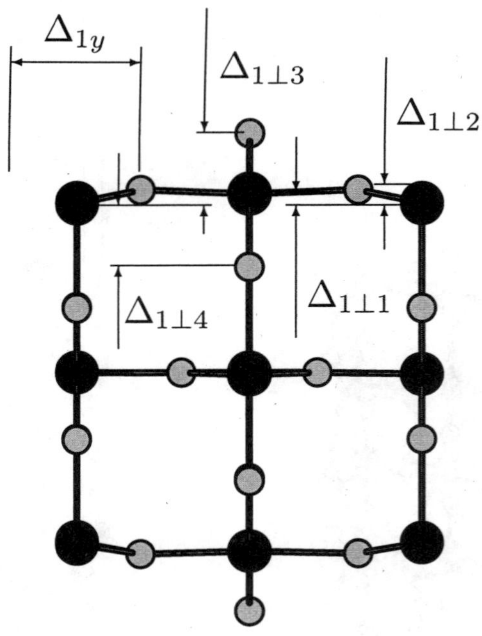

Figure 2: Side view of the Sn_6O_{12} supercell representing slab geometry for the (110) surface, with characteristic distances indicated.

relaxed and stoichiometric cases is not so large in our calculation as in that by Rantala *et al.*

Whereas qualitatively the relaxation pattern remains the same in a larger $Sn_{10}O_{20}$ supercell, actual numbers of relaxation are somehow affected by the supercell size. For instance, both $\Delta_{1\perp 1}$ and $\Delta_{1\perp 2}$ parameters become nearly 0.28 Å. The relaxation goes deep inside the crystal, so that the supercells considered were actually not yet large enough to achieve a convergency towards bulk crystal structure values in the middle of the slab. The same applies to the calculation representing the (001) surface. In this case (see Fig. 1, right panel), the distances between adjacent Sn planes on going inwards change as 1.27. 1.85; 1.48; 1.72 and 1.53 Å, fluctuating around, but only slowly converging to, the bulk value of $c/2$=1.62 Å. With respect to every corresponding Sn plane, the oxygen atoms displace outwards (+) or inwards

Table 1: Relaxation parameters Δ specified in Fig. 2 (in Å) for the (110) surface as calculated by SIESTA, in comparison with earlier calculations by *ab initio* plane-wave method (PW) and by a tight-binding method (TB).

	Sn_6O_{10} (reduced)			Sn_6O_{12} (stoichiometric)		
	SIESTA	TB[a]	PW[b]	SIESTA	TB[a]	PW[b]
$\Delta_{1\perp 1}$	+0.140	−0.05	+0.11	0.172	0.10	0.21
$\Delta_{1\perp 2}$	+0.292	+0.28	+0.40	0.248	0.29	0.22
$\Delta_{1\perp 3}$				1.327	1.37	1.41
$\Delta_{1\perp 4}$	1.207	1.29	1.33	1.239	1.18	1.22
Δ_{1y}	2.430	2.58	2.45	2.453	2.53	2.50

[a] Ref. [17]; [b] Ref. [14]

(−) as +0.33; −0.24; +0.14; −0.10; +0.08 and −0.07 Å (that must become zero in the bulk). Therefore, the treatment of larger supercells, feasible with the SIESTA code, is necessary for the reliable evaluation of surface relaxation.

2 Fe clusters

Magnetic properties of small iron nanoparticles are often unusual and quite different from those of the bulk. The experimental studies become more sophisticated in extracting microscopic information on indirect ways[18], but the difficulties still exist in resolving electronic structure locally, or in establishing accurate morphology of small particles. Here, first-principles simulation schemes, if applied within their limits of accuracy, may be of great help. First-principles studies of clusters have a long history, where the principal subject of study changed several times. Over decades, the clusters served as a model for treating a bulk where translational symmetry was broken, e.g., due to the presence of impurities. Apart from this, the calculations often proclaimed the aim of simulating finite clusters as such, but usually only electronic degrees of freedom have been allowed to relax, with geometrical structure fixed, based on some assumptions. Fujima and Yamaguchi [19] analyzed the variation of magnetization over inner shells of bcc-type Fe_{15} and Fe_{35} clusters. The calculation by Bouarab *et al.*[20], although dependent on a tight-binding Hamiltonian, can be also mentioned in this context.

Non-collinear magnetic structures of small (up to Fe_5) clusters have been recently accessed by Oda *et al.*[21] in a planewave pseudopotential calculation and by Hobbs *et al.*[22] who used projector augmented-wave method[22]. Both approaches were unconstrained in what regards spatial orientation of magnetization and produced close results (albeit disagreeing in some subject, like magnetic ground-state structure of Fe_3). They contested previous results obtained for the cluster geometry and energetics by a (magnetically collinear) Gaussian-orbitals method by Castro *et al.*[23]. The SIESTA code allows an unconstrained (non-collinear over simulation cell) treatment of magnetization; the benchmark results, very closely reproducing those of Refs. [21] and [22] for the Fe_5 cluster, have been reported in Ref. [6]. In the following, only those results of the latter study are cited that deal with larger (Fe_{15} to Fe_{62}) magnetic clusters. For them, fully unconstrained noncollinear treatment of magnetization would become too much time consuming, so a collinear (ferromagnetic or antiferromagnetic) magnetic structure has been assumed.

One should mention that the SIESTA program has been already applied to electronic-structure simulation of Fe nanostructures (small free and deposited clusters, monolayer and nanowire)[24] – primarily, in the view of their magnetic properties and without structure relaxation allowed. In Ref. [6], we intended to go beyond reproducing purely electronic-structure results and concentrated on the performance of SIESTA with respect to reproducing meaningful energetics and structure optimization. The part of Ref. [6] related to the test of calculation setup will be skipped in the present review. Summarizing, generalized gradient approximation, double-ζ singly polarized basis set (according to a classification used, and somehow discussed upon for the case of Fe systems, in Ref. [24]) and the pseudopotential attributing Fe3p functions to the valence panel were found adequate to quantitatively reproduce the hierarchy of energy-volume curves for different structural and magnetic phases of crystalline iron.

While the morphology and magnetic ordering of small clusters remain delicate, com-

Table 2: Relaxed distances from center a and magnetic moments over shells of neighbors M in bcc- and fcc-related Fe clusters. Numbers of neighbors within each shell are given in parentheses.

a (Å)	M (μ_B)	a (Å)	M (μ_B)	a (Å)	M (μ_B)	a (Å)	M (μ_B)
Fe_{35} (bcc)		Fe_{62} (fcc)		Fe_{38} (fcc)		Fe_{62} (fcc), AFM	
(1) 0.0	2.10	(6) 1.857	2.41	(6) 1.827	2.62	(4) 1.767	1.45
(8) 2.345	2.14	(8) 3.082	2.52	(8) 3.281	2.90	(2) 1.877	−2.35
(6) 3.043	3.13	(24) 4.173	2.93	(24) 3.985	3.14	(8) 3.086	−1.82
(12) 3.880	3.21	(24) 5.245	3.21	Fe_{43} (fcc)		(8) 3.883	−2.43
(8) 4.603	3.43			(1) 0.0	2.45	(8) 4.029	3.04
				(12) 2.579	2.52	(8) 4.150	−2.79
				(6) 3.798	3.00	(8) 5.294	3.46
				(24) 4.307	3.22	(16) 5.262	−3.25

putationally unstable and only with difficulties experimentally accessible matters, it is well known from experiment that larger metal nanopaticles exhibit ordered internal structure, with substantial deviations from crystalline behavior being constrained to a (more or less thick) surface layer. The aim of our further study was to simulate structure relaxation and radial distribution of magnetic moments inside Fe particles with several tens of atoms. Calculations for bcc-related Fe_{15} and Fe_{35} clusters, albeit without structure relaxation, have been reported by Fujima and Yamaguchi[19] (in a row of Ni and Cr clusters of comparable size). Therefore in the following primarily the fcc-related clusters are addressed, including Fe_{35} for comparison. These particles either had a central atom, or were centered around an octahedral interstitial, and an unconstrained structure relaxation were allowed after introducing small off-center displacements. The relaxed structure essentially preserved the cubic point symmetry, with the exception of the AFM (of CuAu-type) Fe_{62} particle that developed a slight tetragonal distortion. The (relaxed) radii of atomic shells along with corresponding magnetic moments are presented in Table 2. All these results are obtained in the GGA. The values of local magnetic moments per atom are estimated from the Mulliken population analysis. The values to be compared to in perfect relaxed crystal, according to the SIESTA calculation, are 2.43 μ_B (fcc) and 2.35 μ_B (bcc). The variation of magnetic moments over relaxed spheres of neighbors is shown in Fig. 3 (for particles with a central atom) and Fig. 4 (for particles formed around a central interstitial).

For the (bcc) Fe_{35} cluster, the magnetic moment in the first-neighbors (8 atoms) shell around the central atom was not found to drop down to below 2 μ_B, contrary to what was earlier reported in Ref. [19]. Instead, the magnetic moments grow smoothly towards the surface. This discrepancy was shown to be related to the structural relaxation: with the radii of shells of neighbors fixed at non-relaxed positions corresponding to crystalline lattice constant, the magnetic moments in our calculation come out as 2.31 (on the central atom); 2.21; 2.98; 3.23 and 3.45 (on the surface) μ_B, i.e. in qualitative agreement with the results of Fujima and Yamagichi [19]. The quantitative differences can be due to different exchange-correlation potential (X_α used in Ref. [19]).. Consistently with the results of Ref. [19], the magnetic moments grow in the outer shells and exceed 3 μ_B on the surface. The relaxation (neglected in Ref. [19]) is outwards for the second shell (6 atoms) and inwards in all others. It was found (this applies to the fcc clusters as well) that the inward relaxation is the largest on the surface,

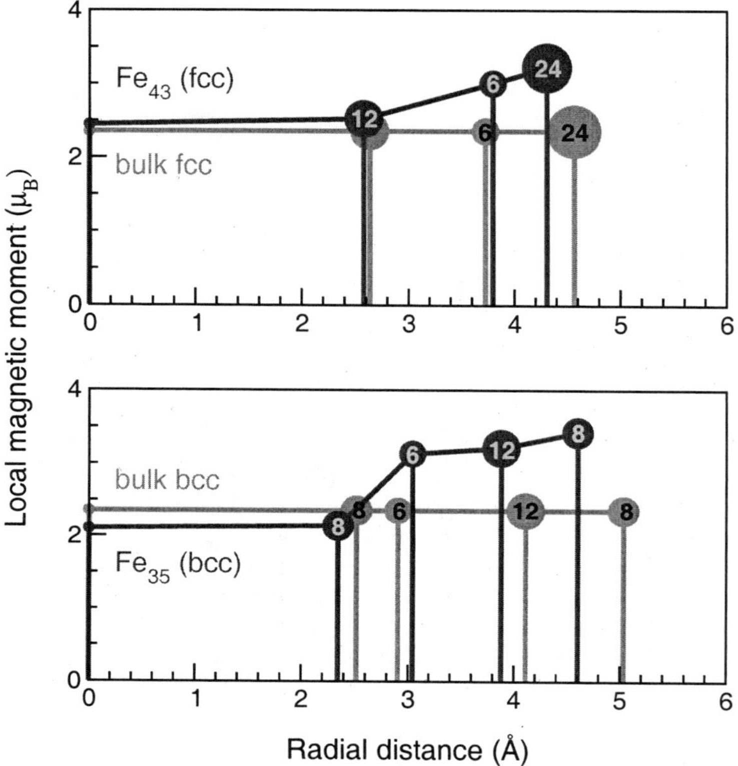

Figure 3: Same as in Fig. 4, for atom-centered Fe_{35} (bcc) and Fe_{43} (fcc) particles. The data for corresponding perfect lattices are shown for comparison.

where the magnetic moment is at most enhanced. Differently from Ref. [19] where the magnetic moments in the Fe_{35} cluster were stabilized from the second shell outwards, the present study found them to increase gradually towards the surface.

In the fcc-related clusters, the internal structure is more densely packed, and a pronounced outward relaxation occurs only for the second shell (6 at.) of the atom-centered Fe_{43} cluster. Comparing this to the result for the (interstitial-centered) Fe_{62} cluster, one can see a general tendency of developing a (quite small) outward relaxation in the subsurface shell, whereas the outer shell is always strongly contracted. Towards further inner shells, the relaxation is rapidly stabilized, and the atomic spacing approaches that in the bulk. As in the case of the bcc cluster, magnetic moments are largely enhanced on the surface (and immediately below it), but rapidly decrease and get stabilized in the deeper shells, without showing any fluctuations. The AFM ordering for the Fe_{62} cluster lifts some degeneracies in the radial distribution of atoms, but otherwise is consistent with the above observation. The total energy per atom is by 0.19 eV higher in the AFM configuration.

Soler *et al.*[25] studied the morphology of small (38 to 75 atoms) "ordered" and "amorphous" gold nanoparticles, using the same calculation scheme as here in order to refine trial geometries, provided by energy minimization in an empirical potential. In all cases "amor-

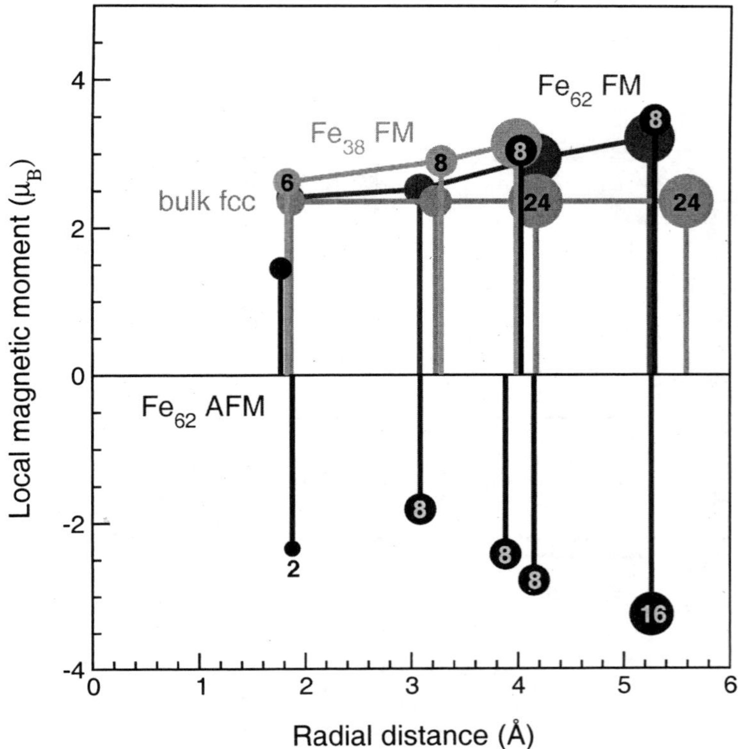

Figure 4: Distribution of local magnetic moments over shells of neighbors in fcc-related nanoparticles Fe_{38} and Fe_{62}, centered around an octahedral interstitial. For the latter particle, results corresponding to FM and AFM ordering are shown. The size of the circle indicates the number of neighbors a in corresponding shell. The data for the fcc lattice are shown for comparison.

phous" particles were found to be more stable, and the subsequent analysis shows the reason for this to be due to high elastic contribution to the total energy of a particle, relaxed in the course of an amorphous-like rearrangement of atoms. In our case, inner shells of Fe particles are also contracted, as compared to the bulk crystal. It would be interesting to probe the effects of amorphization, and their interplay with magnetic characteristics, in an *ab initio* simulation once realistic models for ground-state Fe arrangements become available.

Summarizing, the present results favor a conclusion that the relaxation and magnetic properties of small Fe nanoparticles have certain common features, relatively independent on morphology, magnetic ordering and size. Namely, the structure relaxation is practically confined within 2–3 outer shells, the surface layer relaxes strongly inward, and the magnetic moments on the surface are enhanced to beyond 3 μ_B. The overall magnetic properties of larger nanoparticles must be then primarily governed by the proportion between surface-layers atoms and their deep bulk-like counterparts.

Acknowledgments

The results here reviewed have been obtained in cooperation with P. Ordejón (on the SnO_2 system) and with J. Soler (on the Fe clusters) and with P. Entel and submitted for publication in *Phase Transitions* and *Physical Review B*. The work was supported by the German Research Society (*Sonderforschungsbereich 445*).

References

[1] P. Ordejón, D. A. Drabold, R. M. Martin and M. P. Grumbach, Linear system-size scaling methods for electronic-structure calculations. *Phys. Rev. B* **51** (1995) 1456–1476.

[2] P. Ordejón, E. Artacho and J. M. Soler, Self-consistent order-N density-functional calculations for very large systems. *Phys. Rev. B* **53** (1996) R10441–R10444.

[3] D. Sánchez-Portal, P. Ordejón, E. Artacho and J. M. Soler, Density-functional method for very large systems with LCAO basis sets. *Int. J. Quant. Chem.* **65** (1997) 453–461.

[4] P. Ordejón, Linear scaling ab initio calculations in nanoscale materials with siesta. *Physica Status Solidi (b)* **217** (2000) 335–356.

[5] A. V. Postnikov, P. Entel and P. Ordejón, SnO_2: bulk and surface simulations by an ab initio numerical local orbitals method. (To be published in Phase Trans.).

[6] A. V. Postnikov, P. Entel and J. M. Soler, Density functional simulation of small Fe nanoparticles. (to be published).

[7] A. V. Postnikov, Vibrations in solids and small particles from first-principles calculations. (present proceedings).

[8] J. Watson, K. Ihokura and G. S. V. Coles, The tin dioxide gas sensor. *Measurement Science & Technology* **4** (1993) 711–719.

[9] E. L. Peltzer y Blancá, A. Svane, N. E. Christensen, C. O. Rodríguez, O. M. Cappannini and M. S. Moreno, Calculated static and dynamic properties of β-Sn and Sn-O compounds. *Phys. Rev. B* **48** (1993) 15712–15718.

[10] J. Goniakowski, J. M. Holender, L. N. Kantorovich, M. J. Gillan and J. A. White, Influence of gradient corrections on the bulk and surface properties of TiO_2 and SnO_2. *Phys. Rev. B* **53** (1996) 957–960.

[11] K. Parlinski and Y. Kawazoe, Ab initio study of phonons in the rutile structure of SnO_2 under pressure. *Eur. Phys. J. B* **13** (2000) 679–683.

[12] T. S. Rantala, V. Lantto and T. T. Rantala, A cluster approach for modelling of surface characteristics of stannic oxide. *Physica Scripta* **T54** (1994) 252–255.

[13] I. Manassidis, J. Goniakowski, L. N. Kantorovich and M. J. Gillan, The structure of the stoichiometric and reduced SnO_2(110) surface. *Surf. Sci.* **339** (1995) 258–271.

[14] T. T. Rantala, T. S. Rantala and V. Lantto, Surface relaxation of the (110) face of rutile SnO_2. *Surf. Sci.* **420** (1999) 103–109.

[15] T. T. Rantala, T. S. Rantala and V. Lantto, Electronic structure of SnO_2 (110) surface. *Materials Science in Semiconductor Processing* **3** (2000) 103–107.

[16] J. Oviedo and M. J. Gillan, Reduction and oxidation processes at the SnO_2 (110) surface (2000). Ψ_k-2000 conference "Ab initio (from electronic structure) calculation of complex processes in materials", Schwäbisch Gmünd, Germany, August 22–26, 2000, see http://psi-k.dl.ac.uk/psi-k2000/: Programme and Abstracts, p. 185.

[17] T. J. Godin and J. P. LaFemina, Surface atomic and electronic structure of cassiterite SnO_2 (110). *Phys. Rev. B* **47** (1993) 6518–6523.

[18] D. Gerion, A. Hirt, I. M. L. Billas, A. Châtelain and W. A. de Heer, Experimental specific heat of iron, cobalt, and nickel clusters studied in a molecular beam. *Phys. Rev. B* **62** (2000) 7491–7501.

[19] N. Fujima and T. Yamaguchi, Geometrical magnetic structures of transition-metal clusters. *Materials Science and Engineering A* **217–218** (1996) 295–298.

[20] S. Bouarab, A. Vega, J. A. Alonso and M. P. Iñiguez, Tight-binding study of the ionization of iron clusters. *Phys. Rev. B* **54** (1996) 3003–3006.

[21] T. Oda, A. Pasquarello and R. Car, Fully unconstrained approach to noncollinear magnetism: Application to small Fe clusters. *Phys. Rev. Lett.* **80** (1998) 3622–3625.

[22] D. Hobbs, G. Kresse and J. Hafner, Fully unconstrained noncollinear magnetism within the projector augmented-wave method. *Phys. Rev. B* **62** (2000) 11556–11570.

[23] M. Castro, C. Jamorski and D. R. Salahub, Structure, bonding, and magnetism of small Fe_n, Co_n and Ni_n clusters, $n \leq 5$. *Chemical Physics Letters* **271** (1997) 133–142.

[24] J. Izquierdo, A. Vega, L. C. Balbás, D. Sánchez-Portal, J. Junquera, E. Artacho, J. M. Soler and P. Ordejón, Systematic *ab initio* study of the electronic and magnetic properties of different pure and mixed iron systems. *PRB* **61** (2000) 13639–13646.

[25] J. M. Soler, M. R. Beltrán, K. Michaelian, I. L. Garzón, P. Ordejón, D. Sánchez-Portal and E. Artacho, Metallic bonding and cluster structure. *Phys. Rev. B* **61** (2000) 5771–5780.

Computational Materials Science
C.R.A. Catlow and E.A. Kotomin (Eds.)
IOS Press, 2003

SEMICONDUCTORS

I. Methods of calculating point defects in semiconductors

Peter Deák
Department of Atomic Physics, Budapest University of Technology and Economics
Budafoki út 8., Budapest, H-1111, Hungary (e-mail: p.deak@eik.bme.hu)

Abstract This paper describes how solution of the quantum mechanical problem of a model of the defective semiconductor might help defect engineering in electronics. The relation of the calculated results to experimental data are clarified and comments are given on the characteristics of various models, Hamiltonians and basis sets used in such calculations as well as on the convergence of the results.

1. INTRODUCTION

For the solid state theorist semiconductors are crystals with covalent (apolar or slightly polar) bonds and having, at T = 0K, an empty conduction band (CB) separated by an energy gap from the valence band (VB). In their intrinsic (pure) state they are insulators* but the area selective introduction of special foreign atoms (*dopants*) to various regions of the otherwise almost perfect semiconductor crystal allows the tight spatial control of free carriers. This makes semiconductors the base material of electronic devices. The engine of technological development in the last decades has been the fast and steady advance of solid state electronics. The continuing miniaturization and the drive for ever higher device performance made the control of the material composition necessary down to the nanometer scale. That, together with the fact that the very operation of semiconductor devices are based on quantum phenomena, makes quantum mechanical simulation of atomic processes an ever more important tool of R&D in this field.

Dopant atoms provide free carriers (i.e., become activated) only if they occupy substitutional positions** and become sufficiently immobile there. Obviously, the dopants have to compete not only with the host atoms for lattice positions, they may interact with unintentional *impurities* — which themselves might have influence on the concentration of free carriers. Figure 1 shows the different possible effects of impurities or intrinsic defects. They might behave as dopants themselves *compensating* the effect of the majority dopant, sometimes pinning the Fermi level (Fig.1a). If the impurity level is deep, *trapping* occurs dimin-

* The old restriction in the definition, that the semiconductor gap should be small is obsolete. Intrinsic silicon is a reasonably good insulator at room temperature where electronic devices are usually operated, while some metal oxides show semiconducting behavior at 7-800 °C. It is more appropriate to call any insulator a semiconductor if — at the required temperature of operation — the free carrier concentration can be controlled by the concentration of dopants.

** In a compound semiconductor they are effective only in one of the sublattices.

ishing the free carrier concentration (Fig.1b). Impurities might also outcompete dopants for the substitutional sites *impairing activation* (Fig. 1c), or by forming electrically inactive complexes they may be *passivating* them (Fig.1d). The knowledge of the chemical reactions of dopants and impurities, and the resulting effect on free carrier density and lifetime is of primary importance to designing a device technology. This so-called defect engineering relies heavily on the identification of defects responsible for unwanted changes occurring in the electronic behavior of the semiconductor after various process steps. Point defects (or their complexes consisting of a few atoms) are as yet "invisible"; they manifest themselves only as distinctive signals in various optical, electrical and magnetic resonance spectra. Typically these methods offer only relatively few cues as to the origin of the signal, called the defect *center*. It is the task of quantum mechanical simulations carried out on various possible *models* of the center to find the real composition and configuration of the defect complex. This is done by calculating the thermodynamic and kinetic feasibility and many spectroscopic properties of a variety of complexes in the semiconductor which can be envisaged as model of a given center based on information about the samples the experiments had been carried out on.

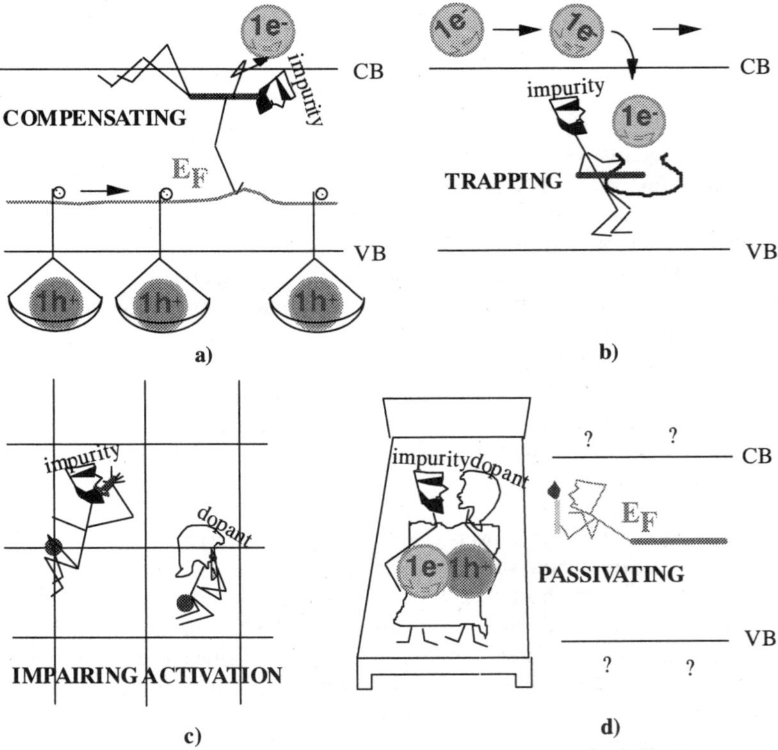

Figure 1. Possible effects of an impurity on the electronic properties of a semiconductor

The present paper is not a lecture on methods of computational physics but rather an account of my personal experience on how to use them in this particular field. As such it is going to be subjective and mostly void of references (except mostly to my own work and the compilation [1] where further reading can be found). In Section 2, I will elaborate on the calculated quantities needed to identify the origin of experimentally observed phenomena. Section 3 contains my comments on the models most often used for defects in semiconductors together with some tests on size convergence. Section 4 gives my views on various Hamiltonians and bases used to solve the quantum mechanical problem of the system. Finally an illustrative example is described in Section 5.

2. TARGET QUANTITIES OF THE CALCULATIONS

Any atomistic computer simulation starts by providing an input describing the atoms in the system under investigation. In case of point defect complexes in solids, the position of most of the host atoms is given by the original crystalline structure (allowing for some *relaxation* due to the defect) but the number, type and position of the dopants and/or impurities, as well as the *reconstruction*[*] of the immediate neighborhood of the crystal is a matter of guesswork based on intuition and experience (see Part II). Experiments mostly provide a hint on the type of foreign atoms involved, often the symmetry, and sometimes even the number of atoms. Still this information is never a direct one and, at best, dependent on the interpretation of the *measured data,* but sometimes it is hardly more than speculation. The *results of the calculation* should be checked against the data — and this decides whether the interpretation (speculation) was right. Therefore, it is always recommendable to try many different configurations and find the energetically most favorable one. Calculation of the total energy and its minimization with respect to the atomic coordinates in the given configuration is a "must" but the chances of finding the global minimum of the energy surface are not always good and there is no "microscopic method" which would provide *a-priori* geometry data. (Sometimes ion beam analysis may give at least some help for checking the result, though.) Still, at least, energetic metastability (especially with respect to dissociation or symmetry lowering) is a necessary condition[**] before other calculated properties can be compared to experimentally observed spectroscopic data. The more of the latter are available on the same center the higher the chance for identification with a model. It should be noted, however, that, on the one hand, the error bar of the very best calculation is still ridiculously large in the eye of a spectroscopist. Therefore, identification is acceptable only if many different data seem to be well approximated by the calculated values, or if trends (on, say, annealing, applied uniaxial stress, isotope substitution) are well reproduced. On the other hand, theorists should be aware that signals seen by different spectroscopies are all too often believed to have the same center as origin, without a conclusive proof.

The main spectroscopic tools used in the research of point defects in semiconductors are infrared absorption (IR) and Raman scattering; optical absorption, photoluminescence (PL) and photoconductivity; thermally stimulated conductivity (TSC) and deep level transient spectroscopy (DLTS); as well as electron-paramagnetic resonance (EPR), electron-nuclear-double-resonance (ENDOR), or optically detected magnetic resonance (ODMR). Description of these methods are out of the scope of the present paper but it is important to understand the relation of the experimental information to the calculated quantities (see Table 1).

---

[*] The word *reconstruction* will be used in the sense of a symmetry lowering displacement of host atoms in answer to the defect complex, in contrast to *relaxation*, where the symmetry of the system is maintained.

[**] A word of caution: thermodynamic stability may not suffice! It is very important to check the information on how the given center arose. For example, stability of an XY complex does not help if X and Y are introduced under circumstances where both are expected to be positively charged and, therefore, to repel each other.

Table 1. Experimental observables which can be elucidated from atomistic quantum mechanical calculations

Total energy (E_{tot}) as a function of the total charge (q) and spin (S) as well as the type (Z_i)and position (x_i) of atoms $$E_{tot}[\ldots,Z_i,\mathbf{x}_i,\ldots,q,S]$$ Forces: $$F=-gradE_{tot}$$ + Automatic minimization of E_{tot} with respect to selected coordinates. + a) Automatic search for a "saddle point" between two minima. b) Solution of Newtonian equations for nuclei	• *Relative stability of different equilibrium configurations, heat of formation, or heat of reactions between (binding energies) between point defects:* **appearance and disappearance of the "center", relative concentration** • *Occupation levels, or "adiabatic" ionization energies (→* **TSC, DLTS***)* • *Activation energies for reactions and diffusion (→* **annealing***)* • *Dynamic simulation of reactions and diffusion*
Dynamic matrix: $m_i\ddot{\mathbf{x}}_i=-\sum_j\frac{\partial^2E_{tot}}{\partial\mathbf{x}_j^2}\mathbf{x}_j\rightarrow\omega_l$	• *Frequencies of local mode vibrations* ω_l *(→***IR, Raman***)*
One-electron energies: $\{\varepsilon_i\}$	• *"Vertical" ionization and excitation energies (→***PL, photoconductivity, optical absorption***)*
One-electron wave function for spin α: $\psi_n^\alpha(\mathbf{r})$	• *Spin distribution, hyperfine tensor (→* **EPR, ENDOR, ODMR***)*

The first task is, as said before, to establish the energetic (meta)stability of the complex. Almost any theory (with the exception of some empirical tight binding schemes and a few quantum chemical methods developed to calculate electronic excitation spectra) yields the total energy of the system, E_{tot}, at T = 0K. E_{tot} can be used to establish the energy difference, ΔE, between different configurations of the same complex (relative stability), or to calculate its binding energy with respect to its isolated constituents in the solid.* Comparing stabilities the charge state must be taken into consideration. In a real semiconductor there is always a reservoir of free carriers with an average energy expressed by the Fermi-level (chemical potential of the carriers), E_F, the position of which may vary between the top of the valence band, E_V, and the bottom of the conduction band, E_C, depending on doping and temperature. The energy of a defect with charge q in the solid is, therefore,

$$E^q=E_{tot}[\ldots,Z_i,\mathbf{x}_i,\ldots,q,S]+qE_F \qquad /1/$$

expressing the fact that the defect can be ionized at the expense of the rest of the solid (inner ionization). The first thing to do is to check the relative stability of the different charge states of a configuration as a function of the Fermi-level (see Figure 2).

It is important to minimize the total energy of each charge state with respect to the coordinates (and also the spin!). The displacement of atoms may lead to energy gain compensating Coulomb-repulsions, so sometimes two electrons or none on the defect might turn out to be more stable than one (e.g., if the E^+ and E^- curves in Figure 2 crossed before they cross E^0): this is called *negative-U* behavior. Different configurations of the same complex usually have different occupation levels and one configuration might turn out to be more stable in one charge state and another in the other (*bistability*). Obviously, heat of reactions should be calculated between entities which are stable at the given position of the Fermi-level, and necessary charge transfer to-and-fro E_F should be accounted for. The Fermi-level positions where the charge state changes are the *occupation levels*. These de-

* The binding energy is a good approximation to the heat of formation, $\Delta H=\Delta E-p\Delta V$, for normal pressures, since the difference in specific volume is negligible compared to errors in ΔE.

scribe the *adiabatic* ionization of the defect while maintaining equilibrium conditions and, therefore, corresponds to the peak positions obtained in TSC and DLTS measurements.

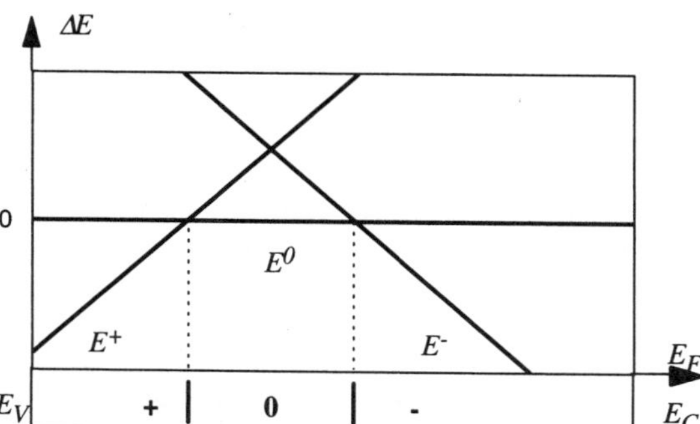

Figure 2. Relative energies of the different charge states of the same complex show which one is in equilibrium as a function of the Fermi-level. The Fermi-level positions where the charge state changes are called occupation levels, and are often incorporated in a band diagram showing the charge state below and above.

By calculating the relative energy of a saddle point configuration between two equilibrium ones, one may have a guess at the activation energy of diffusion. However, extreme caution is advised when comparing to experimental data (coming from Arrhenius-plots taken with depth profiling upon annealing). Apart from the difficulties with calculating saddle-point configurations (typically unsuited to one-electron theories), total energies are calculated at T = 0K, ignoring dynamic effects of the lattice. Unfortunately, molecular dynamics (MD) simulations can – as yet – typically be performed only at temperatures too high or on a timescale too short in comparison with experiment. It should be noted at this point, that the observed dissociation energies (again from exponents of Arrhenius-plots) for a complex can be calculated as a sum of the binding energy and the diffusion barrier for the more mobile constituent. The relative error is appreciably smaller in these cases. It is also worth mentioning that the energy surface for moving the charged defect may differ from that of the neutral, and charging/recharging may occur during diffusion (a fact usually ignored in MD simulations). If the maxima of the energy surface for different charge states do not coincide in such a case, the diffusion may become anomalously fast.

Confirmation of the stability of our complex with respect to doping (Fermi-level position) and temperature (dissociation or out-diffusion) may put it in the ball park of competing models of a given defect center. This, of course, means that theorists, reading experimental papers in search for data to compare their results to, should not skip the "method" section! The circumstances under which the sample was prepared and processed before measurements but also the way the measurement was performed are of vital importance. Again a word of caution: measurements are not always carried out after a systematic series of isochronal anneals, and, therefore, Arrhenius-plots giving the activation energy for appearance or disappearance (i.e., activation energy for diffusion or reactions) are not always available. There is, however, usually a statement regarding the temperature where the center appears or disappears. Multiplying that by the Boltzmann-factor does not give us the required value!!! (For example: oxygen in silicon has a diffusion barrier of 2.5 eV but it is mobile at 300 °C, or the free hydrogen molecule dissociates around 2000 °C despite the 4.7 eV binding energy, due to vibronic excitations!)

To compare the energy of different complexes (different number and type of atoms), one needs a common reference and the suitable thermodynamic quantity is not ΔE but $\Delta G = \Delta H - T\Delta S$. The free enthalpy of formation can be defined as

$$\Delta G_f^q = \left[E^q - TS^q\right] - \sum_i n_i \mu_i \qquad /2/$$

where n_i are the number of atoms of type i in the investigated system and μ_i are their chemical potential in a reservoir which is in equilibrium with the system. (This is usually chosen as close to the real environment pertinent to the process, e.g. annealing or growth, as possible.) The calculation of the entropy of the system is possible (after calculating the vibration modes). However, this possibility is not always available in accessible computer codes and it is hardly ever practical to utilize. In most cases (up to 1000 – 1500 °C) the entropy term in eq. /2/ is, therefore neglected. This does not, typically, impede the comparison of the relative abundance of the complexes but forbids direct comparison with experimentally observed heats of solution. Unless the calculated "energy" of formation is negative – which is often the case – the resulting equilibrium concentrations are still reasonable (give or take an order of magnitude ...). Note, however, that in calculating equilibrium concentration of electrically active complexes, the position of the Fermi-level has to be determined self-consistently at any level of doping and temperature [2].

Once it has been established that the chances for observation of our complex are good under the circumstances the defect center has been seen, we may start calculating the observed quantity. The second derivative of the total energy supplies the dynamic matrix which allows to calculate the frequency and oscillator strength of the localized vibration modes (LVMs) of our defect. These quantities are directly measured by IR and Raman spectroscopy. One should be aware that the accuracy of the measured frequencies is better than 1 cm^{-1} (about 10^{-4} eV!) and any spectrum can show a wealth of bands within 100 cm^{-1} which is the best accuracy theory can hope for. Therefore, assignment can only be attempted if the complex has a number of observable modes and all seem to fit more or less. Alternatively, one can calculate isotope shifts (often known from experiment) or predict splitting (qualitatively) upon uniaxial stress.

Optical methods are very favored in the defect community due to their sensitivity, accuracy and wide energy range. Photon excitation is a non-equilibrium process. (This holds for photoconductivity measurements as well.) That means that the nuclei of the system have no time to respond to the excitation, and the measured energy of the emitted or absorbed photon reflects the (de)excitation energy of the electron system alone. Obviously, photon excitation may cause the (inner) ionization of a defect, which is called *vertical* in contrast to the adiabatic one described above. The difference between adiabatic and vertical excitation may be understood with the help of the configuration coordinate diagram of Figure 3. The horizontal shift of the *E(Q)* curves show that the atoms have different equilibrium positions in the two states. Excitation by photon absorption (*PE*) and deexcitation by photon emission (*PL*) are vertical processes, while the adiabatic (de)excitation energy is the energy difference between the two equilibria. (It should be noted that, during recombination of excitons bound to defects, LVMs may be excited simultaneously. This is manifested by the PL sidebands shifted from the main line by $h\nu_{...}$.)

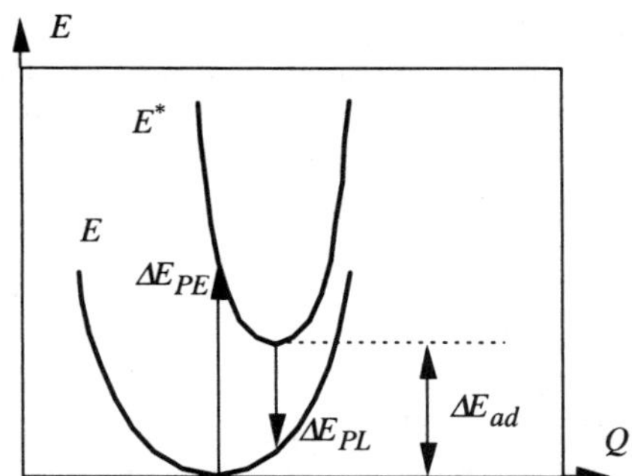

Figure 3. Configuration coordinate diagram showing a system and its excited (ionized) state. For explanation, see the text.

In infinite systems (what the real defective solid approximately is) the electronic excitation energies correspond to energy differences between one-electron levels:

$$\Delta\varepsilon_{ij} = \varepsilon_j - \varepsilon_i \qquad /3/$$

(Note, however, that in finite *models* of a defective solid $\Delta\varepsilon_{ij} = \varepsilon_j - \varepsilon_i - (J_{ij} - 2K_{ij})$ should be used, where J and K are the Coulomb- and exchange-interactions.) Therefore, one-electron levels of a defect with respect to the band edges supply – in good approximation – the defect-to-band electronic transitions (vertical inner ionization) measured optically.* (Obviously, the direction of the transition decides the calculated one-electron levels of which charge state should be taken.) Optical methods being the leading tool in identifying electrically active defects, prediction of one-electron level positions relative to the gap became paramount to understanding the operation of a semiconductor. So, despite of the spectacular weakness of all *ab-initio* one-electron theories in this respect, calculation of one-electron energies seem to be unavoidable necessities in any simulation.** It should be noted though, that many-body effects often play an important role in the excitation (for that matter, even in deciding the spin of the ground state of some defects, e.g. an unreconstructed vacancy or a split self-interstitial). Calculating one-electron states has another advantage. The most valuable experimental information about defects (in their states with S . 0) comes from paramagnetic resonance. The hyperfine tensor is directly related to the spin-distribution from which symmetry, and number and type of atoms in consecutive shells can be deduced. Here, again, the deduction is not unique and subject to errors but the calculated spin distribution can be checked against the measured hyperfine tensor.

3. MODELLING THE DEFECTIVE SEMICONDUCTOR CRYSTAL: CLUSTER (SUPERCELL) SIZE AND CONVERGENCE

Regarding the size and high level of perfection of today's intrinsic semiconductor boules, the usual assumption of solid state physics about an infinite, translationally invariant crystal is quite justified. This allows for the handling of a single unit cell only in the description of the crystal itself. The presence of just one defect, however, ruins the translational symme-

* Defect-to-defect transitions can also be measured. For donor (D) – acceptor (A) recombination the Coulomb attraction between the charged defects shifts the observed PL line. According to the possible relative positions of D^+ and A^-, a series of lines arise characteristic of whether D and A both prefers the same sublattice or not.

** Almost disqualifying *O(N)* methods despite their postulated efficiency.

try, so the explicit treatment of a finite but very large number of atoms would be required. This is circumvented by the application of simplified models of the defective solid. The obvious way is to separate the part of the crystal perturbed by the defect from the still crystalline background, and treat only the former in an explicit manner. On the one hand, the need for economizing calls for a model consisting of as few atoms as possible. On the other, the explicitly treated part of the crystal should, in principle, extend far enough for the effect of the defect to become negligible. In practice, this is only assumed for a manageable group of atoms around the defect — usually called a *cluster*. While the cluster is handled explicitly at any given level of calculation, the problem arises at its surface, i.e. with the representation of the rest of the material. As it is shown schematically in Figure 4, there are two basic possibilities. One either embeds the cluster into a background simulating the rest of the crystal or repeats the cluster periodically in all directions and constructs an artificial periodic superlattice [3]. Sophisticated *embedded cluster models* are based on Green-function techniques, which give rise to serious complications due to the semiconductor gap (where complex solutions enter). Therefore, although these models seem to be most desirable, they are very rarely used for defects in semiconductors.

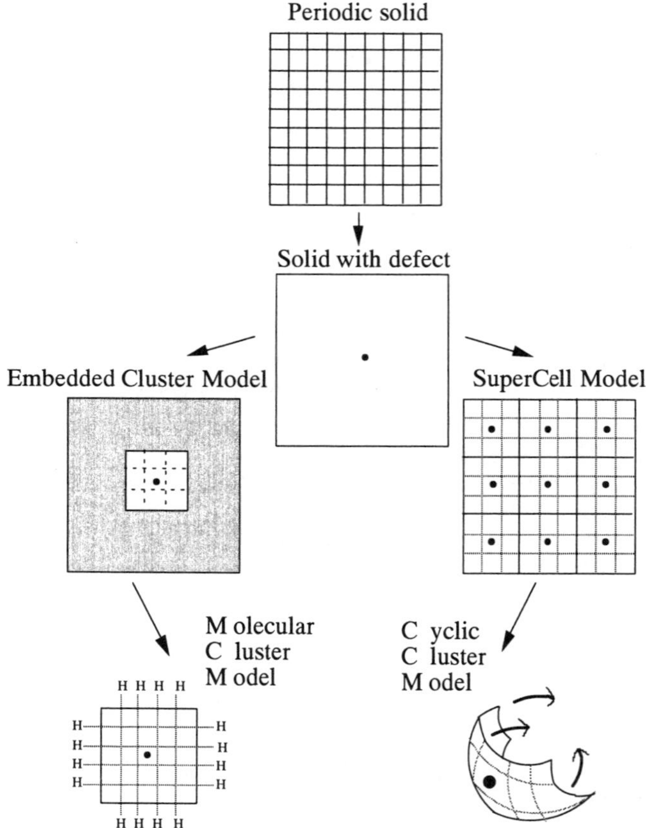

Figure 4. Possible models of a defective solid

The simplest form of embedding is the molecular cluster model (MCM). Its usefulness follows from the covalent nature of semiconductors. Because of that, the crystalline orbitals of electrons in a semiconductor can well be described by localized bonds. (The set of canonical one-electron orbitals may, in principle, be transformed into a set of bonding orbitals localized around two centers to a high degree.) Ideally, the quantum mechanical interaction of the cluster with the background is completely represented by the localized bonds crossing the interface between the two regions. If these bonds could be kept unchanged while cutting off the atoms of the background, the embedding would be perfect (save for long range Coulomb interactions, in case of polar bonds, which can be treated – if needed – by embedding the cluster into a set of background point charges). To achieve that, the dangling bonds on the cluster surface have to be saturated in such a way that the new bonds become equivalent with the original bonds of the crystal. In principle, one can construct monovalent pseudoatoms with parameters fitted to restore the original bonding. Since the fitting is never perfect, in most cases simply hydrogen atoms are used. The hydrogen saturated cluster of atoms is a molecule in the chemical sense (e.g., 5 atoms of diamond saturated with hydrogen atoms form the molecule neopenthane, C_5H_{12}), and this is the origin of the name MCM. The MCM is very popular thanks to its absolute simplicity and to the fact that any standard computer code for solving the Schrödinger equation of a molecule can be applied. In fact, up to now, it is the MCM only which allows the application of real many-body theories. Recent density functional techniques based on numerical basis functions are also restricted as yet to the MCM. The basic assumption of the MCM is reasonable for valence band states. The problem of the MCM arises from the fact that conduction band states (or defect states derived from them) are inherently not of localized nature, i.e. cannot be localized to pairs of atoms. Therefore, the "embedding" idea of the MCM does not work for those states, and calculated quantities like the gap converge to the perfect crystal value extremely slowly (from above). The much too large size of the gap also means that the energy of defect orbitals which are dominated by conduction band admixture are overestimated, sometimes leading to an incorrect occupation sequence. This usually results in a false ground state configuration. There are some additional disadvantages. The symmetry is reduced even without introducing the defect, and the uniformity of the sites is lost. Unless the size of the MCM is really big relative to the defect, the environment of an impurity in different sites may change considerably. This makes comparison of total energies in different configurations problematic. The bonds to hydrogen atoms create a dipole layer on the surface of the MCM, shifting ionization energies and heats of formations in a manner which is hard to account for. Also, this dipole layer induces charge transfer between concentric shells of the cluster, even if it consists of nominally equivalent atoms. This effect can be somewhat relieved by tuning the host-atom – hydrogen distances but eliminating it requires considerable increase in MCM size. Due to the direct space formalism, however, clusters of considerably larger size can be handled in MCM than in supercell models. The geometry, energy of formation and LVMs of defects can, therefore, be calculated to a good accuracy, provided the (true) occupied one-electron defect wave functions are sufficiently well localized.

The alternative to embedded cluster models is to restore the periodicity artificially, i.e., to construct a superlattice with the cluster as unit cell. A cluster chosen in shape of a unit cell (allowing repetition without gaps and overlaps) is called a *supercell*. In the *supercell model* (SCM), the momentum space description of the perfect crystals can be applied without modification. The electron states of the superlattice are characterized by wave vectors $\boldsymbol{K}$ of its Brillouin-zone (BZ), and the one-electron (pseudo) eigenvalue equation has to be solved at all of the allowed $\boldsymbol{K}$ vectors — or at least at a well chosen *special* $\boldsymbol{K}$*-point set*. The set has to ensure that the electron density, calculated as an average of the one-electron wave function squares over the BZ, is well approximated by a weighted sum calculated at the special points (for details see [3]). The standard presently is the use of the Monkhorst-Pack scheme to pick these special $\boldsymbol{K}$ vectors:

$$\boldsymbol{K}_q = \sum_{i=1}^{3} q_i \boldsymbol{B}_i \quad ; \quad q_i = \frac{2p_i - Q_i - 1}{2Q_i}; \quad p_i = 1 \ldots Q_i \qquad /4/$$

where $\boldsymbol{B}_i$ are the reciprocal unit vectors of the superlattice. The one-electron problem has to be solved for those $\boldsymbol{K}_q$ vectors of a $Q_1 \times Q_2 \times Q_3$ set which fall into the irreducible part of the BZ. For a cubic crystal (where all Q values are equal) the smallest set is 2^3. Therefore, depending on the symmetry of the defect, the simultaneous iterative solution of 2 - 8 sets of one-electron equations is necessary — for a supercell! Obviously this influences the choice of the supercell size. Actually, the decoupling of the artificially repeated defects can rarely be achieved. This can be seen in the dispersion of the defect levels as a function of $\boldsymbol{K}$. The position of the "isolated" defect level can be found by a tight binding retrofit of the defect band.

The SCM is arguably the standard method today for calculating defects in semiconductors. The required resources, however, often force researchers to apply the $\boldsymbol{K} = 0$ or Γ approximation, i.e., to solve the one-electron equations only for the center of the BZ. The electron density at the Γ point is not a good approximation for the average over the BZ, unless the supercell size is close to infinite and the BZ shrinks to the immediate vicinity of Γ. This is the limit of non-interacting defects so the approximation can be justified if the supercell size is big enough so the interaction between the repeated defects can be assumed to be negligible.* Since this interaction is artificial anyhow, it is more efficient to disregard it in the first place. This is the basis of the *cyclic cluster model*, CCM (for details see [4]), which is a "crystal" consisting of an appropriately chosen single supercell with cyclic boundary conditions (resulting in the only $\boldsymbol{K}$-point Γ). In practical implementations of the CCM this is achieved by limiting the interaction length to the boundary of the Wigner-Seitz cell around each atom. This can only be done if localized basis functions are used to expand the one-electron states. It is relatively simple for two-center interactions (tight binding and semi-empirical quantum chemical methods [5]) but becomes rather complicated for three-center ones (density functional theory [4]) and has not even attempted yet for four-center interactions (Hartree-Fock-Roothaan theory). Still, the CCM is a very efficient alternative to SCM retaining the conceptual simplicity of the direct space formalism of the MCM without its deficiencies.

The use of both the SCM and the CCM are mostly restricted by the handling of the long range Coulomb interactions. In ionic crystals a defect may cause charge transfers way beyond any practical supercell size. Fortunately, most semiconductor crystals are not too ionic and the screening of any charge is rather effective. Still, in SCM, a problem arises from the fact that the supercell has to have overall charge neutrality (otherwise the repeated charge would give rise to infinite energies). Therefore, a charged defect is always compensated by a jellium of opposite charge spread evenly in the supercell. Unfortunately, while the Coulomb interaction between the repeated charged defects goes — in principle — to zero with increasing supercell size, the interaction of the repeated jellium charges do not. This makes additional corrections necessary which is rather difficult to calculate [7].

---

* The correct description of the extended "crystalline" states is ensured by the fact that the Γ point of the BZ of the perfect superlattice – which is reduced with respect to the BZ of the primitive lattice of the crystal– represents a special $\boldsymbol{k}$-point set of the latter (see ref. [6]).

Given the approximations introduced by any of the models, convergence of the results should always be checked. The effect of the special ***K***-point set in SCM calculations has already been mentioned. In the following I will concentrate on the size of the cluster (be it molecular or a supercell). With the choice of the cluster – in principle – three criteria have to be satisfied:

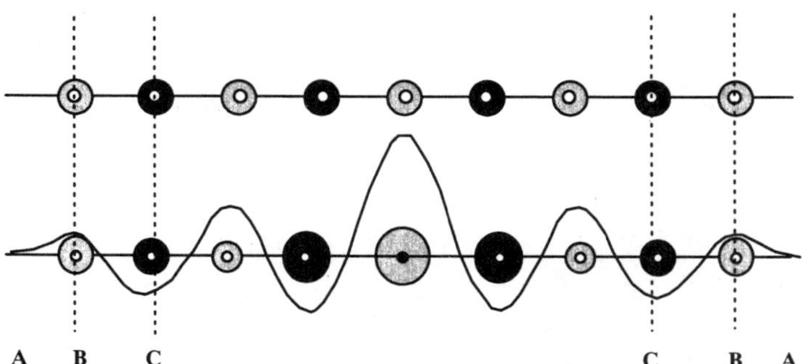

Figure 5. Effects of a substitutional defect in the cluster of a slightly ionic linear crystal. Upper part: perfect, lower part: defective, showing the localized defect wave function. Dark/light gray spheres represent +/- charges.

A/ the amplitude of the localized wave functions,
B/ the deviations of host atom positions relative to the unperturbed case, and
C/ the deviations of charge density

should vanish at the border of the cluster. A schematic representation of these criteria are given in Figure 5 for a one-dimensional crystal with one substitutional impurity. The criteria above are listed in order of increasing stringency in case of a semiconductor. Convergence tests are most often carried out on the vacancy — one of the fundamental intrinsic defects. In the following I will quote values on vacancy formation energies to give a feel for sizes necessary to convergence.*

In SCM calculations the necessity of using a good quality ***K*** set arises due to the interaction of defect wave functions in the superlattice (c.f. criterion A) which manifests itself also in the dispersion of the defect levels between various ***K*** points. It has to be noted, though, that the vacancy formation energy changes less than 0.2 eV going from a 64 to a 216 atom supercell while the dispersion still remains substantial [8]. This fact justifies the use of the $\boldsymbol{K} = 0$ approximation or of the CCM. Obviously, the size of the supercell (cluster) in these cases should be sufficiently large to provide a good description of the extended states by the perfect unit. This is ensured by the fact mentioned before that the Γ point of the supercell represents a special ***k***-point set of the primitive cell. In fact, the Monkhorst-Pack $Q \times Q \times Q$ sets are equivalent to a ***k***-point set represented by Γ of a supercell which is a $Q \times Q \times Q$ times multiple of the primitive unit cell (save for a uniform shift in case of even Q, which, however diminishes the quality of the set). Table 2 shows the quality of Monkhorst-Pack sets of the primitive unit cell compared with that of the implicit application of the special ***k***-point theorem in a $\boldsymbol{K} = 0$ approximation for the perfect supercell.

* Note, that vacancies usually have wave functions which are not well localized. So vacancies can be regarded as "worst case" among single point defects.

Table 2. The quality of explicit or implicit $\boldsymbol{k}$-point summation using Monkhorst-Pack sets of the primitive fcc unit cell or with the Γ point approximation of perfect diamond supercells, respectively, expressed by the first shell-number of lattice vectors, M, for which the Chadi-Cohen condition for special $\boldsymbol{k}$-points is not satisfied.

Monkhorst-Pack set Q^3 for the fcc primitive cell	M	M	Γ point of the diamond supercell with	
			Number of atoms	Unit vectors
2×2×2	8	4	16	fcc 2×2×2
		5	32	bcc 1×1×1
3×3×3	9	9	54	fcc 3×3×3
		4	64	sc 2×2×2
4×4×4	30	15	128	fcc 4×4×4
		8	216	sc 3×3×3
5×5×5	24	24	250	fcc 5×5×5
		17	256	bcc 2×2×2

Experience shows [4] that a 64 atom CCM is already sufficient in diamond to reproduce perfect crystal properties, but not quite enough for silicon. The formation energy of the vacancy [9] converges similarly in the two materials (Table 3).

Table 3. The formation energy of the vacancy in diamond and silicon using the $\boldsymbol{K} = 0$ approximation [9].

$E_f(V)$	C	Si
64 atom supercell	6.97 eV	3.57 eV
216 atom supercell	6.97 eV	3.50 eV

The deviation for lower cell sizes in silicon comes partly from the $\boldsymbol{K} = 0$ approximation (c.f. criterion C), partly from the long range relaxation (c.f. criterion B) as can be seen in Table 4. The change with cell size alone is much weaker than the change in quality of the Γ-point approximation. This shows the precedence of criterion C over B. It also means that application of the $\boldsymbol{K} = 0$ approximation for supercells smaller than 128 atoms are not advisable for any defect in silicon. For SiC or diamond 64 atoms seem to be sufficient but the quality of the $\boldsymbol{K} = 0$ approximation should be checked in all semiconductors. In case of CCM calculations (which may be implemented in an entirely direct space formalism) this is not always possible. However, the pile-up of charge transfer at the boundary of the Wigner-Seitz cell around the defect is a good enough warning for increasing the cluster size.

Table 4. Convergence of the vacancy formation energy in silicon with the improvement of the Γ-point approximation and with the inclusion of long range relaxation, as the cell size increases. The third column gives the difference for each supercell between a Γ-point calculation and a 2x2x2 Monkhorst-Pack set [10]*, while the fourth column* shows the difference between the given supercell and the largest one in a calculation with 4x4x4 Monkhorst-Pack sets [9].

Supercell	M	$\lvert E_f(\Gamma)-E_f(\text{MP:}2\times2\times2)\rvert$	$\lvert E_f(x)-E_f(216)\rvert$
16 (fcc)	4		0.434
32 (bcc)	5	3.05	
54 (fcc)	9		0.190
64 (sc)	4	0.56	0.063
128 (fcc)	15	0.30	0.067
216 (sc)	8	0.04	0.000

* Note that, erroneously, data for the unreconstructed geometry were given in ref. [3].

In case of MCM calculations recent studies [11] on the silicon vacancy show the energy of formation to be convergent well within 0.15 eV at the MCM size $Si_{122}H_{100}$ (9 shells of Si neighbors). Comparing this with the difference of 0.07 eV, going from a 128 to a 216 atom supercell (see Table 4), the MCM is not doing too bad despite of the fact that the vacancy is a defect notorious of having considerable admixture of CB states to its defect states. It should be noted, however, that $Si_{206}H_{158}$ (13 shells of Si neighbors) is the smallest MCM which gives the correct geometry for the divacancy [12]. Despite of this discouraging example, one has to be aware that a large amount of qualitatively and quite often quantitatively correct information has been obtained from 32-atom SCM calculations in several semiconductors.

4. SOLVING THE QUANTUM MECHANICAL PROBLEM FOR THE MODEL: THE HAMILTONIAN AND THE BASIS

The necessity of calculating many different properties with reasonable accuracy raises high demands against the approximation for solving the quantum mechanical many-body problem of our model. Traditionally, a common approximation in atomistic simulations is the adiabatic principle of Born and Oppenheimer, i.e., the assumption that stationary states of the electron system exist at any fixed position of the nuclei. In recent computer codes the conjugate gradient method is often applied, where the nuclear and electronic degrees of freedom are varied simultaneously in search for the energy minimum — thereby "jumping" between Born-Oppenheimer surfaces. Although these methods are very effective, they diminish the chance of finding the global energy minimum of the electronic ground state with respect to the atomic coordinates. Still even in this case the nuclei are handled distinct from the electron system in a classical matter. (MD simulations solve the Newtonian equations of motion for the nuclei, too, using forces derived from the gradient of the quantum mechanical total energy of the electron system.) In should be noted, however, that some defects (especially the ones acting as non-radiative recombination centers in semiconductors) are, in fact, exceptions to the general validity of the adiabatic principle at "normal" temperatures, due to the strong local coupling between nuclear and electronic degrees of freedom. In Jahn-Teller unstable systems a vibronic rather than an electronic wave function should be used, but that is not yet possible in standard computer codes.

For solving the many-electron problem, atomistic computer simulation of defects in semiconductors is usually carried out within the framework of a one-electron theory. To my knowledge, many-electron theories have not yet been implemented into computer codes capable of handling periodic systems, and their excessive demands on computational resources strongly limit the size of the MCM which can be used, too. However, such investigations are always advisable when the occupied and unoccupied localized one-electron states of the defect are close to each other in energy. For example, in case of a two level system with two electrons, one-electron theories occupy the lower lying level with $S = 0$ if there is any difference between the two in energy (no matter how small!!), or both with $S = 1$ if they are exactly degenerate. Generally, the interaction between different electron configurations (i.e., different population of the available close-lying defect levels) may substantially influence the nature of the ground state. E.g., a *biradical*, i.e., a defect with two spatially separated, symmetrically equivalent localized dangling bond will probably be no less stable with both being singly occupied with $S = 0$ than with $S = 1$, not to speak about the "default" result of closed shell (spin-unpolarized) calculations which would occupy just one of the two dangling bonds, eventually leading to artificial decrease in symmetry. The situation may be clarified by using the Multi-Configurational Self-Consistent-Field (MC-SCF) method (see, e.g. ref. [13]).* Calculations with all configurations built from a selected set of defect orbitals in MCM-s up to 50 atoms at fixed geometry are still feasible on medium size work stations.

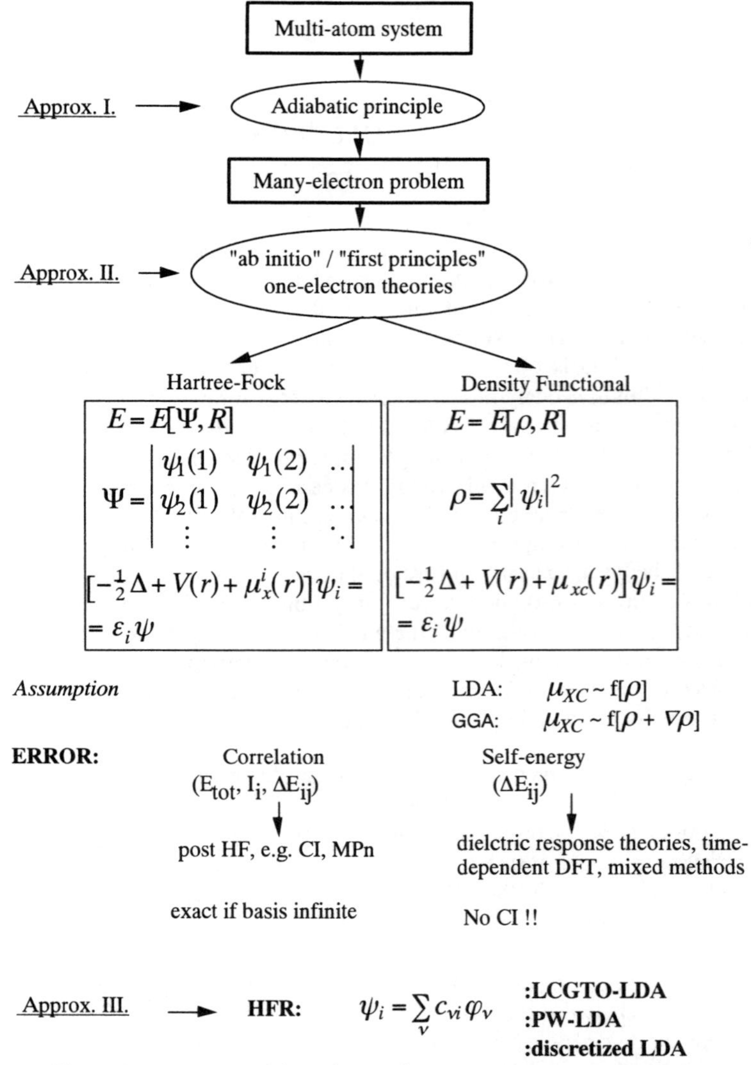

Figure 6. Comparison of the main one-electron theories. For details see the text.

The two main competing one-electron theories, Hartree-Fock (HF) and Density Functional (DF) are compared at the *ab-initio* (sometimes called *first-principles*)** level from the viewpoint of defect calculations in Figure 6. The HF theory is based on the total energy as

---

* This method expands the many-electron wave function in terms of Slater-determinants constructed from one-electron orbitals which are, in turn, linear combinations of localized basis functions. The total energy is obtained using the variational principle. The MC-SCF method is included, e.g. in the package GAMESS [14].

** A warning: these words only mean that no experimental data or fitted parameters are used but do not mean "exact"! Both theories in any practical implementation contain a series of severe approximations!!

functional of the many-electron wave function Ψ at fixed nuclear positions R, which is approximated by a single Slater-determinant built from the one-electron wave functions ψ which are occupied in the ground state. The application of the variational principle leads to a pseudo-eigenvalue-equation for ψ_i with the one-electron eigenvalues ε_i. This HF equation contains the kinetic energy, the Hartree potential $V(r)$ of the nuclei and of the charge distribution represented by the occupied one-electron states, as well as the exact exchange potential, μ_x of the latter. The direct application of the non-interacting one-electron approximation in the total energy functional results in the *correlation* error: two electronic quasiparticles can be at the same point of space, i.e., their motion is not correlated. The correlation error manifests itself in the total energy and it can be quite different in the ground state, in excited states or in charged states, not to speak about the difference between the correlation error for the isolated constituents and the system. As a consequence, HF theory in itself is all but useless in predicting formation, excitation or ionization energies. On the other hand, the correlation error does not depend much on the geometry in the vicinity of the equilibrium of a given configuration, and is always roughly 1 eV per electron pair in a system with "normal" chemical bonds. That means that HF theory gives very accurate equilibrium geometries (in fact, a lot better than DFT) and a qualitatively correct energy order of different configurations (except typically the saddle point configuration where the bonding situation is far from normal). The calculated one electron wave functions and properties derived from them (like hyperfine constants) are also very accurate. In principle, the correlation error of HF theory can be corrected by so-called post-HF treatments, like Moeller-Plesset (MP) perturbation theory (usually applied to second or third order, MP2 or MP3, respectively) or by configuration interaction (CI) correction. (The latter is similar to MC-SCF but the one-electron states building up the additional Slater determinants corresponding to excited configurations are kept fixed in their ground state form.) With post-HF treatment, in principle, exact values can be obtained even for energies, provided the basis used for expanding the one-electron states is infinite and one goes to infinite order of MP or includes an infinite number of configurations in CI. In practice, however, even finite implementations are prohibitively expensive. While HF itself scales with the fourth power of the number of electrons (order N^4), post-HF is typically of order N^5. In fact, I know of no implementation of post-HF into computer codes for periodic systems, so they can only be used for MCMs of moderate size with almost any molecular package (e.g. refs. [14-16]).

DFT on the other hand is based on the total energy as functional of the total electron density, which is given by the sum of squares of the occupied one-electron states. As a consequence, the variational principle leads to a pseudo-eigenvalue equation (the Kohn-Sham equation) which contains, besides the kinetic energy and the Hartree-potential an exchange-correlation potential μ_{xc}.* The problem only is that μ_{xc} is not known! In the most often used practical implementation, the Local Density Approximation (LDA), the assumption ("Ansatz") is made that μ_{xc} is some function of the electron density. The usual approximation holds well in certain asymptotic cases but works more or less as a matter of luck in between. (While the approximation to exchange part is rather serious, it is not really clear, how much of the correlation is, in fact included.) More recently the Generalized Gradient Approximation (GGA) was introduced as an improvement but its effect is often questioned. Still, the LDA works reasonable well for semiconductors and scales only with N^3! That makes it the favorite choice for defect calculations in semiconductors — but at some cost!

---

* This, however, does not make DFT a true many-body theory, as long as the electron density is defined as in Figure 4. Therefore, configuration interaction is not included and this may lead to the problems explained before.

Geometries are non as accurate as those of HF but this is not a quantity of primary importance in this case, so that is no problem. Energy differences between various configurations (even between equilibrium and saddle-point), or heat of formations, though not as accurate as that of post-HF (even at an MP2 level), are reasonably good unless many-body effects become important. The real trouble comes with excited states. Although DFT can, in principle, be extended to states other than the ground state, usage of the ground state one-electron states and exchange-correlation potential derived from them causes double-counting errors and a need for an electron self-energy correction. Rigorous theories like those based on dielectric response (GW-theory), or mixing exact exchange with an LDA type correlation potential are not much more practical for large units than post-HF. There are new attempts to overcome the problem (e.g. in the framework of time-dependent DFT) but these are still far from practical application in large systems. Without a correction, however, the excitation energies are seriously and systematically underestimated: a well converged DFT calculation may show germanium to be a metal! (It is a matter of ongoing dispute whether the "ground state" of the ionized system is well reproduced by LDA and if yes, could the difference between electron affinity and ionization energy of a semiconductor be used as an approximation for the gap.) Therefore, the infamous scissor operator is most often used in many variations. It is generally agreed that the energy of CB states of the perfect crystal can be rigidly shifted: but what about the defect levels? Some researchers do not apply any shift there, some shift acceptor states by the same amount as the CB states but not at all the donor states, other groups scale the shift of defect levels with their energy difference to the CB edge, while my personal preference is [2] to scale the shift with the overlap of the defect wave function with the CB... So, I believe, some skepticism is justified for using the term *ab-initio* in this business. Still, one has to keep in mind that these theories provided undisputed and very valuable help to defect engineering in the past decade!

In fact the use of "first principles" ends when introducing the third approximation necessary for practical implementation: the expansion of the one-electron states on a basis of a finite number of functions. Since the delocalized crystalline and localized defect states pose concurring requirements one is bound to make compromises and, by exercising judgment, empirism enters. Choice of the Hamiltonian and the model of the defective crystal mostly predetermines the choice of the basis.

SCM-LDA calculations allow the use of plane waves (PW) as basis set:

$$\varphi_{\mathbf{k}}(\mathbf{r}) = \sum_{\mathbf{G}}^{\hbar^2(\mathbf{k}+\mathbf{G})^2<E_C} c_{\mathbf{G}} e^{i(\mathbf{k}+\mathbf{G})\mathbf{r}} \qquad /5/$$

i.e., a Fourier-expansion according to the reciprocal lattice vectors **G**. The obvious advantages of a PW basis are the

- good description of delocalized states, and
- easy convergence tests.

The latter can be done simply by brute force: increasing the summation limit until convergence is achieved. The highest kinetic energy allowed this way is called the cut-off energy, E_C, usually expressed in Rydbergs. However, some disadvantages also follow. The PW basis is

• not so good for localized states,
• is inflexible with atom types,
• surface calculations or comparative calculation for atoms, molecules are difficult, and
• calculation of hyperfine data are cumbersome.

Obviously, a relatively high cut-off is necessary to describe the localized states of a defect. Typically, the stronger the reconstruction and the lower the symmetry, the higher cut-off is needed so, for other than the smallest defects the cut-off is practically hardly ever determined by convergence but by the limit on resources. A heavy burden on the PW basis is that it does not allow the full utilization of the advantage of pseudopotentials. The latter are introduced to account for the joint effect of nucleus and core electrons in order to avoid the use of valence electron wave functions which are strongly oscillating in the region of the nuclei and needing, therefore, too high a cut-off in the plane wave expansion. Hydrogen has no core electrons and the nuclei of first-row atoms (with only *s*-electrons in the core) are not completely screened. Therefore, a single one of these among the many host atoms forces the increase of the cut-off, leading to enormous basis sizes (or forces one to use non norm-conserving pseudopotentials). Defect studies are often carried out on the surface which can be modeled by periodically repeated slabs consisting of a layer of the material and a layer of vacuum. The presence of this vacuum layer decreases the BZ so a lot more PWs are needed to reach the same cut-off. The same problem arises when comparative results on "SCM"s containing an isolated atom or molecule become necessary (e.g., to obtain the heat of formation). Finally, PW-SCM calculation of defects without the use of pseudopotentials seem to be as yet unfeasible. The calculation of hyperfine data can, therefore, be done only with very complicated approximations.

Despite all this problems, SCM-PW-LDA still seems to be the most often used tool for determining relative stabilities and occupation levels of defects and are implemented in easily accessible codes like FHI98MD[17], VASP [18] or ABINIT [19].

While LDA and SCM also allows that, many-body or HF theory as well as the MCM and CCM necessarily mean a localized basis set in direct space. The advantages of a localized basis set are:

• good description of localized states,
• flexibility with atom types,
• calculation of surfaces or isolated entities easy,
• calculation of hyperfine data straightforward, and
• population analysis of localized orbitals is a big help in understanding defect phenomena.

The first one is obviously beneficial in describing the localized defect states. The second one means that each atom introduces a custom-tailored set of functions, i.e. the presence of hydrogen, or that of a single transition metal atom do not significantly influence the total size of the basis. Since the basis functions are centered on atoms (or bonds) any amount of "vacuum" in the model does not increase the basis. The calculation of hyperfine data are straightforward even in case of pseudopotentials within a frozen core approximation (see, e.g. ref. [20]). The disadvantages are fewer but important. Localized basis functions

---

* This does not mean, however, that some indication of the level of convergence should not be given in any report.

- are not so good in the interstitial region (unless high order polarization functions are included),
- it is difficult to test the convergence with size of the basis, and
- overcompletness of the basis may lead to numerical instabilities.

The traditional approach in case of HF theory is the Hartree-Fock-Roothaan (HFR) method based on using a linear combination of atomic orbitals (LCAO) for the one-electron wave functions. After early attempts with Slater-type functions the computationally more effective Gaussian-type orbitals (GTOs) became standard (see e.g. ref. [21]). Due to their less appropriate asymptotic behavior, however, several GTOs are needed to describe an atomic orbital:

$$\varphi_{lm}(\mathbf{r}) = \sum_{i} \sum_{\substack{n_1,n_2,n_3 \\ n_1+n_2+n_3=l}} A_l^i(m,n_1,n_2,n_3) x^{n_1} y^{n_2} z^{n_3} e^{-\xi_l^i r^2} \qquad /6/$$

In expanding the one-electron orbitals into such functions the ratio of the coefficients A^i_l can be held fixed (contracted GTOs) or varied freely (independent GTO's). In case of contracted GTOs the change of the valence AOs in response to the ligands can be taken into account by using two (three) contracted set of GTOs for each AO (double or triple zeta basis). Minimal basis sets (only *s*- and *p*-type GTO's for all atoms other than hydrogen) can be used for calculating the geometry and energy of electrically inactive defects but a full valence basis including *p*-orbitals for H and *d*-orbitals for other atoms (usually denoted by stars in the superscript of the abbreviation of standard basis sets) is recommendable otherwise. Since the use of *d*-orbitals more than doubles the basis set, often *s*-type GTOs placed on bond-centers are used instead.*

AO basis sets are designed to described states around the atoms but they are not so successful in the interstitial region of the crystal, unless higher order polarization functions (*d*, *f*) are included [22], although that obviously depends also on the supercell size and the number of primitive Gaussians involved. Inclusion of these polarization functions (extended basis) may cause difficulties with the population analysis (otherwise very useful, among others, in calculating hyperfine constants) and the large number of primitive Gaussians may lead to overcompletness. The convergence of the GTO basis is very difficult to check in individual studies. Standard GTO bases have been carefully tested over the years both in all-electron and pseudopotiential calculations and are available in the HF-based molecular codes GAMESS [14], Gaussian® [15] and HyperChem® [16], but also in molecular DFT codes like DGauss® [23], ALLCHEM [24] and NRLMOL [25]. They are also implemented into the supercell HFR code CRYSTAL [26]. On the other hand, the very effective AIMPRO [21] code, which has been developed specifically for defect studies both in MCM and SCM version, operates with basis sets about which very little is known.

In a matter of speaking, a "mixed" basis is used in the Linearized Muffin Tin Orbital (LMTO) approximation (see, e.g. ref. [27]) where localized functions in the atomic spheres are fitted to spherical waves in the interstitial regions. Besides the possibility of SCM-type calculations, embedded cluster model implementation of LMTO for defect studies also exist [28] but the complexity of the method has so far restricted its application to defects with high symmetry and small reconstruction. The advancement of parallel computing brought another initiative in foreground: to discretize the pseudo-eigenvalue equation on a grid in

* This is very cost-effective but poses series conceptual difficulties in case of, e.g., vacancies.

real space. This can be regarded as the use of a kind of localized numerical basis functions. Besides molecular packages as DMOL® [29] this technique has also been utilized specifically for MCM defect calculations [30], allowing the use of extremely large MCMs.

The use of the methods mentioned in the last paragraph is still very limited while the use of PWs or AOs require enormous basis sizes for full convergence (be it the cell size for PWs, or the number of atoms for AOs which determines the basis). The practical restrictions on the basis set used in a specific calculation are, therefore, usually posed by the available resources and not by convergence. In light of this, one is tempted to turn to more "cost-effective" approximations, so I will give some very personal remarks on them. The order N, or *O(N)*, methods which scale linearly with *N*, are not very welcome in this field, due to the luck of one-electron states and eigenvalues. Besides, their practical implementations are not yet sufficiently accurate. Among the order N^2 methods, the same can be said about strictly *ab-initio* tight-binding theories, while empirical tight-binding (quite accurate in the situation for which it was parameterized) can only be used with extreme caution due to a notorious lack of wide range transferability which leads sooner or later most schemes to disaster. My best experience so far is with the charge-self-consistent, non-orthogonal, all-neighbor SCC-DFTB method [31]. The total energy is calculated as the sum of the band energy (sum of one-electron eiegenvalues) and a repulsive term. The eigenvalues are obtained by using the sum of *ab-initio* LDA atomic potentials in the Hamiltonian on a linear combination of the corresponding AOs (which are expressed on a parameterized Slater basis). By fitting one parameter in the repulsive term to reproduce *ab-initio* LDA results on simple systems, this method is — strictly speaking — semi-empirical in nature, but it is on the boundary of *ab-initio* tight binding methods. As a consequence it is both quite reliable and accurate within about 0.5 eV per individual defect. This is not worse than typical not-well-converged *ab-initio* LDA but allows the treatment of much larger systems, many defects or longer MD simulations.

While DFTB is – to my knowledge – the only viable semi-empirical approximation to DFT, semi-empirical HFR theories are used widely in chemistry and can be used for solids as well (for a review, see ref. [32]). Despite some success [5,33], in the simulation of defects in semiconductors they are by now almost extinct in this field. The reason is more psychological than physical. Semi-empirical quantum chemical methods like AM1 or PM3 encompass a lot more physics than any form of tight-binding theory but rely heavily on an extended set of parameters — which among others ensure the inclusion of ground state dynamic correlation effects in a formally HFR framework which scales with N^2. The problem is that the parameters have been fitted to sets of molecules where organic ones, first row atoms and aromatic bonds dominate over inorganic ones, heavier atoms and saturated bonds. The transferability of the parameters – if they exist at all– to semiconductors is far from perfect (although not too bad [5]) and the solid state community shrunk back so far from the investment into an appropriate parameterization, especially since the assumption of orthogonal orbitals in these early methods causes serious errors in supercells [32]. New versions, like MSINDO [34] have overcome this problem but parameters sufficiently optimized for defect studies in semiconductors are still missing.

5. A LEARNING EXAMPLE: THE HYDROGEN PLATELETS IN SILICON

Finally, I would like to give an example which – I believe – illustrates nicely the principles outlined in Section 2.

In 1987 transmission electron microscopy studies of hydrogen plasma annealed silicon revealed disc-shaped platelets in the (111) plane with diameters between 70 – 500 Å and thickness of about 3.3 Å, i.e., that of a Si double layer [35]. Raman spectra taken from these samples showed hydrogen related bands at 1960 and 2100 cm^{-1}. From their intensity it was deduced that the platelets may contain 1 or 2 hydrogen atoms per unit cell of the (111) plane. Later, it was shown [36] that the stress field of the platelets caused a decrease of 0.13 eV in the gap of silicon, corresponding to a stress of 1 MPa. With so much clue in hand, of course, theorist started to find a model for the structure of these complexes. Van de Walle et al. have suggested a model in which a double layer of silicon atoms in the (111) plane was removed and the remaining dangling bonds saturated by hydrogen [37]. This model seemed energetically quite favorable, since for each broken Si–Si bond a Si–H bond was created, but produced only a single frequency around 2100 cm^{-1}. Zhang and Jackson have, however, proposed hydrogenated stacking faults [38] which gave rise to calculated frequencies close to the observed bands at 1960 and 2100 cm^{-1}. The only problem with this model was that intrinsic stacking faults are supposed to end in dislocations, while none of those were observed around the platelets.

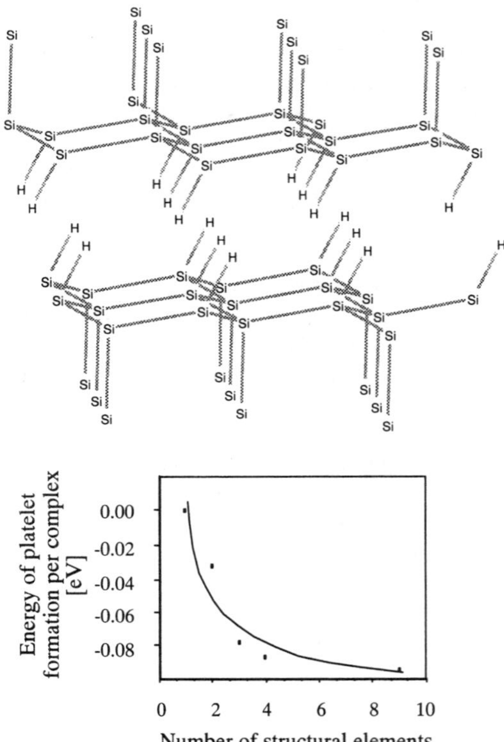

Figure 7. Di-hydrogen model of the platelet and the energy gain on accumulating the structural elements.

Our group has investigated several possibilities but instead of doing the calculation only on an infinite platelet, we have calculated the energy gain on accumulating individual hydrogen defects in the (111) plane. (These calculations have been carried out using a 54 atom (111) slab of silicon as SCM with the $\boldsymbol{K} = 0$ approximation, and a simple semi-empirical Hamiltonian, MINDO/3 [39]).This way we eliminated single bond centered interstitial hydrogen or interstitial H_2 molecules, as well as the H_2^* defect (one of the hydrogen atoms in bond center position binding to one of the silicon atoms, the other binding to the other Si from behind) as structural element of the platelet. Energy gain on accumulation was only obtained for two hydrogen atoms binding to two silicon atoms whose bond was interrupted (Figure 7). The built in stress was calculated to be 1.055 MPa and the resulting gap was lower by 0.15 eV than the one calculated for the perfect slab. The displacement caused by the presence of hydrogen extended to a 3.55 Å wide region of the slab. These data fit the experimental observations nicely but the hydrogen atoms – while rotating almost freely – had only one Raman active stretching frequency at 2122 cm^{-1} and an IR active one at 2095 cm^{-1} — but none around 1960 cm^{-1}.

Then, in 1993, the vibrations related to the hydrogen platelets have been reinvestigated [40] — this time by both Raman and IR spectroscopy. First of all, it was found that the 1960 cm^{-1} band does not correlate after all with the platelets! However, in addition to a 2095 cm^{-1} mode in the Raman spectrum, a 2075 cm^{-1} mode was observed in IR. These vibrations have been assigned to a single complex with C_{3v} symmetry. Considering the accuracy of frequency calculations and the fact that motional averaging at any finite temperature would show an effective C_{3v} symmetry, our model seemed to be in good agreement with these findings. Finally, in 1995, comparing high resolution TEM pictures with the simulated image of the available models, Muto et al. [41] concluded that only our model fits the experimental observation.

Table 5. The calculated energy [eV] of hydrogen defects in silicon with respect to interstitial molecular hydrogen.

	LDA-PW-SCM [42]	MINDO/3-CCM [35]
2 x ($Si–H_{BC}–Si$)	2.2	0.2
$(H_2)_T$	0.0	0.0
H_2^*	0.4	0.1
(Si–H ... H–Si)	1.8	0.2

Can we now triumph? Not really! Table 5 compares the stability of individual hydrogen defects in silicon in our calculation and in a LDA-PW-SCM calculation [42] (both using 32 atom units). As can be seen, our defect, Si–H ... H–Si, is only marginally more stable than two isolated bond-center hydrogen interstitials but less stable than the molecule. Due to this reason the model cannot be accepted, despite some additional evidence regarding vibration frequencies of H_2 molecules trapped in platelets [43], as long as a reaction sequence is not found leading to the creation of the first Si–H ... H–Si defects by, e.g., mediation of vacancies or self-interstitials which are known to dissociate molecular hydrogen [44]. The chances are not bad, since atomic interstitial hydrogen has been observed, too after all, even though they are less stable than molecules!

Acknowledgment. The author is indebted for the support of the grants OTKA-T032174 and FKFP-0289/97.

6. REFERENCES

[1] In: P. Deák, Th. Frauenheim, M. R. Pederson (eds.), Computer Simulation of Materials at Atomic Level. Wiley-VCH, Berlin 2000; also appeared as Vol. **217** of *Phys. Stat. Sol. (b)* (2000).

[2] B. Aradi, A. Gali, P. Deák, J. E. Lowther, N. T. Son, E. Janzén, W. J. Choyke, *Phys. Rev. B* **63**/24 (2001)preprint pp. 245202/1-19.

[3]P. Deák, in ref. [2], pp. 9-23.

[4] J. Miró, P. Deák, C. P. Ewels and R. Jones, *J. Phys: Condens. Matter.* **9** (1997) 9555.

[5] P. Deák, L. C. Snyder, *Phys. Rev. B* **36** (1987) 9619; P. Deák, L.C. Snyder, J. W. Corbett, *Phys. Rev. B* **37** (1988) 6887; P. Deák, L. C. Snyder and J. W. Corbett, *Phys. Rev. B* **45** (1992) 11612.

[6] R. A. Evarestov, in this volume

[7] G. Makov and M. C. Payne, *Phys. Rev. B* **51** (1995) 4014.

[8] S. J. Clark, G. J. Ackland, *Phys. Rev. B* **56** (1997) 47.

[9] A. Zywietz, J. Furthmüller, and F. Bechstedt, *Phys. Stat. Sol. (b)* **210** (1998) 13.

[10] M. J. Puska, S. Pöykkö, M. Pesola, and R. Nieminen, *Phys. Rev. B* **58** (1998) 1318.

[11] S. Ögüt, H. Ch. Kim, J. R. Chelikowsky, *Phys. Rev. B* **56** (1997) R11353.

[12] S. Ögüt, J. R. Chelikowsky, *Phys. Rev. Lett.* **83** (1999) 3852.

[13] P. Deák, J. Miró, A. Gali, L. Udvardi, and H. Overhof, *Appl. Phys. Lett.* **75** (1999) 2103.

[14] GAMESS©, http://www.msg.ameslab.gov/GAMESS/GAMESS.html

[15] GAUSSIAN98®, http://www.gaussian.com

[16] HYPERCHEM®, http://www.hyper.com/sales/electronic/electronic-lite.htm

[17] FHI98MD©, http://www.fhi-berlin.mpg.de/th/fhi98md/

[18] VASP®, http://cms.mpi.univie.ac.at/vasp/

[19] ABINIT©, http://www.pcpm.ucl.ac.be/ABINIT

[20] A. Gali, P. Deák, R. P. Devaty, and W. J. Choyke, *Pys. Rev. B* **60** (1999) 10620.

[21] P. R. Briddon and R. Jones, in ref. [2], 131-172.

[22] G. B. Bachelet, H. S. Greenside, G. A. Baraff and M. Schlüter *Phys. Rev. B* **24** (1981) 4745.

[23] DGauss®, http://www.cachesoftware.com/cache/dgauss/index.shtml

[24] AllChem©, http://www.theochem.uni-hannover.de/AllChem/

[25] M. R. Pederson, D. V. Porezag, J. Kortus, and D. C. Patton, in ref. [2], pp. 197-218.

[26] R. Dovesi, R. Orlando, C. Roetti, C. Pisani, V. R. Saunders, in ref. [2], pp. 63-88.

[27] R. W. Tank and C. Arcangeli, in ref. [2], pp. 89-130.

[28] U. Gerstamnn, M. Armkreuz, H. Overhof, in ref. [2], pp. 665-684.

[29] DMol®, http://www.accelrys.com/cerius2/dmol3.html

[30] J. R. Chelikowsky, Y. Saad, S. Ögüt, I. Vasiliev, and A. Stathopoulos, in ref. [2], pp. 173-196.
J. Bernholc, E. L. Briggs, C. Bungaro, M. Buongiorno nardelli, J.-L. Fattebert, K. Rapcewicz, C. Roland, W. G. Schmidt, and Q. Zhao, in ref. [2], pp. 685-702.

[31] Th. Frauenheim, G. Seifert, M. Elstner, Z. Hajnal, G. Jungnickel, D. Porezag, S. Suhai, and R. Scholz, in ref. [2], pp. 41-62.

[32] P. Deák, in R. Hull (ed.), Properties of Crystalline Silicon, EMIS Data Reviews Series No. 20. INSPEC, London 1999. p. 245.

[33] M. J. Caldas, in ref. [2], pp. 641-664.

[34] B. Ahlswede and K. Jug, J. *Comput. Chem.* **20** (1999) 563; **20** (1999) 572; **21** (2000) 974; **22** (2001) 861.

[35] N. M. Johnson, F. A. Ponce, R. A. Street, and R. J. Nemanich, *Phys. Rev. B* **35** (1987) 4166.
[36] J. Hartung and J. Weber, *Mater. Sci. Eng. B* **4** (1989) 47.
[37] C. G. Van de Walle, P. J. H. Denteneer, Y. Bar-Yam, and S. T. Pantelides, *Phys. Rev. B* **39** (1989) 10791.
[38] S. B. Zhang and W. B. Jackson, *Phys. Rev. B* **43** (1991) 12144.
[39] P. Deák, C. R. Ortiz, L. C. Snyder, and J. W. Corbett, *Physica B* **170** (1991) 223.
[40] J. N. Heyman, J. W. Ager III, E. E. Haller, N. M. Johnson, J. Walker and C. M. Donald, P*hys. Rev B* **45** (1992) 13363.
[41] S. Muto, S. Takeda, and M. Hirata, *Phil. Mag.* **A 72** (1995) 1057.
[42] K. J. Chang and D. J. Chadi, *Phys. Rev. B* **40** (1994) 11644.
[43] B. Hourahine, R. Jones, S. Öberg, P. R. Briddon, Mater.Sci. Eng. B **58** (1999) 24.
[44] S. K. Estreicher and J. L. Hastings, Phys. Rev. Lett. **82** (1999) 815.

Computational Materials Science
C.R.A. Catlow and E.A. Kotomin (Eds.)
IOS Press, 2003

SEMICONDUCTORS

II. Tips for modeling point defects in semiconductors

Peter Deák
Department of Atomic Physics, Budapest University of Technology and Economics
Budafoki út 8., Budapest, H-1111, Hungary (e-mail: p.deak@eik.bme.hu)

Abstract While the previous paper described how to use the available computational tools for identifying defect models with observed defect centers in semiconductors, this one attempts to give ideas on how to construct models to do the calculations on and ways to check the outcome.

1. INTRODUCTION

Quantum mechanical simulation of defect phenomena in semiconductors is becoming an ever more important tool not only for research but for technology development. Its spread is being promoted by the increasing number of computer codes accessible in the public domain and commercially. As the sophistication of theory and the complexity of the codes increases, the interpretation of the results becomes more and more cumbersome. This is especially so because the difficulty of the problems treated grows faster than available computer capacity and, therefore, the necessary amount of approximations (neglect) is always severe. While the access to and use of ever increasing computational resources might cause overconfidence, a healthy distrust of the results is advised. In the previous paper I have shown that only the (necessarily approximate) agreement of a set of calculated and observed physical properties entitles one to identify a theoretical defect *model* with an experimentally detected defect *center*. However, even that is not enough. The results must have "sense" in themselves — the physics must not be lost in the jungle of the computational technique. Unless one understands why the given configuration of atoms should give rise to the calculated properties, the comparison with the experimental data should not even be attempted! In fact, it is profitable to have an *a priori* guess at the expected result of the calculation. Since a starting input has to be provided anyhow, this can save a lot of precious computer time – not to speak of the time of the researcher spent on randomly selected models the "hopelessness" of which could have been seen in the first place.

An invaluable assistant in analyzing the electronic behavior of dopants and impurities in semiconductors is the concept of the "defect molecule", which is based on the realization that reaction of atoms resulting in stable complexes is nothing but chemistry in a solid medium. With that in mind the complete experience about chemical bonding in molecules can be invoked and rely upon. The difficulty lies in the fact that the classical chemical bond can only be described quantum mechanically by means of localized orbitals: unitary transforms of the canonical orbitals obtained from the eigenvalue equation. Still, in the case of semiconductors – and especially in case of their defects — a qualitative correspondence can be established without doing the mathematics. This way of looking at a system makes prediction of the results before and interpretation after the calculation a lot easier. In Section 2, as an introduction to this "trick", the band structure of crystalline and amorphous semiconductors will be interpreted in terms of a linear combination of atomic hybrid orbitals. The role of the bonding – antibonding splitting will be explained and the concept of bonds, strained (long) bonds, dangling bonds and lone pairs will be introduced. In Sections 3-5 examples of "defect molecules" in silicon will be shown to explain, how the symmetry, geometry, energetics, number of electrically active states and their occupancy can be predicted. Finally, a new example is shown explaining the calculated behavior of hydrogen in the anion vacancies of III-nitride semiconductors.

2. THE SEMICONDUCTOR BANDS RESULTING FROM LOCALIZED ATOMIC HYBRIDS

The quantum mechanical problem of electrons is usually presented in form of a (pseudo)-eigenvalue-equation – as described in Section 4 of the previous paper – resulting in the canonical orbitals (eigenstates) of the system. By a unitary transformation of these states one may obtain localized states with well defined chemical meaning, however, these will not be eigenstates anymore, i.e., can only be characterized by an average energy. In case of semiconductors where tetrahedral bonding environment seems to be dominating, it is useful to introduce such a transformation already at the level of atoms. From the canonical *s*- and *p*-orbitals one can mix four equivalent *sp^3-hybrids* (Figure 1) by

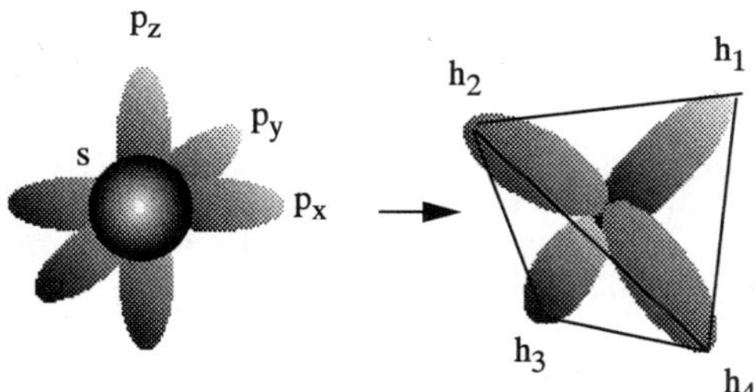

Figure 1. Transformation of the canonical atomic orbitals to sp^3-hybrids

$$\left.\begin{aligned} \varphi_1 &= (s + p_x + p_y + p_z)/2 \\ \varphi_2 &= (s - p_x - p_y + p_z)/2 \\ \varphi_3 &= (s - p_x + p_y - p_z)/2 \\ \varphi_4 &= (s + p_x - p_y - p_z)/2 \end{aligned}\right\} \quad \varepsilon^{sp^3}_{1\ldots4} = \frac{\varepsilon_s + 3\varepsilon_p}{4} \qquad /1/$$

According to simple tight binding theory, the hybrids of neighbor atoms which point toward each other, form a bonding and an antibonding combination, with a splitting characteristic of the binding energy. Say, with φ_1^A of atom A and φ_1^B of atom B pointing toward each other,

$$\frac{\hat{H}\left(\varphi^A + \varphi^B\right) = \varepsilon\left(c_A\varphi^A + c_B\varphi^B\right)}{\begin{aligned}\left\langle\varphi^A\middle|\hat{H}\varphi^A\right\rangle c_A + \left\langle\varphi^A\middle|\hat{H}\varphi^B\right\rangle c_B &= \varepsilon c_A \\ \left\langle\varphi^B\middle|\hat{H}\varphi^A\right\rangle c_A + \left\langle\varphi^B\middle|\hat{H}\varphi^B\right\rangle c_B &= \varepsilon c_B\end{aligned}} \qquad /2/$$

follows after multiplication of the eigenvalue equation by φ^A and by φ^B, and by assuming $<\varphi^A\,|\varphi^B> \approx 0$. Introducing

$$\begin{aligned} \frac{\left\langle\varphi^A\middle|\hat{H}\varphi^A\right\rangle + \left\langle\varphi^B\middle|\hat{H}\varphi^B\right\rangle}{2} &= \frac{\varepsilon^a + \varepsilon^c}{2} - \Delta\varepsilon \\ \frac{\left\langle\varphi^A\middle|\hat{H}\varphi^A\right\rangle - \left\langle\varphi^B\middle|\hat{H}\varphi^B\right\rangle}{2} &= -V_i \\ \left\langle\varphi^A\middle|\hat{H}\varphi^B\right\rangle = \left\langle\varphi^B\middle|\hat{H}\varphi^A\right\rangle &= -V_c \end{aligned} \qquad /3/$$

(where ε and ε are the energy of the isolated hybrids of the anion and cation atoms lowered by the attraction, $\Delta\varepsilon$, of the neighbor atom, and V_c and V_i are the covalent and ionic component of the binding energy), a *bonding* ($\mathrm{sign}(c^A) = \mathrm{sign}(c^B)$) and an *antibonding* state ($\mathrm{sign}(c^A) = -\,\mathrm{sign}(c^B)$) arises at

$$\begin{aligned} \varepsilon_b &= \frac{\varepsilon^a + \varepsilon^c}{2} - \Delta\varepsilon - \sqrt{V_c^2 + V_i^2} \\ \varepsilon_b &= \frac{\varepsilon^a + \varepsilon^c}{2} - \Delta\varepsilon + \sqrt{V_c^2 + V_i^2} \end{aligned} \qquad /4/$$

As can be seen in Figure 2, the binding energy comes partly from the fact that the electron of the cation lowered its energy by charge transfer to the anion (ionic part) and partly by the splitting due to the quantum chemical interaction between the orbitals (covalent part).

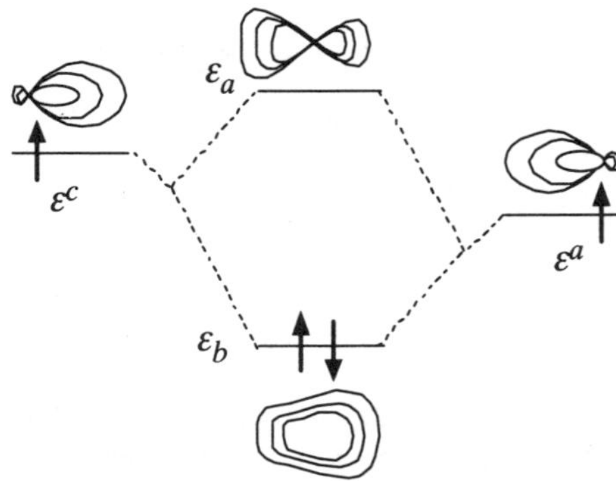

Figure 2. The bonding and antibonding states arising from the overlap of two sp^3 hybrids.

Generally speaking, V_c is the bigger, the smaller V_i is, and $\Delta\varepsilon$ is much smaller than the leading term, V_c or V_i.* In semiconductors the dominating term is V_c which increases as the distance of the atom decreases.

In a crystal there is a large number of equivalent bonds, so there would be four times as many bonding and antibonding states as unit cells, N, all with equal energy. Therefore, according to the Pauli-principle, both the bonding and the antibonding states split up forming the valence band (VB) and the conduction band (CB), respectively, of the semiconductor (Figure 3). Note, that the individual states in the band diagram of Figure 3 do not correspond to the eigenstates appearing in the usual band structure representations, but the band edges are at the right relative energy positions.

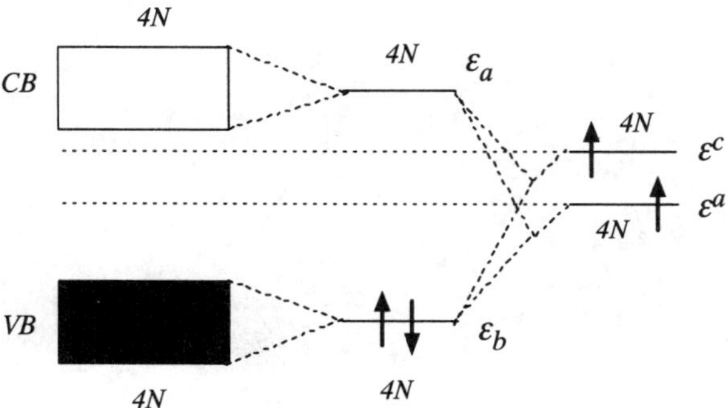

Figure 3. Formation of bands from sp^3-hybrids in a zincblende semiconductor with N unit cells

Note that if bonds are broken somewhere in the crystal as, e.g., in a vacancy, unsaturated hybrids, so-called *dangling bonds* remain with corresponding levels in the gap. The gap, as well as the binding energy of the semiconductors scale inversely with the interatomic distance through the bonding-antibonding splitting: this is shown in Table 1 for group IV elements.

Table 1. The cohesive energy and the fundamental gap of group IV elements as function of the equilibrium lattice constant

	a_0 [Å]	E_b [eV]	E_g (T=0K) [eV]
C_{diam}	3.56	7.37	5.64
Si	5.42	4.55	1.17
Ge	5.66	3.90	0.75
αSn	6.49	3.12	0.07

The crystal of semiconductors are typically built from regularly arranged tetrahedrons where all first neighbor distances are the same. In an amorphous semiconductor, the tetrahedra are randomized. While the majority of the first-neighbor distances are still very close to the "ideal" value in the crystal, a relatively sizeable portion of the bonds is strained, i.e., spans a longer first neighbor distance. About 10-15 % of the atoms are in fact only threefold coordinated (have only 3 neighbors in bonding distance). In case of the *long bonds* the bonding-antibonding splitting is smaller. The bonding and antibonding orbitals corresponding to the randomly distributed long bonds have a weak interaction and split into narrow bands with density of states having a nearly Gaussian line shape. As can be seen in Figure 4 for a homopolar semiconductor, these add up with the bands of "regular bonds", producing

* In the following $\Delta\varepsilon$ will be neglected.

an exponential decay (*tail*) of the density of states toward the gap (instead of the square-root type, characteristic of the normal bands). The sparsely distributed tail-states allow only hopping conduction, so an amorphous semiconductors has a mobility band edge (where the "normal" bonding and antibonding states end) and an optical band edge (where the tail states end) for both VB and CB. Actually, the splitting of the normal bonding and antibonding states into bands is smaller, since the average distance between normal bonds is bigger due to the interdispersion of the long bonds. This makes the mobility gap wider than in a crystalline semiconductor. In addition to the normal and tail states, the large number of dangling bonds causes interaction between them leading to a splitting into a dangling bond band. This might even make the band gap disappear. (That is why amorphous silicon has to be hydrogenated for use in solar cells. Hydrogen saturates the dangling bonds and the arising Si-H bonding and antibonding states are swept out of the gap.)

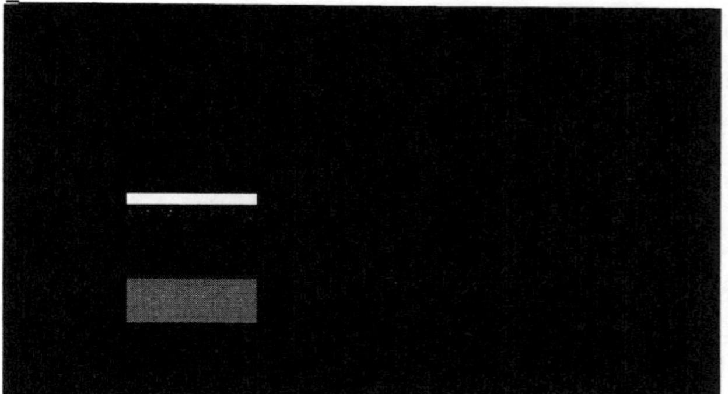

Figure 4. The electronic states of a homopolar amorphous semiconductor.

The concepts of *bonds, long bonds,* and *dangling bonds*, introduced here will be very useful in analyzing the electronic structure of defects. The tetrahedrally bonded semiconductors – as we shall see – have a tendency to force even foreign atoms into an sp^3-hybridization. For example, oxygen is two-fold coordinated in most semiconductors, as it is in water (see Figure 5.a). Its electronic structure can be thought of in terms of four – somewhat distorted – sp^3-hybrids: two participating in bonds with the neighbors (and having a bond angle varying from ~105 to ~165°, depending on environment), while the other two hosting the remaining four (two each) of the six valence electrons of oxygen. Such doubly occupied – and, therefore, chemically inactive – hybrids are called *lone pairs*. (In case of oxygen, they are the reason for twofold coordination.)

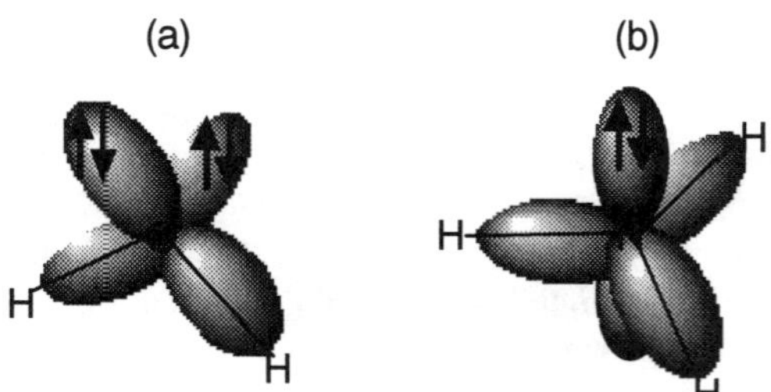

Figure 5. The electronic configuration of oxygen in water (a) and of nitrogen in ammonia (b).

Sometimes the semiconductor does not succeed in molding the impurity into its own picture and likeliness. For example, nitrogen prefers its electron configuration adapted in ammonia (see Figure 5.b): three sp^2-hybrids and a perpendicular p-like lone pair:

$$\begin{aligned}\varphi_1 &= (s + p_x)/\sqrt{2}\\ \varphi_2 &= \left(s - \tfrac{1}{2}p_x + \tfrac{\sqrt{3}}{2}p_y\right)/\sqrt{2}\\ \varphi_3 &= \left(s - \tfrac{1}{2}p_x - \tfrac{\sqrt{3}}{2}p_y\right)/\sqrt{2}\\ \varphi_4 &= p_z\end{aligned} \qquad /5/$$

leading to three-fold coordination.

3. THE NATURE OF DEFECT STATES AND THE ROLE OF RECONSTRUCTION

With the exception of isoelectronic substitutional impurities, defects always introduce extra electron states into the electronic structure of the crystal. In case of a substitutional atom (impurity or dopant) the reconstruction of the host decides the nature of these orbitals. Figure 6 shows the situation for nitrogen and phosphorus – both having 5 valence electrons) in silicon. Phosphorus has a covalent radius comparable to (a little longer than) silicon, so it is forced into sp^3-hybridization and, thereby, into fourfold coordination which uses up four of its valence electrons in four bonds with the silicon neighbors. There is no localized state left for the fifth valence electron of phosphorus, so that one can only find place on the CB edge. In fact, the extra positive charge in the core of phosphorus (with respect to silicon) perturbs the states of the CB edge into forming a series of weakly localized hydrogen-like orbitals with levels close to the CB edge, according to the effective mass theory (EMT). The lowest of these so-called EMT orbitals is s-like and this is occupied by the extra electron (Figure 6.a).

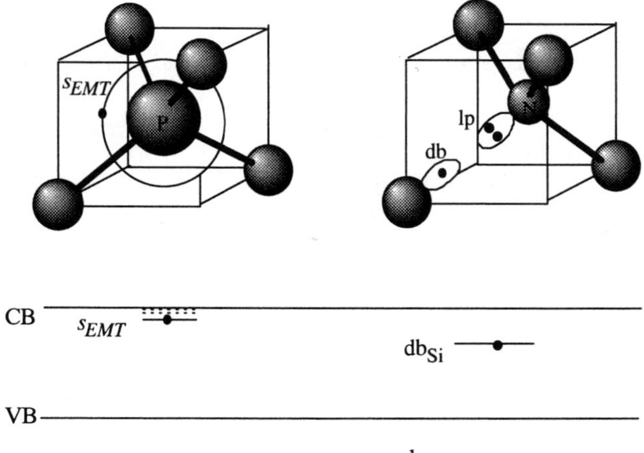

Figure 6. The electronic structure of the on-center phosphorus substitutional (a) and of the off-center nitrogen substitutional (b) in silicon.

Nitrogen, on the other hand, is small: the typical Si–N distance is only 1.78 Å in contrast to the first neighbor distance of 2.35 Å in silicon. Having some "room to move", nitrogen goes *off-center* in the [111] direction (reducing the symmetry, unlike phosphorus which remained *on-center*), and takes on its favored three-fold coordinated configuration as in ammonia. As a consequence, one of the silicon neighbors is left with a dangling bond. The repulsion of the nearby lone pair pushes the energy of the corresponding level somewhat up, but it still remains quite deep in the gap. That is the reason why P is and N is not a shallow donor in silicon. (Note that due to the higher electronegativity of nitrogen, the nitrogen lone pair has a level deep in the silicon VB.) The situation in N is such that there are only two truly localized defect states in the system which do not mix with each other. As a consequence, only one canonical orbital has a strong contribution to the LCAO expansion of each of the localized states. Therefore, these can easily be identified among the eigenstates.

A vacancy is a bit more complicated. There are two ways to asses the situation (see Figure 7).

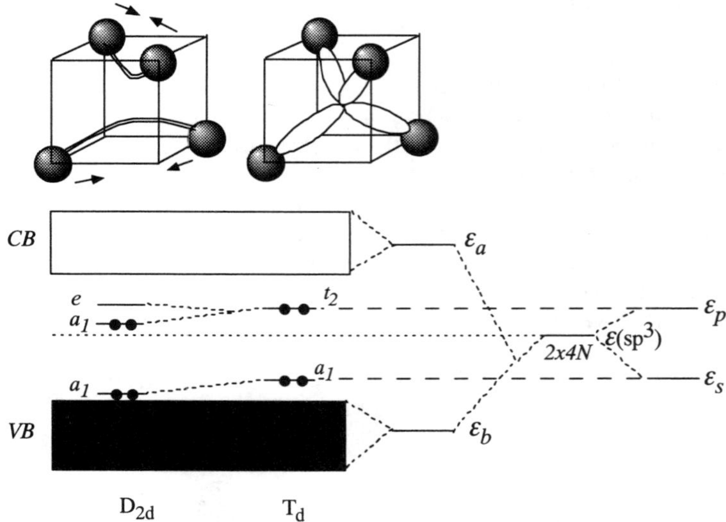

Figure 7. Localized and canonical orbital description of a vacancy

Removing an atom from a (homopolar) semiconductor, one is left with the dangling bonds of the four neighbors pointing toward the center of the vacancy. From these four sp^3-hybrids one can remix one *s*- and one *p*-orbital centered at the position of the missing atom: the first one transforming according to the non-degenerate a_1 representation, the second according to the triply degenerate t_2 representation of the T_d point group. These canonical one-electron states have to be occupied by the four electrons available from the four dangling bonds. There are more than one possible occupations (due to the degeneracy of t_2), i.e., the ground state is degenerate and, therefore, instable. According to the Jahn-Teller principle, the system should undergo a reconstruction so the symmetry-lowering lifts the degeneracy. In this case both a reconstruction to C_{3v} and to D_{2d} symmetry would split t_2 into a non-degenerate a_1 and a doubly degenerate *e* state, thereby allowing unambiguous occupation. To decide between the two without doing the calculation (which should be done without the Born-Oppenheimer approximation, actually) one has to call on chemical intuition. It seems quite reasonable to assume, that the dangling bonds form long bonds pairwise between the neighbor silicon atoms, and members of the pairs move closer to each other leading to D_{2d} symmetry. In fact, from the combination of the two canonical a_1 eigenstates, obtained from such a calculation, one can build two equivalent localized bonding orbitals, while the *e* states can be used to construct the equivalent antibonding ones. In this case, therefore, the energy levels shown in the band diagram cannot be assigned in a one-to-one fashion to the localized states depicted in the defect molecule in the upper left of Figure 7.

4. CONSTRUCTING THE LEVEL DIAGRAM OF A DEFECT MOLECULE

The localized states of the defect can be constructed in advance from constituents of the defect molecule, like one does it in case of molecules using the states of the atoms. As an example I will use the oxygen impurity both in substitutional and interstitial positions in silicon. Figure 8. shows the construction for the substitutional from using the electronic structure of an unreconstructed vacancy and a free oxygen atom.

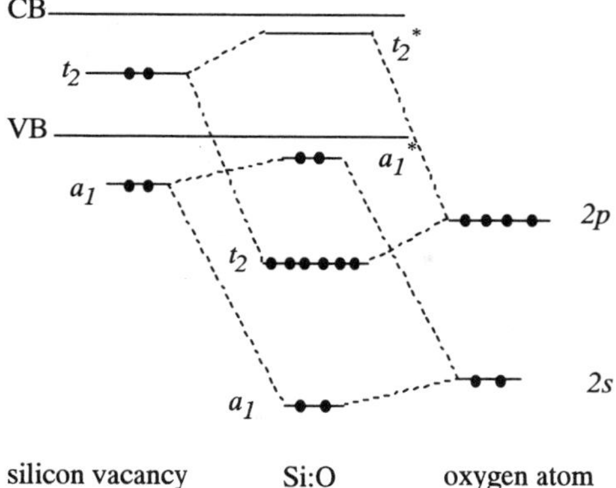

Figure 8. The construction of the electronic structure of unreconstructed substitutional oxygen in silicon from those of the unreconstructed vacancy and free oxygen.

Due to symmetry reasons the t_2 (a_1) states of the vacancy interact only with the *2p* (*2s*) states of the oxygen atom. In both cases a pair of bonding and antibonding states arise (the latter denoted by asterisks). The splitting between the *s*-like states is more ionic, between the *p*-like states more covalent. The size of the splitting and the relative position of the constituent levels can only be guessed at qualitatively, and the position of the resulting defect levels with respect to the band edges comes out only from the actual calculation. However, these are the states one has to look for among the calculated eigenstates! (In this case again, the correspondence is close to one-to-one between the expected localized states and some canonical eigenstates.) Unless they are there and their position roughly corresponds to chemical intuition, the calculation must have had a flow! One knows that oxygen is a lot more electronegative than silicon, so the bonding a_1 state of Si:O should be below the bottom of the VB, and the t_2 one deep in it. Usually the covalent splitting between *p*-type orbitals is bigger (they are more diffuse, so the overlap is stronger) but in this case there is very big ionic splitting between the *s*-type states. Therefore, one expects t_2 between a_1 and a_1^*. Indeed, this is the situation one observes not only in silicon but in diamond and SiC as well. (However, e.g., in the case of sulfur in diamond the sequence of a_1^* and t_2^* is reversed.) The difference is in the positions relative to the band edges. In silicon the situation corresponds to Figure 7 and oxygen is an electron trap (hyper deep acceptor). In diamond both a_1^* and t_2^* are in the gap making oxygen an amphoteric impurity, while, e.g., oxygen on the carbon site in cubic SiC produces both a_1^* and t_2^* in the CB. As a consequence, the two electrons with highest energy occupy no defect states but the CB edge instead, and EMT states are formed — making SiC:O_C an EMT double donor. Obviously, this information can only be obtained from the calculated results, the interpretation of which should be carried out, however, in this manner in every case!

Actually, oxygen in silicon does not remain on-center, connected to the fact that a dynamic Jahn-Teller distortion occurs due to the interaction of the vibronic and electronic degrees of freedom. This is very difficult to visualize and, therefore, to anticipate. Chemical intuition helps again. Oxygen is relatively small compared to silicon (the typical Si-O bond length is 1.65 Å, in contrast to the first neighbor distance of 2.35 Å), so one may expect that chemical rebonding occurs putting oxygen into its favored covalent configuration. (Note that oxygen likes to have an ionic environment as well with higher coordination.) An off-center movement in the [001] direction does the job: oxygen binds to two Si neighbors only, forming also two lone pairs, wile the other two silicon atoms form a long bond. The symmetry lowering splits the canonical triply degenerate states into three non-degenerate ones. Figure 9 shows the participation of the canonical states in the localized orbitals of the defect molecule.

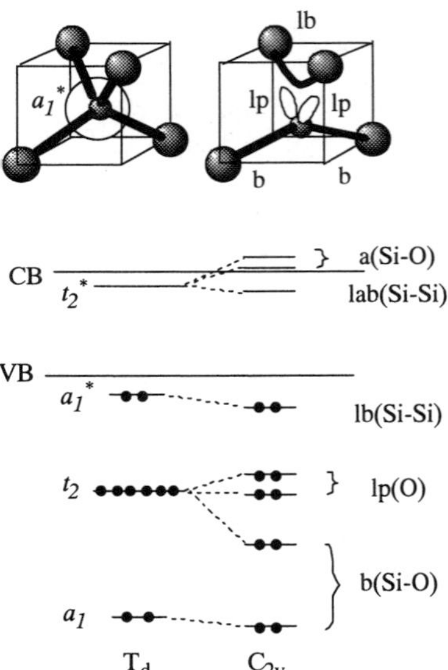

Figure 9. The reconstruction of substitutional oxygen in silicon.

The configuration of the oxygen interstitial is known in silicon: oxygen is in a puckered bond-center position, forming bonds with two silicon atoms whose bonding has been disrupted. In this case it is easy to guess at the resulting electronic structure (Figure 10.a). Oxygen is two-fold coordinated in a distorted sp^3 configuration. The Si–O bonds are stronger than Si–Si bonds, so the bonding-antibonding splitting will be larger, putting the former below the VB edge and the latter above the CB edge. Therefore, there should be one s-like and one p-like eigenstate in both bands from which the (Si–O) bonding and antibonding orbitals can be constructed. Since oxygen is more electronegative than silicon, there must be two eigenstates below the VB edge strongly centered on oxygen, from which the two equivalent lone-pairs can be built. The (Si–O) bonds and the two lone-pairs host all six valence electrons of oxygen. The defect is obviously inactive electrically, as it is, indeed, observed.

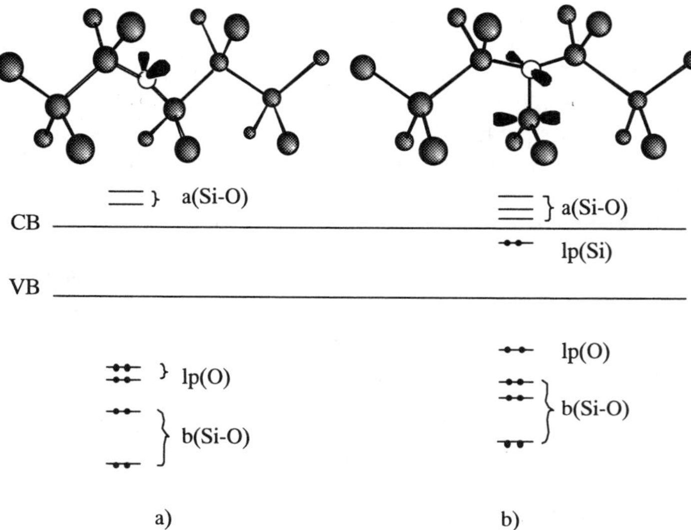

Figure 10. The electronic structure of interstitial oxygen in equilibrium (a), and at the saddle point of diffusion (b).

It is interesting to examine the saddle-point configuration along the diffusion path of interstitial oxygen. Jumping from one bond-center site to the next, the symmetrical position is depicted in Figure 10.b. At this point oxygen is bonded to three silicon atoms at the same time, disrupting two Si–Si bonds. This makes both oxygen and the Si atom on the symmetry axis become three-fold coordinated with sp^2 electron configurations instead of sp^3. There will be three Si–O bonds: a little different from those of the equilibrium case but the corresponding canonical orbitals should still be in the bands. Oxygen will have one lone-pair only, forcing its sixth valence electron into the *p*-like orbital of the three-fold coordinated silicon, making that a lone-pair.* A pure *p*-state of Si lies in the upper part of the gap (c.f. Figure 7) and the repulsion of two electrons on it pushes that even higher. Therefore, the configuration is subject to ionization (has donor character). Actually, this is the only possibility to obtain electrical activity from oxygen in silicon. It is well known that unless irradiated (when Si:O_{Si} forms), oxygen is inactive in silicon as long as it is dispersed as isolated interstitials (Si:O_i). Aggregation of oxygen, however, can create electrically active complexes. All the models consistent with such centers are based on three-fold coordinated oxygen — but of course one has to find complexes which are stable in equilibrium and have properties in correspondence with observations (see, e.g. [1,2]).

* That would mean a $q = -1$ formal charge on that silicon but, actually, due to the higher electronegativity of oxygen the net charge of silicon will be a lot lower.

5. PREDICTION OF THE BEHAVIOR OF HYDROGEN IN VACANCIES

Similarly to the case of dangling bonds in amorphous silicon, hydrogen is known to passivate the vacancies in many semiconductors. The passivation is both chemical and electrical: hydrogen saturates the dangling bonds and the bonding and antibonding Si–H orbitals fall in the bands. This happens also in the case of the vacancy in silicon, Si:V, and the silicon vacancy in SiC, SiC:V_{Si}. Recently it was found, however, that H in the carbon vacancy of SiC (SiC:V_C) does not follow this trend [3]. Instead of forming a normal Si–H bond, it is more favorable for hydrogen to have a symmetric position between two Si atoms neighboring the vacancy. The electron density shows that the two-center long bond between these two Si neighbors has become a three-center bond, enveloping also the H atom. The one-electron level corresponding roughly to this bond is in the valence band. Another level, corresponding to the long bond between the other two silicon atoms is in the lower half of the gap. The electron introduced by the H atom finds place on the antibonding combination of the sp^3 hybrids of its Si neighbors, so that the H atom is at the node of this orbital. As a consequence, the corresponding level lies high in the gap, making the system a hole trap. Therefore, no passivation can be achieved.

Hydrogen becomes similarly over-coordinated as a bond-center interstitial in Si and diamond but that this seems to be unique to elemental semiconductors [4]. SiC appears to be the first example for H entering a three-center bond when trapped in a vacancy. The reason probably is that the Si-Si distance across the carbon vacancy in SiC is about 3.1 Å, while the normal Si-H bond length is ~ 1.5 Å, so the H atom can interact with two Si atoms. One begins to wonder whether this might not occur in other compound semiconductors where the distance between cations is comparable to the double of the typical cation–hydrogen bond length. We have investigated hydrogen in the anion vacancy of the nitrides of the third column of the periodic table [5]. The calculations for different charge states in all of the vacancies resulted in the configurations shown in Table 2 (for q = -2, a three-center bond forms in every case). Analyzing the results in terms of a defect molecule diagram, it seems that a consistent picture emerges.

Table 2. The nature of the cation and the configuration of the hydrogenated anion vacancy in different charge states. $\Delta d_{cc}/d_{cc}$ ($\Delta d_{ch}/d_{ch}$) is the percentage difference between the cation–cation distance in the semiconductor and in the pure crystal of the cation (and the double of the normal cation-hydrogen distance). ΔX is the Pauling electronegativity difference between hydrogen and the cation.

$\Delta d_{ch}/d_{ch}$	$\Delta d_{cc}/d_{cc}$	ΔX	system	q = +2	q = 0
+3 %	61 %	0.1	BN:$(V_N+H)^{0+q}$	H remains on-center	two-center B–H bond
+4 %	32 %	0.3	SiC:$(V_C+H)^{1+q}$	H remains on-center	inward puckered three-center Si–H–Si bond
-1 %	30 %	0.5	GaN:$(V_N+H)^{0+q}$	H remains on-center	linear three-center Ga–H–Ga bond
-6 %	9 %	0.6	AlN:$(V_N+H)^{0+q}$	H remains on-center	outward puckered three-center Al–H–Al bond

The decisive question is how the cation-cation distance across the anion vacancy relates to the ideal bonding distance between the cations. If the former is much bigger than the latter then the dangling bonds of the vacancy "retract" and the cations relax outward. Putting the hydrogen in will simply give rise to a single cation–hydrogen bond. This happens in SiC:V_{Si} and in BN:V_N as the calculations prove it. If the two distances are, however, comparable than a pairing distortion occurs in the vacancy and long bonds will be formed between pairs of the cations. It is convenient to construct the electronic structure of the V+H defect molecule if we start from the reconstructed neutral vacancy, V_C^0, in SiC and the negative one, V_N^-, in the nitrides on the one hand, and a proton, H^+, on the other, as constituents. This way we have four electrons in the vacancy and equivalent long bonds between two pairs of cations. This happens in SiC:V_C, AlN:V_N, and GaN:V_N. The situation is depicted in Figure 11 assuming cubic crystals (apart from small symmetry differences the situation is the same in the hexagonal lattices of AlN and GaN as well). In the actual calculations the orbitals shown here were identified in all cases.* Positions relative to the band edges were not always as in the Figure, though. The differences in the resulting geometries can also be explained.

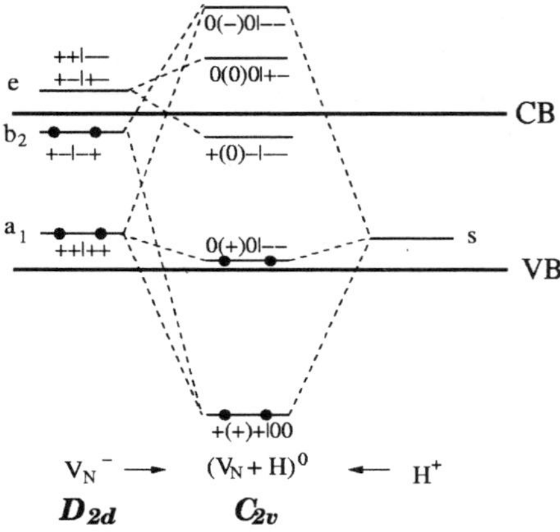

Figure 11. Defect molecule diagram for SiC:$(V_C+H)^{+1}$ and XN:$(V_N+H)^0$, where X = B, Al, and Ga. The sign of the coefficient of the cationic sp^3-hybrids on both sides of H and that of the s-orbital of H is given left of the "|" mark, while the sign of the contribution of the other two cations are given to the right.

The essence is that the H^+ "slips" into the long bond between one of the cation pairs (if there exists any), forming a three-center bond and diminishing the symmetry from D_{2d} to C_{2v}. Differences arise due to the electronegativity difference (ΔX) between H and the cations, as well as due to the relation of the ideal cation-cation distance and the ideal cation – hydrogen distance (see Table 2). In case of SiC:V_C, ΔX is small, so H^+ "owns" little of the two electrons in the three-center bond. Therefore, the Si–H–Si bridge is slightly puckered "inwards", i.e., toward the other long bond, so H can "chip off" a little of those electrons to become less frustrated. In case of GaN, ΔX is bigger, therefore H^+ gets a higher share of the two electrons in the three-center bond, which becomes somewhat ionic making H a strong link in the straight Ga–H–Ga bridge. In case of AlN, ΔX is the largest. H^+ gets most of the two electrons in the three-center bond. Since the other long bond is also well concentrated into the vacancy (the Al–Al distance across the unrelaxed vacancy is closest to the ideal Al–Al bonding distance), the repulsion of the electrons in that one to those around H^+ becomes strong: therefore, the Al–H–Al bridge is puckered outwards. (Still the positive charge of the proton is not completely screened. Therefore, putting one or two electrons into the antibonding combination of the hybrids of the neighboring cations does not destabilize the three-center bond.)

* For that purpose, the real part of the canonical one-electron functions from an LDA-PW-SCM calculation were depicted in the (110) plane.

Starting from neutral $(V_N+H)^0$ in nitrides and positive $(V_C+H)^+$ in SiC, it is easy to understand the states with two electrons less (these systems show negative U behavior, so the intermediate charge state is not stable). The situation for the ++ charge state of the nitrides (3+ for SiC) can be understood in the sense of Figure 8. Taking a positive V_N (or dipositive V_C in SiC) we have an "electron-poor" vacancy, stable in T_d symmetry (assuming cubic crystals). There are no distinct long bonds between the pairs across the vacancy: the four cations share the two electrons on a totally symmetric a_1 orbital. In this situation the *s*-orbital of H^+ sitting on-center simply interacts with this a_1 state and the T_d symmetry remains.

In these cases the use of defect molecule diagrams have not only helped to understand the calculated results, but allowed generalization. In fact, the three-center bond of hydrogen is far from obvious and chances to find the stable outward puckered configuration in case of AlN would have been meager if searching the energy surface at random. The defect molecule interpretation was also leading to important clues in determining which vacancy can be passivated by hydrogen, i.e. it resulted in conclusions of practical importance.

Acknowledgment. The author is indebted for the support of the grants OTKA-T032174 and FKFP-0289/97.

6. REFERENCES

[1] P. Deák, L. C. Snyder, and J. W. Corbett, *Phys. Rev. B* **45** (1992) 11612.
[2] C. P. Ewels, R. Jones, S. Öberg, J. Miro, and P. Deák, *Phys. Rev. Lett.* **77** (1996) 865.
[3] A. Gali, B. Aradi, P. Deák, W. J. Choyke, N. T. Son, and E. Janzén, *Phys. Rev. Lett.* **84** (2000) 4926.
[4] C. H. Chu and S. K. Estreicher, *Phys. Rev. B* **42** (1990) 9486.
[5] B. Szücs, PhD Thesis, Budapest University of Technology and Economics, 2001.
B. Szücs, A. Gali, P. Deák, and Ch. G. Van de Walle, to be published

Modeling of Point Defects, Polarons and Excitons in Ferroelectric Perovskites

E.A.KOTOMIN[a,b], R.I. EGLITIS[c], G. BORSTEL[c], and P.W.M. JACOBS[d]

[a]*Institute of Solid State Physics, University of Latvia, Kengaraga str.8, Riga LV 1063, Latvia*
[b]*Max-Planck Institut für Festkörperforschung, Heisenbergstr. 1, Stuttgart 70569, Germany*
[c]*Fachbereich Physik, Universität Osnabrück, D-49069 Osnabrück, Germany*
[d]*Dept of Chemistry, The University of Western Ontario, London ON N6A 5B7, Canada*

Abstract. We review the results of our recent large-scale computer simulations of point defects, excitons and polarons in ABO_3 perovskite crystals, focusing mostly on $KNbO_3$ and $KTaO_3$ as representative examples. We have calculated the atomic and electronic structure of defects, their optical absorption, and their activation energies for migration. The majority of the results were obtained by means of the quantum chemical method, the Intermediate Neglect of Differential Overlap (INDO) based on the Hartree-Fock formalism, and the shell model (SM). The main findings are compared with the results of *ab initio* Density Functional Theory (FP-LMTO) first-principles calculations.

1. Introduction

Ternary ABO_3 ferroelectric perovskites have numerous technological applications [1]. In particular, $KNbO_3$ crystals are widely used for laser frequency doubling, and doped $KTaO_3$ is prospective for electrically controlled holographic and graded pyroelectric devices. Their properties are influenced by point defects, primarily by vacancies (since as-grown crystals usually are non-stoichiometric). For instance, the second harmonic generation in $KNbO_3$ is reduced by the blue-light-induced infra-red-absorption (BLIIRA) [2]. Despite the fact that such light-induced absorption is well known for many perovskites, the effect has been studied only in a few papers.

Concerning intrinsic point defects in these materials, relatively little is known for $KNbO_3$, and almost nothing for $KTaO_3$ [1,3]. A broad absorption band around 2.7 eV has been observed in electron-irradiated crystals and ascribed tentatively to the F-type centers (oxygen vacancy with one or two trapped electrons, F^+ and F centers, respectively) [4,5]. Transient optical absorption around 1 eV has been associated recently [5], in analogy with other perovskites, with a hole polaron (a hole bound to some defect), whereas IR absorption (around 0.8 eV) with an electron polarons [6]. The ESR study of $KNbO_3$ doped with Ti^{4+} provides strong evidence that holes can be trapped by negatively charged defects [7]. For example, hole polarons were found in $BaTiO_3$ doped with Na or K alkali ions replacing for Ba and thus forming a negatively charged site attracting a hole [8]. Primary candidates for such defects are cation vacancies. In irradiated MgO they are known to trap one or two holes giving rise to the V- and V^0 centers [9] which are also known as bound hole polaron and bipolaron, respectively. Lastly, we want to shed some light on the long-standing debate

[4,10] on the nature of the so-called "green" excitonic luminescence observed in both $KTaO_3$ and $KNbO_3$ around 2.2-2.3 eV.

In this paper, we review the results of our theoretical modeling of radiation-induced point defects (including polarons and excitons) in $KNbO_3$ and $KTaO_3$ crystals. Theory is compared with available experimental data.

2. Methods

To perform large-scale modeling of defects, we used mainly the semi-empirical INDO method modified for ionic and partly ionic solids [11]. The INDO method is based on the Hartree-Fock formalism and allows self-consistent calculations of the excited states of defects, and thus the relevant absorption energies, using the so-called ΔSCF method. The method has demonstrated very good results in previous calculations of defects, both in the bulk and on the surface of many oxide materials [3], including pure $KNbO_3$ and $KTaO_3$ perovskites and defects – Li impurities in $KTaO_3$ [12]. The calculated frequencies of the transverse-optic (TO) phonons at the Γ point in the Brillouin zone (BZ) of cubic and rhombohedral $KNbO_3$ and the atomic coordinates in the minimum energy configuration for the orthorombic and rhombohedral phases of $KNbO_3$ are also in good agreement with experiment thus indicating that a very successful INDO parametrization has been achieved [12]. Frozen-phonon calculations for T_{1u} and T_{2u} modes of cubic $KTaO_3$ are also in good agreement with experiment.

The degree of covalency in the chemical bonding could be seen from the calculated (static) effective charges on atoms (calculated using the Löwdin population analysis): 0.62 *e* for K, 2.23 *e* for Ta and –0.95 *e* for O in $KTaO_3$, which are far from those expected in the purely ionic model (+1 *e*, +5 *e* and –2 *e*, respectively) often used. These charges show slightly higher ionicity in $KTaO_3$ as compared with the relevant effective charges calculated for $KNbO_3$: 0.54 *e* for K, 2.02 *e* for Nb and –0.85 *e* for O atoms.

Along with our INDO method, recently we performed *ab initio* calculations of *F* centers and hole polarons were performed recently employing DFT method as implemented within the full-potential linear muffin-tin orbitals (FP-LMTO) code [13,14], and used 40 atom supercells. Lastly, for modeling defect diffusion, in addition to semi-empirical INDO we employed classical SM [15] where the interatomic interactions are described by the core-core, core-shell and shell-shell pair potentials, representing the shell-model. In this approach each ion has a charged core and electronic shell. The sum of the core and shell charges is equal to the formal charge of the corresponding ion. The spring constant *k* connects the core and the shell of the same ion. The core-shell separation is a measure of the polarization.

The interactions between the cores and between cores and shells of different ions are Coulombic, whereas the interactions between the shells of different ions besides the Coulombic part contain also the short-range potentials accounting for the effects of the exchange repulsion together with the van-der-Waals attraction between them. The short-range Buckingham potentials contain three parameters (D, ρ, and C) per pair of atoms. All parameters are carefully fitted to the lattice structure, elastic and dielectric properties of crystals (see details in Section 7). The use of integer ionic charges does not imply restrictions to ionic materials, in fact the short-range potential effectively takes into account the covalency and charge-transfer effects.

The quantum chemical simulation of all defects has been done within a supercell

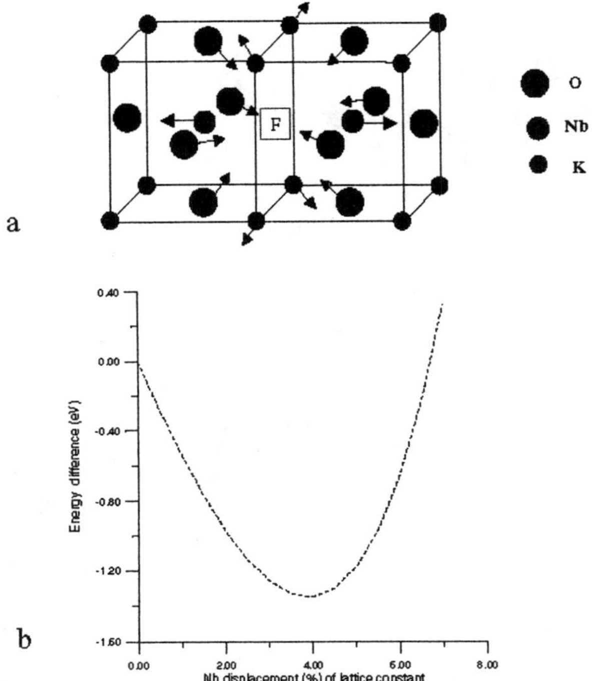

Figure 1. Sketch of lattice relaxation around the F center (a) and the relevant potential energy curve (b) vs. the Nb atom displacement calculated by INDO method.

approach. One (neutral) O atom was removed from the supercell for modeling the *F* center, whereas two atoms, O and K, were removed to simulate the F^+ center. For the modeling of the hole polarons, a K atom has been removed, and the actual type of the hole polaron (one-site and two-site) was set by the symmetry of the local lattice relaxation. In semi-empirical INDO calculations we have used large supercells expanding primitive cells up to 4x4x4 times (320 atoms), and allowed for the relaxation of many atoms surrounding a defect.

3. *F*-type centers [3, 13, 14, 16-18]

In the cubic phase of $KNbO_3$ all O atoms are equivalent and have the local symmetry C_{4v} due to which the excited state of the *F*-type centers should be split into a non-degenerate level and a doubly-degenerate energy level. The optimized INDO relaxation of the two Nb atoms next to the O vacancy was found to be 3.9 % a_0 (lattice parameter) (Fig. 1), very close to the value of 3.5 % found in the *ab initio* study (FP-LMTO method).

The outward INDO relaxation of nearest K atoms and inward displacements of O atoms are much smaller. They contibute only about 20 % to the total relaxation energy of 1.35 eV (considerably larger than 0.5 eV found in *ab initio* calculations). The *F* center local energy level lies 0.6 eV above the top of the valence band. Its molecular orbital contains primarily the contribution from the atomic orbitals of the two nearest Nb atoms. Only 0.6 e resides at the orbitals centered at the vacancy site; hence the electron localization at the defect is much

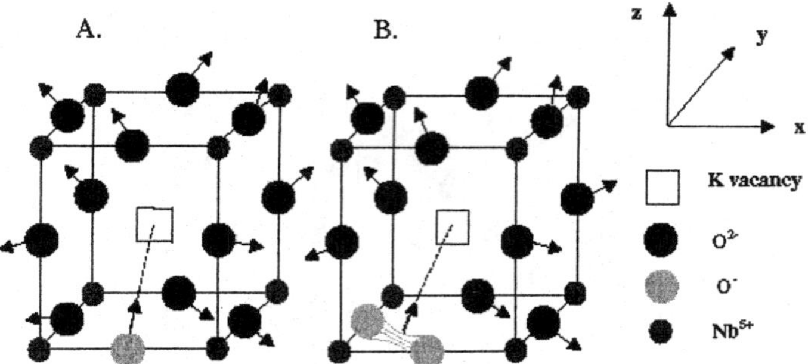

Figure 2. Sketch of the one-site (a) and two-site (b) hole polarons in $KNbO_3$.

smaller than is known for traditional *F* centers in ionic oxides (see [3,9] for a comparison). This difference stems from much smaller Madelung potential at the vacancy in partly covalent $KNbO_3$ (10.5 eV versus 23 eV in MgO).The symmetry analysis of the ground-state wave function associated with the *F* center shows that the major contribution comes from the e_g states centered at Nb neighbors (it is essentially the $3z^2$-r^2 component, with the *z* axis in the direction towards the F center).

For the F^+ center the lattice relaxation energy of 2.23 eV and the Nb displacements of 5.1% are larger than those for the F center due to a stronger Coulomb repulsion between unscreened O vacancy and Nb atoms: a share of the electron density inside the O vacancy decreases to 0.3 e.

The calculated optical absorption energies for the F^+ center are 2.34 and 2.66 eV (and for the *F* center 2.73 and 2.97 eV). That is, the two absorption bands predicted for the F^+ center are shifted to the low-energy side compared to *F* center energies, in agreement with similar defects in ionic oxides [3,9]. However, *both* defects are predicted to have one of the bands around 2.6-2.7 eV, in agreement with the experimental observations [4,5].

4. Polarons

4.1. Hole polarons [3,16,19,20]

Our calculations show that there are two energetically favourable atomic configurations in which a hole is well localized (Fig. 2): one-site and two-site (molecular) polarons. In the former case, a single O^- ion is displaced in our INDO calculations *towards* the K vacancy by 3% a_0. Simultaneously, 11 other nearest oxygens surrounding the K vacancy are slightly displaced outwards from the vacancy. In contrast, in the two-site (molecular) configuration, a hole is shared by the *two* O ions which approach each other by 3.5% a_0 and their center-of-mass shifts towards a vacancy by 2.2% a_0. The relevant lattice relaxation energies are close, 0.4 eV and 0.53 eV, respectively. Despite the fact that the two-site configuration of a polaron is lower in energy, proper incorporation of the electron correlation effects could reverse this delicate energy balance. In turn, *ab initio* FP LMTO calculations gave similar atomic displacements but predicted smaller relaxation energies, 0.12 eV and 0.18 eV, respectively.

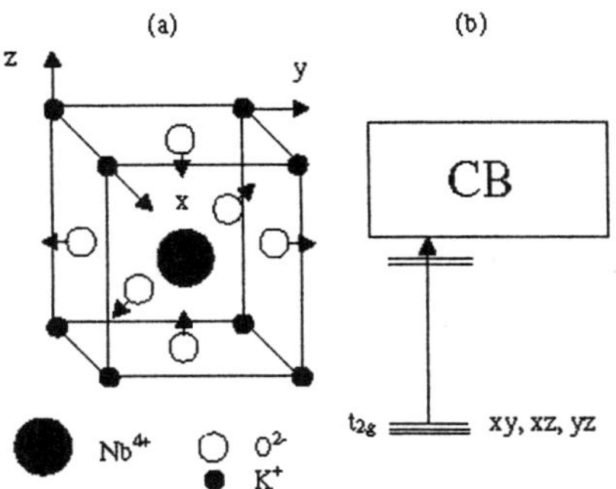

Figure 3. Atomic structure of self-trapped electron: (a) asymmetric O atom relaxation around an electron localized on the central Nb atom. (b) Local energy states within the gap.

In spite of the general observation of a considerable degree of covalency in $KNbO_3$ (see, e.g., [3,12]) and contrary to a delocalized character of the *F* center state, the one-site polaron state remains well localized at a single displaced O atom, with only a small contribution from atomic orbitals of other O ions but not from those of the K or Nb ions. In agreement with general theory for the small-radius polarons in ionic solids [21], the optical absorption corresponds to a hole transfer to the state delocalized over nearest oxygens. The calculated absorption energies for one-site and two-site polarons bound to the K vacancy are close, both around 1 eV which is half the experimental value for a hole polaron trapped near Ti [7]. This shows that the optical absorption energy of small bound polarons could be strongly dependent on the defect involved.

4.2. Electron Polarons [22]

The existence of electron self-trapping in ionic solids was predicted by L. Landau in 1933, but confirmed experimentally by means of ESR only recently; in 1993 for $PbCl_2$ [23] and in 1994 for $LiNbO_3$ [24]. Our quantum chemical INDO calculations argue also for an

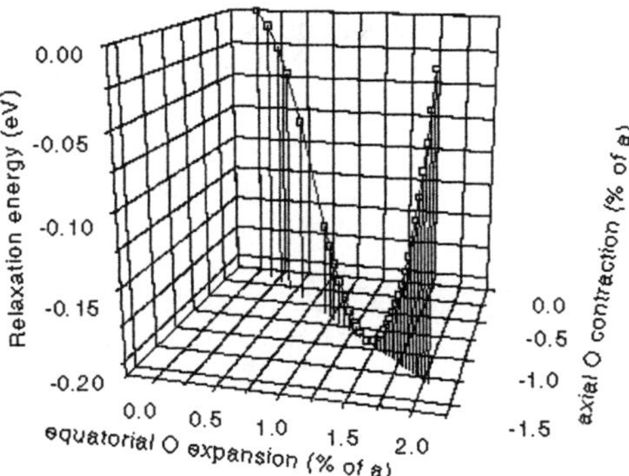

Figure 4. Lattice energy gain in the self-trapped electron due to relaxation of four equatorial O atoms and axial contraction of two O atoms along the z axis.

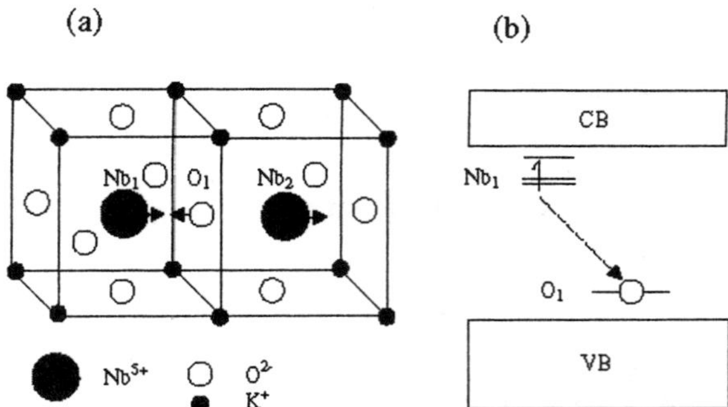

Figure 5. Schematic view of the charge transfer vibronic exciton (a) and its luminescence (b).

existence of the self-trapped *electrons* in $KNbO_3$ and $KTaO_3$, associated with the net lattice relaxation energy of 0.21 eV and 0.27 eV and estimated optical absorption of 0.78 eV and 0.75 eV, respectively. An electron in the ground state occupies t_{2g} orbital of a single Nb^{4+} or Ta^{4+} ion (Fig. 3).

Its orbital degeneracy is lifted by a combination of the breathing and Jahn-Teller modes where four nearest equatorial O atoms are displaced by 1.4% a_0 outwards and two oxygens shift 1% a_0 inwards along the *z* axis (Fig. 4) The energies mentioned above are close to the experimental data on transient optical absorption [6]. As to the energetics of electron self-trapping, calculations show that the bottom of the $KNbO_3$ conduction band has a narrow subband which makes the electron localization energy small enough for the total self-trapping energy balance to be positive.

5. Excitons [10,22,25]

Many ABO_3 ferroelectric perovskites reveal a photoluminescence in the visible range ('green luminescence') with peaks around 2.2 eV in both $KTaO_3$ and $KNbO_3$. Many hypothetical mechanisms have been suggested as the origin of this luminescence, including donor-acceptor recombination, recombination of electron and hole polarons, charge transfer vibronic excitation (CTVE), electron transitions in MeO_6 complexes, etc. We performed quantum chemical modeling of the atomic and electronic structure of the relevant triplet excitons in $KNbO_3$ and $KTaO_3$ and calculated their luminescence energies.

Table 1. Calculated properties of excitons in the two crystals (see Figure 5 (a) for atom labelling). Numbers in brackets give the experimental values.

Crystal	E_{rel} (eV)	Displ. (% of a_o)	Eff. charge (*e*)	Charge transf. (*e*)	E_{lum} (eV)
$KNbO_3$	2.37				2.17 (2.2)
Nb1		2.9	+1.55	-0.47	
Nb2		4.3	+1.92	-0.1	
O1		4.9	-0.20	+0.65	
$KTaO_3$	2.71				2.14 (2.2)
Ta1		3.1	+1.74	-0.49	
Ta2		4.5	+2.12	-0.11	
O1		5.2	-0.24	+0.71	

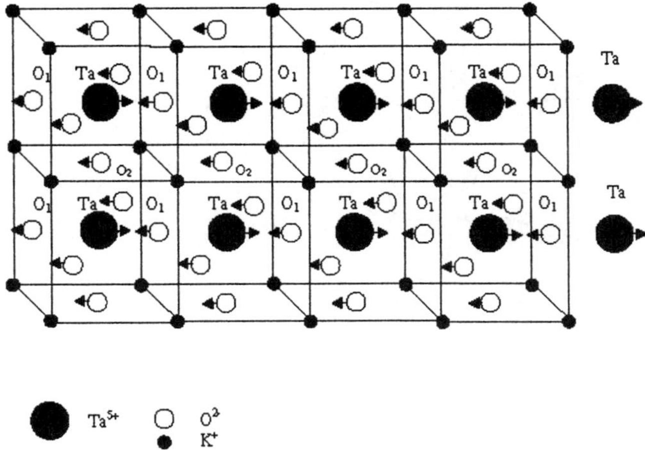

Figure 6. The displacement structure in the ferroelectric state of charge transfer vibronic exciton phase in $KTaO_3$ crystal.

Our calculations give strong support to 'green' luminescence in $KTaO_3$ and $KNbO_3$ crystals occuring as a result of the radiative recombination of the electrons and holes forming the *charge transfer vibronic exciton* (CTVE) rather than due to the electron transitions in MeO_6 complex. The calculated luminescence energy for both $KTaO_3$ and $KNbO_3$ is very close to the experimentally observed 2.2 eV. The CTVE in ferroelectric oxides with partly covalent chemical bonding is usually imagined as a correlated pair of electronic and hole polarons (*bi-polaron*).

We found in our calculations (Fig. 5) that this in fact is a Ta_1-O_1-Ta_2 *triad*, accompanied by a strong vibronic energy lowering (2.71 eV). This is associated with a hole localized mostly on O_1 atom which has a 5.2 % a_0 displacement along the (100) axis towards the nearest Ta_1-ion on which the electronic polaron is mainly localized. This Ta_1 ion is in its turn displaced by 3.1 % a_0 towards the O_1 atom, whereas another Ta_2 ion along the (100) axis on the another side of O_1 atom contains a smaller fraction of the electron polaron density and is displaced by 4.5% a_0 outwards the O atom. The charge redistribution between the triad atoms in $KNbO_3$ is: -0.47 e, 0.65 e, and –0.1 e (compared to the ions of the perfect lattice).

Special attention should be paid to the *new phase* of the CTVE predicted theoretically in ref. [26-28]. The CTVE phase consists of the strongly correlated CTVEs which are located in *each* unit cell of the crystal. That is, the CTVE phase corresponds to a new state of the crystal which is characterized by a new equilibrium charge transfer as well as a new set of equilibrium lattice displacements.

The first step in the theoretical study of the CTVE phase was done [26,27] using semi-phenomenological models. Namely, first a co-operative Jahn-Teller effect type model of CTVE phase was developed. This model deals with point dipoles corresponding to equilibrium displacements in the CTVE cell as well as with electronic degrees of freedom of CTVE active ions assuming the electronic state degeneracy, or pseudo-degeneracy. These two type degrees of freedom directly interact with lattice polarization and deformation in the

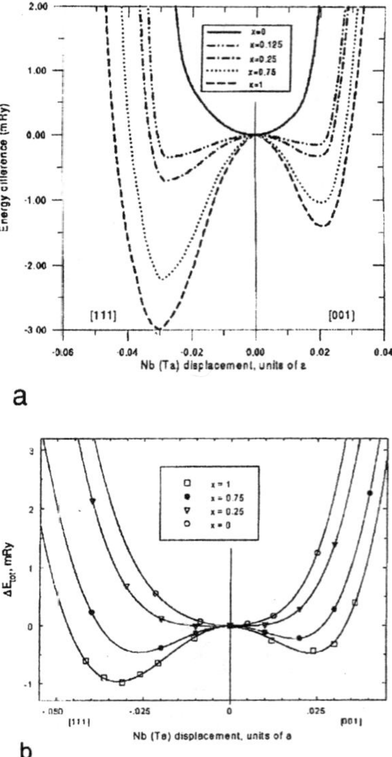

Figure 7. The total energy gain (per cell) *vs.* Nb off-centre [111] and [001] displacement in $KNb_xTa_{1-x}O_3$ solid solution for Nb concentrations x = 0, 0.125, 0.25, 0.75 and 1 as calculated by means of INDO (a) and FP-LMTO methods (b).

linear and quadratic approximations. It was shown that an important role in CTVE formation belongs to a specific vibronic interaction characterized by a direct coupling of charge transfer with lattice distortions. According to these estimates the CTVE phase energy position [26] was deeply in-gap.

However, a direct, model-independent computation of CTVE phase properties was needed. In the present study the computational analysis was performed by means of the semi-empirical Hartree-Fock-type INDO method. This method was previously used in a treatment of both a single CTVE in incipient ferroelectric $KTaO_3$ [25] and in ferroelectric $KNbO_3$ [22] where the existence of CTVE was confirmed.

It is shown in the present work on the basis of INDO-calculations that the novel CTVE *phase* can exist in the well known incipient ferroelectric $KTaO_3$. We have used 3x3x3 supercells in these calculations. The triplet CTVE phase was obtained as a ferroelectric phase with parallel-oriented small polaron electron-hole pairs on O-Ta ions and in-gap states. The total energy lowering per such a O-Ta pair was 2.32 eV. This corresponds to a strong vibronic interaction case. The equilibrium displacements of O-Ta ions in each electron-hole pair are directed towards each other and rather large; 4.33 % for Ta-ion, and 5.62 % for O-

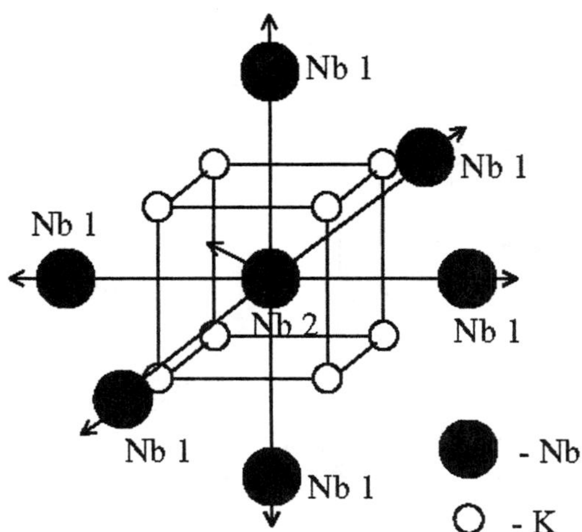

Figure 8. Sketch of the symmetric repulsion of six Nb atoms outwards from the central Nb atom, which as a consequence, is off-center then both in [111] and [001] directions.

ion. Self-consistent charge transfer is also large, -0.77 *e*, accompanied by the equilibrium effective charge of 1.46 *e* for Ta-ion, whereas for O ions they are 0.77 *e* and -0.18 *e*, respectively. Note that ferroelectrically ordered oxygen ions in the intermediate chains relatively to O-Ta CTVE-chains are much less shifted (0.37 %) in the same direction as oxygen ions in these CTVE- chains (Figure 6).

Another model example, characterized by higher bonding covalency, is $KNbO_3$. The parameters of the ferroelectrically ordered CTVE phase were calculated here on the basis of the same approach as for $KTaO_3$. Here the total energy lowering per O-Nb pair is 1.99 eV. This also corresponds to a strong vibronic interaction, which is lower than that for $KTaO_3$ due to a small increase in the bond covalency for $KNbO_3$. The equilibrium displacements of O-Nb ions in each polaronic electron-hole pair are also directed towards each other and remain rather large: 4.15 % for the Nb-ion, and 5.28 % for the O-ion. Self-consistent charge transfer is considerable: -0.70 *e* (the equilibrium effective charge +1.32 *e*) for the Nb-ion, and +0.70 *e* (with an equilibrium effective charge -0.15 *e*) for the O-ion in the CTVE-chains. Oxygen ions in the intermediate chains have 0.35 % shifts which are parallel to the shifts of oxygen ions in the CTVE-chains. This calculation, together with the previous one, confirm the conclusions about a high stability of the CTVE-phase in ferroelectric oxides.

6. KTN Solid Solutions and Impurity Self-assembling

Figure 7 (a) shows the total energy for Nb impurity concentrations at x = 0, 0.125, 0.25, 0.75 and 1 in KTN, as a function of Nb [111] and [001] off-centre displacements. First of all, a test calculation for the pure $KTaO_3$ (x=0) shows no off-center Ta displacement, in good agreement with experiment. In another extreme case of pure $KNbO_3$ (x=1) Nb atoms reveal very clearly the off-center displacements along both the [111] and [001] directions. The magnitudes of these displacements are in good agreement with a recent EXAFS study [29]. According to our INDO calculations, Nb impurities become off-center in $KTaO_3$ already at the lowest studied concentration (x=0.125), in good agreement with another XAFS measurements [30]. The calculated magnitude of 0.146 Å displacement along the [111] direction is very close to the experimental finding of 0.145 Å observed at 70 K [30].

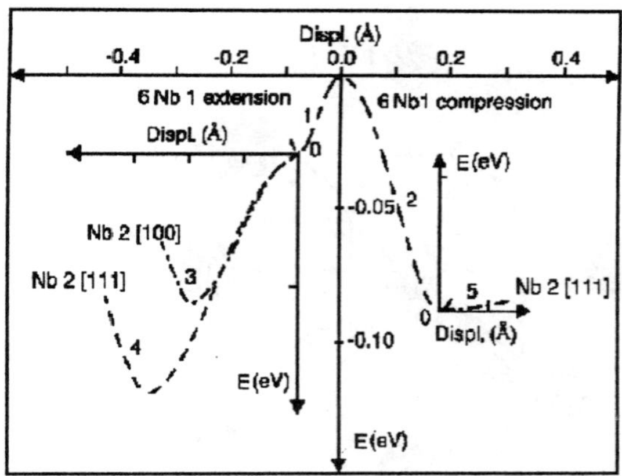

Figure 9. Displacement energy of six Nb atoms relaxed outwards (curve 1) and inwards (curve 2) as shown schematically in Fig. 8. The central Nb atom is off-center in the [001] and [111] direction, (curves 3 and 4), in the case of a symmetric expansion of the cluster, but is on-center (curve 5), in the case of a compression of the cluster.

Fig. 7 (b) shows also the results of the FP-LMTO [31] calculations in which the critical Nb concentration turns out to be much higher than the experimental value.

Next, we simulated self-organization of Nb impurity ions in KTO. In such the Nb-cluster calculations we have extended the primitive $KTaO_3$ unit cell 4×4×4, i.e. 64 times. In order to study cooperative displacements (self-ordering) of Nb impurities in $KTaO_3$, we replaced, in our extended unit cell containing 320 atoms, seven Ta atoms by seven Nb atoms, as is shown schematically in Fig. 8.

In order to find the energy minima of the Nb clusters in $KTaO_3$, we allowed the six Nb atoms to relax towards the central Nb atom (see Fig. 8). The positions of the K and O atoms were kept fixed. The results of our calculations show that the total energy per LUC is lowered by 0.088 eV when the six Nb atoms shift symmetrically by 0.187 Å towards the central Nb atom. Moreover, a uniform outward displacement of the six Nb atoms from the central Nb by 0.073 Å is also favourable and lowers the energy by approximately 0.03 eV (see Fig. 9). In the case, when the six Nb atoms are shifted outwards from the central Nb atom, but the latter moves off-center by 0.27 Å in the [111] direction, a further lowering of 0.09 eV is found to give an overall total energy reduction of approximately 0.12 eV. The central Nb atom also reveals instability in the [100] direction. The shift of the central Nb ion in the [100] direction by 0.192 Å lowers the cluster energy additionaly by 0.056 eV, in the case in which six Nb atoms are shifted outwards the central Nb atom – to give a total energy reduction of 0.086 eV. Nevertheless, the total cluster-structure-induced energy lowering in the ground state, which corresponds to the situation in which six Nb atoms are symmetrically relaxed outwards from the centre and the central Nb ion is displaced off-center in the [111] direction (0.12 eV), turns out to be energetically more favourable.

We have also studied the inverted impurity case: When seven Ta impurities replace seven Nb atoms in $KNbO_3$ a different behaviour is found, compared to seven Nb atoms replacing

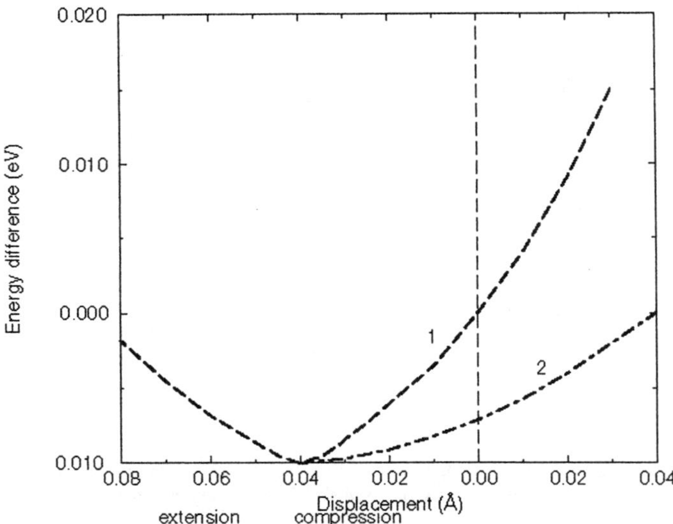

Figure 10. Total energy vs displacement of six nearest neighbour Ta atoms in $KNbO_3$ (see Fig. 8) which are relaxed outwards and inwards with respect to the Ta atom at the origin (curve 1). Curve 2 shows the additional effect of the central Ta atom being displaced along the [111]

seven Ta atoms in $KTaO_3$. The total energy minimum of the former case corresponds to the by 0.04 Å symmetrically extended configuration of the six Ta ions, while the central Ta ion retains its on-center position (see Fig. 10). Such pronounced difference arises due to higher chemical bonding covalency in KNO as compared to KTO.

7. Defect diffusion [32]

We have modeled diffusion of K, O and Nb atoms in $KNbO_3$, employing both the INDO and SM techniques. For this purpose, the SM parameterization was carefully developed, in order to reproduce experimental structural, elastic and dielectric properties, as well as frequences of transverse and longitudinal optic phonons in $KNbO_3$.

Our starting point in SM calculations was the interatomic potentials developed by Donnerberg and Exner [33] which they used to calculate defect formation energies. These Buckingham-type potentials

$$\phi_{ij}(R) = D_{ij} \exp(-R/\rho_{ij}) - C_{ij}/R^6 \qquad (1)$$

where R is the distance between ions of species i and j, ϕ_{ij} is the interaction energy and i,j = 1 for K, 2 for Nb, and 3 for O, give reasonable defect formation energies but have the disadvantage of predicting a negative eigenvalue for the lowest transverse optic mode, which is the soft mode in $KNbO_3$. We therefore modified their potentials to fit ω_{TO1} at the Γ point in the BZ at 710 K (in the cubic phase). Moreover, we found that, as for an earlier investigation of $SrTiO_3$ [34,35], slight adjustments in the Nb–O repulsive parameter D_{23} were all that was required to fit ω_{TO1} over the whole T-range 710–1180 K. However, we found it impossible

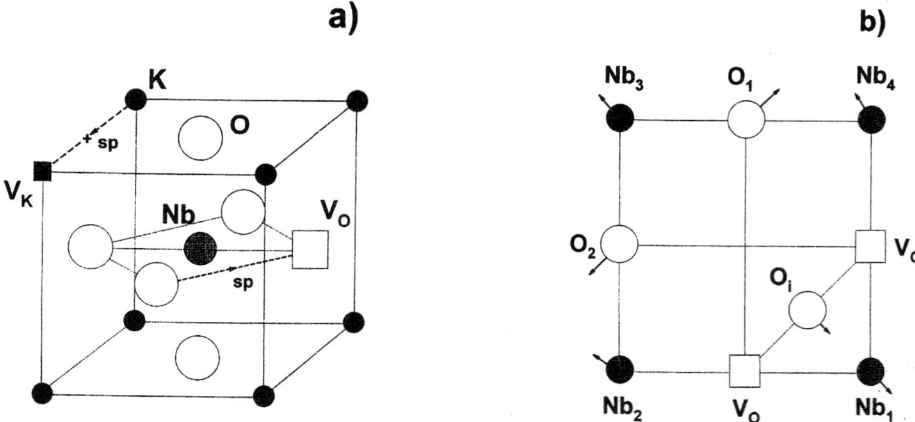

Figure 11. Unit cell of $KNbO_3$ (a) indicating saddle points (sp) for the O and K vacancy (V_o, V_K) migration paths. Displacements of atoms around interstitial Oi ion shown in sketch (b) are given in table ??.

to fit the strong temperature dependence of the static permittivity ε_s as well as that of ω_{TO1} using a conventional SM with temperature-dependent parameters.

Table 2 shows that the temperature dependence of the soft mode could be fitted successfully using a modified potential which has a K–O repulsive parameter D_{13}=600.3 eV, and a temperature-dependent D_{23}. Calculated values of the static permittivity ε_s were about half the values calculated by Fontana *et al* [36], from their IR reflectance data, using the standard Lyddane–Sachs–Teller (LST) relation, but the calculated temperature-dependence $\varepsilon_s(T)/\varepsilon_s(T_r)$, where the reference temperature Tr is 1180 K, agreed quite closely with their values [36] for this ratio (table 2). The frequencies of the other optic modes vary hardly at all with temperature, so calculated and experimental values are compared in table 3 at just two temperatures, 710 K and 1180 K. Calculated frequencies of the Raman mode are 320 cm^{-1} at

Table 2. Temperature dependence of the lattice constant a of cubic $KNbO_3$ [36], of the Nb–O repulsive constant D_{23}, of calculated and experimental values [32] of the frequency of the soft mode, and of the ratio of the static permittivity at T, to that at T_r =1180 K. Experimental values of this ratio are from Fontana *et al* [36].

			ω_{TO1}(cm^{-1})		ε_s(T)/ε_s(Tr)	
T(K)	a(Å)	D_{23}(eV)	Calc.	Expt.	Calc.	Expt.
710	4.0220	1332.10	96.1	96	1.92	2.00
730	4.0225	1333.00	100.2	100	1.84	1.89
803	4.0241	1334.88	106.0	106	1.65	1.67
910	4.0265	1338.00	115.7	116	1.38	1.40
1030	4.0289	1341.60	127.1	127	1.15	1.18
1180	4.0320	1345.44	136.2	136	1.00	1.00

Table 3. Comparison of frequencies of transverse and longitudinal optic modes of $KNbO_3$ calculated from the SM I (see Table 4) with experiment [36].

	TO1		LO1		TO2		LO2		TO3		LO3	
T(K)	Calc.	Expt.	Calc.	Expt.	Calc.	Expt.	Calc.	Expt.	Calc.	Expt.	Calc.	Expt.
710	96	96	488	419	194	198	194	190	492	521	777	826
1180	136	136	489	418	193	204	193	194	492	511	778	815

Table 4. (a) Shell - model potentials used for $KNbO_3$ at 710 K. Buckingham repulsive (Dij), hardness (ρij) and van derWaals (Cij) parameters are defined in Eq.(1). Y is the shell charge (or core charge on K+ when it has no shell) and K is the core-shell spring constant. Potential (I) for unpolarizable K+; (II) for polarizable K+, $i; j$ = 1 for K, 2 for Nb, and 3 for O. (b) Crystal properties calculated from potentials I and II, using the experimental lattice constant at 710 K (4.022 Å).

(a)

Potential	ij	D(eV)	ρ(Å)	C(eV Å^{-2})	i	Y(|e|)	K(eV Å^{-2})
(I)	13	600.3	0.36198	0.0	1	1.0	-
	23	1332.1	0.36404	0.0	2	-2.811	103.07
	33	22746.3	0.14900	27.88	3	-4.496	2100.0
(II)	The only parameters different to those in I are:						
	A_{13} = 640.0 eV; Y_1 = - 2.76 |e|; K_1 = 80.0 eV Å^{-2}						

(b)

Properties	I	II	Expt.	Ref. [33][c]
Cohesive energy (eV)	-175.52	-175.34		
Elastic constants (GPa)				
c_{11}	371	373	255[a]	398
c_{12}	116	118	80[a]	142
c_{44}	127	129	90[a]	142
Permitivities				
static ε_s	115.5	249	246[b]	516
high-frequency ε_∞	1.80	2.22	1.81[b]	1.81
Frequency of lowest transverse optic mode				
ω_{TO1}(cm^{-1})	96	69	96[b]	[d]

[a]Nunes et al [37].
[b]From Fontana et al [36]; the value of ε_s is the lattice contribution to ε_s calculated from the LST reaction. Actual values measured by Yanovskii [38] are much higher due to some further polarization mechanism of unexplained origin [36].
[c]Recalculated for a =4.026 Å, as used in [33].
[d]Negative eigenvalue.

710 K and 318 cm^{-1} at 1180 K, about 12% higher that the experimental value [36] of 280 cm^{-1} in the tetragonal $KNbO_3$ phase at 585 K.

The SM potentials used are given in table 4 along with calculated and experimental values of elastic constants and permittivities. Defect vacancy formation energies are compared in table 5 with the earlier calculations of Donnerberg and Exner [33]. Our modified potentials give formation energies about 3% higher for O and Nb, about 30% higher for K, perhaps indicating some special difficulties in modelling the K^+ ion. Tables 6 and 7 give the main results for the vacancy calculations using both the semi-empirical methods, SM and INDO. The atomic neighbourhood of the K and O vacancies and saddle-points are sketched in figure 11. One can see that atomic displacements agree quite closely. The SM with a polarizable K^+ ion results in only slightly larger K^+ displacements.

Table 5. Calculated vacancy formation energies for $KNbO_3$ (in eV).

Defect	This work, potential I	Ref. [33]
a (Å)	4.022	4.026
V_O	20.44	19.68
V_k	5.76	3.99
V_{Nb}	122.83	119.57

Table 6. Atomic displacements δ around O vacancy (A) and saddle point for the migrating Oi atom (B) (in fractions of the lattice constant multiplied by 100)—see sketches in figures 11 (a) and (b), respectively. SM stands for the shell model, numbers in brackets are results for the rigid K+ ion potential I (table 4), q_{eff} is the effective ionic charge (in |e|). Outward displacements are positive, inward displacements negative; *N* is the number of ions equivalent by symmetry.

			SM	INDO	
configuration	Ion	N	δ	δ	q_{eff} (e)
A	Nb	2	8.70 (8.16)	8.21	1.72
	O	8	-2.35 (-2.80)	-2.44	-0.82
	K	4	1.31 (1.11)	1.73	0.50
B	O_i[a]	1	1.54 (1.25)	1.14	-0.72
	Nb_1	1	8.50(8.10)	7.89	+1.68
	Nb_2	2	7.40(7.14)	7.80	+1.83
	Nb_3	1	1.06(1.70)	Fixed	+1.85
	O_1	2	1.61(1.58)	1.30	-0.81
	K	8	2.20(1.1)	1.52	0.50
Migration energy (eV)			0.67(0.68)	0.79	

[a]With respect to the mid-point of the line joining the two vacancies.

The INDO effective atomic charges presented in table 6 show that: (i) charges for Nb and O are quite different from those expected in a completely ionic model (Nb^{+5}, O^{-2}, K^+) but close to those found earlier for perfect $KNbO_3$ [12], (ii) the effective charges for the *interstitial* O_i and K_i atoms are close to those for regular lattice sites, which justifies the use of the same SM potentials in migration studies as developed for the perfect crystal. The remarkable point is that a SM potential can incorporate so effectively the covalency of Nb–O bonds in $KNbO_3$. The calculated migration energies for O atoms are very similar for the SM with and without a K^+ shell, and are only smaller by 15% than the INDO value where the covalency effects are properly incorporated.

The largest displacements occur for Nb atoms which give the dominant (90%) contribution to the lattice relaxation energy. We are aware of only one experimental estimate of about 1 eV [39] for the O vacancy migration energy which agrees qualitatively with our findings [32]. This demonstrates that O vacancies could be quite mobile in $KNbO_3$ at elevated temperatures. The INDO calculations give the activation energy for the K vacancy migration to be 0.60 eV, which is even smaller than that for the O vacancy. Unfortunately, the SM calculations were unable to locate saddle-point configurations with one imaginary eigenvalue for the migrating K^+ interstitial ion. Lastly, results of the SM calculations for the atomic displacements around Nb vacancy are also presented in Table 7. We faced here the same problem in locating the saddle point, as for the K^+ ion.

In conclusion of this Section 7, we believe that the results presented are likely to be valid also for *ferroelectric* $KNbO_3$ phases as well since the atomic displacements and energy relaxations obtained here for defects are two orders of magnitude larger than those

Table 7. The same as table 6 for Nb vacancy (A), K vacancy (B), and the saddle point of the hopping K+ ion (C) (see figure 11 (a)). We were unable to find the saddle-point energy for Nb^{5+} and K^+ ions with either of the two SM potentials, in Table 4.

			SM	INDO	
Defect	Ion	N	δ	δ	q_{eff}(e)
A	O	6	11.7		
	K	8	-8.2		
	Nb	6	1.3		
B	Nb	8	-1.58 (-1.50)	-2.70	2.01
	O	12	0.73 (0.80)	0.51	-0.83
	K	6	-0.22 (-0.20)	-0.88	0.53
C	O	4		3.8	-0.83
	Nb	4		-2.5	2.01
	K	8		-0.94	0.53
Migration energy for K^+(eV)				0.60	

responsible for the ferroelectric transitions [12]. It would be of great interest to check our findings with firstprinciples calculations, which are now in progress.

Figure 11 shows the equilibrium and saddle point (*sp*) configurations for K and O vacancies. Results obtained independently by means of the INDO and SM agree quite well, in particular, the *migration energy* for the O vacancy was found to be 0.79 eV and 0.68 eV, respectively. This is in agreement with experimental estimate [39]. The migration energy for K vacancy is even smaller than for the O vacancy (0.6 eV). This means that both types of vacancies are mobile at room temperature and thus could form aggregates and complexes with impurities.

8. Conclusions

We demonstrated in this study how fruitful the combination of different methods and techniques (semi-empirical quantum chemical INDO, *ab intio* FP LMTO, and classical SM) techniques can be. Based on our calculations (whose reliability is supported by use of several different techniques), we predict that each of the electronic F and F^+ centers have two absorption bands, one of these bands is around 2.7 eV, which supports the interpretation of the experimental observation for electron-irradiated $KNbO_3$ [4,5]. The ESR experiments are necessary to discriminate between these two defects. Analysis of the electron density redistribution clearly demonstrates that, in the ground state of both defects, the electron density is considerably shifted from the O vacancy towards the two nearest Nb atoms, which is very unusual for ionic oxides (such as MgO) but well known for analogous, so-called E' centers in quartz, and indicates a chemical bonding covalency in *I-V* ABO_3 perovskites.

The calculated optical absorption energies for electron and hole polarons (ca. 0.8 eV and 1 eV) support their preliminary interpretation. Lastly, we attribute the green luminescence around 2.3 eV in both $KNbO_3$ and $KTaO_3$ to the charge-transfer vibronic exciton (a Ta-O-Ta triad). Calculated migration energies for K and O vacancies (0.6 eV and 0.7-0.8 eV) show their high mobility and thus the ability to aggregate in *I-V* ABO_3 perovskites at room temperatures.

Modeling of KTN solid solutions demonstrated that well-parametrized semiempirical methods could serve as a useful tool for the study of tiny self-organization effects in doped semiconductors and perovskites.

Acknowledgments

This study was partly supported by the European Excellence Center for Materials Research and Technology in Riga, Latvia, NATO collaborative grant # PST.CLG.977561, and DFG. Authors are greatly indebted to L. Grigorjeva, J. Maier, D. Millers, V. Vikhnin, A. I. Popov, O.F. Schirmer, and R. T. Williams for many stimulating discussions.

References

[1] P. Günter and J.-P. Huignard (eds.) *Photorefractive Materials and Their Applications,* Topics in Applied Physics, vol. 61, 62 (Springer Verlag, Berlin, 1998); H-J. Donnerberg, *Atomic Simulations of Electro-Optical and Magneto-Optical Materials*, Springer Tracks in Modern Physics, vol. **151**, Berlin, 1999.

[2] L. Shiv, J.L.Sorensen, E.S.Polzik, G. Mizell, *Opt. Lett.*, **20** (1995) 2270.

[3] E.A.Kotomin, A.I.Popov, *Nucl. Instr. Methods B* **141** (1998) 1.

[4] E.R.Hodgson , C.Zaldo and F.Agullo-Lopez, *Solid State Commun.*, **75** (1990) 351.

[5] L.Grigorjeva,D. Millers, E.A.Kotomin,E.S.Polzik, *Solid State Commun.*,**104** (1997) 327. L.Grigorjeva, D.Millers, A.I.Popov, E.A.Kotomin, E.S.Polzik, J.Lumin., **72-74** (1997) 672.

[6] R.T. Williams, L.Grigorjeva, D. Millers, G. Corradi, *Rad. Eff.& Defects in Solids* **155** (2001) 265.

[7] E.Possenriede, B.Hellermann, O.F.Schirmer, *Solid State Commun.,* **65** (1988) 31.

[8]T.Varnhorst, O.F.Schirmer, H.Krose, R.Scharfschwerdt, Th.W.Kool, *Phys.Rev.B*, **53** (1996) 116.

[9]Y.Chen, M.M.Abraham, *J.Phys.Chem.Sol.* **51** (1990) 747; A.E.Hughes, B.Henderson, *In: Point Defects in Solids* (ed. J.H. Crawford Jr and L.M. Slifkin, Plenum Press, New York), 1972.

[10] V. Vikhnin, R.I. Eglitis, E.A. Kotomin, S. Kapphan, and G. Borstel, *Mat. Res.Soc. Symp. Proc. Vol.* **677**, p. AA.4.15.1 (2001)

[11] E.V.Stefanovich, E.K.Shidlovskaya, A.L.Shluger, *phys.stat.sol. (b),* **160** (1990) 529 .

[12] R.I.Eglitis, A.V.Postnikov, G.Borstel, *Phys.Rev.B*, **54** (1996) 2421; R.I. Eglitis, A.V. Postnikov, and G. Borstel, *Phys. Rev. B* **55** (1997) 12976.

[13] R.I.Eglitis, N.E.Christensen, E.A.Kotomin, A.V.Postnikov, and G.Borstel, Phys.Rev. B, 56 (1997) 8599.

[14] N.E. Christensen, E.A. Kotomin, R.I. Eglitis, A.V. Postnikov, D.L. Novikov, S. Tinte M.G. Stachiotti, and C.O. Rodriguez, In*: Proc. NATO ARW on Defects and Surface-Induced Effects in Advanced Perovskites* (Riga, August 1999), Kluwer, 2000, p.3

[15] C.R.A. Catlow and W.C. Mackrodt (eds.) *Computer Simulations of Solids*, Lecture Notes in Physics, vol. 166, Springer Verlag, Berlin, 1982.

[16] E.A. Kotomin, R.I. Eglitis, G. Borstel, *Comp. Mater. Sci.*, **17** (2000) 290.

[17]. E.A. Kotomin, R.I. Eglitis and A.I. Popov, *J. Phys.: Cond. Matt.*, **9** (1997) L 315.

[18] E.A. Kotomin, R.I. Eglitis, G. Borstel, L. Grigorjeva, D. Millers, V. Pankratov, *Nucl. Inst. Meth. B* **166-167** (2000) 299.

[19] R.I.Eglitis, E.A.Kotomin, G.Borstel, *phys.stat.sol. (b)* **208** (1998) 15 .

[20] E.A. Kotomin, R.I. Eglitis, A.V. Postnikov, G. Borstel and N.E. Christensen, *Phys. Rev. B* **60** (1999) 1.

[21] O.F.Schirmer, P.Koidl, H.G. Reik, *phys.stat.sol. (b),* **62** (1974) 385 ; O.F.Schirmer, *Z.fur Physik B*, **24** (1976) 235 .

[22] E.A. Kotomin, R.I. Eglitis, G. Borstel, *J. Phys.: Cond. Matt.*, **12** (2000) L557; R.I. Eglitis, E.A. Kotomin, and G. Borstel, Computational Materials Science **21** (2001) 530-534.

[23] S.V. Nistor, E. Goovaerts, D. Schoemaker, *Phys. Rev. B* **48** (1993) 9575.
[24] B. Faust, H. Müller, O.F. Schirmer, Ferroelectrics **153** (1994) 297.
[25]. V.S. Vikhnin, H. Liu, W. Jia, S. Kapphan, R.I. Eglitis, D. Usvyat, J. Lumin., **83-84** (1999) 109.
[26] V.S. Vikhnin, *Ferroelectrics* **199** (1997) 25-40; *Z. Phys. Chem.* **201** (1997) 201-213.
[27] V. Vikhnin, *Ferroelectrics Lett.* **25** (1999) 27-35.
[28] V.S. Vikhnin, R.I. Eglitis, E.A. Kotomin, S. Kapphan, and G. Borstel, *Mat. Res. Soc. Symp. Proc. Vol.* **677** (2001) AA4.15.1 – AA4.15.6.
[29] V.A. Shuvaeva, K. Yanagi, K. Yagi, K. Sakaue, and H. Terauchi, *Solid State Commun.* **106** (1998) 335.
[30] O Hanske-Petitpierre, Y. Yacoby, J. Mustre de Leon, E.A. Stern, and J.J. Rehr, *Phys. Rev. B* **44** (1991) 6700.
[31] A.V. Postnikov, T. Neumann, and G. Borstel, *Ferroelectrics* **164** (1995) 101.
[32] P.W.M. Jacobs, E.A. Kotomin and R.I. Eglitis, *J. Phys.: Cond. Matt.*, **12** (2000) 569.
[33] H.J. Donnerberg and M. Exner, *Phys. Rev. B* **49** (1994) 3746.
[34] J.E. Crawford and P.W.M. Jacobs, J. *Solid State Chem.* **144** (1999) 423.
[35] P.W.M. Jacobs, *Nuovo Cimento D* **20** (1999) 1187.
[36] M.D. Fontana, G. Metrat, J.L. Servoin and F. Gervais, *J. Phys. C: Solid State Phys.* **17** (1984) 483.
[37] A.C. Nunes, J. Axe, and D. Shirane, *Ferroelectrics* **2** (1971) 291.
[38] V.K. Yanovskii, *Sov. Phys. Solid State* **22** (1980) 1284.
[39] DM Smyth, Ferroelectrics **151** (1994) 115.

Computational Materials Science
C.R.A. Catlow and E.A. Kotomin (Eds.)
IOS Press, 2003

Modelling Point Defects and Diffusion in Earth Materials

Kate WRIGHT
Departments of Chemistry and Earth Sciences University College London and Davy Faraday Research Laboratory, Royal Institution, Albemarle Street, London W1S 4BS, UK.

Abstract Computer simulation techniques ranging from atomistic to full quantum mechanical descriptions of elements are increasingly being applied to the study of Earth materials. The study of defects in minerals can be problematic since the materials of interest often have large unit cells, low symmetry and variable stoichiometry. In many cases, we are interested in sampling pressure, temperature and time scales which are difficult to access in the laboratory. Computational modelling, therefore, can provide important insights into defect behaviour in minerals.

1. Introduction

The study of point defects and diffusion in minerals has applications to a wide variety of problems in the Earth Sciences. These include dating of rocks and determination of their pressure/temperature (P/T) history, studies of deformation and creep in the Earth's mantle, and processes such as phase transitions and the nucleation and growth of minerals. The experimental study of diffusion and defect processes in minerals is difficult and, for the Earth Sciences, often complicated by the need to acquire data at extremes of pressure and temperature, or to measure very slow diffusion rates. Computer simulation methods offer an alternative way to explore defect behaviour in complex minerals at those conditions and timescales that are difficult to access in the laboratory.

Atomistic simulations have been widely and successfully used to model defects and diffusion in geological materials [1-5] both by static and molecular dynamics methods although the latter is only really effective when diffusion is rapid and activation barriers low. More recently it has been possible to calculate defect formation energies in minerals at the quantum mechanical level [6,7] using periodic electronic structure codes as well as QM/MM embedded cluster methods [8,9]. In this chapter, we will briefly review the basics of diffusion theory and then examine how defects can be studied using static lattice and embedded cluster methods. Examples relating principally to the problem of OH defects in minerals will be used to illustrate the above.

2. Defects and diffusion in solids

The diffusion process involves the transport of matter through a solid in response to some driving force or gradient and can be described by Fick's Laws [10]: Fick's first law describes the flow of a diffusing species in terms of the concentration of that species:

$$J = -Ddc / dx \tag{1}$$

where $\boldsymbol{J}$ is the flux (number of diffusing species crossing a unit area s^{-1}), D the diffusion coefficient and $\boldsymbol{c}$ is the concentration of the diffusing species at a point $\boldsymbol{x}$. Fick's second law relates the change in concentration over time (*dt*) to the diffusion coefficient (*D*) along direction *x*.

$$\frac{dc}{dt} = \frac{d}{dx}\left[D\frac{dc}{dx}\right] \tag{2}$$

Experimental diffusion data can be analyzed to give values of D using various solutions to these equations [11], depending on the experimental setup.

Where there is no gradient, then we have "self diffusion" which is dependent on temperature and where diffusion takes place by a random walk process. The driving force for self diffusion is essentially the change in entropy associated with an atomic jump. In crystalline materials diffusion can take place through the bulk lattice, along extended defects such as dislocation cores and along grain boundaries or surfaces. At high temperatures, bulk or volume diffusion will dominate although when grain size is small or at low temperatures and in the presence of fluids, diffusion along surfaces and grain boundaries will be important. Here, however, we will consider only volume diffusion where atoms migrate via point defects.

Diffusion is a temperature dependent process, which can be described by an Arrhenius equation:

$$D = D_0 \exp\left(-Q/_{RT}\right) \tag{3}$$

where D_0 is a pre-exponential factor related to jump frequency and defect concentration (see [10] for full derivation), $\boldsymbol{Q}$ is the activation energy required to make the jump, R the gas constant and T the temperature. Plots of ln D against 1/T over a wide temperature range show non-linear behaviour, with two distinct regimes, as shown in Figure 1.

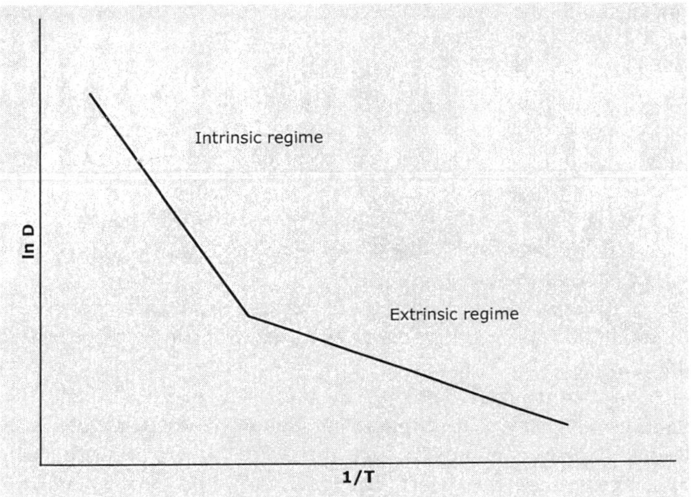

Figure 1. Arrhenius plote showing the extrinsic and imtrinsic regimes.

In the low temperature extrinsic regime, diffusion is characterised by low activation energies. This is because the majority of point defects are generated by impurities so that the activation energy is associated only with migration. As temperature increases, thermally generated defects become dominant and the activation energy becomes higher as it must also include the defect formation energy. The position of the "knee" which marks the transition from extrinsic to intrinsic behaviour is difficult to predict so that the extrapolation of diffusion data beyond the range of experiment can be problematic [12], as can comparison with computed values of activation energy.

The mechanism for bulk diffusion in solids depends on crystal structure, the type of defects present, concentration of impurities etc.. In a close-packed solid, we would expect to vacancies to be the dominant defects present, and thus diffusion is likely to occur via a vacancy mechanism. More open structures can easily accommodate interstitial atoms so that an interstitial mechanism is preferred. In practice, more complex mechanisms such as intersticialcy or ring exchange may dominate where there is cooperative motion involving more than one atom. Finding direct evidence of the diffusion mechanism from experiment is difficult although it can be obtained by isotope-mass-effect methods [13]. This involves the simultaneous diffusion of two isotopes of different masses from which a correlation factor is deduced whose value is associated with a particular mechanism [13].

Computer simulation methods offer an effective alternative to experiment, and can be used to evaluate the most stable defect species and to determine the most likely diffusion mechanism in a range of materials.

3. Methods

In this section, classical atomistic and embedded cluster approached to modelling defects will be described. Other techniques such as those using Hartree-Fock (HF) and Density Functional Theory (DFT) are covered elsewhere in this volume.

3.1 Atomistic methods

The atomistic simulation method is based on the Born model of solids where interatomic potential functions are defined to model the long- and short-range forces acting between atoms or ions in the solid. The effects of oxygen ion polarizibility on the system are included by use of the shell model [14] while directionality of the bonding is described by three-body and four-body terms. The variable potential parameters are derived by fitting to experimental data such as structure, elastic constants and vibrational spectra [15]. Simulations are carried out using standard energy minimization schemes [15] in which the energy of the system is calculated with respect to all atomic coordinates. Thus the equilibrium positions of the ions are evaluated by minimising the lattice energy until all forces acting on the crystal are removed.

In ionic and semi-ionic materials, defects are charged species and cause long-range disruption in the crystal lattice. These long-range perturbations can be effectively modelled using the approach developed by Mott and Littleton [16] where the crystal is divided into two, concentric spherical regions, R1 and R2, as illustrated in Figure 2a. In R1, which contains the defect at its centre, an explicit simulation is carried out to adjust the coordinates of all ions in the region until they are at positions where no forces are acting on them; i.e. they are relaxed around the defect. The radius of R2 is selected so that the forces within this super sphere are relatively weak and the relaxation can be treated essentially as the polarization response to the effective charge of the defect. An interfacial region (R2A) is introduced to deal with interactions between the two regions. The resulting defect energy is then a measure of the perturbation by the defect on the static lattice energy of the crystal.

In this approach, both the zero point energy and the effects of entropy are neglected. However, it has the advantage that single charged defects or small defect clusters can be considered in isolation so as to mimic infinitely dilute concentrations.

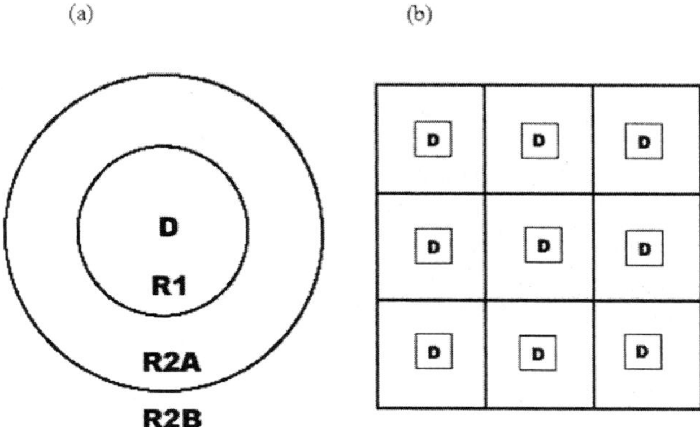

Figure 2. Modelling defect energies using (a) the Mott-Littleton and (b) the supercell methods.

An alternative approach is to use a supercell (SC) method where the defect is part of a periodic 3D repeat unit (Figure 2b). If single defects or low defect concentrations are to be considered then the supercell must be sufficiently large so as to minimize defect-defect interactions. In practice, it is often prohibitively expensive in terms of computational resources to use the SC method and where defect-defect interactions are important the Mott-Littleton (ML) method is more appropriate. However, where high concentrations of defects interacting in a periodic way are to be modelled, then the SC method is more applicable. The advantage of the SC method is that algorithms for the calculation of free energies can be employed [15] thus including the effects of entropy on the defect energy. In addition, the influence of defects on cell volume can be assessed.

The methodology described above can also be used to model atomic migration. For vacancy migration, two vacancies are created: the first is the initial site of the migrating atom and the second the destination site. The potential energy surface between the two sites can be mapped out by fixing the migrating species at different points along the migration pathway and carrying out a relaxation at each point. The highest point on the lowest energy pathway is then the saddle point. Within codes such as GULP [15,17], transition state searching algorithms can be used to automate much of this process. The migration or jump energy (E_{mig}) is then given by:

$$E_{mig} = E_{sp} - E_{in} \tag{4}$$

where E_{sp} is the energy at the saddle point and E_{in} the energy of the initial configuration with the migrating ion at its own site and one vacancy. E_{mig} corresponds to the extrinsic diffusion energy, where the majority of vacancies are formed by impurities. At high temperatures, or in very pure crystals, the majority of defects are formed thermally rather than in response to impurities, provided that formation energies are sufficiently low. In this case to determine the intrinsic diffusion energy, it is necessary to add a fraction of the defect formation energy (n) to the migration energy [12]. Thus Equation (4) becomes:

$$\boldsymbol{E_{dif} = E_{mig} + (E_{def} / n)} \qquad (5)$$

The same approach can be also be used for other diffusion mechanisms involving interstitial species. If we wish to calculate the absolute diffusion coefficient, then an alternative treatment is required by which the pre-exponential factor in Equation 3 can be determined. D_0 can be obtained from reaction rate theory, Greens functions, or dynamical theory. One of the most common approaches uses Vinyard reaction rate theory [18] to calculate the jump frequency of the diffusing species.

Atomistic modelling can therefore provide important information on the nature of defects and their migration in complex minerals. However, classical simulations have the inherent limitations of the potential model, most importantly the fixed ionic charges that allow no redistribution of the electronic charge during defect formation. In many cases, a full description can only be obtained by using quantum mechanical (QM) based methods such as those described in the next section.

3.2 Embedded cluster methods

The embedded cluster method[9,20] uses a nano-cluster (Figure 3), which is divided into regions in an analogous manner to the Mott-Littleton method described above. The central region contains the defect of interest and is treated at the quantum mechanical level of theory. This QM region is normally terminated with atoms described by effective core

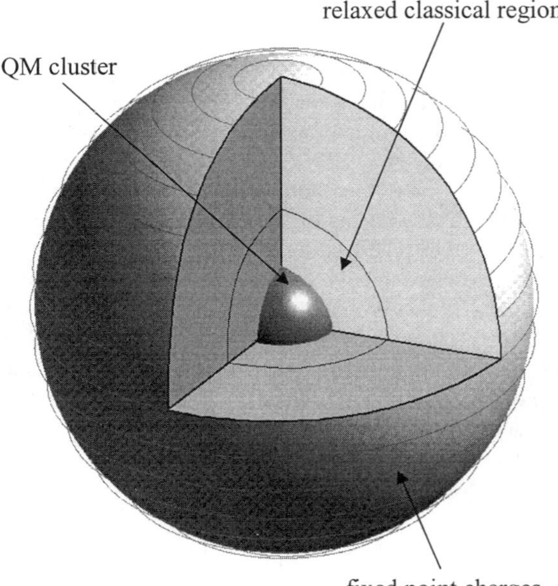

Figure 3. Schematic representation of the nano-cluster used in embedded cluster calculations

pseudopotentials and is embedded in a large (>50Å) array of point charges which represent the bulk crystal. Between the two, is a sphere of MM atoms that are described by interatomic potentials. The inner QM and outer MM atoms are fully relaxed in the course of calculations, while all the point charges in the outer region are kept fixed at their ideal crystalline positions.

The total energy is calculated as a sum of several contributions. The energy of the QM cluster in the electrostatic potential due to the rest of the nano-cluster is calculated using either Hartree-Fock or Density Functional Theory (see other chapters in this volume). Classical ions interact with each other electrostatically and via short-range parameterised potentials as described above. Similar short-range interactions are introduced between the QM cluster atoms and the classical atoms in region I. A correction to the total energy of charged defect species needs to be applied because, in the embedded cluster model, polarisation of the lattice due to the defect charge only occurs within the QM cluster and the relaxed classical region. However, the polarisation of ions at greater distances has a significant contribution to the total energy of the defect, which can be estimated by carrying out a Mott-Littleton calculation with a region 1 size that is the same as the size used in the embedded cluster calculations.

4. Computational studies of defects and diffusion in minerals

4.1 Modelling H defects in minerals

The presence of water and hydrogen containing defects in oxide and silicate minerals is known to have a profound influence on both their chemical and physical properties, Examples include hydrolytic weakening in quartz [20], the role of Brønsted acidic hydroxyl groups in catalytic zeolites [21], and solid state proton conduction in perovskite-type oxides [22]. The presence of water also leads to an increase of oxygen diffusion rates and lowering of activation energies in materials such as feldspars and quartz [23,24].

4.1.1 OH defects in olivine

Olivine $(MgFe)_2SiO_4$ is the dominant mineral in the Earth's Upper Manntle and as such is likely to control the rheology of this section of the Earth's interior. The structure of forsterite, the Mg end member, is illustrated in Figure 4. It has been known for some time that olivine and the other Upper Mantle minerals could contain substantial amounts of water in the form of H defects incorporated into the host lattice. [25-28]. IR spectroscopy has been used to quantify the amounts of water present and to identify the type of defects associated with H incorporation [27-29]. However, due to the complex nature of the spectra, it has not been possible to interpret these data unambiguously. Atomistic simulations of OH defects in forsterite [30], suggested that the dominant defect would be associated with vacancies as the Mg1 site (Figure 4), but also pointed out that if Fe^{3+} were present in the systems, then interstital OH groups could be formed very easily.

Braithwaite et al [8,31] have been able to model H defects in pure forsterite and to calculate vibrational frequencies associated with each configuration, using embedded cluster methods. They use the GUESS code [20], with a 30Å radius nano-cluster in which the central QM region is described using HF theoty and localised basis sets. The nano-cluster is constructed from charge neutral structural units, $[Si(O^{(Si)})_4]$ and $[Mg(O^{(Mg)})_6]$, that represent the local coordination around the Mg and Si atoms so that multipole interactions are minimised.

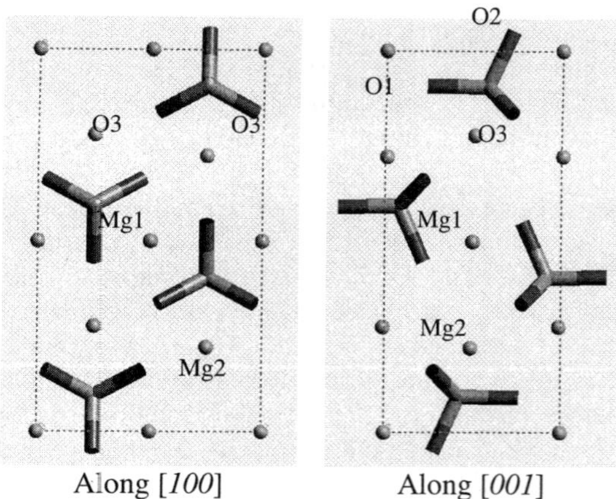

Figure 4. The structure of forsterite viewed along the [100] and [001] directions

Specifically, the defects species they examined were, in Kröger-Vink defect notation, $V_{Mg}^{//}$ and $V_{Si}^{////}$, i.e. the Mg^{2+} and the Si^{4+} vacancy, $H_{Mg}^{/}$ in which a single hydroxyl group is next to an Mg vacancy site, $(2H)_{Mg}^{X}$ where the vacancy has two adjacent hydroxyl groups; and the defect species formed by incremental protonation around a Si^{4+} vacancy, from one hydrogen, $H_{Si}^{///}$, through two, $(2H)_{Si}^{//}$, three, $(3H)_{Si}^{/}$, and finally four, $(4H)_{Si}^{X}$. Isolated hydrogen interstitial sites were defined as $H_{O1}^{\bullet}$, $H_{O2}^{\bullet}$, and $H_{O3}^{\bullet}$, which correspond to protonation at oxygens O1, O2, or O3 respectively (see Figure 4). The calculations were performed with the full relaxation of several hundred atoms surrounding the defect site, and optimized to give relaxed geometries and total energies for each of defect species. When there were a number of initial geometries for a particular defect species, Mott-Littleton calculations were used to search for the best starting configuration. The results [8,31] show that Mg1 vacancies will indeed be important sinks for H^+ species, and also, that reactions leading to the formation of a hydrogarnet type defect is also favourable. However, in the pure forsterite end member, isolated OH species will not readily form. The following reactions are predicted to occur:

$$\boldsymbol{Mg_{Mg}^{X} + H_2O_{(g)} \rightarrow (2H)_{Mg}^{X} + MgO_{(s)}} \tag{6}$$

$$\boldsymbol{Si_{Si}^{X} + 2H_2O_{(g)} \rightarrow (4H)_{Si}^{X} + SiO_{2(s)}} \tag{7}$$

with calculated reaction energies of 5(±30) kJmol^{-1} and –7(±30) kJmol^{-1} respectively for equations (6) and (8).

The optimised geometries for the $(2H)_{Mg}^{X}$, H_{O3} and hydrogarnet defects are shown in Figure 5. For the defects around the Mg vacancy, the optimised structure shows that the defect retains the centre of symmetry at the vacancy site. The most stable site for an isolated OH group is formed by protonation at O3, with the H orientated away from the SiO_4 tetrahedra and contains weak hydrogen bonds, at 2.23Å, The O-H distances in the hydrogarnet defect, $(4H)_{Si}^{X}$, are as short as 0.95Å, where crowding around the Si vacancy site means that hydrogen bonds cannot form as easily.

Calculated vibrational frequencies are obtained from the first and second derivates of the total energy with respect to each bond length. Infra-red (IR) absorption frequencies can be used to 'fingerprint' neutral defect species, especially when compared to polarised measurements. A comparison with experimental data suggests that the concentration of the isolated hydroxyl species in pure forsterite is low, as the observed spectra do not contain peaks that exhibit the expected pleochroic behaviour.

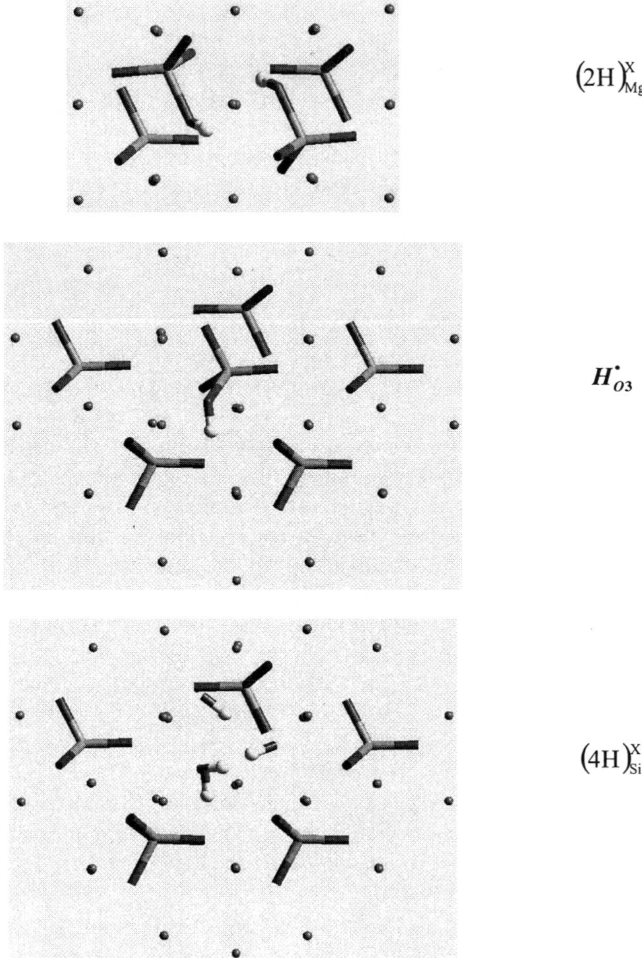

$(2H)_{Mg}^{X}$

$H_{O3}^{\bullet}$

$(4H)_{Si}^{X}$

Figure 5. Optimised geometries from embedded cluster calculations. SiO_4 tetrahedra are shown as sticks, Mg are dark grey and H are shown in white.

The use of embedded cluster methods represents an effective and accurate method for determining the geometries and energetics of complex defects in minerals. The ability to routinely calculate IR frequencies is particularly useful for aiding the interpretation of measured frequencies. In the case of pure forsterite, these methods have shown that H defects will easily be trapped at vacant Si and Mg sites rather than as isolated hydroxyl species [8,31]. This is confirmed by analysis of the energetics of defect incorporation and by comparison of calculated and measured vibrational frequencies.

4.1.2 Oxygen diffusion in silicates

The transport properties of O in silicate minerals have attracted growing interest in recent years because of the importance of the rates of stable isotope exchange, and the control that O diffusion exerts on a number of geological processes. A knowledge of O diffusion coefficients at a range of temperatures, and the resulting activation energies, are critical for an understanding of the kinetics of O isotope exchange between coexisting minerals.

Experimental studies of O diffusion have been performed for a wide range of silicate minerals [see for example 23,24,32,33]. The accumulated body of data suggests that O diffusion in a number of framework silicates under dry conditions is associated with large activation energies (typically 200 - 300 kJ/mol) and slow diffusion rates, whilst diffusion under 'wet', hydrothermal conditions is associated with lower activation energies (~120 kJ/mol) and faster diffusion rates. Such differences are now well established for the feldspar minerals and the diffusion data for quartz shows a similar trend. However, the reasons for these apparent differences are not well understood, although it has been postulated that this phenomena could be due to changes in the migration species, or in the migration mechanism.

Atomistic simulations, using the Mott-Littleton methods [16] have been successfully used to study O and OH diffusion in feldspars [1] and in garnet minerals [34]. Garnet minerals are densely packed with a structure consisting of interconnected octahedra and tetrahedral, which suggests that diffusion will be predominately governed by a vacancy mechanism. Using the approach outlined in 3.1, Wright et al. [34] were able to model O and OH vacancy migration in grossular garnet ($Ca_3Al_2Si_3O_{12}$). Grossular is known to accommodate significant amounts of water where the SiO_4 tetrahedra is replaced by 4(OH) in the hydrogarnet defect. In fact grossular exhibits complete solid solution with its hydrous end member (($Ca_3Al_2(OH)_{12}$). Their results gave calculated migration activation energies of 106 $kJmol^{-1}$ and 383 $kJmol^{-1}$ for O diffusion in the extrinsic and intrinsic regimes respectively and is independent of direction. This value is the same for OH diffusion also, although when hydrogarnet defects are introduced, then the OH extrinsic activation energy is reduced to 67 $kJmol^{-1}$.

Preliminary experimental data for O diffusion in grossular garnet at 850°C and 1050°C gave activation energies of 102 $kJmol^{-1}$ although the results were somewhat ambiguous [35], so that the agreement between experiment and calculation may be fortuitous. The calculations do show, however, that Frenkel defect energies are too high to be the dominant form of point defect and that small amounts of water in the structure are unlikely to influence O diffusion activation energies. The lower migration energy found for hydrous grossular probably relates to the increase in cell size with H_2O content, making movement easier on electrostatic grounds.

Much more data are available for "wet" and "dry" diffusion in the feldspar group of minerals where the lower activation energy and faster diffusion rates for "wet' experiments are well documented [13,23,24]. Wright et al. [1], again using atomistic techniques, carried out a series of calculations on O diffusion in Albite ($NaAlSi_3O_8$) to determine the mechanism of diffusion, the manner of H incorporation in the structure, and to identify the most likely diffusing species under hydrous conditions.

Unlike grossular, albite has an open framework structure based on four-membered rings of SiO_4 tetrahedra that are linked into chains parallel to the *a* axis. One fourth of the tetrahedral sites will be occupied by Al which is neutralised by the incorporation of Na into available voids. This open structure favours Frenkel rather than Schottky defect formation. H can be incorporated into the structure either as interstitial water molecules (E=87 $kJmol^{-1}$) or OH groups(E=70 $kJmol^{-1}$). The computed activation energies for extrinsic and extrinsic migration of O and OH are given in Table 1 and show a distinct diffusional anisotropy as well as a drop in activation energy for OH diffusion. The most favourable migration direction is parallet to *a.* The experimentally determined activation energy for albite under hydrous conditions measured parallel to *c* is 89 $kJmol^{-1}$, which agrees well with the calculated value of 92 $kJmol^{-1}$. Data for other feldspar minerals including the K-feldspar sanidine (see [1 and 24] for details) give activation energies of 109 and 259 $kJmol^{-1}$ for wet and dry O diffusion parallel to *c*, also in excellent agreement with the calculated values.

Table 1. Calculated migration energies ($kJ\ mol^{-1}$) for O and OH interstitial migration in albite.

Migration direction	$E_{mig}\ O^{2-}$	$E_{mig}\ (OH)^-$
a	83	48
b	251	126
c	178	92

The two examples above illustrate the ways in which atomistic computer modelling thechniques can be used to complement experimental observations by providing information on the mechanisms of diffusion; on the species involved, as well on difusional anisotropy.

4.2 Cation ordering in dolomite

Dolomite ($CaMg(CO_3)_2$) is a common carbonate mineral which is found in great abundance in the geological past. However, dolomite formation is almost non-existent in modern day marine environments and is extremely difficult to synthesise under room temperature laboratory conditions. The reasons for this are unknown and hence the formation of ancient dolomite is an area of intense debate. Dolomites (Figure 6) often contain impurities such as Mn, Fe and other divalent cations, which replace either Mg or Ca in the lattice. The site distribution of these impurities is dependent on external factors

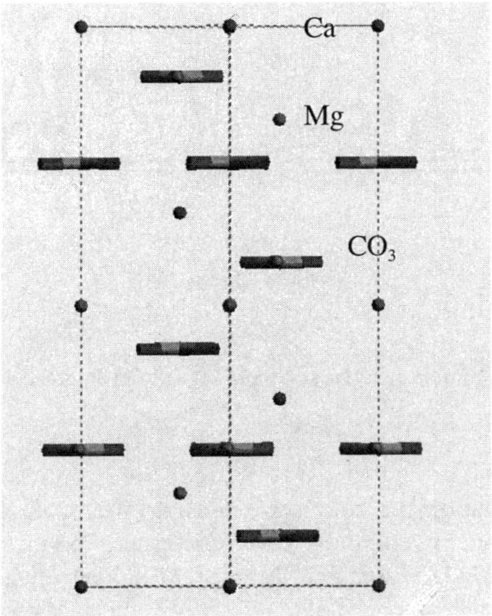

Figure 6. The structure of dolomite.

prevalent at the time of formation such as temperature, pH and fluid composition [36]; and in the case of Mn impurities, the degree of non-stoiciometry and disorder is also thought to influence site distribution [36]. Non-stoichiometry due to Ca enrichment in dolomite is common and characterised by the presence of domains, which differ from the ideal ordered dolomite matrix [37]. Ca-rich dolomites are also considered to be more reactive and less stable than those with an ideal Ca:Mg ratio [38] and thus the presence of Ca excess in dolomite may be a factor in determining the site distribution and concentration of impurities. Understanding such relations could provide clues to the conditions of formation of dolomite in the geological past.

Calculations of the structure of Ca excess dolomite and impurity defect ordering [39] have used both Mott-Littleton and supercell approaches. The results show that Ca excess is most likely incorporated into the structure as stacking fault domains, where the Mg/Ca exchange energy if of the order of 36 $kJmol^{-1}$. This is in good agreement with calorimetric experiments [40], which suggest an enthalpy of Ca/Mg disordering of 35 $kJmol^{-1}$ in dolomite. Results of substitutions indicate a strong preference of impurity elements for Ca rather than Mg sites in stoichiometric dolomite, with substitution energy increasing with increasing ionic radii as shown if Figure 7. However, this is the opposite of what has been inferred from experiment [36,38] where most impurities are found at Mg sites. This has been interpreted as showing that the site preference for impurities will be influenced during growth of dolomite so that impurities get "pinned" at surface Mg sites and are prohibited from exchanging with Ca in the bulk due to kinetic factors [39].

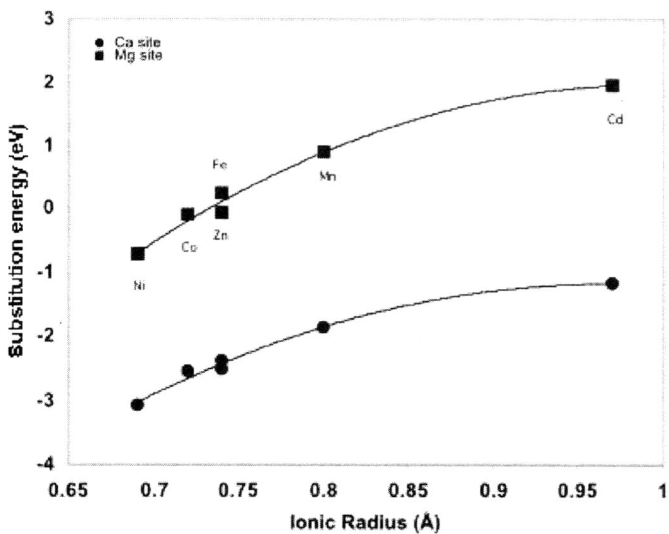

Figure 7. Impurity substitution energies at Ca and Mg sites in dolomite as a function of ionic radius

5. Conclusions

Computer simulation methods can be used to investigate a wide range of defect properties and processes in complex Earth materials. When used in conjunction with experimental observations, they enable us to probe defect phenomena at the microscopic level and enhance our understanding of a variety of important processes.

6. References

[1] K. Wright *et al.*, Water Related Defects and Oxygen Diffusion in Albite. *Contrib. Miner. Petrol.* **122** (1986) 161-166.

[2] N.L. Vocadlo *et al.*, Absolute Ionic Diffusion in MgO – Computer Calculations via Lattice Dynamics. *Phys. Earth Planetary Interiors* **88** 193-210.

[3] K. Wright and R.A. Jackson. Computer Simulation of the Structure and Defect Properties of Zinc Sulphide. *J. Materials Chemistry* **5** (195) 2037-2040.

[4] D.K. Fisler *et al.*, A Shell Model for the Simulation of Rhombohedral Carbonate Minerals and Their Point Defects. *American Mineral.* **85** (200) 217-224.

[5] F. Azough *et al.*, A Computer Simulation Study of Point Defects in Diopside and the Self Diffusion of Mg and Ca. *Mineral. Mag.* **62** (1998) 599-606.

[6] J. Brodholt, Ab initio Calculations on Point Defects in Forsterite (Mg_2SiO_4) and Implications for Diffusion and Creep. *American Mineral.* **82** (1997) 1049-1053.

[7] J.P.Brodholt and K. Refson, An Ab initio Study of Hydrogen in Forsterite and a Possible Mechanism for Hydrolytic Weakening. *J. Geophys. Res. – Solid Earth.* **105** (2000) 18977-18982.

[8] S. Braithwaite *et al.*, Hydrogen Defects in Forsterite: A Test Case for the Embedded Cluster Method. *J. Chem.Phys.* **116** (2002) 2628-2635.

[9] P.V. Sushko *et al.*, Embedded Cluster Calculations of Metal Complex Impurity Defects: Properties of the Iron Cyanide in NaCl. *J. Phys. Condens. Matter* **12** (2000) 8257-8266.

[10] R.J.D. Tilley Defect Crystal Chemistry and its Applications. Blackie, Glasgow,(1987)

[11] J. Crank,. The Mathematics of Diffusion. Oxfoird University Press, (1975)

[12] A. Lasaga The Atomistic Basis of Kinetics: Defects in Minerals In: A. Lasaga and R.J. Kirkpatrick (ed.) Kinetics of Geochemical Processes. Reviews in Mineralogy Volume 8. (1981)

[13] R. Freer *et al.*, Oxygen Diffusion in Sanidine Feldspar and a Critical Appraisal of Oxygen Isotope-Mass-Effect Measurements in Non-Cubic Materials. *Phil. Mag.A* **75**(2) (1997) 485-503.

[14] B.G. Dick and A.W Overhauser, Theory of the Dielectric Constants of Alkali Halide Crystals. *Phys. Rev.* **112** (1956) 90-103.

[15] J.D. Gale, Simulating the Crystal Structures and Properties of Ionic Materials from Interatomic Potentials. In: Molecular Modelling Theory: Applications in the Geosciences. Reviews in Mineralogy Volume 42, (2001) pp 37-59.

[16] N.F. Mott and M.J. Littleton, Conduction in Polar Crystlas.I Electrolytic Conduction in Solid Salts. *J. Chem. Soc. Faraday Trans.* **34** (1938) 485-499.

[17] J.D. Gale, GULP:A Computer Program for the Symmetry Adapted Simulation of Solids. *J. Chem. Soc. Faraday Trans.* **93** (1997) C29-637.

[18] G.H. Vinyard, Frequency Factors and Isotope Effects in Solid State Processes. *J.Phys Chem Solids* **3** (1957) 121-127.

[19] P.V. Sushko *et al.*, Embedded Cluster Approach: Applications to Complex Defects. *Radiation EffectsDefects Solids* **151** (1999) 215-221.

[20] D. Griggs, Hydrolytic Weakening of Quartz and Other Silicates. *Geophys. J. Royal Astron. Soc.* **14** (1967) 19-31.

[21] A.A. Sokol *et al.*, Defect Centres in Microporous Aluminium Silicate Materials. *J. Phys. Chem. B* **102** (1998) 10647-10649

[22] R. Glockner *et al.*, Protons and Other Defects in $BaCeO_3$ – A computational Study. *Solid State Ionics* **122** (1999) 145-156.

[23] S.C Elphick *et al.*, An Ion Microprobe Study of Anhydrous Oxygen Diffusion in Anorthite: A Comparison with Hydrothermal Data and Some Geological Implications. *Contrib. Miner. Petrol.* **100** (1988) 490-495

[24] D. Derdau *et al.*, Oxygen Diffusion in Anhydrous Sanidine Feldspar. *Contrib. Miner. Petrol.* **133** (1998) 199-204

[25] D.R. Bell and G.R. Rossman Water in the Earth's Mantle: The Role of Nominally Anhydrous Minerals. *Science* **255** (1992) 1391-1397

[26] J. Ingrin and H. Skogby Hydrogen in Nominally Anhydrous Upper-Mantle Minerals: Concentration Levels and Implications. *Eu. J. Mineral.* **12** (2000) 543-570

[27] Q. Bai and D.L. Kohlstedt Effects of Chemical Environment on the Solubility and Incorporation Mechanism for Hydrogen in Olivine. *Phys. Chem. Mineral.* **19** (1993) 460-471

[28] A. Beran The Contribution of IR Spectroscopy to the Problem of Water in the Earth's Mantle. In: K.Wright and R. Catlow (eds.) Microscopic Properties and Processes in Minerals. NATO Science Series C, Kluwer, Dordrecht 1999, vol 534 pp 523-538

[29] A. Beran and E. Libowitzky IR Spectroscopy and Hydrogen Bonding in Minerals. In: K.Wright and R. Catlow (eds.) Microscopic Properties and Processes in Minerals. NATO Science Series C, Kluwer,

Dordrecht 1999, vol 534 pp 493-508.

[30] K. Wright and C.R.A. Catlow A Computer Simulation Study of OH Defects in Olivine. *Phys. Chem Minerals* **20** (1994) 515-518.

[31] J.S. Braithwaite *et al.*, Embedded Cluster Calculation of Hydrogen Incorporation in Forsterite. *J. Geophys. Res. – Solid Earth* (2002) Submitted

[32] R. Freer Diffusion in Silicate Minerals and Glasses: A Data Digest and Guide to the Literature. *Contrib. Mineral. Petrol.* **76** (1981) 440-454.

[33] J.P. Brady, Diffusion Data for Silicate Minerals, Glasses, and Liquids. In: T.J. Ahrens (ed.) Mineral Physics and Crystallography Handbook of Physical Constants, AGU, Washington, 1995, pp 269-290

[34] K. Wright *et al.*, Oxygen Diffusion in Grossular and some Geological Implications. *American Mineralogist.* **80** (1995)1020-1025

[35] R. Freer and P.F. Dennis, Oxygen Diffusion Studies 1. A Preliminary Ion Microprobe Injvestigation of Oxygen Diffusion in Some Rock-Forming Minerals. *Mineral. Mag.* **45** (1982) 179-192.

[36] D.N Lumsden *et al.*, Mineralogy and Mn geochemistry of laboratory synthesized dolomite. *Geochim. Cosmochim. Acta* **53** (1989) 2325-2329.

[37] R.J. Reeder, R.J. Carbonates: Growth and alteration microstructures. *Rev. in Mineralogy* **27** (1992) 381-424.

[38] R.J. Reeder, Constraints on cation order in calcium-rich sedimantary dolomite. *Aquat. Geochem.* **6** (2000) 213-226.

[39] K. Wright *et al.*, Impurities and Nonstoichiometry in the Bulk and on the (1014) Surface of Dolomite. *Geochim. Cosmochim. Acta.*, **66** (2002) 2541-2546.

[40] A. Navrotsky *et al.*,.Calorimetric studies of the energetics of order-disorder in the system $(Mg_{1-x}Fe_xCa(CO_3)_2$. *American Mineralogist* **84** (1999) 1622-1626.

Computational Materials Science
C.R.A. Catlow and E.A. Kotomin (Eds).
IOS Press, 2003

Molecular Dynamics Simulations of Silicate Glasses

A.N. Cormack
School of Ceramic Engineering and Materials Science
New York State College of Ceramics
Alfred University, Alfred NY USA

Abstract. The application of molecular dynamics to the modelling of silicate glasses is reviewed. Several examples are used to illustrate the value of these computer simulations in advancing our understanding of the structure of polycomponant silicate glasses. In addition to structure, transport propoerties, in particular alkali ion migration, emerge naturally from the simulations. It thus becomes possible to relate transport properties to structure. Some examples are given of alkali migration mechanisms.

1. Introduction

Glasses are of considerable scientific interest, because of their use in a wide range of technologies, as diverse as fibre optics and traditional container glass, and also to mineralogists because of the role they play in mantle processes. They also attract the attention of metallurgists because of their occurrence in refining processes. Glasses are thus not only of intrinsic interest.

However, the application of traditional structural analysis tools to glassy materials is somewhat more difficult than for crystalline materials, leading to incentives to develop other ways of probing the structure of non-crystalline solids. Computational approaches obviously appeal; however, the same problems arising from the lack of translational periodicity also limit the deployment of computational tools. For this reason, until very recently, because of limitations to computer resources (cpu memory and speed), only classical simulations have been possible; of these, molecular dynamics has had the most success. In this chapter, we will discuss the application of molecular dynamics (MD) to the structure of silicate, and related, glasses. To begin with, the technique itself will be reviewed. For a more complete discussion of molecular dynamics (and other statistical mechanical based techniques), the reader is referred to the book by Allen and Tildesley [1].

2. Molecular Dynamics

Molecular dynamics (MD) is a computational approach to modelling of solids, with roots in statistical mechanics, which includes the effects of both temperature and time. It has been used with some considerable success to probe the structures and the dynamics of ion migration in glasses.

In the present context, the objectives of molecular dynamics are two-fold. Firstly, it is to provide an atomic scale picture of the structure of glasses. From this, an understanding of how the composition will affect the structure. Secondly, since the MD technique is inherently time-dependent, it is to provide an insight into the atomic migration mechanisms that operate in the glasses. Since these latter will be determined by the structure, one may also infer how compositional changes will affect the transport of ions in the glasses.

In principle, MD is straightforward: the evolution of an ensemble of atoms is followed as a function of time. The positions and velocities of each of the atoms in the simulation are determined from classical mechanics, using Newton's second law, and forces that are obtained from an appropriate interatomic potential model. For inorganic systems, such as silicate glasses, the Born model of the solid is commonly used: atoms are treated as point charges with short-range forces acting on them. The forces are found from the derivatives of the interatomic potential, which usually has a Buckingham form:

$$V(r_{ij}) = \frac{q_i q_j}{r_{ij}} + A_{ij}\exp\left(-r_{ij}/\rho_{ij}\right) - \frac{C_{ij}}{r_{ij}^6}$$

Here A, ρ and C are parameters that are chosen to provide the optimum description of known physical properties of analogous crystalline systems. In general, one must consider all the (pair) interactions together, in deciding whether the potential model is adequate. The earliest models used formal oxidation state charges; as a result, it was found necessary to include three-body terms in the potential model in order to get the geometry of the SiO_4 tetrahedron correct. This was because the electrostatic interactions between the Si and O and the oxygens were so string that they tended to flatten out the bond angles. Bond-bending interactions were introduced with a θ_0 that "encouraged" the O – Si – O bond angle to be 109.4°. A number of potential models for silica have recently been developed, using quantum mechanical calculations on small clusters of Si and O; the most commonly used nowadays are those due to TTAM and to van Beest et al., the BKS model. These most recent potential models employ partial charges; that is to say that ions do not have their formal charge, e.g. $Si4^+$, but rather a smaller one. The effect of this is to reduce the dominance of the Coulomb interactions in the potential and, consequently, to eliminate the need for three-body terms. The main problem with these models is that because of the partial charges, none of the previously derived potentials for other cations are usable. Thus new Na – O and Ca – O potentials had to be found that were compatible with, say, the BKS model for silica. This is a time consuming exercise, but has been done with reasonable success by the present author's group.

With so much attention being paid in other quarters, it is well to remember that the potential model, irrespective of the numerical values of the charges used, is really a means to evaluating the forces that act on the ions, as the interatomic separations vary with time. The model itself can say nothing about the degrees of ionicity or covalency of the interatomic bonds. It is sometimes tempting to read too much into the physics apparently represented b y the analytical form of, especially, the short-range components of the potential model; this to be avoided.

The equations of motion are solved in an iterative process: the positions and velocities are updated in small increments of time, known as timesteps. Each timestep corresponds to a Δt of the order of 10^{-15} s. The structure of the glass is obtained by averaging over a large number of timesteps, of the order of thousands. Newton's Laws assume conservation of energy and hence use of the NVE statistical mechanical ensemble is implicit. There are a number of algorithms available for integrating the equations of motion, those due to Verlet and Beeman being the most common. These take into account the fact that the forces acting on the ions will change as the positions of the atoms change.

Most applications of molecular dynamics in which we are interested actually require control of temperature, or use of the NVT ensemble. Temperature control is accomplished conceptually by coupling the system to a heat bath. The temperature is controlled using a thermostat by which the velocities of the ions are managed. The kinetic energy distribution of the ions is indicate by the temperature of the thermostat; essentially the velocities are adjusted, either directly or indirectly, to give a Gaussian distribution of kinetic energies centered on the appropriate temperature.

There are, naturally, some technical issues that complicate the apparently simplicity of MD simulations. The most important one concerns the interatomic potential model (from which the forces entering into the equations of motion are obtained), on which rests the validity of the simulation. A variety of means of determining the potentials is possible [2]; a detailed discussion is outside the scope of this paper. In our work, the potential model is imported from other, crystalline, systems. These are usually the binary oxides, which comprise the components of the glass. This assumes that the potentials are transferable; although this is an approximation, experience suggests that it is an acceptable one. However, more detailed studies may require more accurate potentials, such as the inclusion of ion polarisability. Our experience is that factors other than the accuracy of the potential model play a more significant role in limiting the agreement between the simulated structural data and that obtained experimentally.

One consequence of the limited accuracy of the potential model centers on the Boltzmann relation between energy and temperature: $\Delta E = k_b \Delta T$. In this case, uncertainty in the energy of the simulation leads to an uncertainty in the absolute temperature of the simulation. For a 1,500 atom simulation, the total energy of the simulation box will be of the order of 20,000 eV. Given numerical precision limitations (even to 7 or 8 significant figures), relatively small errors or uncertainties in energy result in large uncertainties in the absolute temperature: $0.025\text{eV} = 300k_b$. Absolute, quantitative temperature dependencies thus may not be possible.

Two other facets of the simulation govern the effective temperature of the simulation. The first relates to the size of the simulation, which is relatively small. Simulation boxes typically contain around 1500 atoms, although they may range from a low of around 400 to 10,000 or more. Only the available computational resources limit the size: Vashista and colleagues [3] have run simulations containing more than 1 million atoms, for example. Nowadays, even using desktop workstations, it should not be necessary to run simulations using less than about 1,500 atoms. The temperature of the simulation is affected by the ability of the atoms in the simulation box to reach their equilibrium configuration.

If equilibrium cannot be achieved then the effective internal (or fictive) temperature will be higher. The periodic boundary conditions used to create an infinite solid will prevent the structure from reaching complete equilibrium, because of the need for structural coherence across the faces of the simulation box. Although larger simulation sizes would seem to be the way around this limitation, it is not yet completely clear how the effective temperature depends on the size of the simulation. Larger simulation sizes would also appear to need longer simulation runs, but again information on this topic is scant.

The second, and perhaps more obvious, aspect of the simulation which influences the effective temperature concerns glass formation procedures, that is, how the glass is created in the computer. The key property here is the so-called "cooling rate", that is how the thermal energy is removed from the glass. Many procedures, which involve taking a simulation from a high temperature, in which the system is in a molten, highly disordered, state to a temperature below the glass transition temperature, do so in stages. Others adopt a nominally continuous procedure. However, even in the case of very slow rates of cooling, only nominal rates of 1K/ps can be practically achieved. These are considerably faster than actual physical rates experienced in the laboratory (or the field) and would appear to imply very high fictive temperatures (or the order of thousands of Kelvin). However, the direct comparison between the simulated and physical cooling rates is not appropriate. One reason for this is the fact that the simulations are performed on an infinite solid, which is effectively superheated (on melting) or supercooled (on forming the glass). The absence of external surfaces really means that these systems are being subjected to mechanical melting [4], for which the melting temperature is somewhat higher, and not true thermodynamic melting. Notwithstanding this issue, the effect of different cooling rates on the simulated structure of silica has been examined by Vollmeyr et al. [5], as well as the present author's group. Small differences in structures obtained with the different cooling rates were observed. However, how these differences relate to differences in density was not clearly established.

Comparison of calculated Q^n distributions for a series of sodium silicate glasses, with those obtained from both annealed and rapidly quenched (splat-cooled) glasses suggest that the simulated glasses have internal temperatures that are only about 200K higher than the target room temperature. The calculated Q^n distribution is very close to that of the splat-cooled glasses indicating that they have the same effective internal temperature. (See Figure 2)

3. Basic Glass Properties

Glass properties may be divided into two principal categories, focusing on either structural or transport-related aspects. One of the roles of MD is to relate the microscopic properties resulting from the simulations to the macroscopic properties that are probed experimentally. Since the range of information available from the simulations is much greater than is available from experiment, points of contact between theory and experiment provide a means of validating the simulations [6] so that one has some confidence in the additional details that are available from the simulations.

3.1 Structure

There have been a number of proposals for the structure of glasses, the most widely accepted being based on Zachariasen's random network model for silica. An alternate view, now largely discounted because of conflicts with experimental data, suggested that glasses were comprised of micro-crystallites which were themselves randomly arranged in space. This idea still persists because some aspects of diffraction data, namely the first sharp diffraction peak, indicate a certain amount of short-range order which could, possibly, be interpreted in terms of micro-crystallites. However, more detailed analyses, and other data are not consistent with this picture.

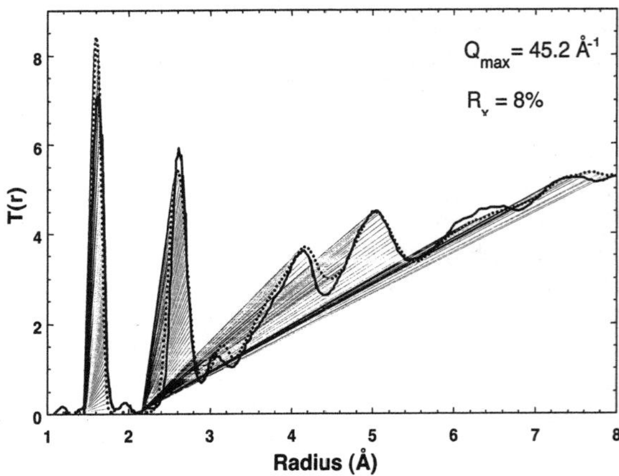

Figure 1. Comparison between experimental (solid) and calculated (dashed) T(r) for vitreous silica.

Whilst Zachariasen's model is generally accepted for silica (and vitreous B_2O_3), how the picture changes when the network is disrupted by the addition of network modifiers, such as alkali or alkaline earth ions, is not so clear. On the one hand, one might postulate that the alkalis would be randomly, but homogeneously, distributed throughout the glass, along with their associated non-bridging oxygen ions [7]. On the other hand, however, is the suggestion that the alkalis (and NBO) are spatially aggregated, so that

NBO ions are shared between more than one alkali ion, since this occurs in comparable crystalline compounds. This picture is represented by the Modified Random Network Model proposed by Greaves [8].

There are basically four ways of analyzing the structural information provided by an MD simulation. Some of these provide for a comparison with experimental data whereas others supplement the insight available from interpreting that data.

The macroscopic description of the structure is embodied in the radial distribution functions. These allow, in principle, a comparison with experiment, since they can be measured directly. The most popular way of measuring them nowadays is neutron scattering [9]. From the scattered intensity, the total distribution function, T(r), is obtained. T(r) is also the sum of the individual pair distribution functions. If the experimental T(r) could be deconvoluted into the individual pdfs, then we could have a reasonably detailed idea of the structure of the glass. Unfortunately, this cannot be done, for a couple of reasons. Firstly, the scattering intensity is isotropic and so T(r) is one dimensional. Secondly, measurements may only be made over a finite region of reciprocal space, thus broadening the experimental peaks and making deconvolution even harder. On the other hand, the pdfs are available from the simulations, so T(r) can be calculated. Concordance between measured and calculated T(r) implies that the simulated structure reflects the details of the actual glass structure. One should bear in mind, however, that there may not be a unique structure giving rise to a particular T(r).

How well the current models reproduce the measured T(r) may be gauged from Figure 1 which compares, for vitreous silica, the latest data of Wright with T(r) calculated from a simulation by Yuan and Cormack [10], using a slightly modified BKS potential. The Wright reliability index [11] is 8%, which is amongst the lowest obtained so far. One also needs to bear in mind that calculated structures have to be quite poor before substantial discrepancies appear between calculated and experimental T(r), largely for the same reasons that make it difficult to deconvolute experimental distribution functions, i.e. their one dimensional nature and peak broadening. The second structural property, which can also be measured (by solid state NMR), focuses more on local structure, is the Q^n distribution, which describes the distribution of non-bridging oxygen ions over the tetrahedra. Here n represents the number of bridging oxygen ions on a tetrahedron, so, for example, in silica, all the tetrahedra are Q^4 species, because each tetrahedron is connected to four other tetrahedra through bridging oxygens. When the network is modified, however, through the introduction of alkali or alkaline earth cations, non-bridging oxygen ions are created, and so, consequently, will be Q^3 species. This presupposes that, because the number of modifier cations is less than the number of silicon ions, that only some of the tetrahedra will have one NBO. Of course, if the distribution of modifiers and NBO is not homogeneous, so that the modifiers are co-ordinated to more than one NBO in their nearest co-ordination shell,

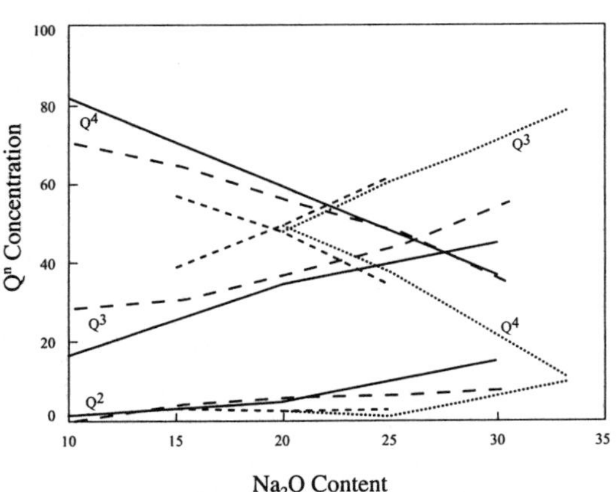

Figure 2. Q^n concentration as a function of soda content for sodium silicate glasses. The solid lines are from MD simulations and the other data are from experiment: (-----) Olivier et al [12]; (- - - -) from Maekawa et al. [13] and (- - - - -) from Buckerman et al. [14]

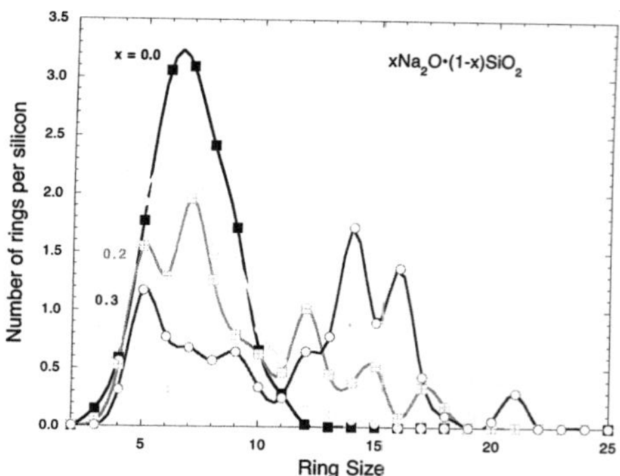

Figure 3. Tetrahedral ring size distribution in sodium silicate glasses

then the possibility exists for Q^2 species. Indeed, as the concentration of modifier cations increases, then so will the probability that Q^2 or even Q^1 or Q^0 species will be created. Figure 2 compares, for a series of sodium silicate glasses, the Q^n distribution calculated from MD simulations with some experimental data. Included are results from both annealed [12,13] and rapidly cooled glasses [14]. It can be seen that the calculated distribution follows that measured for the rapidly cooled glasses, whose fictive temperature is about 200K higher than for the annealed glasses, based on the quench rates associated with the splat-cooling. More advanced NMR techniques allow one to probe the linkages between the different Qn species, thus giving access to the intermediate Range II structure. Clearly this information is also readily obtained from the simulations. Comparison with this data is a real test for the simulations, but has been attempted [15]. The limited agreement found indicates that the structure of the glass at this level is significantly affected by temperature, that is to say that the different effective internal temperature of the glass is apparent.

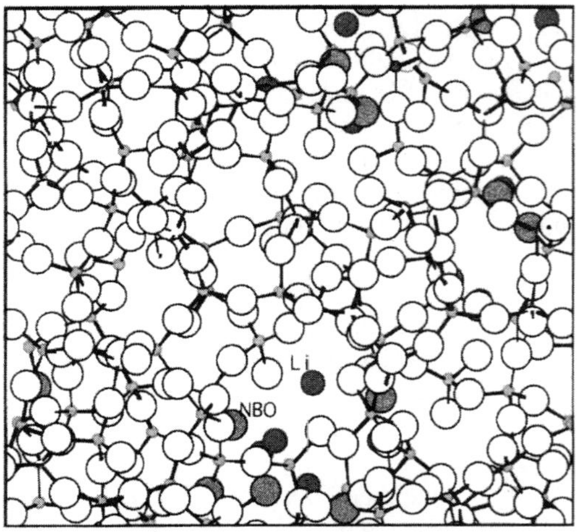

Figure 4. Segregation of Li and NBO in a silicate glass containing 5mol% lithia.

A third way of discussing the network structure of glasses is in terms of the ring-size distribution [16]. In this context, the rings consist of paths which are realized by passing from one tetrahedron to the next through a bridging oxygen ion. A ring is defined as the shortest path that brings one back to one's original starting tetrahedron. In crystalline network-structured silicates, there are rings of only a few discrete sizes. However, in the random network structure of the glass, there is, instead, a distribution of ring sizes.

Yuan and Cormack [16] found, in a series of sodium silicate glasses, that as the sodium concentration increased, the ring size distribution became essentially bimodal, as shown in Figure 3. The number of rings associated with the pure silica network structure (i.e. 4 – 10 member rings) decreased with a corresponding increase in larger sized rings which are associate with the sodium ion distribution. In regions where the sodium has disrupted the network through the introduction of NBO, more tetrahedra must be included in order to create a closed path. Note that it is not really possible to determine ring size distributions experimentally, except for some special small sized rings that may be detected spectroscopically.

The fourth approach to characterizing simulated structures is a visual one. However, in many cases gained is necessarily qualitative and somewhat limited. An early example is shown in Figure 4, in which it can be readily seen that the lithium and NBO ions are aggregated spatially instead of being more homogeneously distributed through the structure. This information, coupled with more quantitative analyses, suggests that the structure of these glasses owes more to the description afforded by the Modified Random network Model, than to simple extensions to Zachariasen's picture.

Increasing the amount of alkali generally causes the alkali/NBO rich regions to get larger. At the kind of alkali loadings used commercially, however, the simple picture is hard to visualise, because the sizes of the alkali-rich and silica rich regions must be rather small. Yuan [17] has developed a way to estimate the sizes of these regions, which turn out to be of the order of 10-15Å, similar in magnitude to the anionic cages proposed for the structure of silicate melts by Bokris and colleagues [18] more than forty years ago.

Now, it is known that the addition of Al reduces the number of NBO in the structure, for the same reasons that NBO are created by alkali addition (as long as the Al acts as a network former, of course). The work of Cao [19] indicates that the addition of Al promotes a Zachariasen like structure: Al prefers Q^4 sites and thus causes a redistribution of the alkali ions. The prediction that Al will occupy Q^4 sites is now being confirmed by O^{17} solid state NMR studies [20]. As the amount of Al increases, so that the ratio of Al to alkali approaches unity, the number of NBO decreases, reaching zero at that ratio. Further addition of Al must lead to a change in co-ordination. Although there has been some suggestion that the Al moves into octahedral sites, this has not been observed in the simulations. Instead, some of the oxygens become three fold co-ordinated, resulting in a change to the network topology. These tri-clusters have been invoked to account for viscosity changes in alumino-silicate melts of geological significance [21].

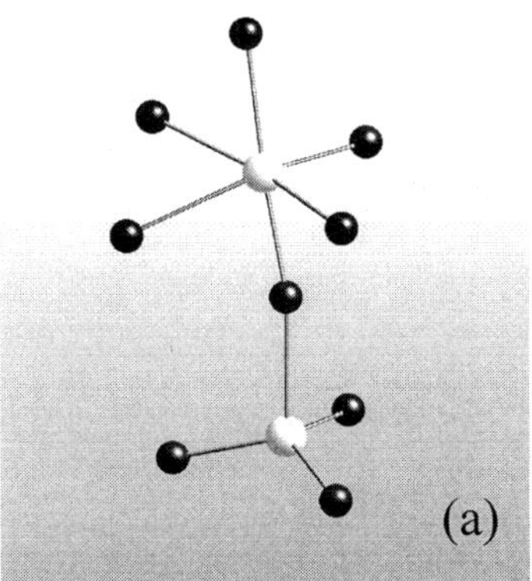

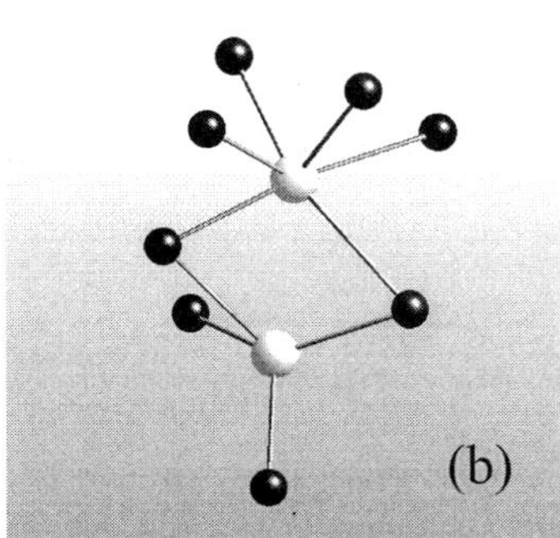

Figure 5. Examples of the linkages between Na-O polyhedra, from a 25mol% sodium silicate glass. (a) corner-shared: one oxygen in common; (b) edge-shared: two oxygens in common.

The addition of Al leads to a re-polymerising of the tetrahedral network structure, making a Zachariasen-like picture more appropriate. LaCourse [22] has introduced the term "transitional structures" to describe the structures of those compositions for which small changes in composition result in large changes in properties. Many commercial glass compositions behave this way, and it is thought that the small compositional changes cause a (significant) change from a modified random network to a Zachariasen like network.

More recently, our attention has been focused on the environment of the network modifier ions, such as the alkalis. Because these species break up, or modify, the tetrahedral network structure, they must be considered an integral part of the overall structure; a complete description of the structure should include descriptions of both the network forming and network modifying components of the glass. In addition, modifiers such as the alkalis are the mobile species, governing properties involving mass and charge transport, which range from chemical durability to ionic conductivity. We have found that in a number of systems, the alkali ion coordination polyhedra are connected through common oxygen ions, in ways which are reminiscent of those encountered in crystalline structures. Some examples of corner-shared and edge-shared Na-O polyhedra, taken from a recent study of a 25mol% sodium silicate glass [23], are given in Figure 5. The connectivity provides for more or less natural pathways for alkali ion migration, as schematically depicted in figure 6.

3.2 Ion Transport

The molecular dynamics technique is, by definition, a dynamic one: that is to say, that time-dependent processes such as diffusion and other mass transport properties emerge naturally from the simulations. Since these processes are governed, ultimately, by the structure of the glass, the simulations are an ideal way to investigate the way in which transport properties are governed by the structure. Here, again, there is a good symbiosis between experiment and theory because, by and large, the former produces macroscopic observations, such as activation energies and magnitudes of diffusivity or conductivity, and the latter provides atomic scale information, such as detailed descriptions of the migration mechanisms.

The compositional dependencies of alkali ion transport in alumino-silicate glasses present a technologically relevant illustration of the insight that MD can provide, since the simulations afford a systematic approach to investigating compositional and, hence, structural influences on ion migration.

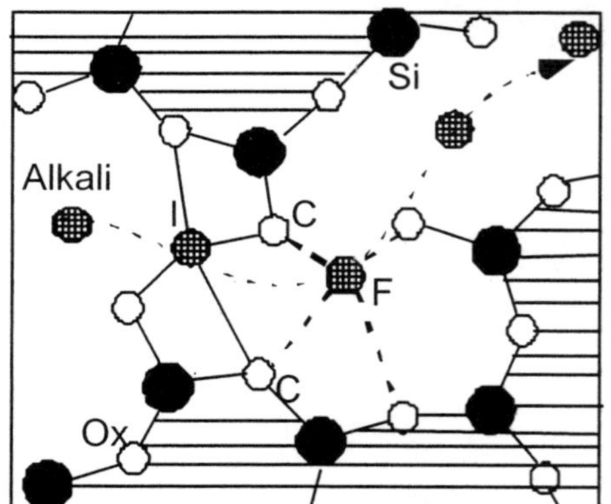

Figure 6. Schematic diagram for ion migration pathways in an alkali silicate glass. I is the initial position of an alkali ion, which jumps to position F after the ion in that site has jumped into an originally vacant site. The arrow indicates the migration path. (Ox = oxygen; Si = Silicon)

When alumina is added to alkali silicate glasses, the overall trend is that these additions decrease the activation energy for electrical conduction, ostensibly because there is a reduction in the number of NBO and so the binding energy between the alkali ion and the network is reduced, freeing the ion for migration. However, the data [24] actually show two extrema, a maximum at about R (= Al/alkali) = 0.2 and a minimum at R = unity. The minimum is understood in terms of a structural change that must occur because further additions of alumina cannot create a negative number of NBO. One argument is that the aluminium ions change co-ordination from tetrahedral to octahedral, although this has not been seen in MD simulations. Rather, there is a change in oxygen co-ordination from two to three, creating so-called tri-clusters [19]. The tri-clusters have been seen experimentally in NMR studies, and have also been used to explain the compositional dependencies of the viscosity of alumino-silicate melts around R = 1.

The maximum, on the other hand, is not expected, although similar behaviour is seen in a number of properties in a number of systems. Our simulations [19] have reproduced this effect, which has been ascribed to a redistribution of alkali ions out of the modifier-rich regions (using the MRN picture) as a result of the fact that Al ions prefer to occupy Q^4 tetrahedral sites and not Q^3 or Q^2 sites. Thus local charge balanced is achieved by having an alkali ion adjacent to the Al (in the Q^4 site). This association leads to an initial increase in the migration activation energy, which only falls when there are sufficient alkali ions to manufacture adequate migration pathways. The preference of Al for Q^4 sites has been observed in NMR experiments.

Probably the most interesting phenomenon involving ion transport is one that has engaged the attention of virtually all glass scientists (and others) at some time or another, namely the Mixed Alkali Effect (MAE). This effect was first observed more than a hundred years ago, but, in spite of a great deal of attention, has still no commonly accepted explanation, even of its most basic aspects. Most suggestions, which have been put forward, invoke an interaction, usually (but not always) between the two different alkali species, which would be absent in the single alkali glasses; the energy associated with this interaction would account for the effect. However, different theories generally propose different kinds of interaction, not all of which are mutually compatible. For example, some propose pairing between like alkali ions whereas some argue for pairing between unlike alkali ions. We will not discus further details of earlier arguments here, but rather will concentrate on what structural and energetic information may be extracted from the simulations.

The general feature of the effect, sometimes also known as the Mixed Mobile Ion Effect, since it is not confined exclusively to alkali ions, but also occurs in systems with non-alkali modifiers (e.g. silver), is as follows. In an alkali containing glass, if some of the alkali is replaced by a different alkali species, then the diffusion coefficients of both species are lowered with respect to the single alkali composition. A remarkable feature of the MAE is its non-linear behaviour. The change in diffusion coefficients, or related property such as ionic conductivity, is highly non-linear, and can be several orders of magnitude. In the alkali silicates, the addition of aluminium also has a marked influence: the deviation from additivity (Vegard's Law) is greater in systems containing Al, and increases with increasing Al content [25].

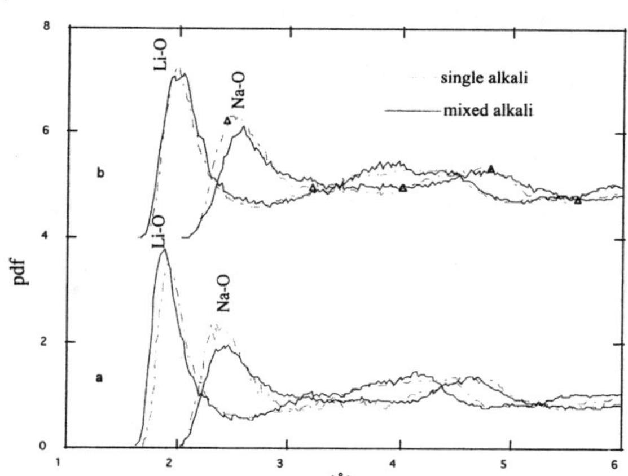

Figure 7. Li-O and Na-O pdfs from single and mixed (a) silicate and (b) alumino-silicate glasses

The MAE offers an excellent example of a compo-sitionally dependent pheno-menon. Not only is there the change in alkali ratio, but there is also the aluminium dependence. Since the transport of alkali ions (or other mobile ions, or, indeed, other modifiers) is ultimately governed by the structure of the glass, a simulated structure, which reproduces the mixed alkali effect, must contain the essential factors underlying the effect. The first simulations to show the mixed alkali effect were those of Huang [26], who also found that there were small, but significant, differences between the alkali ion sites in the single and mixed alkali glasses. Subsequent studies by Balasubramanian and Rao [27] and Smith et al. [28] also reproduced the effect. The structural features underlying the effect may be identified from the simulations. Figure 7 shows the alkali-oxygen pair distribution functions from Na/Li silicate and alumino-silicate glasses [29]. It is observed that the alkali ions retain very similar environments in both single and mixed alkali glasses. Broadly speaking, this means that in the mixed alkali systems, a set of sites becomes inaccessible to one of the alkali species, since the sites occupied by the smaller ions will not be large enough for the larger alkali species. On the other hand, the larger ion sites are not energetically attractive to the smaller ion, a phenomenon familiar to crystal chemists as the "no-rattling" rule [30]. Ion migration will largely be confined to a subset of the total available alkali ion sites in the glass. Thus, for example, Li-Li jumps are much more common than Li-Na jumps. Balasubramanian and Rao [27] reported a similar observation.

At the same time, however, the co-ordination of the alkali ions differs slightly in the mixed alkali systems, compared to those in the single alkali glasses. This is manifested in a change in average site potential. The site potential (similar, say, to the Madelung site potential in crystals) for the smaller ion decreases, in a non-linear manner, as the concentration of the larger ion increases. In mixed alkali silicates, the site potential of the larger species increase slightly, but at a slower rate than the decrease in the site potential of the smaller species. In the alumino-silicates, the site potential of both alkali species decreases. The effect of the decrease in site potential is to increase the activation energy of migration: the alkali ions are more tightly bound to the silicate network.

These observations can be related to ion – ion interactions, which, as we have noted, are a common feature of many theories of the mixed alkali effect. Our results show that the interactions can be considered to act between the different kinds of alkali ions; however, they are not direct interactions as such, but rather are mediated through the oxygen ions in the network. The "excess" interaction energy may be expressed as

$$\varepsilon_{\text{int}} = \varepsilon_{ab} - (\varepsilon_{aa} + \varepsilon_{bb})/2,$$

where ε_{ij} is the energy of the interaction between i and j. The site potential may be used to relate the structural and energetic aspects of the interaction. We may write

$$\varepsilon_{\text{int}} = (V_a - V'_a + V_b - V'_b)/2,$$

where V_i and V'_i are the site potentials of species i in the single and mixed alkali glasses, respectively.

glass	ε_{int} (eV)
experimental: Ag: $Na_2O \cdot 4SiO_2$ (Inman et al.)	-0.063
calculated: $(Na,K)_2O \cdot 3SiO_2$	-0.021
calculated: $(Na,Li)_2O \cdot 3SiO_2$	-0.075
calculated: $(Na,Li)_2O \cdot Al_2O_3 \cdot 2SiO_2$	-0.181

Table 1. Excess interaction energies calculated as described in the text. The experimental data is from Inman et al. (J. Non-Cryst. Solids, 191, 209 (1995))

The values for the interaction energy obtained in this way compare well with those extracted from experimental data, as may be seen from the table. To some extent, the agreement may be serendipitous, but it does allow the interaction between the different alkali species to be understood in terms of how the tetrahedral network packs around the alkalis. In a single alkali silicate, all of the alkalis compete equally for the attention of the network oxygens, resulting in a particular distribution of alkali-oxygen bond lengths. In the mixed alkali glass, the two different species compete differently for the framework oxygens, because their field strengths are not the same (their radii are different). The packing of the coordinating oxygens around the alkalis becomes more efficient, especially for the smaller species. The alkali-oxygen bond length distribution narrows and the site potential becomes more negative. The interaction, which may be considered an alkali-alkali interaction, nevertheless involves ions in the tetrahedral framework and is effectively an indirect sort of interaction, mediated by the network structure.

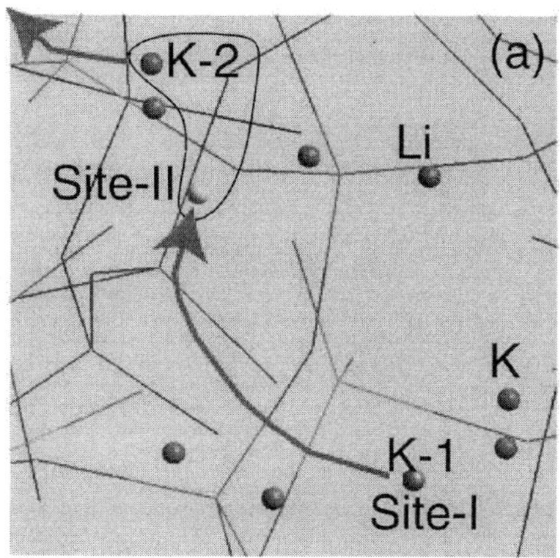

Figure 8. In this migration mechanism, K-1 jumps to site–II, after K-2 jumps to an initially vacant position. Although the positions of K-2 and K-1 in site-II are not exactly the same, they are co-ordinated to the same set of oxygen ions (not shown for clarity). After the jump, site-I is empty.

4. Alkali Migration Mechanisms

The mechanisms by which the alkali ions migrate in the glass structure may also be deduced from a detailed analysis of the simulations. The mechanisms observed are, similar to those found in crystalline materials, being, by and large, 'jump and reside' mechanisms, rather than being fluid-like motion. That is to say that the ions spend considerably more time localized in a well-defined environment, than they do jumping between sites. However, the details are composition-dependent, to the extent that different behaviour is observed for small alkali concentrations (1mol%) than for moderate (25mol%) alkali concentrations. The differences are obviously due to the differences in the structure of the glasses, in particular, in the way in which the alkali content is accommodated.

At moderate alkali loadings, 25 mol%, the migration of the alkali ions is usually correlated. That is to say that isolated, single ion hops, or jumps, are not usually seen. Rather, the jump of one alkali from one site to another involves the concomitant jumps of other ions at the same time. The correlations are apparent in the observation that the alkali ions predominantly jump into sites which were previously occupied by other alkali ions. The earlier mixed alkali glass simulations [29] also suggested that ions jump between well-defined sites, in that case between sites associated with the same type of alkali. The alkalis are found to migrate predominantly within NBO rich regions in the structure. This may be simply understood from the observations regarding the connectivity of alkali ion coordination polyhedra and the fact that the majority of oxygens coordinating to the alkali ions are non-bridging in character. Recent work of Park and Cormack [31], from analyses of mixed Li-K silicates, provided the first detailed description of alkali ion jump mechanisms in an alkali silicate glass.

They monitored the changes in the oxygens which were co-ordinated to particular cations at the beginning and end of the jump process. In one case, involving a potassium ion, shown in figure 8, the cation K-1, in Site-I, was seen to jump into an adjacent, neighbouring site, Site-II, displacing the K ion (K-2) from that site. The displaced K-2 cation jumped into a position that was vacant at that time, and the site from which the K-1 cation moved thus became vacant. The neighbouring sites (Site-I and Site-II) were found to share co-ordinating oxygen ions. Since a vacant site (the one into which K-2 jumped) moved to Site-I, this process can essentially be described as a vacancy mechanism, although the sites are not as ordered as in crystalline materials. In comparison to the usual crystalline vacancy mechanisms, the interesting feature of this mechanism is that it involves the co-operative migration of two cations, so that the net displacement of the "vacancy" is more than a single K – K distance, in contrast to that expected for a vacancy jump in a crystal. We note, however, that in some fast-ion conductors, similar co-operative motions of ions have been observed.

In the same simulations, the direct exchange of two Li ions was observed (see figure 9). Again, it was found that the Li ion sites shared co-ordinating oxygen ions, just as they would in a crystalline structure. Direct exchange mechanisms are not commonly reported in oxide systems, but are considered to be important mechanisms for diffusion in metals.

In the vacancy-like mechanism involving the potassium ion, it was not clear whether the K-1 ion "pushed" the K-2 ion out of its initial site (I), or whether the K-2 ion left its position before K-1 began its jump into the Site-I. A more detailed examination [23] of sodium ion jumps in a 25mol% sodium silicate glass found that the site becomes vacant before it is re-occupied, by the departure of its original occupant, *before* the new occupant begins its jump. That is to say, that the departing ion is not "pushed", but rather leaves of its own accord. It is the vacation of the site that allows the new occupant to begin to move into it. In the sodium silicate study, it was again found that multiple alkali ions were involved in the migration process. It was also argued, again, that the mechanism was more properly described as a vacancy-like mechanism because the sodium ions are an integral part of the structure. Without them, the structure would be different, so they are not interstitial species and cannot behave like them.

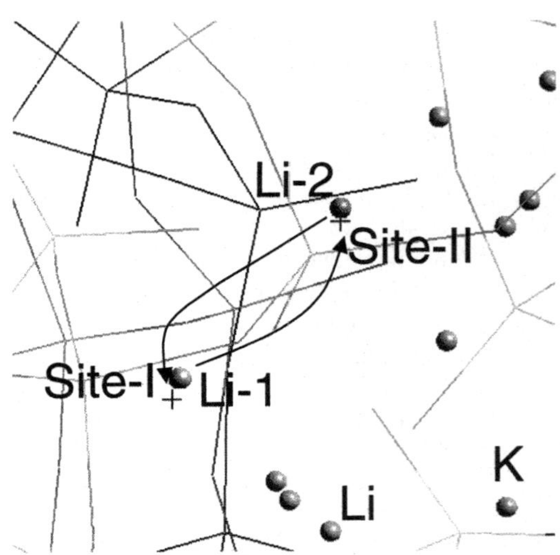

Figure 9. Exchange process involving two Li ions. The spheres indicate the initial positions of Li-1 and Li-2 (in sites I and II, respectively) and the crosses indicate the positions of the two ions after exchanging sites.

On the other hand, the extent to which alkali ions alter the silica network structure when they are only present in very small amounts, such as 1mol%, is much less clear. We have performed some simulations which suggest that the behaviour of the alkali ions is quite different under these circumstances, not in the least because sodium-oxygen co-ordination polyhedra are not connected at these small alkali concentrations. The role of the non-bridging oxygen ions in these structures seems to be important, although more work is necessary to quantify this observation.

5. Summary and conclusions

We have discussed the usefulness of molecular dynamics simulations in probing the structure of silicate glasses. The analysis of the simulated structures provides both contact with experiment (so the validity of the simulations can be assessed) and additional characterization that provides further insight into the structures of these important materials. In addition, atomic migration mechanisms, which are controlled by the atomic structure, may also be elucidated. The technique is not confined to silicates, however; any inorganic system for which a suitable interatomic potential model can be found, is amenable to this approach.

6. Acknowledgements

The New York State College of Ceramics and the Center for Glass Research at Alfred University are thanked for their generous financial support. The ASI Directors and NATO are also thanked for their kind invitation and support to present this work.

7. References

1. Allen, M.P. and Tildesley, D. *Computer Simulation of Liquids*, OUP, Oxford, UK. (1987)
2. Bush, T.S., gale, J.D., Catlow, C.R.A. and Battle, P.D., J. Mater. Chem. **4**, 83 (1994)
3. Vashista P., Nagano, A., Kalia, R.K., and Ebbojo, I. J. Non-Cryst. Solids, **182**, 59. (1996)
4. J. Wang, J. Li, S. Yip, D. Wolf, S. Phillpot, Physica A **240**, 396 (1997).
5. Vollmeyr, K, Kob, W., and Binder, K. Phys. Rev. **B54**, 15808. (1996)
6. Cao, Y., Cormack, A.N., Clare, A.G., Bachra, B., Wright, A.C., Sinclair, R.N. and Hannon, A.C. J. Non-Cryst. Solids, **177**, 317 (1994)
7. Warren B.E., and Biscoe, J. J. Amer. Ceram. Soc. **21**, 49(1938); ibid. 259.
8. Greaves, G.N. J. Non-Cryst. Solids, **71**, 203. (1985)
9. Wright, A.C. (1993) in: Simmons, C.J. and El-Bayoumi, O. (eds.) *Experimental Techniques in Glass Science*, Amer. Ceram. Soc., Westerville, OH, p.205.
10. Yuan, X and Cormack, A.N. J. Non-Cryst. Solids, 283, 69 (2001)
11. Wright, A.C. J. Non-Cryst. Solids, **157**, 254. (1993)
12. Olivier, L., Glock, K., Thomas, B. and Jäger, C., Glastechn. Ber. Glass Sci. Technol. **71C**, 174 (1998)
13. Maekawa, H., Maekawa, T., Kawamura, K. and Yokokawa, T. J. Non-Cryst. Solids, **127**, 53 (1991)
14. Buckman, W., Muller-Warmuth, W. and Frischat, G. H. Glastech. Ber. **65**, 18, (1992)
15. Olivier, L., Yuan, X., Cormack, A.N., and Jager, C., J. Non-Cryst. Solids **293-295**, 53 (2001)
16. Yuan, X., and Cormack, A.N. Comp. Mater. Sci., Accepted (2001)
17. Yuan, X., and Cormack, A.N. Ceram. Trans. **82**, 281-6 (1998).
18. Bockris, J. O'M., Lowe, D.C. Proc. Roy. Soc. **A226**, 423. (1954); Bokris, J. O'M, MacKenzie, J. and Kitchener, J.A. Trans. Faraday Soc. **51**, 1734. (1955)
19. Cao, Y. and Cormack, A.N. p. 137 in: Jain, H. and Gupta, D. (eds.) *Diffusion in Amorphous Materials*, TMS, Warrendale, PA,. (1994)
20. Dirken, P.J., Kohn, S.C., Smith, M.E., and van Eck, E.R.H. Chem. Phys. Lett. **266**, 568. (1997)
21. Topliss, M.J., Dingwall, D.B. and Lenci, T. Geochimica and Cosmochimica Acta, **61**, 2605 (1997)
22. LaCourse, W.C., and Cormack, A.N. (1998) in: Clare, A.G., and Jones, L.E. (eds.) Advances in the Fusion and Processing of Glass II, Ceram. Trans. **82**, The Amer. Ceram. Soc., Westerville, OH.
23. Cormack, A.N., Du, J., and Zeitler, T.R., Phys. Chem. Chem. Phys. **4**, 3193 (2002)
24. Isard, J.O., J. Soc. Glass Tech., **43**, 113T (1958)
25. Lapp, J.C., and Shelby, J.E., J. Non-Cryst. Solids **95/96** 88 (1987)
26. Huang, C., and Cormack, A.N. p.31 in: Pye, L.D., LaCourse, W.C., and Stevens, H.J. (eds.) *The Physics of Non-Crystalline Solids*, Taylor and Francis, London, (1981).
27. Balasubramanian, S., and Rao, K.J. J. Chem. Phys. **97**, 35. (1993)
28. Smith, W., Gillan, M.J., and Greaves, G.N. J. Chem. Phys. **103**, 3091. (1995)
29. Cormack, A.N., and Cao, Y., Molecular Eng. **6** 227 (1996)
30. Megaw, H. *Crystal Chemistry: A Working Approach*, Saunders, Philadelphia, (1973)
31. B. Park and A.N. Cormack, J. Non-Cryst. Solids, **255** 112-121 (1999).

Computational Materials Science
C.R.A. Catlow and E.A. Kotomin (Eds.
IOS Press, 2003

Molecular Mechanism of Ethylene Epoxidation on Silver: State of the Problem and Theoretical Approaches

Georgii M. Zhidomirov*, Vasilii I. Avdeev, Andrei I. Boronin
Boreskov Institute of Catalysis, Novosibirsk 630090, Russia

Abstract. A new concept of the oxygen species epoxidizing ethylene on silver is presented. The epoxidizing oxygen is formed on the defects of the partially oxidized metal silver surface. Oxygen saturates the bulk of silver at high temperature ($T > 500$ K) and pressure of the reaction medium, and the whole subsurface layer becomes highly defective. We consider a cluster model of the defect structure surface AS_V, including a silver atom vacancy and the subsurface oxygen atoms

Calculations were performed in the framework of DFT approach. It was shown that the subsurface oxygen atoms tend towards self-association and formation of quasi-molecular oxygen structures inside of the vacancy space. Adsorption of the oxygen atom on site AS_V also provides stabilization of the surface quasi-molecular ("ozonide") form, $AS_V + O \rightarrow AS_d\text{-}O$.

We discuss the experimental XPS, UPS, IR, and Raman spectroscopy providing evidence in favor of stabilization of the quasi-molecular oxygen forms on the reactive silver surface. A theoretical interpretation of the experimental data is based on the proposed model of associative oxygen forms.

1. Introduction

The oxidized metal silver is a unique catalyst for the selective oxidation of ethylene to ethylene oxide. Since the discovery of this reaction in 1933 by Lefort [1] silver is the only industrial catalyst used for ethylene epoxidation. This reaction also occurs on some other metals as Au, Pt, Pd, Ni, and Cu, but the activity of silver is higher by several orders of magnitude. In spite of considerable efforts of researches and a great number of experimental and theoretical studies, an understanding of the mechanism of this reaction at the molecular level presents essential difficulties [2]. The main question is associated with the nature of active oxygen on the silver surface. Ethylene epoxidation requires oxygen possessing electrophilic properties. Such properties are typical for the chemically adsorbed oxygen species containing bonds O-O. However, according to the literature, peroxide and superoxide adsorbed oxygen forms are readily desorbed from the regular silver surface at high temperatures. Therefore, the associative (molecular) oxygen forms are not sufficiently thermally stable.

Most researchers hold to the idea that active oxygen should be an atomic adsorption form. But such forms are typical oxide-like ones with an essential polarity of the $Ag^{+}\text{-}O^{2-}\text{-}Ag^{+}$ structure and nucleophilic properties of oxygen [3,4,5]. From the standpoint of inorganic chemistry, electrophilic properties of atomic oxygen are not typical for the

* Corresponding author. E-mail: zhi@catalysis.nsk.su; Fax: +7 3832-343056

oxygen/silver system, because there is no chemical compounds with the Ag-O covalent bonding; all compounds have a high ionic degree. The most stable compounds have a degree of silver oxidation of Ag^{1+}, where oxygen is nucleophilic one.

It was suggested [6-9] that electrophilic properties of the adsorbed atomic oxygen can increase due to influence of subsurface oxygen (see scheme 1a) but adsorbed oxygen is proved to exist as nucleophilic one on the regular silver surface. It readily attacks bond C-H, and catalysis is ruled towards deep oxidation (see scheme 1b). Direct calculations, in principle, qualitatively confirm this suggestion [10,11,12]. Thus, it is possible to understand the mechanism of halogen promoters (see scheme 1c). At the same time there is no strong evidence that this effect is really enough for the needed increase of electrophility of adsorbed oxygen. Besides it is difficult to provide a reasonable interpretation of some recent spectroscopic data using the atomic oxygen adsorption forms on the silver catalyst. So we would like to return to the discussion of an alternative of quasi-molecular atomic oxygen adsorption forms of silver taking into account a highly defect structure of the surface of the working catalyst.

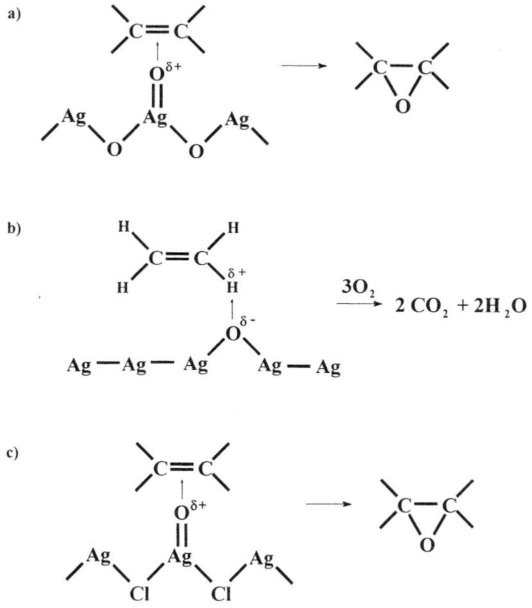

Scheme 1.

First we would like to briefly consider some experimental and theoretical studies on the oxidized metallic silver. Then our cluster model will be presented to compare the properties of oxygen on the regular and defect oxidized silver surfaces. The ensuing parts contain calculations of the possible forms of adsorbed oxygen and ethylene on the oxidized regular and defect Ag(111) surface. The concluding part is devoted to a calculation of the reaction pathway of ethylene epoxidation.

A VG ESCALAB High-Pressure electron spectrometer was used to record photoemission spectra. X-ray photoelectron spectra (XPS) were taken using either MgK_α ($h\nu$ = 1253.6 eV). or AlK_α ($h\nu$ = 1486.6 eV) radiation. UP-spectra were taken using a helium discharge lamp, and two main resonance lines: HeI ($h\nu$ = 21.2 eV) and

HeII (hν = 40.8 eV). The PSILON, CALC software and standard graphic packages served for the data acquisition and spectral analysis.

A polycrystalline foil and electrolytic silver powder were used as the samples. Electrolytic silver was prepared by high temperature silver plating from silver nitrate and silver chloride [13].

Calculations were performed using the GAUSSIAN98 program package [14]. Core electrons of silver were included into the effective potential LANL1 [15] with the corresponding LANL1MB or LANL1DZ basis [16]. All calculations were performed with the Density Functional Theory (DFT) [17] using the Becke three-parameter exchange functional within the gradient corrections [18] and the Lee-Yang-Parr correlation functional [19] (method B3LYP).

2. State of the problem: review of experimental and theoretical studies

Numerous studies using a number of spectral methods show that different species of adsorbed oxygen are formed on the metal silver surface during hard interaction of silver with oxygen in the gas phase. The appearance of such species and their concentration depend on the conditions of preparation and the sample pretreatment, temperature and pressure of oxygen, time of exposure of the sample in the oxygen medium.

It is substantial that such treatment always results in restructuring of the regular surface and formation of defects in the catalyst subsurface layer. The degree of the defect surface heterogeneity determines in turn the state of oxygen. Clearly, the results obtained for different-type surfaces may differ significantly. Therefore, it seemed important to study oxygen species generated on the surface of the carefully cleaned silver foil, which was preliminary treated by ion bombardment and/or annealed under vacuum in order to obtain surfaces with different degrees of the structure irregularity.

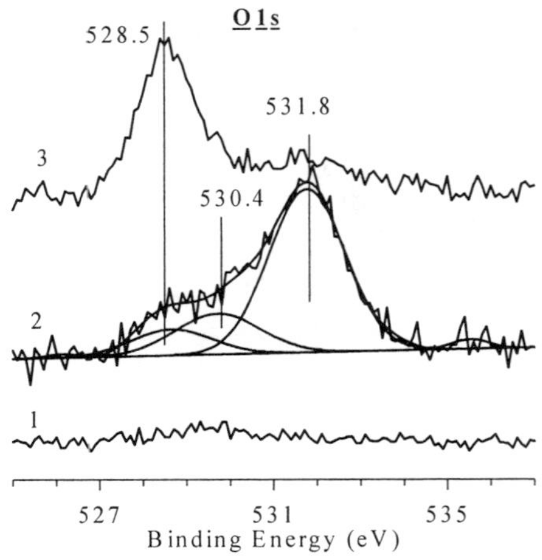

Fig. 1. Adsorption of O_2 on the polycrystalline silver surface. Curve 1: initial pure surface after ion bombardment at 300 K. Curve 2: adsorption of O_2 at 330 K, the exposure is 10^6 L; Curve 3: adsorption of O_2 at 420 K, the exposure is 10^6 L.

If oxygen adsorption occurs at 300 K on the foil treated by ion bombardment without its further annealing, the XPS spectrum exhibits O1s, which is characterized by three peaks with E_b(O1s) = 528.5, 530.4, and 531.8 eV, moreover, the peak with E_b(O1s) = 531.8 eV (Fig. 1) prevails.

Because of heating of the adsorption layer to 420 K (or oxygen adsorption at T = 420 K), the high-energy peaks disappear. As a result, spectrum O1s has only one state with E_b(O1s) = 528.5 eV. A comparison with the literature data permits one to attribute the peak with E_b(O1s) = 528.5 eV to atomic (dissociative) oxygen [20-24] and the peaks with E_b(O1s) = 530.4 eV and 531.8 eV may be attributed to the molecularly absorbed oxygen. Actually, the correlation of these peaks with the UPS spectra permits one to consider this conclusion to be justified (Fig. 2).

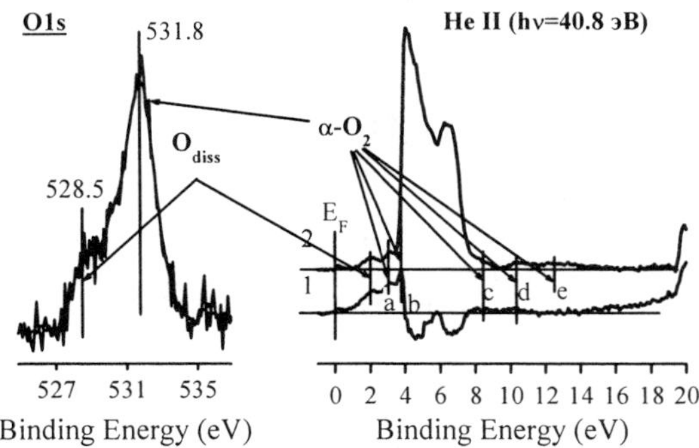

Fig. 2. O1s and UPS (HeII) spectra obtained after O_2 adsorption on the polycrystalline silver surface: (a) O1s spectrum was taken after adsorption at 330 K, the exposure was 10^6 L; (b) the corresponding UP spectrum (curve 2); curve 1 – the difference spectrum (spectrum of clean surface is subtracted from spectrum obtained after oxygen adsorption).

UPS makes it possible to extract several peaks induced by oxygen adsorption under such conditions: 2.1; 3.1; 3.8; 8.4, and 10.2 eV. The peak with E_b = 2.1 eV is unambiguously attributed to the dissociation state, whereas the other peaks (a, b, c, d, e) are associated with the molecularly adsorbed oxygen as two-fold (E_b(O1s) = 530.4 eV) and one-fold (E_b(O1s) = 531.8 eV) peroxide.

Therefore up to T ~ 420 K, stabilization of molecular oxygen forms can occur on the polycrystalline silver surface. However such temperature is not sufficient to provide epoxidation (T = 550-570 K) even with regard for high pressure in the reactor [23]. However, these data point to the primary importance of defects in stabilization of the oxygen molecular forms on the silver surface.

The second important factor in the stabilization of molecular oxygen species is the degree of oxidation of the defect surface. Because the bulk and surface of the electrolytic silver has the highest defectness degree (among the known polycrystalline silver materials) both of bulk and surface, it is intensively studied in the whole world. Compared to the supported silver catalysts (which also consist of a defect structure), XPS and UPS provide information concerning the state of oxygen in the electrolytic silver, because there is no oxygen from support masking the adsorption oxygen on silver.

The effect of oxygen within the range of the "epoxidation" temperatures results in three phases which are associated with the below oxygen species: oxygen adsorbed on the surface, subsurface oxide-like oxygen and oxygen dissolved in the volume. It was established that the subsurface and dissolved oxygen are in equilibrium. Thus, the oxide-like defects were obtained and the state of oxygen on the surface had to be elucidated. The analysis shows that stabilization of the quasi-molecular oxygen on the oxide-like phase of the defect silver can be obtained to temperatures up to 500 K.

Fig.3a shows a typical spectrum O1s, taken after the oxygen action at elevated pressure. Because the shape of spectra is complex, their deconvolution is necessary for analyzing the oxygen species. One clearly sees that the spectrum splits into a number of components which are characterized by E_b varying over a rather wide range 528-538 eV.

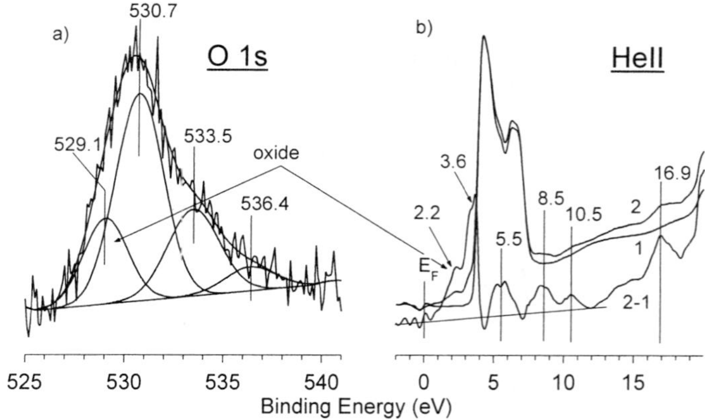

Fig. 3. O1s (a) and UPS (b) spectra obtained from the electrolytic silver surface after treatment in O_2 at 1000 Pa and 420 K. Curve 1 is obtained after surface cleaning; curve 2- after oxygen treatment.; curve 2-1 is the difference spectrum between curves 2 and 1.

The component with a low value of E_b(O1s) ~ 529 eV characterizes the ion type of oxygen bond with silver, that is the nucleophilic state and it unambiguously belongs to oxide Ag_2O [24,25]. The components with E_b(O1s) > 531 eV either correspond to the uncharged oxygen atoms or quasi-molecular species adsorbed on the defect surface. Thermally stable uncharged oxygen atoms have not been described in the literature. That is why we are inclined to believe that both Ag-O and O-O bonds can form on the defect and oxidized surfaces.

On correlating spectra O1s and UPS data (Fig. 3), one actually observes the peaks below the 4d band, which indicates that the oxygen adsorbed in the structure is molecular. In addition, the peaks with E_b(O1s) > 531 eV also give an evidence the molecular nature of adsorption. Molecular peaks in the UPS are not recorded if the sample is heated to 500 K. On the other hand, according to Figs. 4 and 5, the peaks with E_b(O1s) > 531 eV in the O1s are observed up to 700-800K. Disappearance of the molecular peaks in UPS is most likely associated with the subsurface localization of oxygen atoms. Since UPS spectroscopy is more surface sensitive than XPS, the subsurface oxygen visible in XPS is not detected in the UPS spectra.

The analysis of spectra O1s in Figs. 4-5 shows that integral peak consists of 4 components (curves 1 and 2). The peak with the lowest bonding energy (528.3 eV) is unambiguously associated with structure of the surface oxide as Ag_2O. The other peaks, regarding the adsorption conditions and temperature of spectra recording, act in a cymbate manner as if belonging to one species.

Therefore, the adsorbed oxygen species observed on the defect surface of electrolytic silver is a quasi-molecular one containing three non-equivalent oxygen atoms. One oxygen atom with E_b(O1s) =529-530 eV characterizes most probably the localization of oxygen atoms in the subsurface region near the surface structural defect. The peaks with E_b(O1s) = 531-535 eV characterize the oxygen atoms belonging the quasi-molecular form. The nature of the peak in the region E_b(O1s) = 536 eV has not been determined.

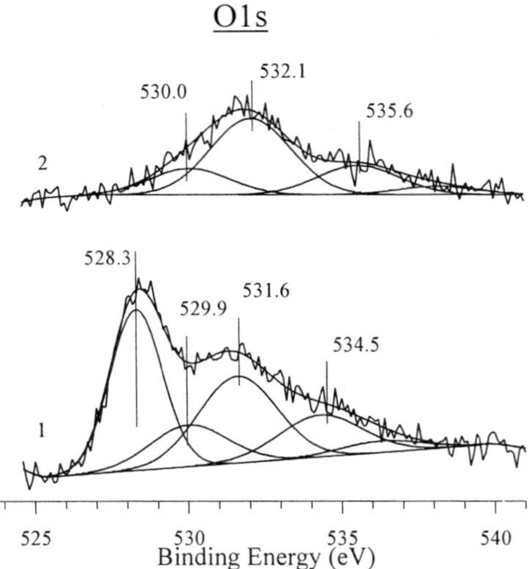

Fig. 4. O1s spectra obtained from the samples of electrolytic silver surface after treatment with O_2: Curve 1 was recorded at P = 3000 Pa, T= 500 K, during 1 hour, then at T = 420 K for 20 min; curve 2 was taken after heating of the sample to 570K (curve 1).

A ratio of peak intensities (that is a ratio of non-equivalent atoms in the oxygen form structure) is 1:2:1. Note that the above oxygen adsorbed on the silver surface is rather thermally stable. As Fig. 4 suggests, heating of the adlayer (curve 2) to 570 K results in desorption of the oxide state (E_b(O1s) =528.3 eV). The structure containing 3 peaks slightly changes, but it retains as a whole. The structure of high energy peaks does not change even after heating to 720 K in vacuum (Fig. 5, curve 1). What's more, a severe treatment in O_2 at T = 720 – 770 K (Fig. 5, curves 2,3) provide an increase of peaks at the range of E_b = 529-535 eV.

The results of spectral deconvolution show that a ratio between the peak areas with E_b(O1s) ~ 532 eV and ~535 eV does not almost change, whereas the peak with E_b(O1s) = 529.4 eV becomes more pronounced. According to [26,27], the thermally stable oxygen state, characterized by E_b(O1s) ~ 529.0 – 529.5 eV in the XPS spectra and denoted as O_γ [26], is formed on the significantly restructured surface under oxygen action at high temperatures. It is proposed that oxygen O_γ– reflects the structure of the most thermally stable layered oxide Ag_2O. [28]. One can assume that the stability of the oxygen states with E_b(O1s) ~ 532-535 eV is provided because of O_γ, which oxidizes the defect sites of the restructured silver surface. Such oxygen is most likely situated in the sub-surface layers of silver.

Therefore, in order to provide fixation of the thermally stable quasi-molecular oxygen species, the silver surface should be both defective and oxidized. For corroboration of this conclusion, it seemed to be important to study the oxygen states adsorbed on the surface or fixed by the lattice of Ag_2O. Because of thermodynamic reasons, the preparation of silver oxide requires very high pressure and not high temperature, which is very difficult to realize if the vacuum equipment is used. For this reason, we have activated the gas-phase oxygen with a microwave discharge directly in the preparation chamber of the electron spectrometer [29].

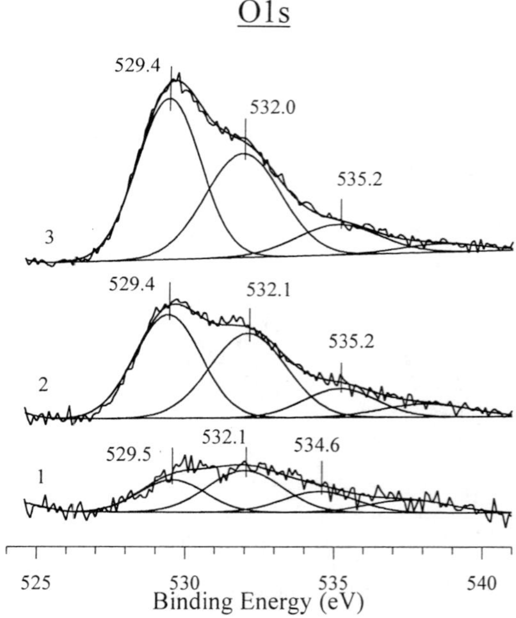

Fig. 5 Spectra O1s were recorded from the electrolyte silver surface treated in O_2. Curve 1: the sample annealed in vacuum at 720 K; curve 2: the sample was treated in oxygen at P = 1000 Pa, T = 720 K; curve 3: the sample was treated in oxygen at P = 1000 Pa and T = 770 K.

Let us consider in detail the results of XPS and UPS study of oxygen species on the oxidized surface of Ag(110) under O_2 microwave excitation [30]. The special conditions of the oxidation were chosen because the thermal oxidation of silver surface requires high-pressure conditions of O_2 treatment and elevated temperatures to increase the rate of oxidation. This imposes a substantial limitation regarding the UHV equipment. In order to prepare oxidized silver in vacuum the silver surface was subjected to a microwave O_2 discharge. Such oxygen treatment provides strong oxidation of silver at low temperatures [24,29].

Fig. 6 shows XPS and UPS spectra of the oxidized silver surface. Two peaks are observed in spectrum O1s (Fig. 6a, curve 1), the first peak with E_b(O1s)= 529.2 eV is associated with ion O^{2-} (Ag_2O phase), but the peak with E_b(O1s) = 532.1 eV could be explained reasonably by a quasimolecular form. It should be noted that the last form disappears after ethylene exposure (see Fig. 6a, curve 2), which indicates its activity in the reaction with ethylene. As silver is treated with exited oxygen, the UPS spectrum becomes very peculiar (Fig. 6b, curve 1). These features disappear after interaction between silver and ethylene, which is to say that this UPS spectrum is associated with E_b(O1s) = 532.1 eV form. To extract all features induced by this oxygen species, the difference spectrum (Fig. 6, curve 1-2) was analyzed, the positive part of it was related to new oxygen features. The multiple peak structure of UPS spectrum definitely excludes the atomic oxygen state.

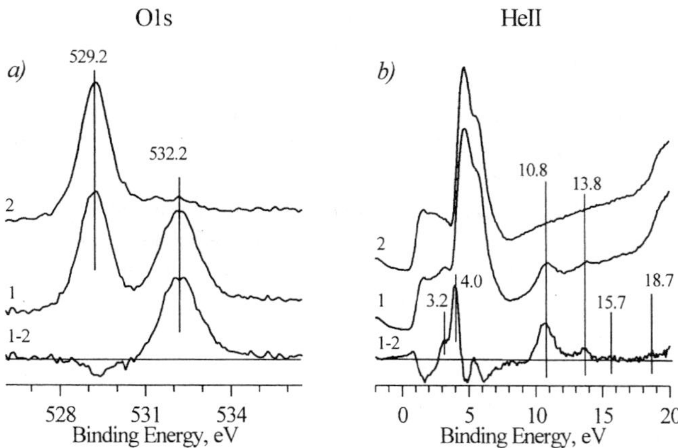

Fig. 6. O1s (a) and UPS (HeII) (b) spectra recorded before (curves 1) and after (curves 2) ethylene interaction (the exposure of C_2H_4 is about $2x10^8$ L). The adsorbed oxygen layer was obtained by oxygen microwave excitation at 420 K. Curves (1-2) are the difference spectra between curves 1 and 2.

An application of photoelectron spectroscopy with angle resolution providing variations of the sample surface thickness, permitted us to find out that localization of the quasi-molecular oxygen occurs directly in the oxide lattice layers but not on the Ag_2O surface. We propose that on the formation of the oxide lattice under hard conditions of oxygen plasma, a part of the overstoichiometric oxygen inside the oxide layers exists as a quasi-molecular form. Localization probably occurs between the layers of silver ions.

Thus, high-temperature treatment of the surface with oxygen provides dissolution of oxygen into the bulk and, as a result, the progressive reconstruction of the surface, which dramatically changes the catalyst morphology. As a result, structural defects inevitably form and oxygen species appear at "high temperature" (T > 500 K). Such species differ from those usually found on the regular surfaces. The binding energy of 1s electrons are within a range of E_b(O1s) = 531-533 eV. The states with E_b(O1s) > 531 eV are typical for the associative oxygen species, but not for the dissociative ones. Joyner and Roberts [31] have observed E_b(O1s) = 532.5 eV at high oxygen pressure on the polycrystalline silver. Ayyoob and Hegde [32] have registered E_b(O1s) = 533.2 eV on the silver foils treated with potassium. Au et al. [33] have shown that the oxygen state on Ag(110) with E_b(O1s) = 532 eV is active for ethylene epoxidation. Grant and Lambert [34] have obtained E_b(O1s) = 532.6 eV by treating Ag(111) with an oxygen flow. According to Bao et al. [35], $E_b(O_{1s})$ = 532.2 eV for electrolytic silver. Boronin et al. have registered E_b(O1s) = 531.9 eV [36] and 532.0 eV [37] for the silver film obtained by thermal decomposition of Ag_2O. In the above papers, the oxygen states were ascribed to the molecular oxygen species.

Vibration spectroscopy is a very important source of structural information about unusual oxygen species formed at high temperature and pressure upon actual catalysis. At present, SERS (Surface enhanced Raman spectroscopy) appears to be the main investigation method. An application of SERS requires roughness of the surface. As shown by Pettenkofer et al. [38], the roughened surface always contains defects even at low temperatures.

The problem was studied in detail by Millar et al. [39,40,41]. In [40], two samples (A and B) of polycrystalline silver, differing much in their surface morphology, have been studied. Sample A had a rather high density of surface defects such as grain boundaries, terraces, etc. Sample B was comparatively smooth and rounded. The Raman spectra were recorded at 295 – 873 K. Sample A exhibited bands at 240, 345, 485, 644, 782, 960 cm^{-1}. As temperature grew from 300 to 673 K, the intensity of all modes, excepting mode 485 cm^{-1}, decreased. On the contrary, the intensity of a band at 485 cm^{-1} increased. In addition, it was shown that if ethylene was introduced into the reaction mixture at 473 K, the modes at 644 and 782 cm^{-1} lost their intensity, and new bands appeared in the range of 200-300 cm^{-1}. Five sub-bands were revealed at 196, 236, 264, 288 and 308 cm^{-1}. Sample B exhibited only two intensive bands at 485, 800 cm^{-1} and a shoulder at 370 cm^{-1}.

Kondarides et al. [42] have studied oxygen species on the supported catalysts Ag/α-Al_2O_3, Ag/quartz, and Ag/ZrO_2 at atmospheric pressure in a wide temperature range (300-673 K). All samples give almost identical bands at 240, 345, 815, 870, and 980 cm^{-1}, but their intensities differ significantly. So mode 980 cm^{-1} is most intensive for Ag/α-Al_2O_3, while modes 815/817 cm^{-1} exhibit poorly resolved shoulders. On the contrary, for the Ag/ZrO_2 , mode 980 cm^{-1} exhibits low intensity, while mode 817 cm^{-1} is the most intensive in the SER spectra. For the Ag/quartz sample, mode 980 cm^{-1} grows in intensity when C_2H_4 is introduced into the reaction mixture, while mode 815 cm^{-1} looses its intensity. The authors assume that this regularity is provided by formation of carbonate structures.

Wu et al. [43,44] have observed oxygen species on Ag(111) with bands within 600-640 cm^{-1}, which are stable up to 400 K. These species are ascribed as molecular ones, and their thermal stability is explained by presence of defects on the surface. Bao et al. [45] have studied Ag(111) and Ag(110) exposed to mixture O_2/C_2H_4 and referred three bands at 632, 802 and 954 cm^{-1} to the specific atomic oxygen species. Deng et al. [46] have also registered three bands in this range for electrolytic silver (624, 808, 1078 cm^{-1}). Despite the fact that the catalysts are different, the origin of modes is most likely the same. However, bands at 624 and 1078 cm^{-1} were ascribed to molecular species, while the band at 808 cm^{-1} was ascribed to atomic oxygen species.

In conclusion, we would like to note that there are difficulties in interpreting the high frequency oxygen vibration bands (> 600 cm^{-1}) as the bands associated with atomic molecular models, which could confirm such supposition.

The investigation of the nature of oxygen adsorption forms on silver was a subject of a number of theoretical studies [47-55] mostly performed in the cluster approximation. Rosch and Menzel [47] have simulated the atomic oxygen adsorption on silver for the Ag_4-O system using the DFT method and showed that one should expect the appearance of two additional emission peaks below the Fermi level near the 4d band of silver. The Hartree-Fock (HF) calculations and qualitative analysis of the density of states (DOS) for a moderate-size cluster Ag_{26}-O confirmed the two-peak ultraviolet photoelectron spectra (UPS) structure [48]. The results obtained by these authors together with those reported in [50] have shown the importance of the electron correlation energy for correct estimation of oxygen adsorption energy on silver. The HF approximation yields the bonding energy value E(Ag-O) = 9 kcal/mol [48], which is significantly below the experimental value (80 kcal/mol).

Selmani et al. [49] performed the analysis of the molecular adsorption of oxygen on Ag_4, Ag_{10}, and Ag_{16} clusters simulating the Ag(110) face by the SCF-X_α-SW method. The computation of a number of oxygen adsorption states was performed for three possible orientations of an O_2 molecule with respect to the surface. A comparison of the calculated energies of one-electron levels with the XPS and UPS spectra led the authors to the conclusion that the bridge adsorption of O_2 molecules oriented parallel to the Ag(110) surface was most probable. The computations also showed DOS to have a complex structure below the 4d band, which is typical only for molecular oxygen species.

An extremely important role of subsurface oxygen forms in the formation of epoxidizing oxygen has been already mentioned. However, the mechanism of its effect is still unknown. Some aspects of this problem have been studied by the DFT method for a set of clusters Ag_4, Ag_{10}, and Ag_{24} simulating the Ag(110) surface [10,11]. In particular, it was shown that the subsurface oxygen, first, weakened the Ag-O bond on the surface, then, increased the ethylene adsorption energy. Higher adsorption energy is known to favor the epoxidation reaction and, on the contrary, hinder the combustion.

A very important attempt to determine the properties of epoxidizing oxygen was undertaken in [4]. Both the chemisorption of O and O_2 and epoxidation of olefins were analyzed by the GVB-CI method on the Ag(111) surface simulated by an Ag_3 cluster. The authors [4] noted that one of the states of adsorbed oxygen $^2\Sigma$ (surface atomic oxyradical anion) was the most preferable form for epoxidation. This form of oxygen was the most electrophilic (it had the lowest effective charge q(O) = - 0.20). The role of electronegative promoters was also studied in this paper and interpreted in a manner slightly different from the one suggested by other authors [10,11].

Jorgensen and Hoffmann [52] have studied general aspects of the atomic and molecular adsorption of oxygen on silver by the extended Huckel method and suggested several models for oxygen insertion into the π-bond of ethylene. The atomic form of oxygen was preferred for epoxidation. It was also noted that DOS of molecular forms had, at least, two peaks below the 4d band, which is in a qualitative agreement with the results of [49]. In a series of theoretical studies by Nakatsuji et al., summarized in review [5], the main intermediates in the ethylene epoxidation reaction with participation of atomic and molecular forms of oxygen were calculated using the dipped adcluster model (DAM). Their calculations predict that the adsorbed molecular superoxide is an oxygen form active in this reaction.

3. A cluster model and computational details

The structure of the surface with strongly bonded forms of oxygen is unknown. Since such structure is formed at high pressures and temperatures when oxygen dissolution in the bulk is significant, it is necessary to begin with determining the properties of subsurface oxygen O_{ss}. The subsurface oxygen structures are usually described in terms of Ag_2O oxide. There is a rather justified assumption that under the real catalysis conditions ($T > 500$ K, $P \sim 1$ bar), unusual oxide-like species (similar to Ag_2O) are produced on the defect silver surface [56,57]. The oxidized state of the surface was obtained by introducing the subsurface oxygen atoms into the model. The model face should have atomic size defects. As a defect we have chosen cationic vacancy V. We also used silver oxide Ag_2O as a structural prototype.

The minimum cluster $AS_v \rightarrow Ag_{12}$-3O (shown in Fig. 7) satisfies all the requirements mentioned above [58,59].

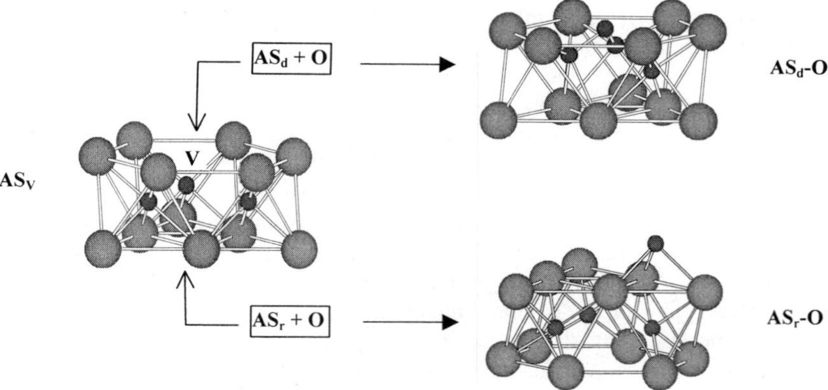

Fig. 7. Cluster model $AS_V = Ag_{12}$-3O for oxidized surface Ag(111). The regular surface is simulated by six silver atoms with adsorption site AS_r. The defect surface, including cation vacancy V, is also simulated by six silver atoms with adsorption site AS_d. Three oxygen atoms between the first and the second silver layers are simulated as subsurface oxygen. The atomic oxygen adsorption on the defect surface (AS_d-O) associates surface oxygen and subsurface oxygen yielding the associative oxygen of metal ozonide type Ag-O-O-O-Ag. AS_r-O is 3-fold atomic oxygen adsorption on the regular surface Ag(111).

The first layer of AS_V consists of 6 silver atoms with cationic vacancy V in the center. The second layer consists of three oxygen atoms. The third layer, consisting of 6 silver atoms, simulates the regular surface. Since defect V in the top metal silver layer has no appreciably effect on the adsorption properties of the bottom (regular) silver layer, we may use such cluster to calculate ethylene and oxygen adsorption on the regular surface with adsorption site AS_r as well as on the defect surface with adsorption site AS_d (see Fig. 7). In the optimized structure each oxygen atom is bonded to two silver atoms from the first layer (Ag1) and with three silver atoms from the second layer (Ag2), bond lengths being r(Ag1-O) = 2.13 Å, r(Ag2-O) = 2.16 Å, r(Ag2-O) = 2.20 Å [59] In the regular lattice of the layered oxide, oxygen occupies the octahedron positions, bond length being r(Ag-O) = 2.16 Å. Therefore, the limited size of a cluster has no essential effect on the oxygen layer localization as compared to that in Ag_2O. That is why the properties of oxygen in model AS_v will be similar to the properties of oxide oxygen O_{ox}.

3.1. Oxygen chemisorption on the regular and defect Ag(111) surfaces

On the regular site AS_r oxygen adsorbs on the 3-fold site with a bond length of r(Ag-O) = 2.32 Å. The situation, however, changes dramatically for oxygen adsorption on the defect site AS_d . Here adsorbed oxygen associates with two oxide oxygen atoms into structure Ag-O-O-O-Ag (see Fig.6, AS_d-O). As bond O-O forms, it changes the nature of bond Ag-O. Oxide oxygen is modified into subsurface oxygen Ag-O_{ox}-Ag → Ag-O_{ss}-O_{ep}. We use here the term "subsurface" for the oxygen located under the surface and entering a bridge bond Ag-O_{ss}-O_{ep} with one of silver atoms on the surface and associative oxygen O_{ep}. The central oxygen in chain O_{ss}-O_{ep}-O_{ss} shows electrophilic properties. Oxide oxygen O_{ox} on the contrary exhibits pronounce nucleophilic properties.

Therefore, the AS_d-O complex includes oxygen of three types such as electrophilic oxygen O_{ep}, oxide oxygen O_{ox}, and subsurface oxygen O_{ss}. A cationic vacancy is demanded for formation of associative oxygen species. The adsorbed oxygen transforms bonds Ag-O into O-O yielding a united associative group Ag-O_{ss}-O_{ep}-O_{ss}-Ag.

According to calculations [59], the binding energies of atomic oxygen on the defect and regular surfaces are practically the same (E(AS_d-O) = 61.2 kcal/mol and E(AS_r-O)= 62.7 kcal/mol, respectively). This means that regarding the thermodynamics one is not able to distinguish adsorption on the regular surface for the defect one. However, since local environment of O_{ep} essentially differs from that of oxygen O_{ss} and oxygen O_{ox}, these two types of adsorbed oxygen essentially differ by their spectral parameters (IR-, Raman-spectra, XPS, UPS). According to XPS one may expect an increase of binding energy for the 1s-electron to the values typical for molecular oxygen species E_b(O1s) ~ 532 eV. At high temperature (> 400 K), one may expect bands at 600, 1000 cm^{-1} in the IR spectra, which are typical for the molecular oxygen species registered at low temperatures [38].

Assuming that the oxygen association is either O_2 or O_3 complexes, we have performed the DFT calculations of energy distribution of the valence band electrons [30]. Fig. 8 presents the state density (DOS) obtained by the Lorentz broadening of the discrete energy spectrum with parameter γ = 0.4 eV for simple models Ag_2-O_2 and Ag_2-O_3. The Ag_2-O_2 structure simulates a surface oxy-metal cycle, which is similar to peroxide species on silver. The model Ag_2-O_3 is similar to an ozone molecule bonded to silver via terminal atoms. The calculations show that in both cases DOS consists of multiple peaks with different structures. Because of the interaction between oxygen 2s electrons and silver 4d electrons, a 2s-level splits either in two (Ag2-O_2) or three (Ag2-O_3) components.

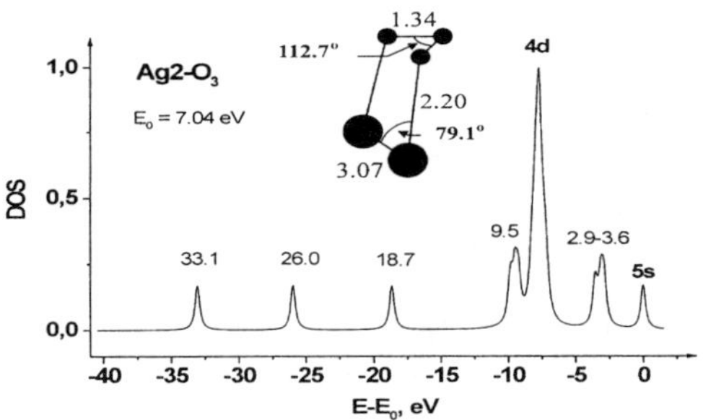

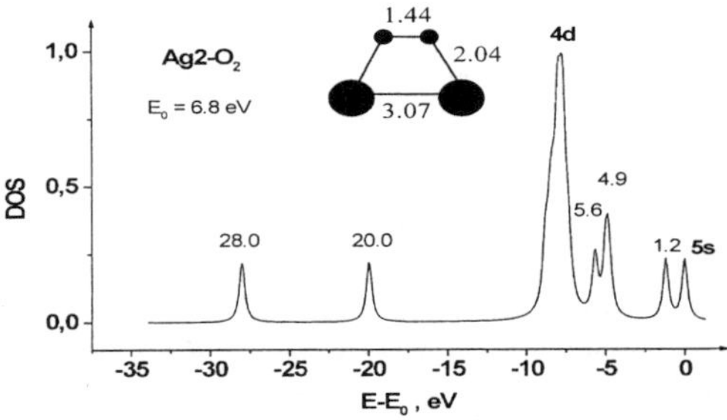

Fig.8. The DOS of the valence band of peroxide and ozonide species of oxygen on silver for simple models Ag2-O_2 and Ag2-O_3.

The interaction of 2pσ electrons with the Ag4d levels induces an additional structure below a 4d band for the ozonide model (peaks near 9.5 eV) and above a 4d band for the peroxide model (peaks with 5.6 eV, 4.9 eV). Other peaks near the Fermi level are not characteristic of the associative oxygen species. Thus, to distinguish the type of oxygen association, it is necessary to analyze the energy range below the Ag4d band. Since the UPS were obtained with HeII radiation, the maximum spectral range, 0.0 eV - 19.6 eV, is open for analysis. This implies that the characteristic range below Ag4d band is about 8 - 19 eV. The calculations of the Ag2-O_2 model show that a single feature (peak with E_b = 20.0 eV, Fig. 8) could appear in UPS (HeII). However, a peak with E_b = 28 eV cannot appear in the HeII spectrum, because the binding energy is too high. Even if we take into account the final state effects of the photoionization process, it cannot provide a shift to lower BE by more than 8 eV. For ozonide species (model Ag2-O_3), not less than two peaks are expected to appear in UPS (peaks with 9.5 eV and 18.7 eV). Besides, it is not improbable that peaks around E_b = 9.5 eV may split into three components upon transition from a small cluster to the real surface. This effect will obviously increase the number of peaks below the Ag4d band. We suggest that three experimental peaks with E_b = 10.8, 13.8, and 15.7 eV (see Fig. 5) will correlate with the calculated unsplitted peak with E_b = 9.5 eV. The other peak with E_b = 18.7 eV can also be correlated. This qualitative analysis makes it possible to attribute the multiple peak structure of the UPS (Fig. 6, curve 1-2) to ozonide species.

Now we shall try to estimate binding energy $E_{1s}(O)$ within the framework of our model. One may obtain a theoretical estimation of $E_{1s}(O)$ from the bound one-electron eigenstates $\varepsilon_i(1s)$ by a scale shift $E_{1s}(O_i) = k^* \varepsilon_i(1s)$. Scale factor k is determined from experimental value $E_{1s}(O_{ox})$ = 529.2 eV [30] and value $\varepsilon_i(1s\ )$ for the AS_v site [59] which contains only oxide oxygen O_{ox}, $k = E_{1s}(O_{ox})/\varepsilon_{1s}(O_{ox})$. Taking this condition into account, one may independently estimate $E_{1s}(O)$ for all types of oxygen entering the AS_d-O center. For oxide oxygen, $E_{1s}(O_{ox})$ = 529.4 eV. This value is close to $E_{1s}(O)$ = 529.0 eV for bulk oxide Ag_2O [60]. For subsurface and electrophilic oxygen, theoretical values $E_{1s}(O_{ss})$ = 531.1 eV and $E_{1s}(O_{ep})$ = 532.8 eV.

3.2. Ethylene adsorption on the regular and defect oxidized Ag(111) surfaces

Fig. 9 shows the calculated structure of ethylene adsorbed on the regular AS_r-C_2H_4 and defect AS_d-O-C_2H_4 surface Ag(111) [59]. The subsurface oxygen atoms partially oxidize the surface silver atoms. Therefore, ethylene is stabilized in a π-complex on cation Ag^{+q}. According to calculations, the binding energy $E_\pi(C_2H_4)$ = 14.2 kcal/mol. Yang et al. [61,62] performed similar calculations for the binding energy of an isolated cation Ag^{1+} modified by anions F, Cl, Br, I. Regarding the anion, $E_\pi(C_2H_4)$ ranges within 10.9-11.8 kcal/mol.

For π-C_2H_4 an oxygen precovered Ag(110), the experimental formation heat value is Q(π-C_2H_4) = 10 kcal/mol [63]. The estimates of supported oxide Ag_2O/Al_2O_3 give Q(π-C_2H_4) = 13 kcal/mol [64]. Using microcalorimetry, Carter et al. [65] have determined the average heat of ethylene adsorption on zeolite Ag-X, Q(Ag-C_2H_4) = 18 kcal/mol. Structure AS_d-O_{ep}-C_2H_4 is arranged so that solely electrophilic oxygen O_{ep} is accessible for interaction with ethylene. Oxygen atoms O_{ox} and O_{ss} are inserted into the subsurface lattice and are not accessible for ethylene owing to spatial restrictions. Unlike π-C_2H_4 adsorption on the regular surface, when there is a slight activation of π-bond C=C, ethylene adsorption on the defect surface elongates this bond to r(C-C) = 1.49 Å. The lengths of bond r(O_{ep}-O) = 1.47 and 1.48 Å are typical for the peroxide oxygen forms. Therefore, metal-ethylene-peroxide-cycle (MEOOC) forms such as Ag-O_{ss}-O_{ep}-C_2H_4-Ag involve electrophilic O_{ep} and subsurface oxygen O_{ss}.

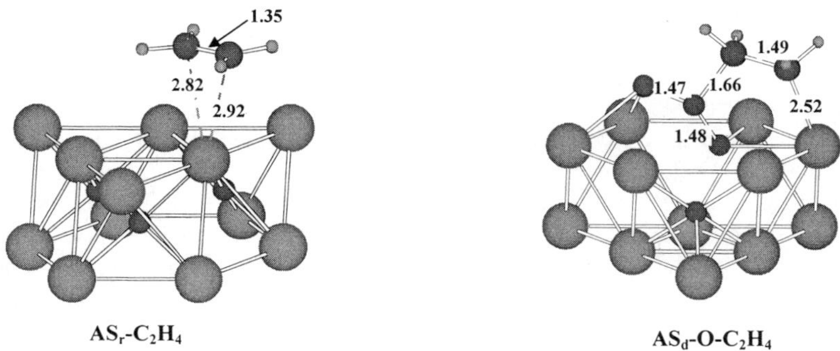

Fig. 9.The optimized structure of ethylene adsorbed on the regular oxidized surface Ag(111) (π-C_2H_4 complex: AS_r-C_2H_4) and defect surface (AS_d-O-C_2H_4). The interaction of electrophilic oxygen with ethylene produces metal-ethylene-peroxide-cycle (MEOOC).

There is a formal similarity between MEOOC and surface oxo-metal cycle (SOMC) Ag-O-C_2H_4-Ag, studied in detail by Barteau et al. [66,67] and Salazar et al. [68]. In both cases there is a full opening of double bond C=C. We assume that SOMC forms from MEOOC as a possible intermediate at the later stages of ethylene epoxidation. The calculated binding energy of ethylene is $E_d(C_2H_4)$ = 10 kcal/mol. Therefore, the values of ethylene binding energy on the regular $E_\pi(C_2H_4)$ and defect $E_d(C_2H_4)$ silver surfaces are approximately equal. However, the structure of ethylene on the regular surface differs significantly from that on the defect oxidized Ag(111).

3.3 Infrared and Raman spectra of oxygen and ethylene on the defect surface

The isotope technique is a very powerful method for the identification of adsorbed oxygen species. The question arises if it is possible to distinguish atomic and molecular adsorbed oxygen forms in similar experiments? As reported by Boreskov et al [69,70], kinetic data permit one to establish a correlation between the rate of ethylene oxidation and the rate of isotope exchange $^{16}O/^{18}O$ on the silver surface. It has been assumed that the adsorbed oxygen layer is uniform. This assumption served as a ground in favor of the "atomic" epoxidation mechanism. However, as we have already mentioned above, it is not possible to distinguish kinetically dissociative and associative oxygen forms.

Later van Santen and de Groot performed a very interesting and fine experiment with isotopes $^{16}O/^{18}O$ [8]. The silver surface, preliminary exposed to $^{16}O_2$, was treated with a mixture of $^{18}O_2/C_2H_4$. First, the ethylene oxide with ^{16}O appeared among the reaction products, then the ethylene oxide with ^{18}O added. The above regularities, as the authors believe, also confirm the "atomic" mechanism of epoxidation. According to the suggested above electrophilic oxygen concept, the associative oxygen species on the defect planes are thermodynamically more favorable than the dissociative species [58]. According to [69,70], a uniform isotope composition $^{16}O/^{18}O$ forms on the surface at the steady state. Therefore, the appearance of ethylene oxide with ^{16}O at the initial experimental stages [8] indicates that the surface state is not steady yet.

McBreen and Moskovits [71] studied ethylene and oxygen interactions with Ag/SiO_2 catalysts using SERS. For mixtures O_2/C_2H_4 they revealed bands at 240, 676, and 995 cm^{-1}, which were stable at T > 400 K. The authors ascribed the band at 240 cm^{-1} to the Ag-C or Ag-O_2 stretching vibration, while bands at 676 and 995 cm^{-1} were attributed to molecular oxygen species. Upon isotope replacement $^{16}O_2 \rightarrow {}^{18}O_2$, the "molecular" bands shifted to 655 and 960 cm^{-1}, the isotope shifts were $\Delta\nu$ = 20 and 35 cm^{-1} respectively. If one assumes that the bands at 676 and 995 cm^{-1} originate owing to the stretching vibrations of bond O-O, it is expected that isotope shifts $\Delta\nu$ = 39 and 57 cm^{-1} should be considerably larger. That is why the authors of [71] suggested that subsurface $^{16}O_2$ should be present in large amounts and create a uniform adsorbed layer $^{16}O^{18}O$ through a fast exchange with $^{18}O_2$. In this case they have a good agreement with the experimental shift value.

Wang et al. [72], however, consider this discrepancy as the main argument in favor of the "atomic" interpreting of bands in the range of 600-1000 cm^{-1}. They used in situ Raman spectroscopy to study oxygen species on the polycrystalline silver and their interaction with ethylene, CO_2, H_2O, and methanol. The experimental conditions (atmospheric pressure and a temperature range of 300-773 K) were close to the real catalyst conditions. The authors also obtained small shifts $\Delta\nu$ = 20-25 cm^{-1} at isotope replacement $^{16}O_2 \rightarrow {}^{18}O_2$. However, this contradiction may be easily explained within our "molecular" concept if molecular species form on the surface not through the O_2 fixation from the gas phase but through the association of subsurface ^{16}O and surface atomic oxygen ^{18}O.

As we have already noted, the conditions for the associative oxygen to form are such that the adsorption site in the defect vicinity should always contain a mixture of isotopes $^{16}O/^{18}O$. Table 1-I shows the calculated frequency values for oxygen adsorption on Ag(111) for normal isotope ^{16}O. Site AS_d-O (Fig. 6) was used for modeling [59]. Two frequencies at 1000 cm^{-1} belong to the symmetric and asymmetric stretching vibrations of bond O-O in the functional group O_{ss}-O_{ep}-O_{ss}. The deformation and wagging vibrations of this group provide bands at 672 cm^{-1} and 238 cm^{-1}. The deformation vibration at 672 cm^{-1} may widely vary depending on the defect type and its structure. Three bands in the range of 450-400 cm^{-1} are assigned to the stretching of bonds Ag-O_{ox} in the Ag-O_{ox}-Ag group, while frequencies at 350-400 cm^{-1} refer to vibrations of bond Ag-O_{ss} in group Ag-O_{ss}-O_{ep}.

According to Table 1, isotope mixing of $^{16}O_2$ and $^{18}O_2$ in structure Ag-O-O-O-Ag causes isotope shifts in a wide range. So at full replacement $^{16}O \rightarrow {}^{18}O$, isotope shift $\Delta\nu$ = 57 - 58 cm^{-1} for the band at 1000 cm^{-1} (Table 1-II).

Table 1. Basic vibration modes (cm^{-1}) for associative oxygen on Ag(111) (column I) and their isotope shifts at replacement $^{16}O \rightarrow {}^{18}O$. Calculated frequencies are scaled by 0.91

	I	II	III	IV	V	VI
Assignment	AS_d-O	^{18}O-^{18}O-^{18}O $^{18}(O_{ox})$ $(\nu^I-\nu^{II})$	^{16}O-^{18}O-^{16}O $^{16}(O_{ox})$ $(\nu^I-\nu^{III})$	^{18}O-^{16}O-^{18}O $^{18}(O_{ox})$ $(\nu^I-\nu^{IV})$	^{16}O-^{18}O-^{18}O $^{16}(O_{ox})$ $(\nu^I-\nu^{V})$	^{16}O-^{16}O-^{16}O $^{18}(O_{ox})$ $(\nu^I-\nu^{VI})$
w(O-O-O)	238	12	12	0	12	0
ν(Ag-O_{ss})	365	18	1	17	14	0
ν(Ag-O_{ss})	371	19	1	17	6	1
ν(Ag-O_{ss})	396	20	0	20	10	0
ν(Ag-O_{ox})	455	20	0	21	0	21
ν(Ag-O_{ox})	483	24	0	24	0	24
ν(Ag-O_{ox})	492	25	0	25	0	22
δ(O-O-O)	672	37	7	30	24	0
ν(O-O_{ep})	1019	57	25	29	61	0
ν(O-O_{ep})	1024	58	30	27	34	0

ν = stretching, δ = deformation, w= wagging

If the subsurface layer is formed by ^{18}O and the adsorbed oxygen consists of ^{16}O (Table 1-IV), the corresponding shift is 27 - 29 cm^{-1}. The other mixed isotope compositions also show 12 - 34 cm^{-1} shifts. Note that the isotope composition III unambiguously shows that bands at 238, 672, 1019, and 1024 cm^{-1} are coupled with the associative oxygen species, while the isotope composition VI confirms that modes at 455, 483 and 492 cm^{-1} correspond to the oxide oxygen vibrations in Ag-O_{ox}-Ag. Thus, three types of oxygen such as subsurface oxygen, oxide oxygen, and electrophilic oxygen are characterized by the most intensive bands: $\nu(O_{ss})$ = 370 cm^{-1}, $\nu(O_{ox})$ = 480 cm^{-1} and $\nu(O_{ep})$ =240, 670, 1000 cm^{-1}. The detailed calculations for the bands within 1000 cm^{-1} show (Table 1, II-VI) that shifts $\Delta\nu$ = 25-30 cm^{-1}. Only in the case of full exchange $^{16}O \rightarrow {}^{18}O$, the shifts indeed increase to $\Delta\nu \sim 60$ cm^{-1} (Table 1, II).

Therefore, low shift values $\Delta\nu$ at exchange $^{16}O \rightarrow {}^{18}O$ can not be used as a criterion for atomic or molecular oxygen species, but they reflect the fact that the conditions for the formation of "high temperature" and "low temperature" molecular oxygen species are different. At low temperature ($T < 200$ K) on the regular surfaces, the molecular species form through O_2 fixation from the gas phase. This case is probably realized in matrix Ar/Ag-O_2 [73] for which $\Delta\nu$ = 30-62 cm^{-1}. The similar situation is observed in matrix Ar/Ag-O_3 [74]. For a full replacement $^{16}O_3 \rightarrow {}^{18}O_3$, shift $\Delta\nu$ = 50 cm^{-1}, and $\Delta\nu$ = 10-38 cm^{-1} depending on the isotopes composition like ^{16}O-^{18}O-^{16}O, though in both cases only the associative oxygen is present in the Ar/Ag-O_3 complex in the form of metal ozonides. At high temperatures ($T > 300$ K), molecular species form through the association of surface and subsurface atomic species to produce isotope mixtures $^{16}O/^{18}O$. These species are stabilized thus providing the lowered values of isotope shifts.

The introduction of ethylene into the reaction mixture essentially complicates the spectrum, but as it will be shown below, some particular features of the frequency spectrum with isotope oxygen mixtures will be retained. Since a preliminary treatment of the silver surface with an oxygen flow is required to adsorb ethylene, one may expect the appearance of frequencies in the spectra, which originate from the ethylene-oxygen mixture.

Fig.10 shows the theoretical spectrum for adsorption of mixture C_2H_4/O_2 on the AS_d-site [59].

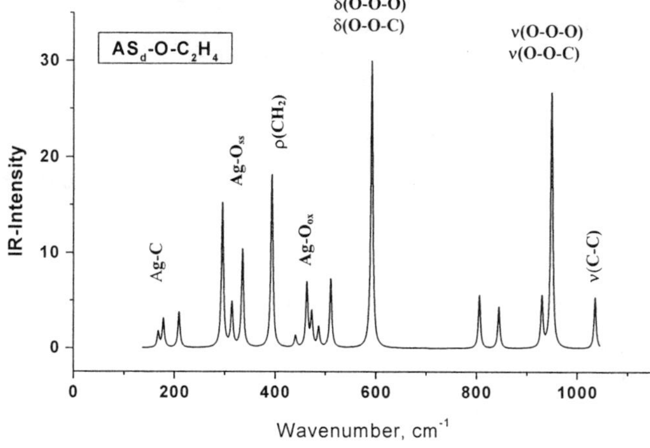

Fig. 10. IR spectrum of the ethylene-oxygen mixture adsorbed on the defect oxidized surface Ag(111) obtained using the Lorenz broadening $\Delta\nu$ = 5 cm^{-1} of theoretical frequencies.

We would like to pay special attention to the bands provided by vibrations of structural unit Ag-O_{ss}-O_{ep}-C-C-Ag. Two frequencies $\nu = 169\ cm^{-1}$ and $\nu = 179\ cm^{-1}$ are interpreted as symmetric and asymmetric stretching vibrations of the Ag-C bond in group Ag-CH_2-Ag. Two group of frequencies at 300 cm^{-1} and 450 cm^{-1} are provided by stretching vibrations of the Ag-O bond. Vibrations of bond Ag-O_{ss} are responsible for bands at 300 cm^{-1}. The stronger bond Ag-O_{ox} produces bands at 450 cm^{-1}.

Note that we have never found atomic oxygen species at frequencies higher, than 600 cm^{-1}. Our calculations show that the range 600-1000 cm^{-1} is related to associative groups O_{ss}-O_{ep}-O_{ss} and O_{ss}-O_{ep}-C. Symmetric ν_s(O-O_{ep}-C) = 930 cm^{-1} and asymmetric ν_a(O-O_{ep}-C) = 949 cm^{-1} stretching vibrations are the superposition of two fundamental modes of bonds O-O and O-C. The largest contribution to frequency $\nu_a = 949\ cm^{-1}$ is given by vibrations of bond O-C and that to frequency $\nu_s = 930\ cm^{-1}$, by vibrations of bond O-O. Two deformation modes δ(O-O_{ep}-O) = 511 cm^{-1} and δ(O-O_{ep}-C) = 592 cm^{-1} are also related to the associative groups. We assume that the most interesting modes at 600 cm^{-1} and 1000 cm^{-1} are the fingerprints of the associative oxygen species formed on the defect silver surface at high temperatures.

Thus, the experimental IR, Raman, XPS, and UPS data, obtained under the real catalysis conditions, and the theoretical estimations of the electronic structure of oxygen and ethylene species adsorbed on the oxidized silver surface, confirm the existence of the thermally stable associative oxygen species on the silver defect surface.

4. Theoretical study of reaction mechanism of ethylene epoxidation

Based on the calculation results, we present the below scheme of formation of the active sites on the silver surface. As oxygen is dissolved into the volume at high temperature (T ~ 500 K) and pressure of the reaction mixture, the surface is reconstructed to give rise to formation of defects as cation vacancy V and intergrain boundaries. Two new types of active sites stabilize in the neighborhood of defects. The first type AS_V includes vacancy V on the surface Ag(111) and the adjoining subsurface oxygen atoms. This site contains O_{ox} solely. As the vacancy captures an additional oxygen $AS_V + O \rightarrow AS_d$-O, the second site AS_d-O forms and is associated with the subsurface oxygen atoms O_{ss} into ozonide type structures such as -Ag-O_{ss}-O_{ep}-O_{ss}-Ag-, which contains electrophilic oxygen O_{ep}. Choosing sites AS_V and AS_d-O, we assumed that the source of all active oxygen forms under the reaction conditions (high temperature and pressure with respect to oxygen) was oxygen dissolved in volume O_v, which stabilized on the defects during segregation into the subsurface silver layer $O_v \rightarrow O_{ss}$. Since the presence of subsurface oxygen is requisite for the selective ethylene oxidation [7], the active site of epoxidation includes also a defect (cation vacancy V) and subsurface oxygen O_{ss}:

(a) -Ag-O_v-Ag-O_v-Ag $\rightarrow$ -Ag-O_{ss}-V-O_{ss}-Ag-Ag-

(b) O + -Ag-O_{ss}-V-O_{ss}-Ag- $\rightarrow$ -Ag-O_{ss}-O_{ep}-O_{ss}-Ag-

Stage (a) describes formation of structural defect AS_V: Ag-O_{ss}-V-O_{ss}-Ag . Stage (b) simulates formation of the other site ($AS_V + O \rightarrow AS_d$-O) which forms by the capturing by vacancy V of atomic (either adsorbed or lattice) oxygen. In the neighborhood of the cation vacancy, the group of oxygen atoms associates to form a stable quasi-molecular form -O_{ss}-O_{ep}-O_{ss}-, containing electrophilic oxygen O_{ep}. During the reaction, the concentration of electrophilic oxygen decreases. The extended defects as intergrain boundaries can act as original channels providing transportation and continuous supplement of electrophilic oxygen. The presence of electrophilic oxygen suggests its high reactivity in the reaction of ethylene epoxidation.

The oxygen transportation to the defect zone, where site AS_V is located, transforms this site into a new site $AS\text{-}O_{ep}$, which contains electrophilic oxygen. We propose that insertion of this oxygen form into the double bond of ethylene reduces site AS_V and the cycle is repeated:

$$AS_V + O + C_2H_4 \rightarrow AS_d\text{-}O + C_2H_4 \rightarrow AS_V + C_2H_4O. \qquad \text{(I)}$$

The interaction between ethylene and site AS_d-O results in a number of stable intermediates INT and transition states TS for the most important reaction of cycle (I):

$$AS_d\text{-}O + C_2H_4 \rightarrow INT_i \rightarrow TS_i \rightarrow C_2H_4O + AS_v. \qquad \text{(II)}$$

According to the calculations there are three intermediates in reaction (II), which are associated with three transition states.

4.1. Structures of ethylene-oxygen intermediates on the AS_d-O center

The first intermediate INT_1 is an ethylene-oxygen complex $AS_d\text{-}O\text{-}C_2H_4$ (its structure was discussed above, see Fig. 9). In Fig. 11 the optimal structures of two intermediates are shown.

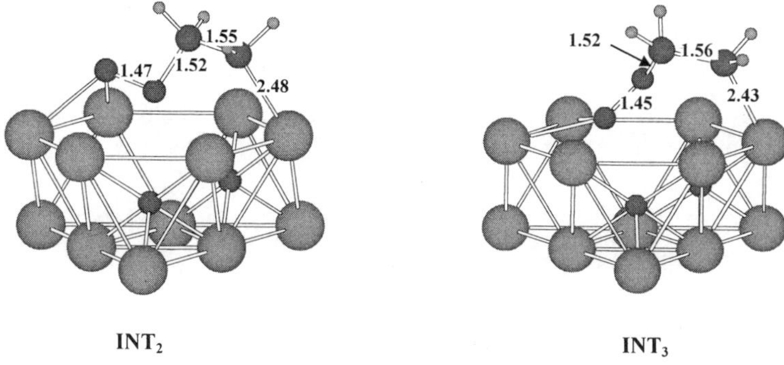

INT_2 INT_3

Fig. 11 The optimized structure of ethylene-oxygen intermediates on the oxidized surface Ag(111) with vacancy V (in approximation B3LYP/LAN1MB). The Ag12 structure skeleton was frozen at r(V-Ag) = r(Ag-Ag) = 3.07 Å, the distance between two silver layers was taken to be 2.47 Å.

It is important to mention that we did not detect ethylene adsorbed on site AS_d-O as a π complex as a stable structure. Such π-complexes of ethylene form on the other sites, which is adjacent to AS_d-O and contain nucleophilic oxygen (their structure is described above). The interaction between ethylene and AS_d-O results in an immediate opening of the double bond C=C and stabilization of structure INT_1 (Fig. 9). Therefore, the first step of reaction $AS_d\text{-}O + C_2H_4 \rightarrow$ INT1 is not an activated process. It should be noted that ethylene forms the bond only with the central oxygen atom of chain $Ag\text{-}O_{ss}\text{-}O_{ep}\text{-}O_{ss}\text{-}Ag$. The second atom of carbon forms a bond with the nearest atom of silver from the first layer. Therefore, the metal-ethylene-peroxide-cycle (MEOOC) of type $Ag\text{-}O_{ss}\text{-}O_{ep}\text{-}C_2H_4\text{-}Ag$ is formed. It includes electrophilic O_{ep} and subsurface O_{ss}. The electron structure of INT_1 was studied in detail in [59].

As one end oxygen atom of MEOOC is dissolved into the volume, the other structure INT_2 stabilizes. In structure INT_2, the peroxide oxygen form with r(O-O) = 1.47 Å is a component of the metal cycle. The bond length r(C-O) = 1.52 Å and bond C-C is slightly longer compared to that in INT_1. There are two O_{ox} in INT_2. During reaction (II), INT_2 is the most stable intermediate. If one atom of the peroxide oxygen diffuses into the volume, the third structure INT_3 stabilizes.

4.2. Structures of transition states (TS) on the AS_d center

Transition from one intermediate to another is an activated process. The energy of activation is determined from the energies of transition states (TS).

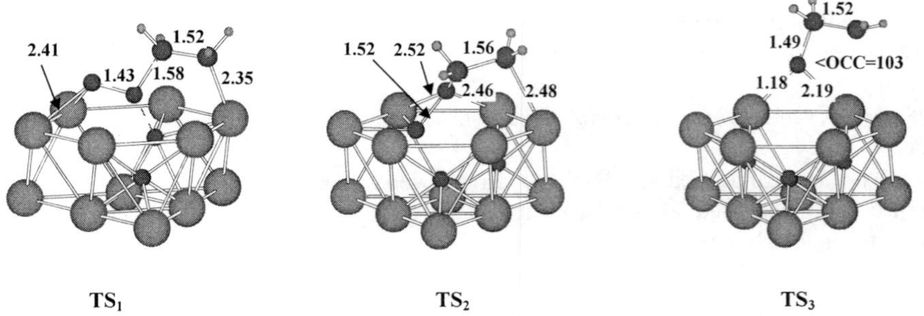

Fig 12. Structures of transition states on the reaction (II) path

Fig.12 shows optimal structures of three TS of reaction (II). The expanded form of reaction (II) is presented as Scheme 2:

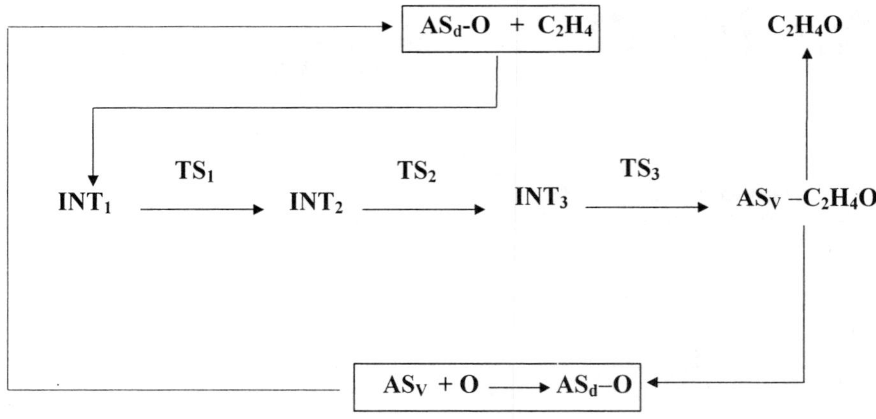

Scheme 2

Adsorbed ethylene oxide C_2H_4O-AS_V is the last intermediate on the reaction path. The desorption of ethylene oxide restores the structure of the active site with a cation vacancy (see Scheme 2). The cycle of ethylene epoxidation is closed by reaction $AS_v + O \rightarrow AS_d$-O.

Fig.13 shows the energy profile corresponding to Scheme 2.

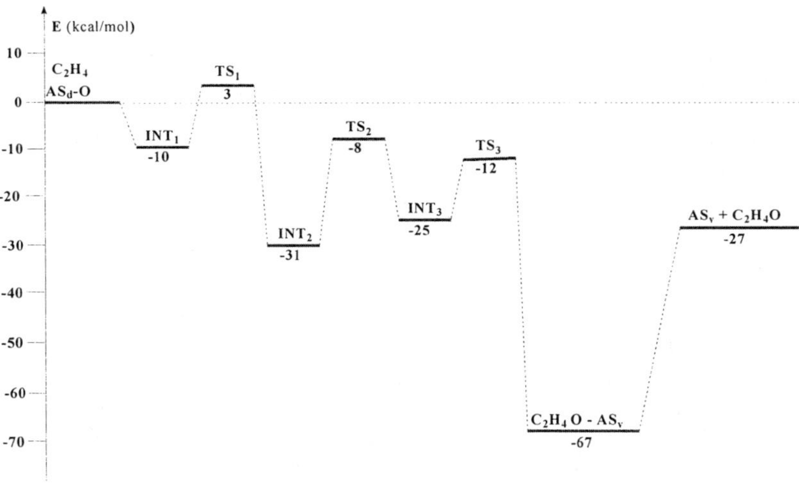

Fig. 13. Energy profiles for the mechanism of quasi-molecular oxygen epoxidation on site Ag-O-O-O-Ag

As Fig. 13 suggests, the stage $INT_1 \rightarrow TS_1 \rightarrow INT_2$ is the activated process with $E^* = 13$ kcal/mol. However, ethylene from the gas phase can immediately produce stable complex INT_2 with $E^* = 3$ kcal/mol. Structure TS_1 describes the transition from "ozonide" oxygen form to peroxide one. One bond O-O of "ozonide" form increases from r(O-O) = 1.48 Å in INT_1 to r(O-O) = 1.83 Å in TS_1. In INT_2 this bond is fully disrupted and oxygen diffuses to the subsurface layer and takes on the properties of oxide oxygen.

Stage $INT_2 \rightarrow TS_2 \rightarrow INT_3$ describes structural changes in a zigzag cycle Ag-O-O-C-Ag. This stage is required to provide minimal energy expenditures for breaking bond O-O in zigzag Ag-O-O-C at the final stage of formation of adsorbed ethylene oxide: $INT_3 \rightarrow TS_3 \rightarrow C_2H_4O\text{-}AS_V$.

As follows from Fig. 13, energies TS_2 and TS_3 are lower than the energy of initial reagents $E(AS_d\text{-}O + C_2H_4)$. A rather high value of the reaction heat $INT_3 \rightarrow C_2H_4O\text{-}AS_V$ (67 - 25 = 42 kcal/mol) is determined by high adsorption heat of ethylene oxide on the defect surface. We propose that two stages $INT_3 \rightarrow C_2H_4O\text{-}AS_V$ and $AS_V + O \rightarrow AS_d\text{-}O$ are conjugated, which permits one to close the cycle of conversion of ethylene into ethylene oxide (see Scheme 2).

5. Discussion

The concept of the active site of ethylene epoxidation on silver includes modification of the surface with subsurface oxygen as a necessary step (formation of vacancy and surface oxide Ag_2O). The simplest model of such sites AS_v and AS_d-O (Fig. 7) shows one probable way of formation of electrophilic oxygen O_{ep}. To provide formation of electrophilic oxygen, the below conditions should be met:

1. High degree of oxygen solubility into the silver volume. This is possible at high pressures with respect to oxygen and T > 300 K. Under such conditions, subsurface oxygen, defects (steps, adatoms, breaks, and vacancies) are formed on the surface.

2. A considerable weakening of Ag-O-Ag bonds and high oxygen mobility. This is possible at high concentration of oxygen in the subsurface layer. In this case, the associated species of oxygen is energetically preferable to the atomic one [58].

3. The presence of atom-size defects on the surface makes it possible to obtain high oxygen concentration in the neighborhood of defects. This is also required for evolution of nucleophilic oxygen to electrophilic one. The calculations suggest that there could be a direct interaction between electrophilic oxygen and ethylene.

Analysis of electron structure AS_d-O shows that only electrophilic oxygen O_{ep} is open for interaction with ethylene on such site. The other oxygen forms are inaccessible for the direct interaction with ethylene (though their presence is necessary). Therefore, structure AS_d-O is such that the reaction route involving participation of atomic oxygen responsible for complete ethylene oxidation is excluded. The calculated reaction route (Fig. 13) proceeds through a number of stages with low activation energies. At the final stage, the ethylene oxide desorption restores cation vacancy (site AS_V), and the following oxygen adsorption, occurring on this site with an activation energy $\Delta E(AS_V\text{-O}) = 61.2$ kcal/mol, regenerates electrophilic oxygen. Therefore, the formation of oxygen associates on the structural defects solves the problem of stable electrophilic oxygen. As discussed above, the stages of ethylene oxide desorption and formation of site AS-O are conjugated. As a result, reaction $2C_2H_4 + O_2 \rightarrow 2C_2H_4O$ is performed with high efficiency and selectivity (S = 100 %) [58].

An introduction of an idea of quasi-molecularity of the oxygen form being active on epoxidation permits one to understand and to discuss a very important aspect of epoxidation. It seems improper to discuss formation of oxygen species under action of only one oxygen form. It would be correct to consider the action of reaction medium $C_2H_4 + O_2$. Ethylene in the gas phase provides more efficient modification of the surface structural and chemical properties. Structural modification (reconstruction) proceeds at lower temperature than under the action of only oxygen. From the standpoint of chemical modification, the structural modification is accompanied by formation of carbon/carbonate species and thermally stable OH groups [26,75]. Using the quasi-molecular approach, one can consider elemental carbon and OH groups as the particles that favor disordering and reconstruction of the surface stabilizing defects, that is, strictly speaking, required for quasi-molecular forms of oxygen. Simulating a structure of the quasi-molecular oxygen species where the end oxygen atoms are changed to carbon atoms, one can conclude (from the general chemical standpoint) that electrophility of the central oxygen atom would not change and thermal stability of carbon-oxygen quasi-molecules, which can be signed as carbonate species, would only increase. Thus, under conditions of real catalysis where the surface of silver is not metallic (it would be proper to consider it as a layer of different chemical compounds), different associative forms can play the role of intermediate species.

Let's consider the relation between "molecular" and "atomic" conceptions of the mechanism of ethylene epoxidation. Based on the conception of electrophilic oxygen one can unite both conceptions. If one assumes that electrophilic oxygen atom is in quasi-molecular structure on the silver surface, one should speak about the molecular approach in epoxidation. Based on the results of our work, we consider the quasi-molecular form as adsorption of atomic oxygen on the oxidized parts of defects, moreover, it makes no difference whether the atomic oxygen is formed from the gas phase or from silver bulk.

Consequently, there is no longer contradiction between the information obtained by kinetic methods (the layer is homogeneous) and spectroscopy methods, which should provide information concerning different (non-equivalent) oxygen states. From the standpoint of energy, the oxygen atom sorbed on the oxy-like site is more easily bound than any atomic forms adsorbed directly on silver, and the process of epoxidation becomes more favorable. In addition, the quasi-molecular as a unit is more stable than any peroxide or superoxide species. There is an energetic compromise in a way between strength of bonds Ag-O and O-O. If the strength of bond Ag-O is high, the association of bonds O-O is not favorable. If the strength of bonds Ag-O is considerably low, thermal stability of the whole quasi-molecular structure can be insufficient and even lower than that of the peroxide complexes.

A structural defect as vacancy is of principal importance for formation of the quasi-molecular form, because only in this case the steric restrictions for association are removed. Association needs space and the vacancy provides this space. It is proposed that the mechanism of formation can be of two kinds. The first version is associated with local overheating of the surface, evolution of the hot silver adatom and diffusion flows on the surface under conditions of the reaction medium. It is used for determination of the most optimal defect structures stabilized by oxide-like subsurface oxygen. One may assume that after removal of the reaction medium, such surface can be annealed to regular structures with a loss of the necessary hole-like structure. The second version is based on the preliminary preparation of the specific defect structure of the whole volume of particles, which is thermally stable and favorable for formation of quasi-molecular species under oxygen action. The second version is probably realized in the technology of preparation of highly selective supported catalysts.

6. Conclusion

We propose a new concept of the ethylene epoxidizing oxygen species on silver. High temperature and pressure of mixture C_2H_4/O_2 increases the concentration of subsurface oxygen. Under these conditions, high mobility of the surface silver atoms inevitably produces many surface defects such as steps, craters, and cationic vacancies. The defects dramatically change the adsorption properties of the silver surface.

Only nucleophilic oxygen is formed on the regular surface. As the defect surface is saturated with oxygen, the Ag-O bonds become weaker, and oxygen atoms are stimulated to associate into stable quasimolecular species Ag-O-O-O-Ag. The latter are similar to metal ozonides and contain electrophilic atom. Therefore, the oxygen adsorption on the defect site transforms nucleophilic oxygen into electrophilic one. The suggested concept of electrophilic oxygen activity on the ethylene epoxidation on silver makes it possible to explain all available experimental data consistently from common positions.

Acknowledgment

This study was supported by the Russian Foundation for Basic Research (RFBR), grants 99-03-32419a and 00-15-97441.

References

[1] T.E. Lefort, French Patent 729952 (1931), Societe Francaise de Catalyse Generale.
[2] A.I. Boronin, V.I. Avdeev, S.V. Koshcheev, K.T Murzakhmetov, S. F. Ruzankin, G.M. Zhidomirov, Kinet. Catal. 40 (1999) 653, (English).
[3] M.L. McKee, J. Chem. Phys. 87 (1987) 3143.
[4] E.A. Carter, W.A. Goddard III, Surf. Sci. 209 (1989) 243
[5] H. Nakatsuji, Prog. Surf. Sci. 54 (1997) 1.
[6] R. B. Grant, R.M. Lambert, Langmuir, 1 (1985) 29.
[7] R. B. Grant, R.M. Lambert, J. Catal. 92 (1985) 364.
[8] R.A. van Santen, C.P.M. De Groot, J. Catal. 98 (1986) 530.
[9] R.A. van Santen, H.P.C.E. Kuipers, Adv. Catal. 35 (1987) 265.
[10] P.J. van der Hoek, E.J. Baerends, Surf. Sci. 221 (1989) L791.
[11] P.J. van den Hoek, E.J. Baerends, R.A. van Santen, J. Phys. Chem. 93 (1989) 6469.
[12] R.A. van Santen, M. Neurock, Catal. Rev.-Sci. Eng. 37 (1995) 557.
[13] C.N. Samoilenko, L.A.Petrov, A.I.Boronin et al. Zh. Fiz. Khim 75 (2001) 1757 (in Russian).
[14] M.J. Frisch, G.W. Trucks, H.B. Schlegel et al. Gaussian 98, Revision A.7, Gaussian, Inc., Pittsburgh PA, 1998.
[15] P.J. Hay, W.R. Wadt, J. Chem. Phys. 82 (1985) 270.
[16] W.R. Wadt, P.J. Hay, J. Chem. Phys. 82 (1985) 284.
[17] R.G. Parr, W. Yang, Density-Functional Theory of Atoms and Molecules, Oxford University Press, New York, 1989.
[18] A.D. Becke, Phys. Rev. A 33 (1986) 2786; J. Chem. Phys. 98 (1993) 5648.
[19] C. Lee, W. Yang, R.G. Parr, Phys. Rev. B37 (1988) 785.
[20] M.K. Rajumon, K. Prabahakaran, C.N.R. Rao, Surf. Sci. Lett. 233 (1990) L237.
[21] R.W. Joyner, M.W. Roberts, Chem. Phys. Lett. 60 (1979) 459.
[22] J.S. Hammond, S.W. Gaarenstrom, N. Winograd, Anal. Chem. 47 (1975) 2193.
[23] E.A. Ivanov, A.I. Boronin, S.V. Koscheev, G.M. Zhidomirov, React. Kinet. Catal. Lett. 66 (1999) 265.
[24] L.H. Tjeng, M.B.J. Meinders, J. Vanelp, J. Ghijsen, G.A. Sawatzky, R.L. Johnson, Phys. Rev. B41 (1990) 3190.
[25] A.I. Boronin, S.V. Koscheev, O.V. Kalinkina, G.M. Zhidomirov. React. Kinet. Catal. Lett. 63 (1998) 291.
[26] X. Bao, M. Muhler, Th. Schedel-Niedrig, R. Schlogl, Phys.Rev. B54 (1996) 2249.
[27] J.H. Wang, W.L. Dai, J.F. Deng et al. Appl. Surf.Sci. 126 (1998) 148.
[28] S.S. Kabalkina, S.V. Popova, N.R. Serebryanaya, L.F. Vereshchagin, Dokl. Akad. Nauk (DAN) SSSR, 152 (1963) 853. (in Russian).
[29] M. Bowker, Surf. Sci. 155 (1983) L276.
[30] A.I. Boronin, S. V. Koscheev, K. T Murzakhmetov, V. I. Avdeev, G.M. Zhidomirov, App. Surface Science 165 (2000) 9.
[31] R.W. Joyner and M.W. Roberts, Chem. Phys. Lett. 60 (1979) 459.
[32] M. Ayyoob, M.S. Hegde, Surf. Sci. 133 (1983) 516.
[33] C.T. Au, S. Singh-Borani, M.W. Roberts, R.W. Joyner, J. Chem. Soc. Faraday Trans. I, 79 (1983) 1779.
[34] R. B. Grant, R.M. Lambert, Surf. Sci. 146 (1984) 256.

[35] X. Bao, S. Dong, J. Deng, Surf. Sci. 199 (1989) 493.
[36] A.I. Boronin, S.V. Koscheev, V.F. Malakhov, G.M. Zhidomirov, Catal. Lett. 47 (1997) 111.
[37] A.I. Boronin, S.V. Koscheev, G.M. Zhidomirov, J. Electron Spectrosc&Relat.Phenom. 96 (1998) 43.
[38] C. Pettenkofer, I. Pockrand, A. Otto, Surf. Sci. 135 (1983) 52.
[39] G. J. Millar, J. B. Metson, G. A. Bowmaker, R. P. Cooney, J. Catal. 147 (1994) 404.
[40] G. J. Millar, J. B. Metson, G. A. Bowmaker, R. P. Cooney, J. Chem. Soc., Faraday Trans. 91 (1995) 4149.
[41] G. J. Millar, M. L. Nelson, P.J.R. Urwins, Catal. Letts. 43 (1997) 97.
[42] D. I. Kondarides, G. N. Papatheodorou, C. G. Vaynas, X. E. Verykios, Ber. Bunsenges. Phys.Chem. 97 (1993) 709.
[43] K. Wu, D. Wang, J. Deng, X. Wei, Y. Cao, M. Zei, R. Zhai, X. Cao, Surf. Sci. 264 (1992) 249.
[44] K. Wu, D. Wang, X. Wei, Y. Cao, X. Cao, J. Catal. 140 (1993) 370.
[45] X. Bao, B. Pettinger, G. Ertl, R. Schlogl, Ber. Buns. Phys. Chem. 97 (1993) 322.
[46] J. Deng, X. Xu, J. Wang, J.Catal. Lett. 32 (1995) 159.
[47] N. Rosch, D. Menzel, J. Chem. Phys. 13 (1976) 243.
[48] R.L. Martin, P.J. Hay, Surf. Sci. 130 (1983) L283.
[49] A. Selmani, J.M. Sichel, D.R. Salahub, Surf. Sci. 157 (1985) 208.
[50] T.H. Upton, P. Stevens, R.J. Madex, J. Chem. Phys. 88 (1988) 3988.
[51] K. Broomfield, R.M. Lambert, Mol. Phys. 66 (1989) 421.
[52] K.A. Jorgensen, R. Hoffmann, J. Phys. Chem. 94 (1990) 3046.
[53] V. I. Avdeev, S.F. Ruzankin, G.M. Zhidomirov, J. Struct. Chem. 38 (1997) 519, (English).
[54] M.R. Salazar, C. Saravanan, J.D. Kress, A. Redondo, Surf. Sci. 449 (2000) 75.
[55] B.Shen, Z. Fang, K. Fan, J. Deng, Surf. Sci. 459 (2000) 206.
[56] S.V. Tsybulya, G.N. Kryukova, S.N. Goncharova, A.N. Shmakov, B.S. Bal'zhinimaev, J. Catal. 154 (1995) 194.
[57] B. S. Bal'zhinimaev, V. I. Zaikovskii, L. G. Pinaeva, A. V. Romanenko, G. V. Ivanov, Kinet. Catal. 39 (1998) 714, (in English).
[58] V.I. Avdeev, A.I. Boronin, S.V. Koscheev, G.M. Zhidomirov, J. Mol. Catalysis A: Chem. 154 (2000) 257.
[59] V.I. Avdeev, G.M. Zhidomirov, Surf. Sci. 492 (2001) 137.
[60] L.H. Tjeng, M.B.J. Meinders, J. van Elp, J. Ghijsen, G.A. Sawatzky, R.L. Johnson, Phys. Rev. B 41 (1990) 3190.
[61] H.Y. Huang, J. Padin, R.T. Yang, J. Phys. Chem. B 103 (1999) 3206.
[62] H.Y. Huang, J. Padin, R.T. Yang, Ind. Eng. Chem. Res. 38 (1999) 2720.
[63] C. Backx, C.P.M. de Groot and P. Bilon, Appl. Surf. Sci. 6 (1980) 256.
[64] A.A. Efremov, Yu.D. Pankratiev, A.A. Davydov, G.K. Boreskov, React. Kinet. Catal.. Lett. 20 (1982) 87.
[65] J.L. Carter, D.J.C. Yates, P.J. Lucchesi, J.J. Elliott and V. Kevorkian, J. Phys. Chem. 70 (1966) 1126.
[66] G.S. Jones, M. Mavrikakis, M.A. Barteau, J.M. Vohs, J. Am. Chem. Soc. 120 (1998) 3196.
[67] M. Mavrikakis, D.J. Doren, M.A. Barteau, J. Phys. Chem. B, 102 (1998) 394.
[68] C. Saravanan, M.R. Salazar, J.D. Kress and A. Redondo, J. Phys. Chem. B 104 (2000) 8685.

[69] G.K. Boreskov, A.V. Khasin, T.S. Starostina,
Dokl. Akad. Nauk (DAN) SSSR, 164 (1965) 606 (in Russian).
[70] G.K. Boreskov, A.V. Khasin,
J. Res. Inst. Catalysis, Hokkaido Univer. 16 (1968) 477.
[71] P. H. McBreen, M. Moskovits, J.Catal. 103 (1987) 188.
[72] Chuan-Bao Wang, G. Deo, I. E. Wachs, J. Phys. Chem. B 103 (1999) 5645
[73] D.E. Tevault, R.R. Smardzewski, M. W. Urban, K. Nakamoto,
J. Chem. Phys. **77** (1982) 577.
[74] D.E. Tevault, R.A. DeMarco, R.R. Smardzewski, J. Chem. Phys. **75** (1981) 4168.
[75] V.I. Bukhtiyarov, A.I. Boronin, I.P. Prosvirin, V.I. Savchenko,
J. Catal. 150 (1994) 268.

Computational Materials Science
C.R.A. Catlow and E.A. Kotomin (Eds.)
IOS Press, 2003

Atomistic Measures of Materials Strength and Deformation

Ju Li, Wei Cai, Jinpeng Chang, Sidney Yip
Department of Nuclear Engineering
Massachusetts Institute of Technology, Cambridge, MA 02139, USA

Abstract. Multiscale materials modeling has emerged as a significant concept as well as a unique approach in computational materials research. We examine here the role of atomistic simulations in modeling the structural responses of solids to thermal and mechanical loading, particularly with regard to the mechanistic understanding derived from the atomic-level details generally not available from experiments. Theoretical strength is defined through elastic modes of instability, or more generally, through the onset of soft vibrational modes in the deformed lattice. Molecular dynamics (MD) simulation of stress-strain response provides a direct measure of the effects of small-scale microstructure on strength, as illustrated by results on single crystal, amorphous, and nanocrystalline phases. In the kinetics of melting, we distinguish between the thermoelastic process of mechanical melting which is homogeneous and sets the upper limit of metastability of the lattice, and the free-energy driven process of thermodynamic melting which involves nucleation and growth and is therefore heterogeneous. In the kinetics of defect mobility we study dislocation dynamics by direct MD simulation and a mesoscale (kinetic Monte Carlo) method which couples kink mechanism energetics to the experimentally measured dislocation velocity. In the area of crack propagation, brittle-ductile transition and crack-tip plasticity are two well-known problems in fracture mechanics that have been looked at from the atomistic standpoint. From a discussion of these case studies one may gain some appreciation of the capabilities, as well as limitations, of atomistic simulations to provide physical insight in computational materials research.

1. Introduction

Understanding materials behavior at the atomic level has long been a grand challenge to scientists and engineers across many disciplines. Currently there is widespread interest in identifying fundamental problems in materials modeling which combine scientific challenges with technologically relevant applications [1]. To provide a basis for such inquiries, we discuss here a particular focus on the molecular understanding of mechanical behavior, in the context of a multiscale approach to materials theory and simulation [2,3]. The aim of this chapter, written in the spirit of a set of lecture notes, is to discuss how strength and deformation at the atomistic level can be probed through structural instability and modes of dynamical response to critical loading, either by heating or an applied stress. In examining several case studies we hope the readers will feel stimulated to draw analogies between fundamental issues which sometimes are considered only separately, such as thermal versus mechanical responses, elastic and plastic deformations, and homogenous and heterogeneous processes. By noting the contrasts and parallels between the individual topics discussed, we believe it is possible to gain some understanding of the role of atomistic simulations in probing complex systems phenomena in the materials research arena.

2. Limits to Strength: Structural Instabilities

The theoretical basis for describing the mechanical stability of a crystal lattice lies in the formulation of stability conditions which specify the critical level of external stress that the system can withstand. Lattice stability is not only one of the most central issues in elasticity, it is also fundamental in any analysis of structural transitions in solids, such as polymorphism, amorphization, fracture, or melting. In these notes our goal is to discuss the role of elastic stability criteria at finite strain in elucidating the competing mechanisms underlying a variety of structural instabilities, and the physical insights that may be gained by probing stress and temperature induced structural responses through atomistic simulations.

Born has shown that by expanding the internal energy of a crystal in a power series in the strain and requiring positivity of the energy, one obtains a set of conditions on the elastic constants of the crystal that must be satisfied to maintain structural stability of the lattice [4,5]. This then leads to the determination of ideal strength of perfect crystals as an instability phenomenon, a concept which has been examined by Hill [6] and Hill and Milstein [7], as well as used in various applications [8]. That Born's results are valid only when the solid is under zero external stress has been explicitly pointed out in a later derivation by Wang et al [9] invoking the formulation of a Gibbs integral. Further discussions were given by Zhou and Joos [10] and by Morris and Krenn [11], the latter emphasizing the thermodynamic basis of the concept of theoretical strength by showing that the conditions of elastic stability, based on Gibbs' original formulation [12], are identical to the results of Wang et al. in which the loading mechanism fixes the Cauchy stress. A consequence of these investigations is that theoretical strength should be considered a property which can be affected by the symmetry and magnitude of the applied load, rather than an intrinsic property of the material system only. In this respect the study of theoretical limits to material strength using atomistic models, including first-principles calculations [13], promises to yield new insights into mechanisms of structural instability.

While the stability criteria say nothing about the final state toward which a structurally unstable system will evolve, nevertheless they can be invaluable in interpreting molecular dynamics simulation results. In the context of simulating the outcome of a virtual strength test, quantitative predictions can be made of the maximum deformation (strain) the lattice can sustain, and the competition between different modes of instability can be analyzed. In this section we will first give a brief derivation of elastic stability at finite strain to bring out in a direct manner the interplay between the intrinsic response to deformation and the effect of external work. Then we note that vibrational instability in the form of soft phonon modes is an extension of this concept, and that direct molecular dynamics simulation can be used to probe both. In the last part of this section we consider briefly how strength is affected by the microstructure of the material. Since a crystal attains its ideal (maximum) strength in the absence of any defect, it is the presence of any microstructural features, such as disorder or interfaces that will lower the strength.

2.1 Elastic Stability Criteria

Consider a perfect lattice undergoing homogeneous deformation under an applied stress τ, where the system configuration changes from X to Y = JX, with J being the deformation gradient or the Jacobian matrix. The associated Lagrangian strain tensor is

$$\eta = (1/2)(J^T J - 1) \quad (1)$$

Let the change in the Helmholtz free energy be expressed by an expansion in η to second order,

$$\Delta F = F(X,\eta) - F(X,0)$$

$$= V(X)[t(X)\eta + (1/2)C(X)\eta\eta] \quad (2)$$

where V is the volume, t the conjugate stress which is also known as the thermodynamic tension or the second (symmetric) Piola-Kirkhoff stress, and C the fourth-order elastic constant tensor. For the work done by an applied stress τ, which is commonly called the Cauchy or true stress, we imagine a virtual move near Y along a path where $J \to J + \delta J$ which results in an incremental work

$$\delta W = \oint_S \tau_{ij} n_j \delta u_i dS$$

$$= V(Y)\frac{\tau_{ij}}{2}\left(\frac{\partial u_i}{\partial Y_j} + \frac{\partial u_j}{\partial Y_i}\right)$$

$$= V(Y) Tr(J^{-1}\tau J^{-T}\delta\eta) \quad (3)$$

The work done over a deformation path ℓ, $\Delta W(\ell)$, is the integral of δW, given by Eq.(3), over the path. To examine the lattice stability at configuration X, we now consider the difference between the increase in Helmholtz free energy and the work done by the external stress,

$$\Delta G(Y,\ell) = \Delta F(X,\eta) - \Delta W(\ell)$$

$$= \int_\ell g(Y) d\eta \quad (4)$$

where

$$g(Y) = \frac{\partial F}{\partial \eta} - V(Y) J^{-1}\tau J^{-T} \quad (5)$$

One may also interpret ΔG in the spirit of a virtual work argument. If the work done by the applied stress exceeds that which is absorbed as the free energy increase, then an excess amount of energy would be available to cause the displacement to increase and the lattice would become unstable.

We regard ΔG as a Gibbs integral in analogy with the Gibbs free energy, the appropriate thermodynamic potential in the (NTP) ensemble. However, notice that ΔG is in general dependent on the deformation path through the external work contribution. This means that strictly speaking it is not a true thermodynamic potential on which one can perform the usual stability analysis. Nevertheless, -g(Y) can be treated as a force field in deformation space for the purpose of carrying out a stability analysis [9]. Suppose the lattice, initially at equilibrium at X under stress τ, is perturbed to configuration Y with corresponding strain η. A first-order expansion of g(Y) gives

$$g_{ij}(\eta) = V(Y)B_{ijkl}\eta_{kl} + \ldots \tag{6}$$

where, by using $V(Y) = V(X)\mathbf{det}|J|$, one obtains

$$B_{ijkl} = C_{ijkl} - \left[\frac{\partial(\det|J|J^{-1}_{im}\tau_{mn}J^{-1}_{nj})}{\partial \eta_{kl}}\right]_{\eta=0, J=I}$$

$$= C_{ijkl} + \Lambda_{ijkl}(\tau) \tag{7}$$

with

$$\Lambda_{ijkl}(\tau) = (1/2)[\delta_{ik}\tau_{jl} + \delta_{jk}\tau_{il} + \delta_{il}\tau_{jk} + \delta_{jl}\tau_{ik} - 2\delta_{kl}\tau_{ij}] \tag{8}$$

δ_{ij} being the Kronecker delta symbol for indices i and j. The physical implication of Eq.(6) is that in deformation space the shape of the force field around the origin is described by B. The stability condition is then the requirement that all the eigenvalues of B be positive, or

$$\mathbf{det}|A| > \mathbf{0} \tag{9}$$

where $A = (\mathbf{1}/\mathbf{2})(B^T + B)$, with B being in general asymmetric [9], In cases where the deformation gradient J is constrained to be symmetric, as in certain atomistic simulations at constant stress, one can argue that the condition det|B| > 0 is quite robust [9]. Thus, lattice stability is governed by the fourth-rank tensor B, a quantity which has been called the elastic stiffness coefficient [14]. It differs from the conventional elastic constant by the tensor Λ which is a linear function of the applied stress. The foregoing derivation shows clearly the effect of external work which was not taken into account in Born's treatment. In the limit of vanishing applied stress one recovers the stability criteria given by Born [4,5].

In the present discussion we will consider only cubic lattices under hydrostatic loading in which case the stability criteria take on a particularly simple form,

$$K = (1/3)(C_{11} + 2C_{12} + P) > 0$$

$$G' = (1/2)(C_{11} - C_{12} - 2P) > 0 \tag{10}$$

$$G = C_{44} - P > 0$$

where C_{ij} are the elastic constants at current pressure P, P > 0 (< 0) for compression (tension). K is seen to be the isothermal bulk modulus, G' and G the tetragonal and rhombohedral shear moduli respectively. The theoretical strength is that value of P for which one of the three conditions in Eq.(10) is first violated. A simple demonstration showing that the external load must appear in the stability criteria is to subject a crystal to hydrostatic tension by direct atomistic simulation using a reasonable interatomic potential. In this case one finds the instability mode is the vanishing of K, whereas the Born criteria, Eq.(10) with P set equal to zero, would predict the vanishing of G' [9].

It is worth mentioning that the six components of the eigenmodes of deformation corresponding to the three zero eigenvalues of det(B) are $(1,1,1,0,0,0)\delta\eta$, $(\delta\eta_{xx},\delta\eta_{yy},\delta\eta_{zz},\mathbf{0},\mathbf{0},\mathbf{0})$ with $\delta\eta_{xx}+\delta\eta_{yy}+\delta\eta_{zz}=\mathbf{0}$ in the order indicated in Eq.(10) [9]. The deformation when the bulk modulus vanishes (spinodal instability) preserves the cubic symmetry, while for the tetragonal shear instability the cubic symmetry must be broken.

The connection between stability criteria and theoretical strength is rather straightforward. For a given applied stress $\underline{\tau}$ one can imagine evaluating the current elastic constants to obtain the stiffness coefficients B. Then by increasing the magnitude of $\underline{\tau}$ one will reach a point where one of the eigenvalues of the matrix A (cf. Eq.(3)) vanishes. This critical stress at which the system becomes structurally unstable is then a measure of theoretical strength of the solid. In view of this, one has a direct approach to strength determination through atomistic simulation of the structural instability under a prescribed loading. If the simulation is performed by molecular dynamics, temperature effects can be taken into account naturally by following the particle trajectories at the temperature of interest.

Under a uniform load the deformation of a single crystal is homogeneous up to the point of structural instability. For a cubic lattice under an applied hydrostatic stress, the load-dependent stability conditions are particularly simple, being of the form

$$B=(C_{11}+2C_{12}+P)/3>0,\quad G'=(C_{11}-C_{12}-2P)/2>0,\quad G=C_{44}-P>0, \qquad (11)$$

where P is positive (negative) for compression (tension), and the elastic constants C_{ij} are to be evaluated at the current state. While this result is known for some time [15-17], direct verification against atomistic simulations showing that the criteria do accurately describe the critical value of P (P_c) at which the homogeneous lattice becomes unstable has been relatively recent [9, 18-22]. One may therefore regard P_c as a definition of theoretical or ideal tensile (compressive) strength of the lattice.

Turning now to molecular dynamics simulations we show in Fig. 1 the stress-strain response for a single crystal of Ar under uniaxial tension at 35.9K. At every step of fixed strain, the system is relaxed and the virial stress evaluated. One sees the expected linear elastic response at small strain up to about 0.05; thereafter the response is nonlinear but still elastic up to a critical strain of 0.1 and corresponding stress of 130 MPa. Applying a small increment strain beyond this point causes a dramatic stress reduction (relief) at point (b). Inspection of the atomic configurations at the indicated points shows the following.

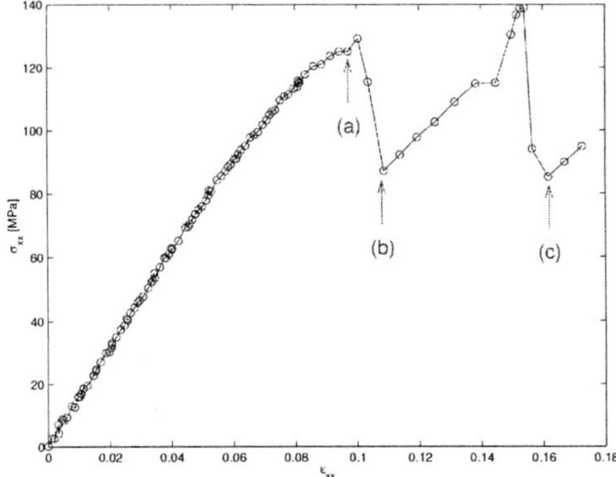

Fig. 1. Atomistic stress-strain response of a single crystal of Ar under aniaxial tensile deformation at constant strain at a reduced temperature of 0.3 (35.9K). Simulate data are indicated as circles and solid line is drawn to guide the eye.

At point (a) several point defect like inhomogeneities have been formed; most probably one or more will act as nucleation sites for a larger defect which causes the strain energy to be abruptly released. At the cusp, point (b), one can clearly discern an elementary slip on an entire [111] plane, the process being so sudden that it is difficult to capture the intermediate configurations. Figuratively speaking, we suspect that a dislocation loop is spontaneously created on the (111) plane which expands at a high speed to join with other loops or inhomogeneities until it annihilates with itself on the opposite side of the periodic border of the simulation cell, leaving a stacking fault. As one increases the strain the lattice loads up again until another slip occurs. At (c) one finds that a different slip system is activated.

2.2 Soft Modes

One may regard the stability criteria, Eq.(5), as manifestation in the long wavelength limit of the general condition for vibrational stability of a lattice. The vanishing of elastic constants then corresponds to the phenomenon of soft phonon modes in lattice dynamics. Indeed one finds that under sufficient deformation such soft modes do occur in a homogeneously strained lattice. To see the lattice dynamical manifestation of this condition, we apply molecular dynamics to relax a single crystal sample with periodic boundary condition at essentially zero temperature for a specified deformation at constant strain. The resulting atomic configurations are then used to construct and diagonalize the dynamical matrix. Fig. 2 shows two sets of dispersion curves for the Lennard-Jones interatomic potential describing Ar which has fcc structure, one for the crystal at equilibrium (for reference) and the other when the lattice is deformed under a uniaxial tensile strain of 0.138 which is close to the critical value [23]. One can see in the latter a Γ'-point soft mode in the [011] direction. Similar results for deformation under shear or hydrostatic tensile strain would show soft modes Γ'-point in the [111] direction and Γ-point in the [100] direction respectively. All these are acoustic zone-center modes, therefore they would correspond to elastic instabilities.

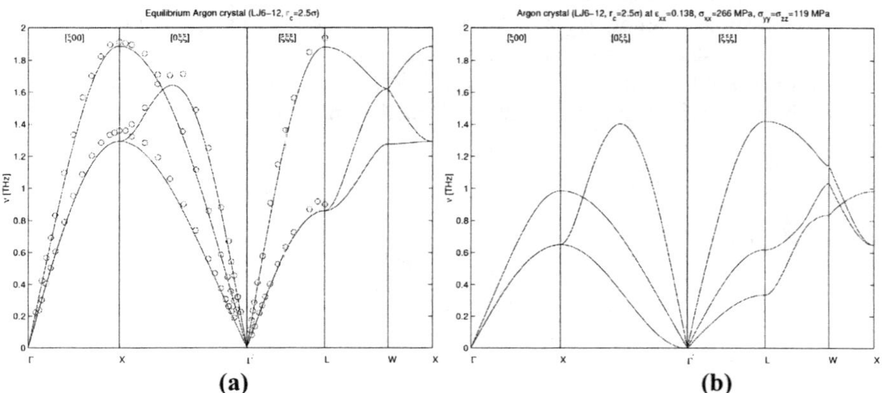

Fig. 2. Phonon dispersion curves of single crystal of Ar as described by the Lennard-Jones potential (lines), (a) comparison of results for equilibrium condition with experimental data (circles), (b) results for uniaxial tension deformation at strain of 0.138 (corresponding stresses of 266 MPa and 119 MPa along the tensile and transverse direction).

For a more complicated lattice such as SiC in the zinc blende structure, one would find that soft modes also can occur at the zone center [23]. The overall implication here is that lattice vibrational analysis of a deformed crystal offers the most general measure of structural instability, and this again demonstrates that strength is not an intrinsic property of the material, rather it depends on the mode of deformation.

2.3 Microstructural Effects

Fig. 1 is a typical stress-strain response on which one can conduct very detailed analysis of the deformation using the atomic configuration available from the simulation. This atomic-level version of structure-property correlation can be even more insightful than the conventional macroscopic counterpart simply because in simulation the microstructure can be as well characterized as one desires. As an illustration we repeat the deformation simulation using as initial structures other atomic configurations which have some distinctive microstructural features. We have performed such studies on cubic SiC (3C or beta phase) .which has zinc-blends structure, using an empirical bond-order potential [24] and comparing the results for a single crystal and prepared amorphous and nanocrystalline structures.[23] Fig. 3 shows the stress-strain response for under hydrostatic tension at 300K. At every step of fixed strain, the system is relaxed and the virial stress evaluated. Three samples are studied, all with periodic boundary conditions, a single crystal (3C), an amorphous system that is an enlargement of a smaller configuration produced by electronic-structure calculations [25], and a nanocrystal composed of four distinct grains with random orientations (7810 atoms). As in Fig. 1, the single-crystal sample shows in Fig. 3 the expected linear elastic response at small strain up to about 0.03; thereafter the response is nonlinear but still elastic up to a critical strain of 0.155 and corresponding stress of 38 GPa. Applying a small increment strain beyond this point causes a dramatic change with the internal stress suddenly reduced by a factor of 4. Inspection of the atomic configurations (not shown) reveals the nucleation of an elliptical microcrack in the lattice along the direction of maximum tension. With further strain increments the specimen deforms by strain localization around the crack with essentially no change in the system stress.

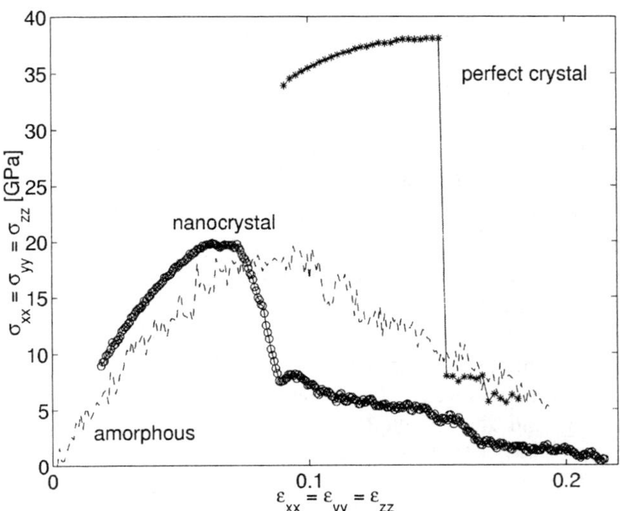

Fig. 3. Variation of virial stress at constant strain from MD simulations of SiC (3C) under hydrostatic tension at 300K in perfect crystal, amorphous, and nanocrystalline phases.

The responses of the amorphous and nanocrystal SiC differ significantly from that of the single crystal. The former shows a broad peak, at about half the critical strain and stress, suggesting a much more gradual structural transition. Indeed, the deformed atomic configuration reveals channel-like decohesion at strain of 0.096 and stress 22 GPa. Another feature of the amorphous sample is that the response to other modes of deformation, uniaxial tension and shear, is much more isotropic relative to the single crystal, which is perhaps understandable with bonding in SiC being quite strongly covalent and therefore directionally dependent. For the nanocrystal, the critical strain and stress are similar to the amorphous phase, except that the instability effect is much more pronounced, qualitatively like that of the single crystal. The atomic configuration shows rather clearly the failure process to be intergranular decohesion. These observations allow us to correlate the qualitative behavior of the stress-strain responses with a gross feature of the system microstructure, namely, the local disorder (or free volume). This feature is of course completely absent in the single crystal, well distributed in the amorphous phase, and localized at the grain boundaries in the nanocrystal. The disorder can act as a nucleation site for structural instability, thereby causing a reduction of the critical stress and strain for failure. Once a site is activated, it will tend to link up with neighboring activated sites, thus giving rise to different behavior between the amorphous and nanocrystal samples.

3. Mechanistic Aspects (Kinetics) of Melting

In 1939 Born set forth a simple criterion for crystal melting by postulating that melting should be accompanied by the loss of shear rigidity.[26] Expressed in terms of the shear modulus G for a cubic crystal, the melting point T_m is that temperature at which

$$G(T_m) = 0 \qquad (12)$$

A year later he extended this stability concept to lattice deformation [4] by deriving the conditions for mechanical stability which in the case of cubic crystals are given by

$$C_{11} + 2C_{12} > 0, \quad C_{11} - C_{12} > 0, \quad C_{44} > 0 \qquad (13)$$

These can be obtained from stability criteria, (11), derived in Sec. 2.1 under the condition of no external stress. In this section we will examine the basis on which Born's two criteria may be considered to be valid. Shortly after (12) was proposed, experimental results obtained on NaCl single crystals were presented showing that the two shear constants, C_{44} and $(C_{11} - C_{12})$, have *nonzero* values at the melting point.[27] Moreover, it was not clear how this criterion could explain the existence of latent heat and volume change in a first-order thermodynamic phase transition. In contrast, the stability criteria (13) seem to be generally accepted, with neither stringent tests having been performed nor qualifications concerning its possible limitations discussed. The challenge of ascertaining whether such criteria are capable of predicting the actual onset of an instability is considerable. The difficulty, on the theoretical side, has been that stability analyses have been formulated in different ways [6,7], and few explicit calculations of elastic constants at the critical condition have been reported to make possible an unambiguous test. On the experimental side, competing effects frequently render the determination of the triggering instability uncertain. Thus, while the shortcomings of (12) are well known, the use of (13) to define structural resistance to thermal agitation has gone unnoticed.

3.1 Mechanical Melting – Limit of Metastability

Our interest here is to test (13) through molecular dynamics simulation of melting instead of testing (12) using experimental data. By performing simulation of isobaric heating to melting at zero pressure of a perfect crystal without surfaces or defects of any kind, we achieve an unambiguous test since without an external stress (13) would be equivalent to (11). As we will see below, simulation shows that at the onset of melting one of the shear constants indeed vanishes, although it is $(C_{11} - C_{12})$ rather than C_{44}. The observed melting temperature, or equivalently the critical lattice strain, is in remarkable agreement with the predictions based on the stability criteria. Since the system is initially a defect- and surface-free lattice, the homogeneous melting observed here is to be distinguished from the conventional melting which is a free-energy based heterogeneous process of nucleation and growth. The latter process, if not kinetically suppressed in simulation by eliminating all defects and surfaces, would set in at a lower temperature, the conventional melting point of the material, and preclude the melting process associated with an elastic instability. Allowing for these modifications, the melting and stability criteria proposed by Born are reconciled. The qualification which is nontrivial is that the concept of thermoelastic mechanism of melting indeed applies to a form of melting, but it is melting in the sense of mechanical stability against thermal excitation as opposed to the conventional thermodynamic process which is always defect mediated and therefore heterogeneous.

Given that the generalized criteria (11) obviously reduce to Born's results in the limit of zero load, then (13) is a valid description of lattice stability in the special case of a cubic crystal being heated to melting at zero pressure. For the simulation we use an interatomic potential model for Au [28] (details of the potential are of no interest in this discussion) and a simulation cell containing 1372 atoms with periodic border conditions imposed in the manner of Parrinello and Rahman.[29] A series of isobaric-isothermal simulations (with velocity rescaling) are carried out at various temperatures. At each temperature the atomic trajectories generated are used to compute the elastic constants at the current state using appropriate fluctuation formulas.[30]

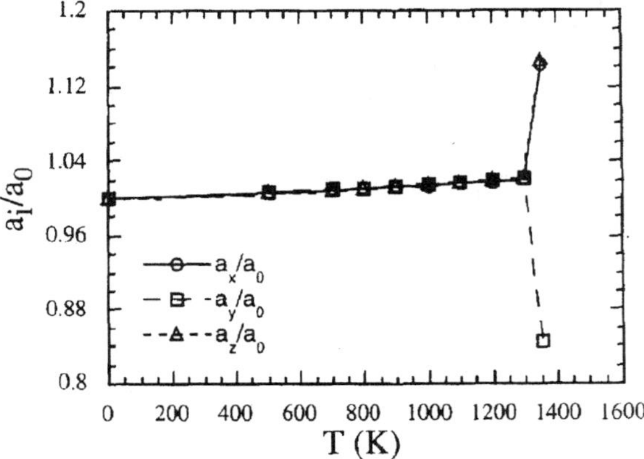

Fig. 4. Variation of lattice strain a/ao with temperature along three Cartesian directions in the simulation of an isobaric (P=0) heating process.

Fig. 4 shows the variation with temperature of the lattice strain a/a_0 along the three cubic symmetry directions.[31] The slight increase with increasing temperature merely indicates the lattice is expanding normally with temperature, and the results for the three directions are the same as they should be. At T = 1350 K one sees a sharp bifurcation in the lattice dimension where the system elongates in two directions and contracts in the third. This is a clear sign of symmetry change, in the present case from cubic to tetragonal. To see whether the simulation results are in agreement with the prediction based on (7), we show in Fig. 5 the variation of the elastic moduli with temperature, or equivalently the lattice strain since there is a one-to-one correspondence as indicated in Fig. 3; the three moduli of interest are the bulk modulus $B_T = (C_{11} + 2C_{12})/3$, tetragonal shear modulus $G' = (C_{11} - C_{12})/2$, and rhombohedral shear modulus $G = C_{44}$. On the basis of Fig. 4 one would predict the incipient instability to be the vanishing of G', occurring at the theoretical or predicted lattice strain of $(a/a_o)_{th} = 1.025$. From the simulation at T = 1350 the observed strain is $(a/a_o)_{obs} = 1.024$. Thus, we can conclude that the vanishing of tetragonal shear is responsible for the structural behavior.

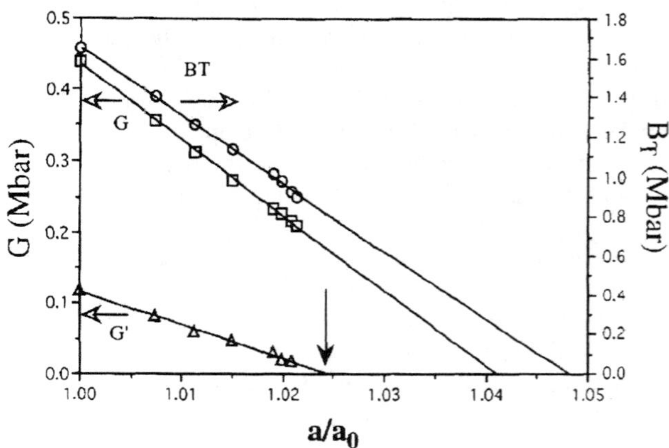

Fig. 5. Variation of BT, G, G' with lattice strain a/ao in the simulation of Fig. 4.

For more details of the system behavior at T = 1350 K we show in Fig. 6 the time evolution of the lattice strain, the off-diagonal elements of the cell matrix H, and the system volume. It is clear from Fig. 6(a) that the onset of the G' = 0 instability triggers both a shear (cf. Fig. 6(b)) and a lattice decohesion (Fig. 6(c)), the latter providing the characteristic volume expansion associated with melting. This sequence of behavior, which has not been recognized previously, implies that the signature of a first-order transition, namely, latent volume change, is not necessarily associated with the incipient instability. Our results also provide evidence supporting Born's picture of melting being driven by a thermoelastic instability [26], later reinterpreted by Boyer [32] to involve a combination of loss of shear rigidity and vanishing of the compressibility. Moreover, it is essential to recognize that this thermoelastic mechanism can only be applied to the process of mechanical instability (homogeneous melting) of a crystal lattice without defects, and not to the coexistence of solid and liquid phases at a specific temperature (heterogeneous melting).[33,34]

It is perhaps worthwhile emphasizing again what the combination of stability analysis and molecular dynamics simulation has contributed to the understanding of Born's two criteria. That the stability criteria (13) are valid only under vanishing external load is quite clear, both theoretically and in simulation studies. Since it is often advantageous to be able to predict a priori the critical stress or strain for the onset of instability, the availability of (11) could facilitate more quantitative analysis of simulation results. Although our results for an fcc lattice with metallic interactions show that homogeneous melting is triggered by G' = 0 and not (6), nevertheless, they constitute clear-cut evidence that a shear instability is responsible for *initiating* the transition. The fact that simulation reveals a sequence of responses apparently linked to the competing modes of instabilities (cf. Fig. 6) implies that it is no longer necessary to explain all the known characteristic features of melting on the basis of the vanishing of a single modulus. In other words, independent of whether G'=0 is the initiating mechanism, the system will in any event undergo volume change and latent heat release in sufficiently rapid order (on the time scale of physical observation) that these processes are all identified as part of the melting phenomenon. Generalizing this observation further, one may entertain the notion of a hierarchy of interrelated stability catastrophes of different origins, elastic, thermodynamic, vibrational, and entropic.[35]

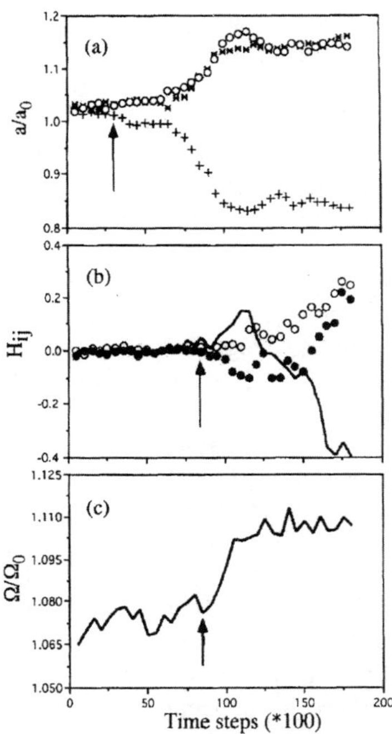

Fig. 6. Time responses of (a) lattice strain along three initially cubic directions, (b) off-diagonal elements of the cell matrix H, H12, H13, H23, (c) normalized system volume. Arrow indicates the onset of Born instability in (a), shear instability in (b), and lattice decohesion in (c).

Finally it may be mentioned that in several studies over the last few years , the stability criteria (11) have lead to precise identifications of the elastic instability triggering a particular structural transition. In hydrostatic compression of Si, the instability which causes the transition from diamond cubic to β –tin structure is the vanishing of $G'(P) = (C_{11} - C_{12} - 2P)/2$.[19] In contrast, compression of crystalline SiC in the zinc blende structure results in an amorphization transition associated with the vanishing of $G(P) = C_{44} - P$.[22] This is discussed further in the next section. For behavior under tension, crack nucleation in SiC [20] and cavitation in a model binary intermetallic [21], both triggered by the spinodal instability, vanishing of $B_T(P) = (C_{11} + 2C_{12} + P)/3$, are results which are analogous to the observations reported here. Notice also that in the present study a crossover from spinodal to shear instability can take place at sufficiently high temperature.[9]

3.2 Thermodynamic Melting

In the preceding simulations we encountered a melting transition which occurred in a perfect crystal that has no surfaces, because the MD simulation was carried out with periodic boundary conditions. A critical temperature was reached at which a bifurcation took place with the cubic simulation cell suddenly becoming tetragonal, and the shear modulus G' vanished. The system responses triggered by this event, along with the atomic configurations, all show that the crystal has melted in a homogeneous manner; it may seem logical to thus conclude that the critical temperature is the melting point of the material. However, such an interpretation would prove to be hasty. We now show that when the heating-to-melting simulation is applied to a crystal with an initial defect such as a free surface or an interface, the crystal-to-liquid transition then occurs through a defect-nucleated process at a temperature *lower* than the critical temperature previously observed. Thus there exist two types of melting transitions, a heterogeneous process of nucleation and growth which corresponds to the conventional melting phenomenon, and a homogeneous process of mechanical collapse of the lattice. Henceforth we will refer to the two as *thermodynamic* and *mechanical* melting respectively, with corresponding melting points denoted as T_m and T_s. While the significance of the former needs no comment, the latter is much less well recognized. We have seen that T_s is the highest temperature at which the crystal can remain structurally stable. Since T_m is always lower than T_s, the region $T_m < T < T_s$ is the temperature range of superheating. It also follows that in this region the crystal is in a metastable state, or in others words, T_s is the upper limit of metastability, and in a sense the thermal analogue to the ideal strength of the crystal.

Despite the extensive efforts in studying the phenomenon of melting [36] certain aspects of this fundamental transition were not clarified until recently. One basic question that was raised [37] is the role of surfaces or interfaces in the mechanism of melting. From the standpoint of thermodynamics, melting occurs at the temperature at which the solid and liquid phases coexist, as expressed by the equality of the Gibbs free energies. However, thermodynamics says nothing about how melting occurs, or how long the process will take. These are issues pertaining to the kinetics of the phenomenon. Thus one can ask whether our thermodynamic picture of melting is one which is consistent with the kinetics [38]. This is a question that can be addressed by molecular dynamics simulation in that simulation provides a method to calculate the free energies of the solid and liquid phases [40, 41], as well as to directly observe the actual melting process at the molecular level [38, 42].

In any simulation study of melting it is essential to recognize that the melting point of the simulation model, T_m, can be quite different from the known melting point of the real substance. How well these two temperatures agree is, in fact, a useful indication of how realistic is the interatomic potential function on which the model is based. We will examine the question of the interplay between thermodynamics and kinetics in the particular case of a simulation model of silicon based on the empirical potential model developed by Stillinger and Weber [39]. For this potential free energy calculations have been reported by Broughton and Li [40] which gives the melting point T_m at 1691 $\pm$ 30 K. Given that the experimental melting point is 1683 K, the excellent agreement between 'theory' and experiment should be regarded as somewhat fortuitous.

Once T_m for the model is known, one then has the proper reference temperature from which to investigate the onset of melting. By taking a perfect crystal model of silicon, composed of 704 atoms in a cell with periodic boundary conditions, and heating it up to T_m using MD, it is found that over a reasonable period of simulation the system shows no indications of any onset of structural disordering. This apparent stability persists up to temperatures well beyond T_m; it is only when T reaches 2500 K that the crystal is observed to suddenly undergo significant disordering over a period of 0.18 ps. These simulation results are perplexing at first sight. Why did the simulation model not melt at T_m as predicted by thermodynamics? Should the disordering at 2500, hereafter denoted as T_s be interpreted as the onset of melting?

To answer these and other questions, another set of simulations is performed using a simulation cell which has free surfaces in one direction, and periodic border in the other two directions. The important point to note is that we are now going to interpret the temperature of 2500 K when the crystal collapsed as the mechanical melting temperature T_s . Since the crystal cannot remain stable this point, there is no reason to do any simulation at temperatures above T_s . So the foregoing simulations are useful in arriving at this upper limit. We can therefore regard T_s as the 'theoretical strength' against thermal agitation (heating), just like the theoretical strength σ_c against mechanical deformation which we have discussed in Sec. 2.1.

The simulation runs in the case of a free surface are made in the temperature range above what we think should be the value of T_m as given by the free-eneergy calculation, but always below T_s . It is observed that structural disordering, which has all the features of local melting, begins invariably at the surface and then spreads toward the interior of the simulation cell. For a quantitative measure of the local disorder, we divide the cell into equal slices along the direction of the surface normal, and calculate the static structure factor $S(\underline{K})$ for each individual slice, with $\underline{K}$ chosen to be a reciprocal lattice vector with orientation parallel to the surface. From the profile of $S(\underline{K})$ obtained at various intervals during the simulation, we can locate the melt-crystal interface, and by following this interface in time we determine its velocity of propagation v(T) at a fixed temperature T. This procedure can be repeated for several temperatures to arrive at a temperature variation of the interfacial velocity. One can now ask what is the temperature at which the interface no longer moves.

As shown in in Fig. 7, the five data points in our particular study extrapolate to a temperature of 1710 K at zero interface velocity. The meaning of this extrapolation is

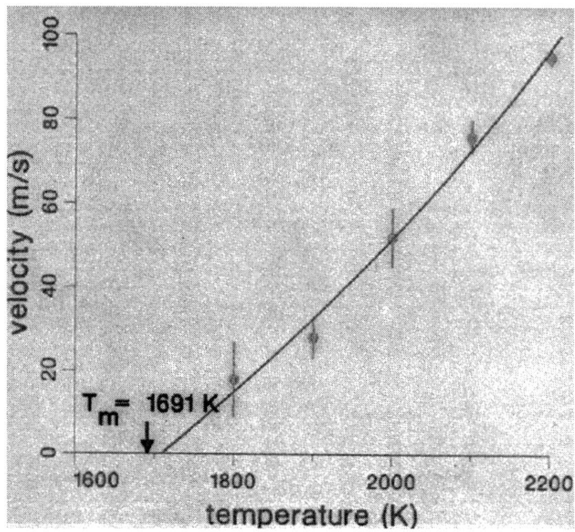

Fig. 7. Variation of velocity of crystal-melt interface with temperature showing the extrapolation to a coexistence temperature in the limit of vanishing velocity.

simple but well worth appreciating. When the velocity of the melt-crystal vanishes, it clearly means that the two phases - the melt (liquid) on the defect side of the interface and the crystal on the bulk side - are 'in equilibrium' in the sense that neither side want to expand into the other, in other words, the two phases are 'in coexistence'. This then is the operational realization of the thermodynamic definition of crossing of the crystal and liquid free-energy curves. If this argument is valid, then one can expect the extrapolated temperature to be the temperature for thermodynamic melting. We note that 1710 is well within the uncertainties in the present study to the free-energy calculation value of 1691, which itself has uncertainties estimated to be 30K.

Similar studies of defect-induced melting also have been carried out using a bicrystal simulation cell representing a grain boundary, and another cell representing a crystal with voids of various sizes [38]. In both cases, extrapolation of the melt-crystal interface velocity leads to essentially the same value of T_m. The conclusion which one can draw from this series of simulations is that there exist in every material two types of melting, thermodynamic and mechanical.

(i) Thermodynamic melting at T_m requires a surface or other defect nucleaton site for the formation of a liquid layer which them propagates into the crystalline bulk at a velocity which depends on the degree of supheating. This process is heterogeneous.

(ii) Mechanical melting at T_s ($> T_m$) is a homogeneous process; it is the upper limit of metastability.

As a final comment we note that recent studies also have been extended to grain boundaries where no premelting has been found although local disordering does take place at $T < T_m$ [43].

3.3 Solid-State Amorphization

When a homogenous, defect-free lattice is driven to structural instability by hydrostatic compression, two types of responses generally can be expected. The crystal can undergo a polymorphic transition to another lattice structure, or a transition to a disordered state, known as solid-state amorphization. Molecular dynamics simulations of compression loading on Si [19] and cubic SiC (β-phase) [22] using essentially the same many-body interatomic interaction model have shown that the former undergoes a transition from diamond cubic to β-Sn tetragonal structure, while the latter undergoes amorphization. The behavior of stability criteria in these two studies are shown in Fig. 8, where one sees that the two transitions involve different instability modes, the vanishing of the tetragonal shear modulus G' and the rhombohedral shear modulus G, respectively. The potential models from which the elastic constants are calculated are of the same bond-order form proposed by J. Tersoff for covalent crystals [24]. In both cases, the critical strains predicted in Fig. 8 agreed with what was observed in the direct simulations. The question then arises as to what is the underlying cause of the different structural consequences of shear instability.

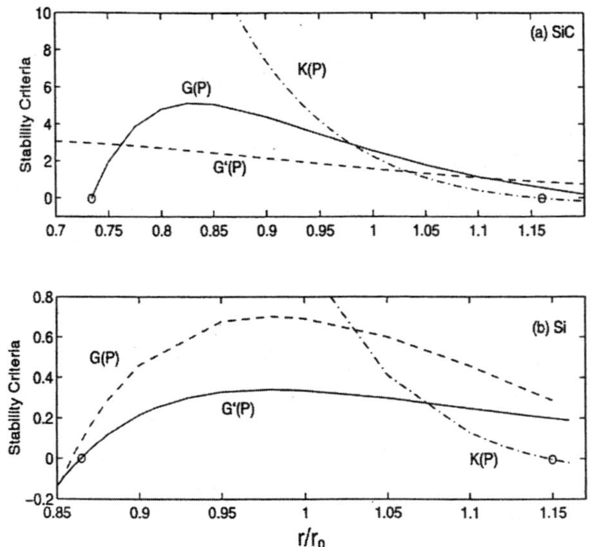

Fig. 8. Variation of elastic moduli, K, G, and G', with lattice strain under hydrostatic loading at 0 K, (a) β -SiC, and (b) Si. $r/r_0 = 1$ denotes the condition of zero stress.

It is apparent that an obvious difference between the two lattices is that one is an elemental system while the other is a binary (AB) compound. Thus in SiC there are chemical ordering effects which are not present in Si. Since in the context of chemical ordering a distinction is made between atomic size effects and chemical bonding effects, it is of interest to assess which effect is more responsible for the observed amorphization. For this analysis one can manipulate the description of interatomic interactions to *intentionally* suppress one effect or the other. Two modified forms of the Tersoff potential model have been produced, in one variant chemical bond preference is suppressed through an adjustment of the interaction between atoms of different species (model I), and in another variant size effects are suppressed by adjusting the bond-order parameter and cross interaction at the same time to leave heat of mixing unchanged (model II) [33]. The relevant physical properties of the Tersoff potential for β -SiC and its two modifications are shown in Table I. It is clearly seen that elimination of chemical bonding preference has little effect (model I), whereas all three elastic constants are significantly altered in the absence of atomic size difference. Although both $C_{11} - C_{12}$ and C_{44} are appreciably reduced, the lowering of the former is more drastic such that in model II the instability mode becomes the vanishing of G'. Thus one may deduce that not only is the presence of size effects responsible for the rhombohedral shear instability in β -SiC, but also their absence allows the tetragonal shear to vanish first in Si. To explicitly verify that these interpretations are correct, a simulation of model II under compression was carried out, indeed revealing a transition from zinc-blende to rock salt structure triggered by a tetragonal shear instability. This is an illustration of the use of modified or manipulated interatomic interaction in simulation; it can be a potentially very useful device for isolating cause-and-effect in probing complex phenomena.

Table 1. Comparison of properties of SiC given by the Tersoff potential and two idealizations, models I and II.

	Tersoff	Model I	Model II
a [A]	4.32	4.34	4.32
E [eV]	-6.19	-6.03	-6.19
B [Mbar]	2.25	2.18	2.25
C_{11}	4.36	4.19	3.31
C_{12}	1.20	1.17	1.72
C_{44}	2.56	2.42	1.61
C_{11} - C_{12}	3.16	3.02	1.59

We have demonstrated that in terms of the competition between instability modes, in this case the vanishing of the two shear moduli, one can gain some insight into the underlying nature of polymorphic and crystal to amorphous transitions. With regard to the experimental implications of our results on β-SiC, we note that amorphization of β-SiC single crystals induced by electron irradiation have been reported [44], the data revealing chemical disordering to take place below a critical temperature of 340 C. On the other hand, the structural transition in β-SiC under compression is found by X-ray diffraction to be polymorphic, from zinc-blende to a rock salt-type structure at 100 GPA [45]. The reason that the simulation predictions do not match precisely with the experimental findings can be attributed to two factors. First is that the empirical classical interatomic potential description is likely not adequate to correctly resolve competing mechanisms involving subtle effects of chemical bonding. Secondly, the role of crystal defects in controlling the experimental observations has not been quantitatively assessed, while for the simulations one knows for sure that no defects were initially present. These uncertainties aside, it is noteworthy that both amorphization and polymorphic transitions have been observed in β-SiC . Apparently, under the relatively 'gentle' driving force of pressure the latter, associated with G'=0, prevails over the former which entails G=0. The driving force induced by electron irradiation is the destabilizing effect of point defect production; under this condition β-SiC undergoes amorphization rather than transforming to another crystal structure.

Even though in β-SiC pressure-induced amorphization appears to be precluded by a polymorphic transition, several experimental studies of this phenomenon in AB compounds can be cited to provide further insights into the kinetics of competing transitions. X-ray measurements show that Nb_2O_5 becomes amorphous at 19.2 GPa at 300 K which is novel because the oxide is simultaneously reduced in the process [46]; the competing polymorphic transition is believed to be kinetically impeded. In BAs a transformation from zinc blende to amorphous structure was observed at 125 GPa, just slightly above the calculated equilibrium transition pressure to the rock salt phase, and interpreted as signifying a kinetically frustrated process [47]. In more complicated systems, such as $CaSiO_3$ an $MgSiO_3$ perovskites, it has been conjectured that stress-induced amorphization arises from the near simultaneous accessibility of multiple modes of instability [48]. The amorphization of $\alpha-$*quartz* (SiO_2) under pressure is a particularly well-known case where molecular dynamics simulation gives a transition pressure in agreement with experiment [49]. The physical mechanism underlying the elastic instability was first identified as the softening of a phonon mode [50]; later a dynamic instability associated with a soft phonon mode at one wave vector was found [51]. These developments are not surprising in view of our discussions in Section 2. It is interesting that the dynamic instability in $\alpha-$*quartz* precede the elastic instability, occurring at 21.5 GPa and 25 GPa respectively.

4. Single Dislocation Dynamics

Dislocations, being the carriers of crystal plasticity, play a fundamental role in any consideration of lattice deformation [52]. For an overview of current atomistic and mesoscale studies of single and multiple dislocations, one may refer to a survey by Bulatov and Kubin.[53] Here we will discuss rather briefly two problems concerning single dislocation mobility to illustrate the kind of mechanistic issues of interest in this active area of simulation research. The first is an-going MD investigation of edge dislocation motion in a metal. This will show the information that direct atomistic simulation can provide. The second problem, an example of multiscale modeling, is the simulation of the dislocation velocity in a semiconductor using a kinetic Monte Carlo approach with kink activation energies determined by atomistic calculations.

4.1 MD Simulations

We are presently conducting MD simulation of moving a pair of edge dislocations in a single crystal of bcc Mo by applying a shear stress [54]. The interatomic potential we use is an effective-medium approximation proposed by Finnis and Sinclair.[55] The set-up of the simulation cell and the application of stress are described in Fig. 9 . Because of the use of periodic boundary conditions, the edge dislocation appears as a dipole. The geometry is such that the two dislocations are arranged to glide on the {112} plane under an applied shear stress σ_{xy}. A method of comparing relative registry of atoms on two adjacent rows is used to locate the dislocation core during its glide. The dislocation profile, shown in Fig. 10, suggests that while double kink nucleation is quite predominant, there is very little kink spreading or migration.

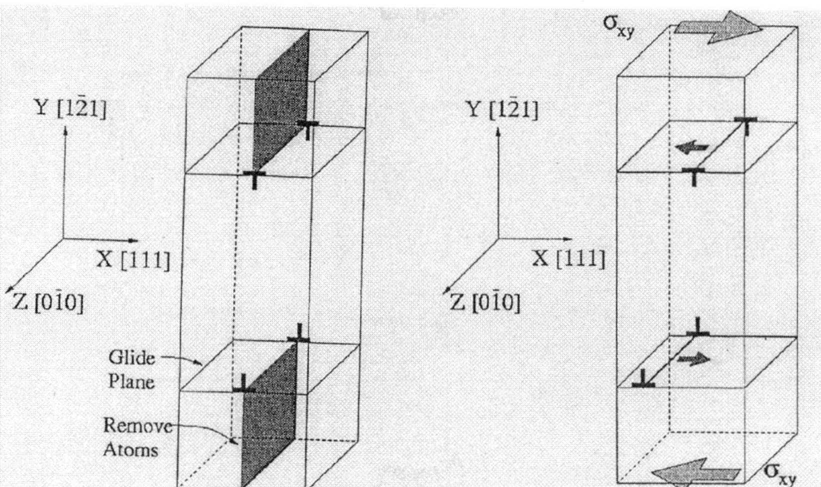

Fig. 9. Simulation cell set-up for MD study of edge dislocation motion in a bcc lattice driven by an applied shear stress. Because of periodic boundary conditions, the cell contains a pair of dislocations (dipole) which will glide in opposite directions. A dislocation is formed by removing a plane of atoms as indicated.

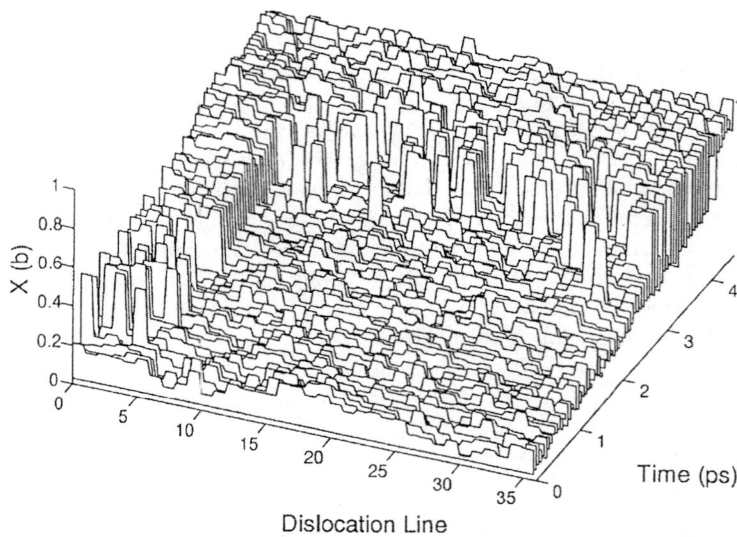

Fig. 10. Profile of an edge dislocation line showing its displacement in the x-direction (upward) as a function of time.

The stress and temperature dependence of the resulting dislocation velocity are shown in Figs. 11 and 12 respectively. In Fig. 11 the simulation results at 77K are seen to lie in the high-stress high-velocity region, while the experimental data are available only in the low-stress and low-velocity regime.

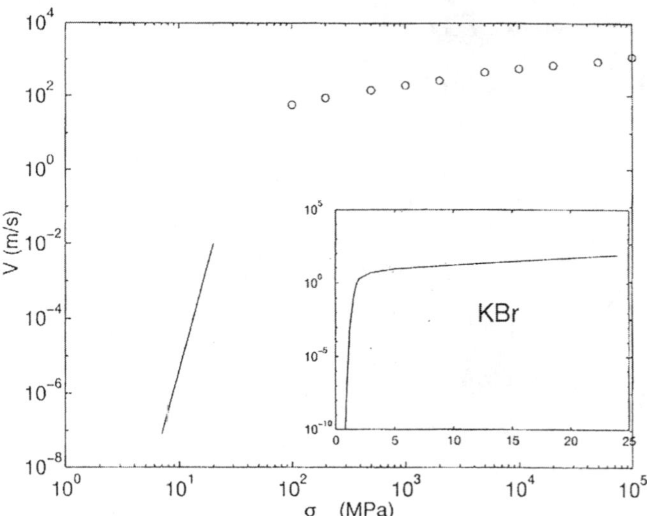

Fig. 11. Variation of velocity with applied shear stress of an edge dislocation in bcc Mo showing the simulation results (circles) and available experimental data (line). Inset shows experimental data on KBr displaying the cross-over behavior suggested in the case of Mo.

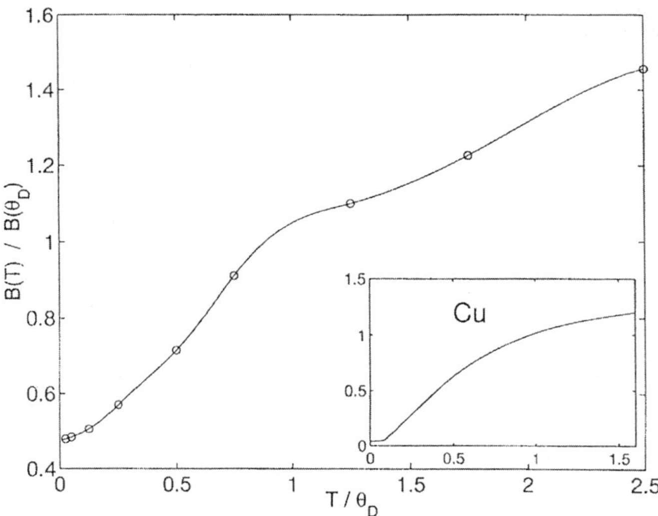

Fig. 12. Variation of dislocation mobility with temperature, simulation results for Mo are shown as circles while the solid line is drawn only to guide the eye. Inset shows experimental data for Cu similarly normalized.

The inset is a plot of the experimental dislocation velocity of KBr where a sharp increase of the velocity with increasing stress has been observed. Thus there is some basis for anticipating that the experimental curve for Mo will join onto the simulation results. The temperatue dependence of the drag coefficient is shown in Fig. 12. From the increase of the drag coefficient with temperature one can conclude that the dislocation mobility is governed by the phonon drag mechanism.

4.2 Mesoscale Modeling

Covalent bonding in silicon and other semiconductors gives rise to a high barrier to dislocation motion [55]. To quantify this behavior in terms of the underlying atomistic mechanisms, much theoretical effort has been expended on obtaining accurate activation parameters for kink nucleation and migration. However, despite the recent progress and a considerable body of experimental observations, how dislocations actually move from one Peierls valley to another under stress is still an open question. The fundamental difficulty is that interpretation of the available data has been hampered by the lack of a theoretical description which is sufficiently free of ad hoc assumptions and capable of relating dislocation mobility behavior to the underlying kink mechanisms.

We consider here such a description by adopting a kinetic Monte Carlo treatment of kink nucleation, migration and annihilation processes along with full elastic interactions between the dissociated partial dislocations [56]. The formulation is designed to produce the overall dislocation movement as the cumulative effect of a large number of individual kink events, requiring for input only the kink formation and migration energies available from atomistic calculations [57]. Using this model we show that current atomistic data on activation energies give the velocities of a dissociated screw dislocation which bracket the experimental data; in the process we obtain bounds on the values of the activation energies to guide future studies. In focusing on the variation of dislocation velocity with applied stress, we show that the coupling between the two dissociated partials can be characterized in terms of compatibility between the strong Peierls barrier and the combined action of the stacking fault attraction and the elastic repulsion between dislocation segments which also influences the average separation distance between the partials X_o.

We identify two compatibility scenarios of particular interest, when X_o is an integral or half-integral multiples of the period of the Peierls barrier, and demonstrate that these two conditions give rise to distinct velocity variations with stress. When the average separation is commensurate with the period (integral X_o), the dislocation mobility is low at low stress and increases super-linearly below a critical stress value. In the case when X_o is half-integral, no threshold behavior is observed and the velocity is essentially linear in the stress. Both types of variations have been observed experimentally. Previous attempts to rationalize the threshold behavior have resulted in the postulate of a random distribution of weak obstacles or ``dragging points" on the dislocation line [58-60], a notion which we are able to clarify with the present description.

Our model deals with a screw dislocation with Burger's vector $(a/2)[1\bar{1}0]$ dissociated into two 30° partials in the (111) plane. Each partial is represented by a piecewise straight line composed of alternating horizontal (H) and vertical (V) segments. The length of H-segments can be any multiple of , while the V-segments are all of the same length, the kink height h. The stacking fault bounded by the two parallel partials has a width measured in multiples of h. The simulation cell is oriented with the partials running horizontally so that periodic boundary condition can be applied in this (z) direction. The dislocation glides in the vertical direction, upward being along $<11\bar{2}>$ (x), as a result of co-migration of the two partials which in turn follows from the elementary kink events.

In each simulation step, a stochastic sampling is carried out to determine which event will take place next: a kink pair nucleation on the H-segments in the upward (downward) direction or a kink (V-segment) translation to the left (right) by an amount b. The rates of these elementary events are calculated based on the energetics of the corresponding kink mechanisms. For example, for a kink pair nucleation mechanism three energy terms are considered, a formation energy obtained by relaxing the atomic configuration of the kink pair using an empirical potential or first principles methods, an energy bias which favors the reduction of the stacking-fault area, and the elastic interaction between a given segment and all the other segments and applied stress (so-called Peach-Koehler interaction). Accordingly, the nucleation rate for an embryonic double-kink (width one b) on a partial with Burgers vector $\underline{b}_p$ and under stress σ_{ij} is calculated as

$$j_{kp} = \omega_o \exp\left(-\frac{E_k + (\pm\gamma_{SF} - \sigma{:}\underline{b}_p)A/2}{k_B T}\right) \tag{14}$$

where ω_o is the pre-exponential ``frequency" factor, which we set equal to the Debye frequency. $\gamma_{SF}$ is the stacking fault energy, with '+' or '-' sign for the leading and trailing partials respectively. A is the area swept out by the dislocation during kink pair nucleation, with``+" or ``-" sign for upward or downward nucleation respectively. The factor of 1/2 appears because we assume that in the saddle point configuration the dislocation has swept out half of the total area A. k_B is the Boltzmann's constant and T is the temperature. Similar expressions, with appropriate modifications, hold for the kink migration rates. The remaining barrier terms, such as E_k, are imported directly from atomistic simulation data.

We will rely on atomistic calculations to provide values for the kink formation and migration energies. Based on various calculated and experimental estimates available at present, a reasonable set of values to take for (E_k and W_m) would be around (0.7 eV, 1.2 eV). The dslocation velocity results obtained by using the kinetic Monte Carlo method just described are shown in Fig. 13 (solid line) along with two sets of experimental data (open symbols) [61,62].

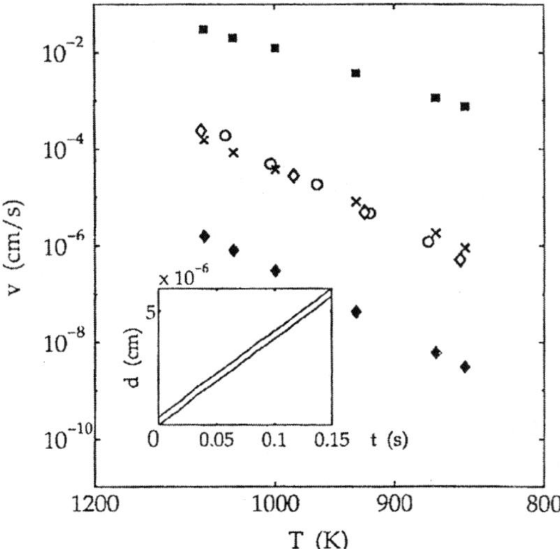

Fig. 13. Temperature variation of screw dislocation in silicon at a shear stress of 10 MPa, kinetic Monte Carlo results are denoted by crosses, experimental data denoted by open circle and diamond. Two other sets of closed symbols are also simulation results using the formation and migration energies obtained by a classical potential (square) and tight-binding method (diamond) respectively. Inset shows simulated instantaneous positions of the two partials.

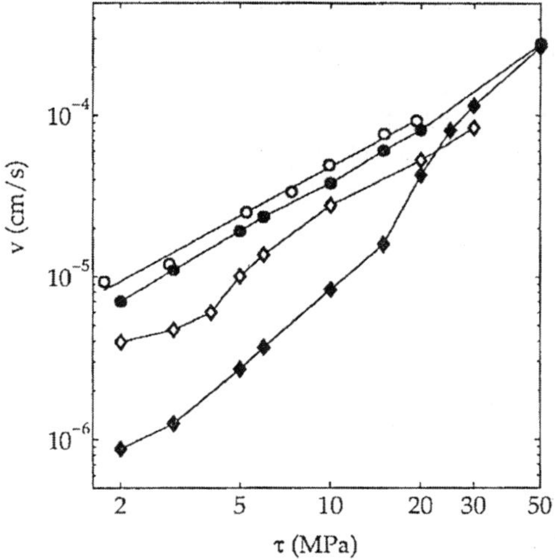

Fig. 14. Variation of screw dislocation velocity with stress in silicon at 1000K, kinetic Monte Carlo results for a commensurate case (Xo=10h) and a non-commensurate case (Xo=10.5h) are shown as closed diamonds and circles respectively, while experimental data are denoted by open symbols.

In Fig. 14 we show the comparison of predicted stress dependence of the dislocation velocity at T = 1000 K, the open symbols, circles and diamonds, refer to the data of Imai and Sumino [62] and George [61] respectively, whereas the closed symbols denote the kinetic Monte Carlo simulations. One can see that depending on whether X_o has integral or half integral value, the velocity varies smoothly (circle symbols) or shows a starting stress behavior (diamonds) [56].

5. Atomistics of Crack Propagation

A crystal is said to be intrinsically brittle if an existing crack is able to propagate along a crystallographic plane in a cleavage manner when the solid is under stress. If a material is intrinsically brittle, then it is capable of undergoing a transition to ductile behavior, emission of dislocations with crack-tip blunting, at a characteristic temperature T_{BD}. This transition has been and continues to be a formidable challenge to our theoretical understanding of fracture mechanics. In this section we will describe two atomistic simulation studies of crack-tip extension, first a demonstration that brittle-ductile transition can be observed in a dynamical scenario, then followed by a characterization of dislocation emission in the stress field at a crack tip.

5.1 Brittle-Ductile Transition

Before discussing the simulation of brittle-to-ductile transition we consider first the set up of the simulation cell containing a crack tip. Fig. 15 shows a schematic of a macroscopic elastic body containing an atomically sharp crack of length 2c subject to an uniaxial tensile loading σ. If we focus on a small region containing one of the crack tips, then we have a simulation cell composed of N atoms (N is around 4000 in the study we will discuss). It is understood that the cell is large enough that the region beyond the cell border can be reasonably well treated as an elastic continuum. Thus we can throw away the elastic medium beyond the border and represent the effects of this medium on the interior atoms by a traction imposed at the cell border [63]. For the third direction, that coming out of the plane of the paper, we will use periodic boundary condition, so the problem is one of plane strain. To demonstrate that this treatment of the border condition does not introduce appreciable numerical artifact, we first study the simpler problem of determining the critical stress for cleavage fracture.

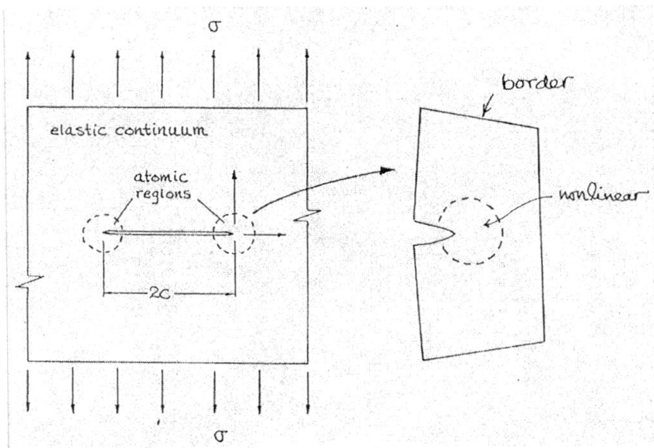

Fig. 15. Schematic showing an atomistic region containing the crack-tip, treated in MD simulation, embedded in an elastic surrounding which can be of macroscopic dimensions. The external stress is applied far from the crack tip; its effect is represented by an appropriate border condition for the simulation cell.

It is well known in fracture mechanics that the Griffith criterion provides a good overall estimate of this critical stress. In the case of an anisotropic lattice the criterion takes the form [64],

$$K_{IG} = \sqrt{2\gamma}\,\{(s_{11}s_{22}/2)^{1/2}[(s_{22}/s_{11})^{1/2} + (2s_{12} + s_{66})/2s_{11}]^{1/2}\}^{-1/2} \qquad (15)$$

where K_{IG} denotes the critical value of the stress intensity factor for pure tensile loading (mode I), sij are the compliance constants with indices 1, 2, and 6 referring to the directions of plane strain, crack propagation (normal to crack front), and normal of the crack plane, respectively. In essence this result is a consequence of imposing the equilibrium condition between strain-energy release rate and twice the surface energy of the crack surface [65]. The Griffith condition has been tested in a number of simulation studies on crack propagation, with varying results. We have shown that it is satisfied to within about 5% for three different crack-tip configurations [66]. This is a significant improvement over previous studies which found deviations from the Griffith condition ranging from 20% to a factor of 3; it is achieved through a combination of better treatment of border conditions, taking into account crystal anisotropy, and more careful handling of the numerical data.

Fig. 16 shows the MD simulation results on $\alpha - Fe$, obtained using an Embedded-Atom-Method potential, . On the left hand side one sees the development of a crack tip under mode I loading at 10% above the Griffith critical value K_{IC} after simulation at the indicated temperature for 3000 time steps. Except for the different temperatures, all three simulations are identifical. In the simulation at T = 300 K the crack has advanced in a cleavage manner (crack tip position at t = 0 is indicated by the cross) with a smooth crack surface profile. The middle result at T = 400 K is quite different; here one sees a significant deformation in the upper part of the crack surface. This is a signature of plastic deformation or dislocation emission. At T = 500 K the deformation at the crack tip is so severely that the crack has blunted and is opening up rather than moving forward (to the right). These results constitute the direct observation of a crack tip undergoing a brittle-to-ductile transition [67]. The right hand side of Fig. 16 shows what happens if one continues the simulation at T = 500 K out to longer times. One sees that at time step 5000 a step appears in the upper border of the simulation cell. This step is created by the movement of a dislocation from the crack tip up to the border.

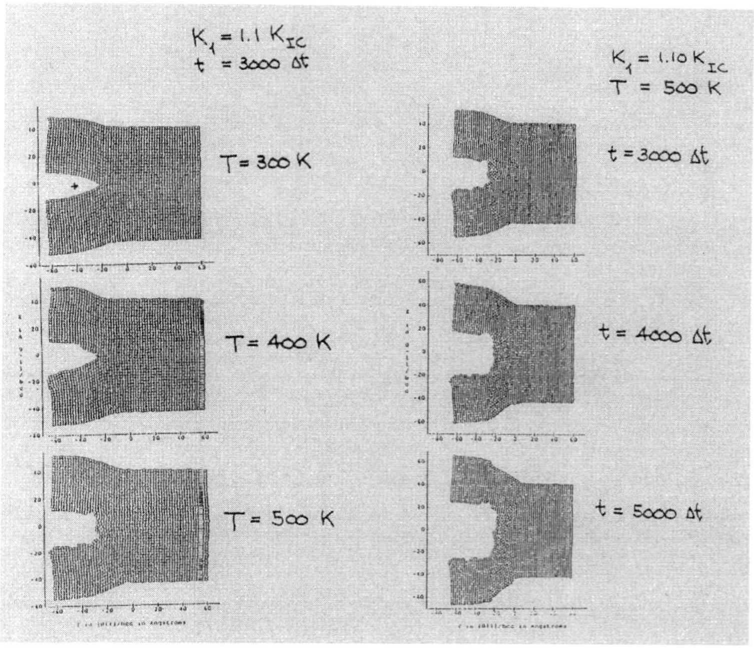

Fig. 16. Evolution of instantaneous configurations of an initially sharp crack (tip position indicated by a small cross) under mode I loading, observed afer the same period of simulation at three temperatures (left side) and at the same temperature but three different periods (right side).

5.2 Crack Tip Plasticity

In the context of understanding materials strength from a multiscale point of view, crack-tip plasticity is a problem that can benefit significantly from the linking of different simulation techniques [68,69]. A central issue that should be resolved at the atomistic level is under what condition a dislocation will be emitted at the crack tip; this is a process that can be realistically captured by MD simulation. At the microstructure level, questions concerning the distribution of dislocations in the plastic zone ahead of the crack tip and their shielding effects on crack extension are particularly relevant. These problems are more properly treated by mesoscale simulation techniques such as discrete dislocation dynamics (DDD) and finite-element method (FEM).

The process of dislocation nucleation at a crack tip is a fundamental problem in the mechanical behavior of stressed crystalline solids at finite temperatures, particularly in the context of understanding the brittle-to-ductile transition [70,71]. Because the crack-tip region is inherently nonlinear and extends over essentially molecular dimensions, the problem would appear to be ideally suited for atomistic simulation [66,67]. From the standpoint of continuum analysis, criteria for predicting brittle or ductile behavior [71,72] are usually based on the concept of the energy-release rate (actually an energy per unit area) for cleavage decohesion, G_{cleav}, and dislocation nucleation, G_{disl}. A ductile material is then characterized by $G_{disl} < G_{cleav}$, where $G_{cleav} = 2\gamma$, the energy of the two crack surfaces [65]. Recent progress has been achieved by introducing an interplanar potential associated with rigid block sliding in a homogeneous lattice [71] Using heuristic models for this potential, which is equivalent to specify the stress-displacement constitutive relation on the slip plane, the critical configurations for dislocation nucleation from the crack tip can be calculated, and in this manner estimates of the brittle-to-ductile transition temperature can be made [73,74].

In a 3D MD simulation of a microcrack in an fcc lattice under tensile loading (plain strain mode I) using the Lennard-Jones potential, the process of dislocation emission from a crack tip has been isolated, and an appropriate stress-displacement relation extracted from the atomic-level displacement and stress fields [75]. The geometric setup, with (221) being the crack plane, is shown at the top of Fig. 17. The simulation cell dimensions are 3a, 60a, and 120a in the x-, y-, and z-directions respectively (86,400 atoms), with lattice parameter a $= 1.56\sigma$, σ being the length parameter in the potential. Periodic conditions are imposed in the x- and y-directions, while surface traction is applied on the z-borders. Simulations are carried out at very low temperatures.

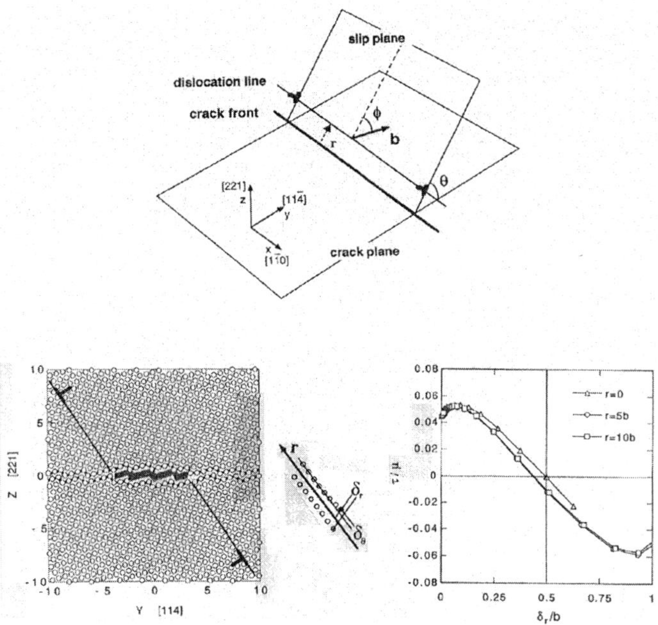

Fig. 17. Geometrical arrangement of a dislocation with Burgers vector b emitted from a crack tip (top). Atomic configuration at the threshold of of dislocation emission along the slip planes (left). Definition of shear and opening components of displacement (center). Shear-displacement curves for one of the dislocations at simulation times corresponding to position r=0, 5b, and 10b, with displacement in unit of b and stress in unit of shear modulus (right).

In the lower left part of Fig. 17 we show an instantaneous atomic configuration during simulation in which one can see that the crack tips have not advanced from their initial positions (shaded region), while two stacking faults on the $(1\bar{1}1)$ planes at $\theta = 54.7^o$ have propagated into the bulk. At one end these stacking-fault strips are bounded by a moving partial dislocation, with Burgers vector b = [112]/6 and $\phi = 0^o$; at the other end they are bounded by a step on the (221) crack surface. This result demonstrates that MD simulation is able to capture the dynamic process of dislocation *emission*, including the *nucleation* from the crack tip and the *motion* away from the crack.

The atomistic results that are most relevant to continuum model formulation are not the atomic positions given by the simulation, rather they are the shear and opening displacements, δ_r and δ_θ, which are components obtained by decomposing the atomic-level displacement field along the slip plane, as indicated in the lower inset in Fig. 17. In the lower right part of the figure we show the variations of displacements δ_r with position r (see the schematic at the top) at various durations of simulation. At any given simulation time during slip along the $(11\bar{1})$ plane, δ_r decreases monotonically with increasing r; this is to be expected since shear displacement should be the largest just in front of the crack tip. As time increases, all the displacements are seen to increase; however, once δ_r reaches a value of b, the dislocation core has moved past that particular atomic site, therefore no further displacement occurs at that site.

Knowledge of the instantaneous atomic positions allows one to calculate the internal stress field, either from its mechanical definition of force across a plane which is a tedious task, or from the virial expression for the atomic-level stress [76]. While the latter method is considerably simpler, its validity in the immediate vicinity of a structural inhomogeneity is open to question. Once the stress field is determined, it can be resolved into shear and tensile components, τ and σ, on the slip plane. With the resolved displacement and stress components in hand, one can cross plot and thus generate shear and tensile stress-displacement curves, $\tau(\delta_r)$ and $\sigma(\delta_\theta)$. Various $\tau(\delta_r)$ results obtained at different distances between the crack tip and the dislocation are shown in Fig. 15.

In contrast to the present direct determination of the shear stress-displacement relation, one can assume a certain form for $\tau(\delta_r)$ and fit the parameters to any known property. As a simple approximation Rice proposed to take for $\tau(\delta_r)$ the Frenkel-Peierls sinusoid, and moreover, identified its integrated area between δ_r=0 and the first zero of $\tau(\delta_r)$ with a quantity γ_{us}, which he called the "unstable-stacking-fault energy", given by the maximum energy encountered during the rigid-block like sliding along the slip plane of one half of a perfect crystal relative to the other [71]. This approximation implies that the actual nonuniform displacement field on the slip plane in the presence of a nucleating dislocation may be replaced by a uniform displacement distribution corresponding to the rigid-block sliding. For further discussions of using the results of Fig. 15 to examine the validity of the approximation, the interested reader should see ref [75].

6. An Outlook on Multiscale Materials Modeling

It is perhaps worthwhile to point out again that the importance of materials technology in our society combined with the advent of high performance computing has given rise to the current focus on multiscale materials modeling as a new component of computational materials research. This in turn provides exciting opportunities for the investigation of longstanding seemingly simple but fundamentally enduring issues where atomistic inputs can lead to significant new insights. In this chapter we have discussed illustrative results on strength and deformation which clearly bring out the important role of defect microstructure at the atomistic level. Three more examples may be briefly cited to emphasize that this type of development will be of growing interest.

For the predictive simulation of crystal plasticity a key question is how can quantitative knowledge of the dislocation core and atomistic methods for analyzing dislocation interactions be used to develop a mechanism-based description of strain hardening? In a large-scale molecular dynamics simulation of crack propagation, it was observed that a Lomer-Cottrell lock, formed by two intersecting Shockley partials, can be destroyed by the action of another partial nearby [77]. The critical stress for unzipping (destroying) the junction was estimated to be about 15 MPa, which is consistent with values previously deduced from experimental data, thus demonstrating the feasibility of linking molecular dynamics and dislocation dynamics simulations and eliminating the need for intermediate experimental input.

Atomistic simulation can play a similar role in our understanding of brittle-ductile behavior in fracture. The current continuum approach is to introduce an interplanar potential, equivalent to specifying a stress displacement constitutive relation on the slip plane, associated with rigid block sliding in a homogeneous lattice. Through this potential one may introduce certain atomic-scale effects and therefore make the approach a continuum-atomistic hybrid. As discussed in Sec. 5.2, this strategy has been implemented, in part, in a molecular dynamics simulation in which dislocation emission from a crack tip under an applied tensile stress has been isolated and analyzed to produce the stress-displacement relation on activated slip plane before and after nucleation [75]. The results revealed clearly the essential role of surface steps in the nucleation event and the effects of crack-tip shielding once the dislocation is emitted.

Another emerging and significant challenge for atomistic simulations is plastic flow in polycrystalline solids where the understanding of how crystal grains deform on decreasing length scales is needed [78]. It has been found recently from simulations of deforming nanocrystals that the well-known Hall-Petch effect where strength increases with decreasing grain size (in the micrometer range) does not hold in nanocrystals where the grain size is only a few nanometers; instead, a reverse Hall-Petch behavior was observed which could be attributed to small-scale sliding in the grain boundaries. This suggests the existence of a critical size separating the two types of behavior, which in turn implies that as grain size is reduced across this critical value, the deformation mechanism changes from the intragranular processes of dislocation nucleation and pile-up to intergranular sliding. A similar transition may occur in nanometer-scale electronic components, such as epitaxial films and quantum dots. In these heterostructures, the lattice mismatch between different materials induces strain, and how it relaxes during fabrication will determine the material structure, and hence its properties. Again, one expects competition between processes mediated by misfit defects such as dislocations, and surfaced based processes such as roughening or morphological changes.

We conclude by returning to the opening theme of our discussion and noting from the Summary of the Report on National workshop on Advanced Scientific Computing, July 1998 [1], that success in any grand challenge enterprise will require not only science and technology, but also integration and partnership among its participants. We believe the field of materials science will be continually enriched as it expands to include scientists and engineers from a multidisciplinary range of background and interest, from physics and engineering, to chemistry, and to biology.

Acknowledgment

This work was supported by AFOSR (F49620-00-10082), NSF (DMR-9980015), Lawrence Livermore National Laboratory (ASCILevel 2 grant), and Honda R&D, Ltd. We acknowledge longstanding collaborations with V. Bulatov, D. Wolf, A. S. Argon, J. F. Justo, M. Tang, J. Wang, F. Cleri and S. Phillpot.

References

[1] Report of the National Workshop on Advanced Scientific Computing, July 30-31, 1998 (National Academy of Sciences), available from http://www.er.doe.gov/production/octr/mics/index.html
[2] Special Issue of J. Computer-Aided Mater. Design, vol. 3, (1996); special issue of Current Opinion in Solid State and Mater. Sci., vol. 3, no. 6 (1998).
[3] G. H. Campbell et al., Mater. Sci. Eng. A**251,** 1 (1998).
[4] M. Born, Cambridge Philos. Soc. **36**, 160 (1940).
[5] M. Born and K. Huang, *Dynamical theory of Crystal Lattices* (Clarendon, Oxford, 1956).
[6] R. Hill, Math. Proc. Camb. Phil. Soc. **77**, 225 (1975).
[7] R. Hill and F. Milstein, Phys. Rev. B **15**, 3087 (1977).
[8] A. Kelly and N. H. Macmillan, *Strong Solids* (Clarendon, Oxford, 1986), 3rd ed..
[9] J. Wang, J., Li, S. Yip, S. Phillpot, D. Wolf, Phys. Rev. B **52**, 12627 (1995).
[10] Z. Zhou and B. Joos, Phys. Rev. B **54**, 3841 (1996)..
[11] J. W. Morris and C. R. Krenn, Philos Mag. A **80**, 2827 (2000).
[12] J. W. Gibbs, in *The Scientific Papers of J. Willard Gibbs, Vol. 1: Thermodynamics* (Ox Bow Press, Woodbridge, Conn. 1993), p. 55.
[13] J. W. Morris, C. R. Krenn, D. Roundy, M. L. Cohen, in *Phase Transformations and Evolution in Materials*, P. E. Turchi and A. Gonis, eds., (TMS, Warrendale, 2000), p. 187.
[14] D. C. Wallace, Thermodynamics of Crystals, Wiley, New York (1972).
[15] T. H. K. Barrons and M. L. Klein, Proc. Phys. Soc **85**, 523 (1965).
[16] W. G. Hoover, A. C. Holt, D. R. Squire, Physica **44**, 437 (1969).
[17] Z. S. Basinski, M. S. Duesbery, A. P. Pogany, R. Taylor, Y. P. Varshni, Can. J. Phys. **48**, 1480 (1970).
[18] J. Wang, S. Yip, S. Phillpot, D. Wolf, Phys. Rev. Lett. **71**, 4182 (1993).
[19] K. Mizushima, S. Yip, E. Kaxiras, Phys. Rev. B **50**, 14952 (1994).
[20] M. Tang and S. Yip, J. Appl. Phys. **76**, 2716 (1994).
[21] F. Cleri, J. Wang, S. Yip, J. Appl. Phys. **77**, 1449 (1995).
[22] M. Tang and S. Yip, Phys. Rev. Lett. **75**, 2738 (1995).
[23] J. Tersoff, Phys. Rev. B **39**, 5566 (1989).
[24] J. Li, Ph.D. Thesis, MIT (2000).
[25] G. Galli, F. Gygi, A. Catellani, Phys. Rev. Lett. **82**, 3476 (1999).
[26] M. Born, J. Chem. Phys. **7**, 591 (1939).
[27] L. Hunter and S. Siegel, Phys. Rev. **61**, 84 (1942).
[28] S. M. Foiles, M. I. Baskes, M. S. Daw, Phys. Rev. B**33**, 7983 (1986).
[29] M. Parrinello and A. Rahman, J. Appl. Phys. **52**, 7182 (1981).
[30] J. R. Ray, Comput. Phys. Rept. **8**, 109 (1988).
[31] J. Wang, J. Li, S. Yip, D. Wolf, S. Phillpot, Physica A **240**, 396 (1997).
[32] L. L. Boyer, Phase Transitions **5**, 1 (1985).
[33] J. F. Lutsko, D. Wolf, S. R. Phillpot, S. Yip, Phys. Rev. B**40**, 2841 (1989).
[34] D. Wolf, P. R. Okamoto, S. Yip, J. F. Lutsko, M. Kluge, J. Mater. Res. **5**, 286.
[35] J. L. Tallon, Nature (London) **342**, 658 (1989).
[36] A. R. Ubbelohde, *Molten States of Matter: Melting and Crystal Structure* (Wiley, Chichester, 1978).
[37] R. W. Cahn, Nature **323**, 668 (1986).
[38] S. R. Phillpot, J. F. Lutsko, D. Wolf, S. Yip, Phys. Rev. B**40**, 2831(1980); S. R. Phillpot, S. Yip, D. Wolf, Computers In Physics **3**, no. 20, p. 20 (1989).
[39] F. H. Stillinger and T. A. Weber, Phys. Rev. B**31**, 5262 (1985).
[40] J. Q. Broughton and X. P. Li, Phys. Rev. B**35**, 9120 (1987).
[41] M. de Koning, A. Antonelli, S. Yip, Phys. Rev. Lett. **83**, 3473 (2000); J. Chem. Phys., submitted.
[42] J. Solca, A. J. Dyson, G. Steinbrunner, B. Kirchner, H. Huber, Chem. Phys. **224**, 253 (1997).
[43] T. Nguyen, S. Yip, P. S. Ho, T. Kwok, C. Nitta, Phys. Rev. B**46**, 6050 (1992).
[44] H. Inui, H. Mori, A. Suzuki, H. Fujita, Philos. Mag. B **65** (1992) 1.
[45] M. Yoshida, A. Onodera, M. Ueno, K. Takemura, O. Shimomura, Phys. Rev. B **48** (1993) 10587.
[46] G. C. Serghiou, R. R. Winters, W. S. Hammack, Phys. Rev. Lett. **68** (1992) 331.
[47] R. G. Greene, H. Luo, A. L. Ruoff, Phys. Rev. Lett. **73** (1994) 2476.
[48] M. Hemmati, A. Chizmeshya, G. H. Wolf, P. H. Poole, J. Shao, C. A. Angell, Phys. Rev. B **51** (1995), 14841.
[49] J. S. Tse and D. D. Klug, Phys. Rev. Lett. **67** (1991) 3559.
[50] N. Binggeli, N. R. Keskar, J. R. Chelikowsky, Phys. Rev. B **49** (1994) 3075.
[51] G. W. Watson and S. C. Parker, Phys. Rev. B **52** (1995), 13306.

[52] J. P. Hirth and J. Lothe, *Theory of Dislocations* (McGraw Hill, New York, 1968).
[53] V. V. Bulatov and L. P. Kubin, Current Opinion in Solid State & Mater. Sci. **3**, 558 (1998).
[54] J. Chang, V. V. Bulatov, S. Yip, J. Computer-Aided Mater. Design **6**, 165 (1999).
[55] M. S. Duesbery and G. Y. Richardson, Crit. Rev. Solid State **17**, 1 (1991).
[56] W. Cai, V. V. Bulatov, J. F. Justo, A. S. Argon, S. Yip, Phys. Rev. Lett. **84,** 3346 (2000)
[57] V. V. Bulatov, S. Yip, A. S. Argon, Philos. Mag. A **72**, 453 (1995).
[58] V. Celli, M. Kabler, T. Ninomiya, R. Thomson, Phys. Rev. **131**, 58 (1963).
[59] H. J. Möller, Acta Metall. **26**, 963 (1978)).
[60] H. Alexander, in *Dislocations in Solids*, F. R. N. Nabarro ed. (North Holland, Amsterdam, 1986), vol. 7, p. 113.
[61] A. George, J. Phys. (Paris) **40**, 133 (1979).
[62] M. Imai and K. Sumino, Philos. Mag A **47**, 599 (1983).
[63] B. deCelis, A. S. Argon, S. Yip, J. Appl. Phys. **54**, 4864 (1983)
[64] G. C. Sih, G. C. and H. Liebowitz, in *Fracture: An Advanced Treatise*, H. Liebowitz, ed. (Academic, New York, 1968), vol. 2, p. 67.
[65] A. A. Griffith, Philos. Trans. Roy. Soc. A **221**, 163 (1920).
[66] K. S. Cheung, S.Yip, Phys. Rev. Lett. **65**, 2804 (1990).
[67] K. S. Cheung and S. Yip, Modell. Simul. Mater. Sci. Eng. **2**, 865 (1994).
[68] E. Kaxiras and S. Yip, Current Opinion in Solid State & Mater. Sci. **3**, 523 (1998).
[69] F. E. Abraham, J. Q. Broughton, N. Bernstein, E. Kaxiras, Computers In Phys. **12**, 538 (1998).
[70] A. S. Argon, Acta metall. **35**, 185 (1987).
[71] J. R. Rice, J. Mech. Phys. Solids **40**, 239 (1992).
[72] J. R. Rice and R. Thomson, Phil. Mag. **29**, 73 (1974).
[73] A. S. Argon, G. Xu, M. Ortiz, Mat. Res. Soc. Symp. Proc. **409**, 29 (1996).
[74] G. Xu, A. S. Argon, M. Ortiz, Phil. Mag. A **75**, 341 (1997).
[75] F. Cleri, S. Yip, D. Wolf, S. R. Phillpot, Phys. Rev. Lett. **79**, 1309 (1997).
[76] K. S. Cheung, and S. Yip, J. Appl. Phys. **70**, 5688 (1991).
[77] V. Bulatov, F. Abraham, L. Kubin, B. Devincre, S. Yip, Nature **391**, 669 (1998).
[78] S. Yip, Nature **391**, 532 (1998).

Computational Materials Science
C.R.A. Catlow and E.A. Kotomin (Eds.)
IOS Press, 2003

Integrated Approach to Atomistic Simulation of Film Deposition Processes

Alexander A. BAGATUR'YANTS
Photochemistry Center, Russian Academy of Science, ul. Novatorov 7a, Moscow 117421, Russia

Anatoli A. KORKIN
Advanced Modeling & Simulation, SPS Motorola Inc, Mesa, AZ 85202, USA

Konstantin P. NOVOSELOV
AOZT Soft-Tec, Nakhimovskii prosp. 34, Moscow, 117218 Russia

Leonid L. SAVCHENKO
Motorola Labs, Physical Science Research Laboratories, Tempe, AZ 85284, USA

Stanislav Ya. UMANSKII
N.N. Semenov Institute of Chemical Physics RAS, Kosigina 4, Moscow, 117977, Russia

Abstract. The review describes an integrated approach to atomistic simulation of CVD processes using silicon nitride CVD from a gas-phase mixture of SiH_2Cl_2 and NH_3 as an example. The mechanisms and kinetics of gas-phase reactions and the mechanism of Si_3N_4 film growth are studied theoretically based on *ab initio* calculations of potential energy surfaces and surface structures. The transition state and RRKM theories are used for calculations of the corresponding rate constants. The gas-phase composition is predicted in a wide temperature–pressure range using the kinetic reaction scheme and the corresponding chemical mechanism proposed. A detailed mechanism is also developed for surface film growth processes. Finally, a kinetic Monte Carlo scheme is applied to describing the process of Si_3N_4 film growth under conditions of CVD from dichlorosilane and ammonia. The results of simulations give evidence that the kinetic Monte Carlo approach is very promising for predicting film growth processes and structural properties of the resulting film.

1. Introduction and General Methodology

Thin semiconductor films (and other nanostructured materials) are widely used in many applications and, especially, in microelectronics. Current technological trends toward ultimate miniaturization of microelectronic devices require films as thin as less than 5 nm, that is, containing only several atomic layers [1]. The structure, properties, and deposition mechanism of such films cannot be understood well without modelling at an atomic level. It is very important for technological applications that the results of such atomistic modelling will be directly connected with macroscopic characteristics. On the other hand, processes and structures essential for thin film growth involve various time and spatial scales.

Therefore, the problem of thin film simulation is a typical example when the use of so-called multiscale (or mesoscale) simulation is of key importance. Various strategies of multiscale modelling as applied to CVD and other film growth processes has been recently described in [2–7].

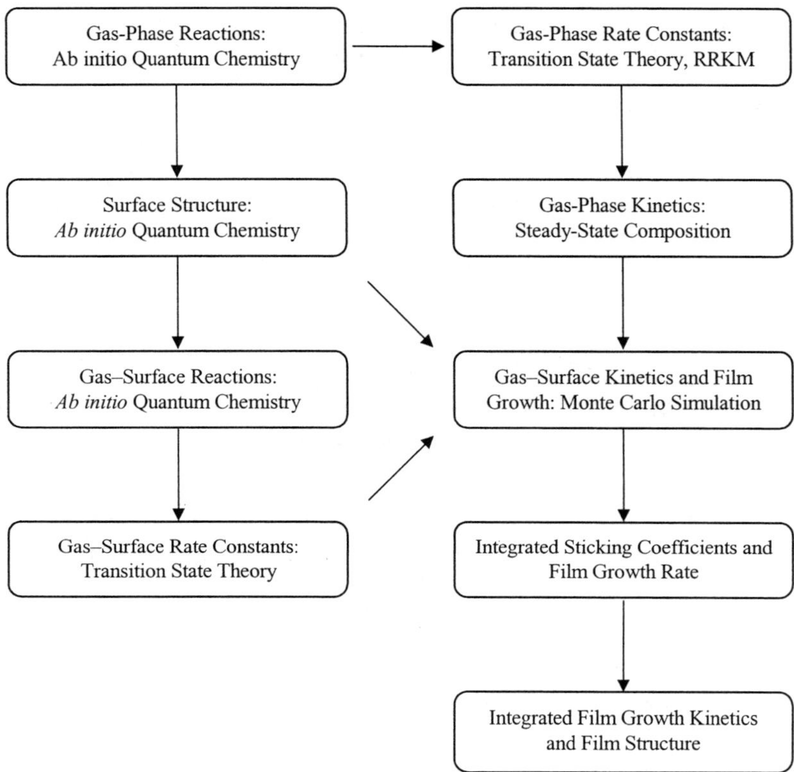

Fig. 1. General scheme of integrated simulation approach.

In this work, we present a general strategy for the multiscale simulation of thin film growth that starts from *ab initio* calculations of elementary steps and structures and eventually leads to the level of reactor modelling. When considering a chemical vapour deposition process, one should take into account gas-phase reactions of reaction components in the CVD reactor and gas–surface reactions. Next, it is necessary to describe film growth on a realistic time and spatial scale, that is, as a sequence of elementary gas–surface and surface reactions (including surface diffusion, if necessary) on a relatively large surface area. Finally, the results of these simulations (overall reaction constants and film characteristics) can be used in the subsequent reactor modelling and in detailed calculations of film structure and properties, including defects and impurities.

Our approach can be illustrated by the schematic diagram in Fig 1. Gas-phase reactions, surface structures and gas–surface reactions are treated at the *ab initio* level, using either cluster or periodic calculations for surface structures, when appropriate. The results of these calculations are used for calculating reaction rate constants within the transition state (TS) or Rice–Ramsperger–Kassel–Marcus (RRKM) theory for bimolecular gas-phase reactions or for unimolecular and surface reactions, respectively. The structure and energy characteristics of various surface groups can also be extracted from the results of *ab initio* calculations. Based on these results, a chemical mechanism can be constructed for both gas-phase reactions and surface growth. The film growth process is modelled within the kinetic Monte Carlo (kMC) approach, which provides an effective separation of fast and slow processes on an atomistic scale. Finally, the results of Monte Carlo simulations can be used in kinetic modelling based on formal chemical kinetics and in detailed *ab initio* (or semiempirical) calculations of film structure and properties. The approach will be illustrated by a detailed application to the chemical vapour deposition (CVD) of silicon nitride.

Silicon nitride is a material of great technological importance because of its electronic and optical properties (high dielectric constant and large band gap), mechanical strength and hardness, and exceptional thermal and chemical stability. Therefore, silicon nitride films are widely used not only in microelectronics as a dielectric material for the fabrication of integrated circuits and dynamic random access memory devices but also in solar cells and for mechanical and optical applications [8–15].

Low-pressure chemical vapour deposition (CVD) using the reaction of dichlorosilane (DCS, $SiCl_2H_2$) and ammonia (NH_3) is one of the most common processes for obtaining silicon nitride thin films [14, 15]. The properties and quality of these films are determined by their structure and stoichiometry, which, in turn, strongly depend on deposition conditions such as temperature, pressure, and gas-phase composition [9]. Therefore, an improved comprehensive understanding of the film growth mechanism and kinetics is very useful for selecting the best deposition conditions.

Much experimental and theoretical effort has been directed toward studying the mechanism and kinetics of the CVD of silicon nitride films from DCS and NH_3 and related processes [16–27]. Theoretical calculations have also been reported on some gas-phase reactions occurring in the CVD of silicon nitride [28–32]. Although these studies have provided some insight into the growth mechanism, a comprehensive model of silicon nitride deposition has not yet been developed. A microscopic approach to formulating the chemical mechanism of CVD processes seems very promising [33, 34], though its consistent application to both gas phase and surface reactions requires very sophisticated computational methods.

Here, we will consider the use of the integrated approach formulated above for the description of silicon nitride CVD from DCS and ammonia. This approach will include (1) *ab initio* quantum chemical calculations of gas-phase reactions that determine the composition of the gas phase in the reactor, (2) *ab initio* quantum chemical cluster calculations of surface reactions that determine the main steps of film growth, and (3) kinetic Monte Carlo simulation of film growth at an atomistic level.

2. Atomistic Simulation of Silicon Nitride CVD from Dichlorosilane and Ammonia

2.1. *Mechanism of Gas-Phase Reactions: Theoretical Formulation*

In the recent theoretical study of the CVD of silicon nitride from DCS and NH_3 based on the application of the B3LYP DFT method [27], it has been demonstrated that the direct reaction between DCS and NH_3 makes an essential contribution to the overall mechanism of gas-phase reactions under typical CVD conditions. It has been shown that, at 700° and lower temperatures, the bimolecular Si–N bond formation reaction

$$SiCl_2H_2 + NH_3 \leftrightarrow SiH_2(NH_2)Cl + HCl \quad (1)$$

is faster than the unimolecular decomposition reactions of $SiCl_2H_2$. These decomposition reactions

$$SiCl_2H_2 \leftrightarrow SiHCl + HCl, \quad (2)$$

$$SiCl_2H_2 \leftrightarrow SiCl_2 + H_2, \quad (3)$$

which result in the formation of chloro- and dichlorosilylenes, may explain the deposition nonuniformity observed experimentally.

A more elaborated theoretical approach to calculations of gas-phase molecules and transition states has been used in [35]. The range of molecules under consideration has been extended in order to analyze systematically substitution effects on the calculated reaction and activation energies. A comprehensive open-end kinetic scheme of gas-phase chemical reactions including all significant stoichiometric and nonstoichiometric reactions has been constructed. This scheme could be further refined and augmented. All gas-phase rate constants for these reactions have been calculated using the transition state theory (TST) and Rice–Ramsperger–Kassel–Marcus (RRKM) theory [36]. The reaction scheme developed was used for calculating the steady-state composition of the gas phase in a CVD reactor at various temperatures and pressures.

Under real CVD conditions, when a cold-wall reactor with a sufficiently large reactor volume is used, the surface reactions should comprise only a small fraction relative to gas-phase processes (see below, section 3.4). The number of active surface sites per unit volume in a typical CVD reactor is much smaller than the concentration of gas-phase molecules, whereas rate constants for gas surface reactions should be of the same order as for the corresponding gas-phase reactions. Therefore, this restricted description can give reliable results for the steady-state composition of the gas mixture in a CVD reactor and can be further used in the simulation of surface reactions, film growth, and the overall CVD process [35].

2.2. Computational details: Quantum chemistry, reaction rates and reactor kinetics

Quantum chemical calculations are performed using the GAMESS program package [37] with energy gradients with respect to nuclear displacements calculated analytically and second derivatives calculated numerically using the analytical gradients Geometry optimizations were performed using Møller–Plesset second-order perturbation theory (MP2) [38]. We use the McLean–Chandler basis sets (MC-311G(d,p)) for the second-row atoms [39] and the 6-311G(d,p) basis sets for N and H atoms. Vibrational frequencies are calculated for all the stationary points found on the potential energy surface (PES). All energies given below include the ZPE corrections calculated at the MP2/MC-311G(d,p) level for gas-phase reactions.

It is well known that the MP2 method mainly describes the effects of dynamic electron correlation. The perturbation theory is generally less reliable if the molecular system of interest has low-lying excited configurations, which can be the case for some systems in our study. Many of the reactions under consideration involve silylenes, which are analogues of carbenes and have a small gap between their highest occupied MO (HOMO) and lowest unoccupied MO (LUMO). The formation of carbenoid systems is a highly endothermic reaction, which proceeds via a late transition state. This transition state also has a small HOMO–LUMO gap and requires a highly correlated treatment. Since the LUMO gives a substantial contribution to the electron correlation energy, accurate calculations must also take into account the quasi-degenerate character of the ground state in these systems.

In light of these considerations, the energy of all stationary points in gas-phase reactions were refined by single-point calculations using the multiconfigurational quasi-degenerate perturbation theory (MCQDPT) implemented in the GAMESS program package [37]. In this case, the one-electron functions used in the single-point MCQDPT calculations were determined using the CASSCF method with two orbitals included in the active space (HOMO + LUMO in the HF ground-state configuration). The ground-state energy was further determined by the multireference second-order quasi-degenerate perturbation theory, which was constructed on the calculated CASSCF one-electron functions with the ground-state CASSCF wave function taken as the zero-order approximation. In this way, both dynamic and nondynamic correlation effects were properly taken into account.

In our recent theoretical study of the CVD of silicon nitride from DCS and NH_3, the B3LYP DFT method was used [27]. The applicability of the DFT methods for systems with high correlation effects (carbenoid-type systems) should be clarified. The reliability of DFT methods with respect to various systems pertinent to vapour phase growth processes was verified in [40]. It was found that DFT methods were generally successful in predicting the structure and vibrational frequencies. However, it was also reported that these methods and the B3LYP approach in particular produced significant errors in the calculated heats of formation and heats of reaction for some bridge-bonded organometallic aluminium compounds [40, 41]. The activation barriers obtained for some representative reactions at the MP2/MC-311G(d,p), MCQDPT//MP2/MC-311G(d,p) (this work) and B3LYP/6-31G(d,p) [27] levels are compared in Table 1. The B3LYP/6-31G(d) activation barriers are generally in a rather close correspondence with the MCQDPT//MP2 values. The mean difference between the two sets of data is only ~7 kJ/mol with the largest discrepancy equal to +14.6 kJ/mol for the reverse reaction R2. On the other hand, the results obtained at the MP2 level exhibit much larger deviations from both other sets, reaching even 25–30 kJ/mol in some cases. The single-point MCQDPT approach provides a reasonable compromise between the accuracy and computational cost and can be recommended for the evaluation of electron correlation contributions to the relative energies of stationary points on a PES in mass-scale *ab initio* calculations.

Table 1. Comparison of activation barriers calculated for forward E_A(f) and reverse E_A(r) gas-phase reactions (kJ/mol) at the MP2/MC-311G(d,p) (I), MCQDPT//MP2/MC-311G(d,p) (II) (this work) and B3LYP/6-31G(d,p) (III [27]) levels

Reaction	E_A(f)			E_A(r)		
	I	II	III	I	II	III
$SiH_4 \leftrightarrow SiH_2 + H_2$	254.7	242.4	234.0	20.2	19.4	12.1
$SiH_2Cl_2 \leftrightarrow SiCl_2 + H_2$	330.5	295.0	303.0	183.9	167.4	182.0
$SiH_2Cl_2 \leftrightarrow SiHCl + HCl$	304.5	276.2	272.0	38.4	33.9	20.9
$SiH_3(NH_2) \leftrightarrow SiH_2 + NH_3$	307.2	283.3	283.0	39.7	41.6	35.1
$SiH_2(NH_2)Cl \leftrightarrow SiHCl + NH_3$	310.1	283.1	285.0	56.1	54.4	54.8
$SiH(NH2)Cl_2 \leftrightarrow SiCl_2 + NH_3$	303.9	280.7	274.0	90.0	84.0	92.9
$SiH_2Cl_2 + NH_3 \leftrightarrow SiH(NH_2)Cl_2 + H_2$	159.8	143.7	155.0	227.2	218.8	216.0
$SiH_2Cl_2 + NH_3 \leftrightarrow SiH_2(NH_2)Cl + HCl$	46.9	31.4	36.0	34.7	21.3	18.0

Transition state calculations of the rate constant *k(T)* are performed using the relationship [42]

$$k(T) = \frac{k_B T}{2\pi\hbar} \frac{Q^{\neq}}{\prod_i Q_i} \exp(-\frac{E_0}{k_B T}). \qquad (4)$$

Here E_0 is the reaction barrier equal to the difference between the ground state energies (including zero-point vibrational energies) of the transition complex and reagents, $Q^{\neq}$ is the partition function of the transition complex and Q_i is the partition function of the reagent i. The partition functions are represented in the form

$$Q = Q_{tr} Q_{rot} Q_{rot}^{(int)} Q_{vib}, \qquad (5)$$

where Q_{tr} is the translational partition function (mol/cm^3), Q_{rot} is the partition function of the overall rotations, $Q_{rot}^{(int)}$ is the partition function of internal rotations and Q_{vib} is the vibrational partition function. Tunnelling and over-barrier reflection are not taken into account. The TST formula gives the rate constant of the direct bimolecular reactions and of unimolecular and association reactions in the limit of high pressures. The equation for the canonical rate constant in the conventional transition state theory involves a symmetry number, which was taken into account in accordance with [43]. The effects of multi-channel reactions are accurately taken into account in accordance with [36]. The statistical factors (reaction path degeneracy) are evaluated according to the Schlag and Haller approach [44]. For the estimation of the collision frequency, we employ the hard sphere approximation. The hard-sphere diameters for the reactant and carrier (bath) gas molecules are estimated as the maximum distance between the pair of atoms in the molecule plus the van der Waals radii of the corresponding atoms.

The rate constant k_{uni} of a unimolecular reaction is determined as the lowest eigenvalue of the master equation [45]

$$\lambda g(E) = \int_0^\infty J(E,E')g(E')dE'. \qquad (6)$$

The integral kernel $J(E,E')$ assumes the form

$$J(E,E') = [M]\{\omega\delta(E-E') - R(E,E')\} + \bar{k}(E)\delta(E-E'), \qquad (7)$$

Here $[M]$ is the concentration in the bath gas in mol/cm^3, E, E' are total molecular energies,

$$\omega = N_A \sqrt{\frac{8k_B T}{\pi\mu}} \pi d^2, \qquad (8)$$

where $\mu = \frac{m m_M}{(m+m_M)}$ is the reduced mass of the reacting molecule with the mass m and the bath gas molecule with the mass m_M, d is the diameter of their gas-kinetic collisions, $R(E,E')$ is the rate constant of collision-induced transitions $E' \rightarrow E$ and $\bar{k}(E)$ is the effective microscopic rate constant of unimolecular transformation at energy E. The master equation (6) was solved by transforming it to the finite system of the linear algebraic eigenvalue equations.

The rate constants k_{rec} of recombination reactions A + B → AB reverse to unimolecular decomposition AB → A + B are calculated using the detailed balance condition

$$k_{rec}(T) = \frac{Q_{AB}}{Q_A Q_A} \exp\left(\frac{\Delta H_0^o}{k_B T}\right) k_{uni} \qquad (9)$$

,

where Q_{AB}, Q_A, and Q_B are the partition functions of AB, A and B, and ΔH_0^o is the enthalpy of decomposition at T=0 K.

Gas-phase kinetics in a CVD reactor is modelled based on the well-stirred reactor (WSR) model [46]. The rates of gas–surface reactions leading to film growth are expressed in terms of sticking coefficients. These are defined as the ratio of the number of molecules of species i captured by the substrate surface per unit time to the number of their collisions with the surface per unit time. The kinetics of species inside the reactor within the WSR model is described by the following system of ordinary differential equations deduced by averaging over the reactor volume [46]:

$$\frac{dC_i}{dt} = \sum_{k=1}^{K} \nu_{ik}(R_k - R_{-k}) - \left(I^* + \gamma_i \sqrt{\frac{RT}{2\pi m_i}} S_S\right)\frac{C_i}{V} + \frac{I_i^*}{V}\frac{P}{RT} \qquad (10)$$

where $i = 1, \ldots, N$ are the numbers of species; $k = 1, \ldots, K$ are the numbers of chemical reactions; C_i is the molar concentration of the i-th species; t is time; ν_{ik} is the stoichiometric coefficient of the i-th species in the k-th reaction; I_i^* is the inlet flow rate of the i-th species; $I^* = \sum_{i=1}^{N} I_i^*$ is the gas mixture inlet flow rate; V is the reactor volume; P is the gas mixture pressure; R is the universal gas constant; T is temperature; γ_i is the sticking coefficient of the i-th species ($0 \le \gamma_i \le 1$); m_i is the molar mass of the i-th species; S_s is the substrate area; and R_k and R_{-k} are the k-th forward and reverse reaction rates, respectively, which are determined from the following expressions [46]:

$$R_k = q_k \prod_{n=1}^{N} C_n^{\|-\nu_{nk}\|} \tag{11a}$$

$$R_{-k} = q_{-k} \prod_{m=1}^{N} C_m^{\|\nu_{mk}\|} \tag{11b}$$

where q_k and q_{-k} are k-th forward and reverse reaction rate constants, respectively, and $\|\nu_{ik}\|$ and $\|-\nu_{ik}\|$ in Eqs. (11a) and (11b) are determined according to the following rule [46]:

$$\|\nu_{ik}\| = \nu_{ik} \text{ and } \|-\nu_{ik}\| = 0 \text{ for } \nu_{ik} \geq 0 \text{ (for products)}, \tag{12a}$$

$$\|\nu_{ik}\| = 0 \text{ and } \|-\nu_{ik}\| = \nu_{ik} \text{ for } \nu_{ik} \leq 0 \text{ (for reactants)}. \tag{12b}$$

Differential equations (12a) and (12b) are solved numerically either by the fifth-order Runge–Kutta method with time step size control (the explicit method) or by the implicit method [47]. The relative importance of various parameters involved in reactor modelling (such as rate constants, activation energies, thermodynamic constants, transport coefficients, initial conditions, and operating conditions) can be evaluated by sensitivity analysis [48, 49]. Here, we used sensitivity analysis in order to determine the reactions whose rate constants should be calculated with higher accuracy. It is achieved by estimating the response of each species concentration to a small variation in each reaction rate constant.

Calculations were performed for typical process conditions for silicon nitride CVD from dichlorosilane and ammonia in a cold-wall reactor [27]: pressure, P = 10–100 Torr, wafer temperature, T_W = 800–1200 K; reactor volume, V_R 3500 cm^3; wafer area, S_W = 314 cm^2; inlet flow rates $I^*(Si_3N_4)$ = 30–90 sccm, $I^*(NH_3)$ = 300 sccm, and $I^*(N_2)$ = 7 slm (sccm and slm mean standard cubic centimetre per minute and standard litre per minute, respectively.

2.3. Gas-phase reactions: activation energies and complex formation

A reaction mixture in Si_3N_4 CVD from DCS and NH_3 and related processes may contain various chloro-, amino-, and mixed chloroaminosilanes. Reactions in which these molecules can participate at temperatures typical for the process conditions specified above can be classified into two main types: dissociation of silanes with the formation of silylenes and bimolecular substitution reactions between silanes and NH_3. Some of these reactions were considered in the literature [27, 28, 29, 32].

The reactions can be further classified according to their products. Silanes can dissociate to silylenes with the abstraction of H_2, HCl, and NH_3 (it was shown previously that the decomposition channels of chlorosilanes with the abstraction of an H or Cl atom and with the elimination of Cl_2 are highly endothermic [28, 29]). The corresponding thermal decomposition barriers for H abstraction from chlorosilanes estimated at the G2 level of theory [50] were found to be about 380 kJ/mol. The heats of Cl abstraction slightly vary, depending on the molecule: 456 kJ/mol (SiH_3Cl), 465 kJ/mol (SiH_2Cl_2), and 470 kJ/mol ($SiHCl_3$). The elimination of Cl_2 requires more then 500 kJ/mol in any case [28]. These values can be compared with the activation barriers 318.0 and 309.6 kJ/mol calculated at the same level for H_2 and HCl elimination from DCS, respectively [28, 29]. For reactions of the same order (unimolecular reactions, in the given case), the difference of more than 60 kJ/mol in activation energies offers enough reason for excluding these reactions from consideration.

Here, in the designations of reactions, the Roman numeral will indicate the reaction type (number I corresponds to decomposition reactions, II to bimolecular substitution reactions with ammonia, and III to some additional reactions with the formation of a Si–Si bond). Reactions that correspond to different transformation channels of the same system of reagents are grouped together and marked with the same Arabic numeral indicating the major reactant.

Different letters (a, b, c) correspond to transformation channels with the elimination of H_2, HCl, and NH_3, respectively. In the case of additional reactions (series III), different letters (a, b) designate the insertion of silylenes into a Si–H or Si–Cl bond, respectively. The forward and the reverse directions of the same reaction scheme will be marked additionally with symbols (f) and (r). The energy barriers and reaction energies calculated for gas-phase reactions are presented in Table 2.

Table 2. Activation barriers for forward E_A(f) and reverse E_A(r) gas-phase reactions and reaction energies ΔE (kJ/mol) calculated at the MCQDPT//MP2/MC-311G(d,p) level

No.	Reaction	E_A(f)	E_A(r)	ΔE
(I.1a)	$SiH_4 \leftrightarrow SiH_2 + H_2$	242.4	19.4	223.0
(I.2a)	$SiH_3Cl \leftrightarrow SiHCl + H_2$	263.6	81.7	182.0
(I.2b)	$SiH_3Cl \leftrightarrow SiH_2 + HCl$	294.3	13.3	281.0
(I.3a)	$SiH_2Cl_2 \leftrightarrow SiCl_2 + H_2$	295.0	167.4	127.6
(I.3b)	$SiH_2Cl_2 \leftrightarrow SiHCl + HCl$	276.2	33.9	242.4
(I.4b)	$SiHCl_3 \leftrightarrow SiCl_2 + HCl$	272.5	73.4	199.1
(I.6a)	$SiH_3(NH_2) \leftrightarrow Si(NH_2)H + H_2$	245.6	110.4	135.1
(I.6c)	$SiH_3(NH_2) \leftrightarrow SiH_2 + NH_3$	283.3	41.6	241.7
(I.7a)	$SiH_2(NH_2)Cl \leftrightarrow Si(NH_2)Cl + H_2$	286.3	180.1	106.2
(I.7b)	$SiH_2(NH_2)Cl \leftrightarrow Si(NH_2)H + HCl$	255.6	34.2	221.3
(I.7c)	$SiH_2(NH_2)Cl \leftrightarrow SiHCl + NH_3$	283.1	54.4	228.7
(I.8b)	$SiH(NH_2)Cl_2 \leftrightarrow Si(NH_2)Cl + HCl$	263.2	74.1	189.1
(I.8c)	$SiH(NH_2)Cl_2 \leftrightarrow SiCl_2 + NH_3$	280.7	84.0	196.7
(I.10a)	$SiH_2(NH_2)_2 \leftrightarrow Si(NH_2)_2 + H_2$	284.4	180.5	103.8
(I.10c)	$SiH_2(NH_2)_2 \leftrightarrow SiH(NH_2) + NH_3$	286.7	94.3	192.4
(I.11b)	$SiH(NH_2)_2Cl \leftrightarrow Si(NH_2)_2 + HCl$	259.0	56.5	202.6
(I.11c)	$SiH(NH_2)_2Cl \leftrightarrow Si(NH_2)Cl + NH_3$	287.7	111.8	175.9
(I.12c)	$SiH(NH_2)_3 \leftrightarrow Si(NH_2)_2 + NH_3$	291.2	128.0	163.3
(II.1a)	$SiH_4 + NH_3 \leftrightarrow SiH_3(NH_2) + H_2$	187.5	211.5	–24.0
(II.2a)	$SiH_3Cl + NH_3 \leftrightarrow SiH_2(NH_2)Cl + H_2$	177.8	231.9	–54.2
(II.2b)	$SiH_3Cl + NH_3 \leftrightarrow SiH_3(NH_2) + HCl$	80.3	44.1	36.2
(II.3a)	$SiH_2Cl_2 + NH_3 \leftrightarrow SiH(NH_2)Cl_2 + H_2$	143.7	218.8	–75.1
(II.3b)	$SiH_2Cl_2 + NH_3 \leftrightarrow SiH_2(NH_2)Cl + HCl$	31.4	21.3	10.1
(II.4a)	$SiHCl_3 + NH_3 \leftrightarrow Si(NH_2)Cl_3 + H_2$	171.1	226.9	–55.7
(II.4b)	$SiHCl_3 + NH_3 \leftrightarrow SiH(NH_2)Cl_2 + HCl$	72.0	52.0	20.0
(II.5b)	$SiCl_4 + NH_3 \leftrightarrow Si(NH_2)Cl_3 + HCl$	108.2	95.9	12.2
(II.6a)	$SiH_3(NH_2) + NH_3 \leftrightarrow SiH_2(NH_2)_2 + H_2$	166.5	224.9	–58.5
(II.7a)	$SiH_2(NH_2)Cl + NH_3 \leftrightarrow SiH(NH_2)_2Cl + H_2$	157.5	230.4	–72.8
(II.7b)	$SiH_2(NH_2)Cl + NH_3 \leftrightarrow SiH_2(NH_2)_2 + HCl$	62.0	27.6	34.4
(II.8a)	$SiH(NH_2)Cl_2 + NH_3 \leftrightarrow Si(NH_2)_2Cl_2 + H_2$	183.9	251.8	–67.9
(II.8b)	$SiH(NH_2)Cl_2 + NH_3 \leftrightarrow SiH(NH_2)_2Cl + HCl$	91.8	72.8	19.0
(II.9b)	$Si(NH_2)Cl_3 + NH_3 \leftrightarrow Si(NH_2)_2Cl_2 + HCl$	115.4	108.1	7.3
(III.1)	$SiH_2(NH_2)Cl + SiH_2Cl_2 \leftrightarrow SiH_2Cl(NH)SiH_2Cl + HCl$	75.5	64.7	10.8
(III.2)	$SiH(NH_2)Cl_2 + SiH_2Cl_2 \leftrightarrow SiHCl_2(NH)SiH_2Cl + HCl$	95.4	88.4	7.0
(III.3a)	$SiH_2Cl_2 + SiH(NH_2) \leftrightarrow SiHCl_2–SiH_2(NH_2)$	48.3	173.5	–125.2
(III.3b)	$SiH_2Cl_2 + SiH(NH_2) \leftrightarrow SiH_2Cl–SiH(NH_2)Cl$	46.4	187.1	–140.7
(III.4a)	$SiH_2Cl_2 + SiHCl \leftrightarrow SiHCl_2–SiH_2Cl$	32.2	189.8	–157.6
(III.4b)	$SiH_2Cl_2 + SiHCl \leftrightarrow SiHCl_2–SiH_2Cl$	33.9	191.4	–157.6
(III.5a)	$SiH_2Cl_2 + SiCl_2 \leftrightarrow SiHCl_2–SiHCl_2$	91.8	203.5	–111.7
(III.5b)	$SiH_2Cl_2 + SiCl_2 \leftrightarrow SiH_2Cl–SiCl_3$	72.7	188.5	–115.8
(III.6)	$SiH_2Cl_2 + SiH_2Cl_2 \leftrightarrow SiHCl_3 + SiH_3Cl$	205.7	196.5	9.1
(III.7)	$SiH_2Cl_2 + HCl \leftrightarrow SiHCl_3 + H_2$	189.4	259.2	–69.8
(III.8)	$SiH_2Cl_2 + H_2 \leftrightarrow SiH_3Cl + HCl$	256.5	175.6	80.8
(III.9)*	$SiCl_2 + SiH_3Cl \leftrightarrow SiHCl_2–SiH_2Cl$	80.8	212.9	–132.1

(*) MP2/MC-311G(d,p) calculations.

In agreement with [27], for all bimolecular (A + B) reaction systems we found local minima on their potential energy surfaces that correspond to molecular complexes A–B between free reagents. However, only some of the complexes have substantial formation energies. The complex formation energies larger than 8 kJ/mol are presented in Table 3. The largest formation energies (about 100 kJ/mol) were found for complexes of silylene, chlorosilylene, and dichlorosilylene with ammonia. All the complexes with H_2 and HCl have rather small formation energies, and the only $Si(NH_2)H{\bullet}HCl$ complex has the formation energy larger than 8 kJ/mol. In general, the formation of intermediate complexes may affect the resulting kinetics of the corresponding reactions. However, the relatively low formation energies of these complexes suggest that these kinetic effects can be neglected at least at rather high experimental temperatures of silicon nitride CVD.

Table 3. Complex formation energies ΔE_c (kJ/mol) calculated at the MCQDPT//MP2/MC-311G(d,p) (II) level

Complex	ΔE_c
Silylenes with HCl	
$Si(NH_2)H + HCl \rightarrow Si(NH_2)H{\bullet}HCl$	–8.5
Silylenes with ammonia	
$SiH_2 + NH_3 \rightarrow SiH_2{\bullet}NH_3$	–99.9
$SiHCl + NH_3 \rightarrow SiHCl{\bullet}NH_3$	–96.3
$SiCl_2 + NH_3 \rightarrow SiCl_2{\bullet}NH_3$	–88.5
$Si(NH_2)Cl + NH_3 \rightarrow Si(NH_2)Cl{\bullet}NH_3$	–44.6
$Si(NH_2)_2 + NH_3 \rightarrow Si(NH_2)_2{\bullet}NH_3$	–9.9
Silanes with HCl	
$SiH_2(NH_2)Cl + HCl \rightarrow SiH_2(NH_2)Cl{\bullet}HCl$	–19.0
$SiH(NH_2)Cl_2 + HCl \rightarrow SiH(NH_2)Cl_2{\bullet}HCl$	–24.2
$Si(NH_2)Cl_3 + HCl \rightarrow Si(NH_2)Cl_3{\bullet}HCl$	–16.5
$SiH_2(NH_2)_2 + HCl \rightarrow SiH_2(NH_2)_2{\bullet}HCl$	–32.6
$SiH(NH_2)_2Cl + HCl \rightarrow SiH(NH_2)Cl_2{\bullet}HCl$	–17.1
Silanes with ammonia	
$SiH_3Cl + NH_3 \rightarrow SiH_3Cl{\bullet}NH_3$	–16.6
$SiH_2Cl_2 + NH_3 \rightarrow SiH_2Cl_2{\bullet}NH_3$	–37.8
$SiHCl_3 + NH_3 \rightarrow SiHCl_3{\bullet}NH_3$	–14.9
$SiH_3(NH_2) + NH_3 \rightarrow SiH_3(NH_2){\bullet}NH_3$	–9.2
$SiH_2(NH_2)Cl + NH_3 \rightarrow SiH_2(NH_2)Cl{\bullet}NH_3$	–21.0
$Si(NH_2)Cl_3 + NH_3 \rightarrow Si(NH_2)Cl_3{\bullet}NH_3$	–15.5

The decomposition reactions of silanes have rather high activation barriers (240–295 kJ/mol) whereas the reverse insertion reactions of silylenes mostly have rather low activation barriers (13–180 kJ/mol). Activation barriers of these reactions, as well as their energies, strongly depend on the substituents at Si that stabilize silylenes in the following sequence: H < Cl < NH_2. Bimolecular reactions between silanes and ammonia that proceed with the elimination of molecular hydrogen and the corresponding reverse reactions (that is, hydrogenolysis of aminosilanes) require a high activation barrier. Similar reactions between chlorosilanes and NH_3 with the elimination of HCl (amination of chlorosilanes) have a rather low activation barrier, increasing in the sequence SiH_2Cl_2 < $SiH_2(NH_2)Cl$ < $SiHCl_3$ < SiH_3Cl < $SiH(NH_2)Cl_2$.< $SiCl_4$ < $Si(NH_2)Cl_3$. Structures of typical transition states for silylene insertion reactions and for bimolecular reactions between silanes and ammonia are presented in Fig. 1.

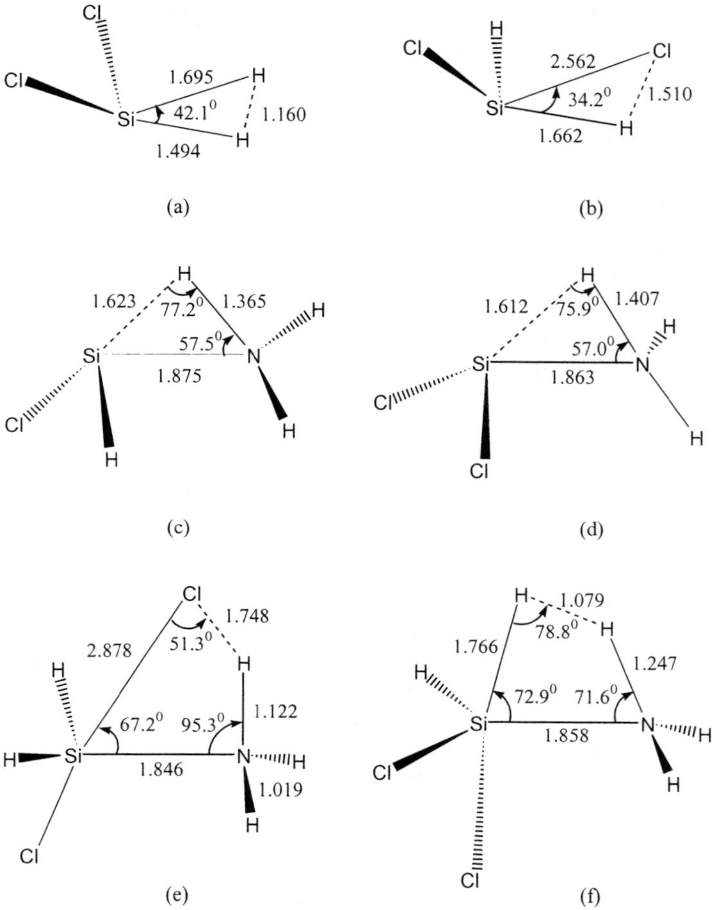

Fig. 2. Typical transition states: silylene insertion ↔ silane decomposition reactions (a) $SiCl_2 + H_2 \leftrightarrow SiCl_2H_2$, (b) $SiHCl + HCl \leftrightarrow SiCl_2H_2$, (c) $SiHCl + NH_3 \leftrightarrow SiH_2(NH_2)Cl$, (d) $SiCl_2 + NH_3 \leftrightarrow SiH(NH_2)Cl_2$ and bimolecular substitution reactions (e) $SiH_2Cl_2 + NH_3 \leftrightarrow SiH_2(NH_2)Cl + HCl$ and (f) $SiH_2Cl_2 + NH_3 \leftrightarrow SiH(NH_2)Cl_2 + H_2$

According to our computations, the reaction between NH_3 and HCl leads to the H-bonded molecular complex $H_3N...HCl$ with a stabilization energy of 30.5 kJ/mol. This reaction may occur only in cold zones of the reactor but not in the high-temperature reaction zone. The ionic form of the complex that formally corresponds to the ion-pair $(NH_4^+)(Cl^-)$ structure is thermodynamically unstable; therefore, we may believe that crystalline ammonium chloride microclusters can result from thermodynamic transitions of growing molecular clusters of $H_3N...HCl$.

2.4. *Gas-phase reactions: mechanism formulation and reaction rate calculations*

The rapid thermal CVD of silicon nitride (Si_3N_4) films from a mixture of DCS and NH_3 is a complicated process involving a number of concurrent gas-phase and surface reactions [16–20, 27]. The temperature, the pressure, and the residence time of molecules in the reactor may affect the extent of chemical transformations of initial reagents in the gas phase. Therefore, the dominant reaction paths may strongly depend on the conditions maintained in the reactor.

We will consider typical reaction conditions and will use the preliminary knowledge of most probable reactions to reduce the number of reactions that must be considered.

The residence time and the depth of conversion of reacting molecules in the gas phase are inversely proportional. Therefore, the reaction scheme should include all the significant reactions with the participation of the initial reagents and the subsequent reactions with the participation of the major products. When the residence time diminishes or the temperature is lowered, the depth of conversion diminishes and the number of consecutive reactions to be considered in the mechanism can be reduced. Below, we will list reactions in the sequence of their expected kinetic significance.

For a gas mixture containing DCS, NH_3, and an inert carrier gas, the starting set will include only the following reactions:

$SiCl_2H_2 + NH_3 \leftrightarrow SiH_2(NH_2)Cl + HCl$, (II.3b)
$SiCl_2H_2 + NH_3 \leftrightarrow SiH(NH_2)Cl_2 + H_2$, (II.3a)
$SiCl_2H_2 \leftrightarrow SiHCl + HCl$, (I.3b)
$SiCl_2H_2 \leftrightarrow SiCl_2 + H_2$, (I.3a)
$SiH_2Cl_2 + SiH_2Cl_2 \leftrightarrow SiHCl_3 + SiH_3Cl$. (III.6)

Reactions of bond breaking, e.g., the dissociation of Si–H, Si–Cl and N–H bonds, require higher temperatures than those typical for CVD or other sources of energy (see, for example, [28, 29]). According to previous results [27, 35], it could be expected that reaction (II.3b) is kinetically more significant than reaction (II.3a), and, respectively, reaction (I.3b) is kinetically more significant than reaction (I.3a). Based on the energy characteristics of the process, we may discard ammonia decomposition reactions $NH_3 \leftrightarrow NH_2 + H$ and $NH_3 \leftrightarrow NH + H_2$. The same is true for Cl–H exchange reactions between DCS and NH_3 resulting in chloroamines. The latter are relatively unstable, which results in rather high endothermicity of such reactions, and preliminary test calculations showed that these reactions are also highly endothermic and can be discarded with certainty.

The number of potentially important reactions that involve possible products of the above reactions is considerably greater than the above reaction set. Some of these reactions were rejected based on the same arguments used above to justify the elimination of their simpler analogues. Additional reactions were discarded because they were found too highly endothermic in preliminary estimations of their reaction energies. Finally, we selected the following set of reactions:

$SiH_2(NH_2)Cl + NH_3 \leftrightarrow SiH_2(NH_2)_2 + HCl$, (II.7b)
$SiH_2(NH_2)Cl + NH_3 \leftrightarrow SiH(NH_2)_2Cl + H_2$, (II.7a)
$SiH_2(NH_2)Cl + SiH_2Cl_2 \leftrightarrow SiH_2ClNHSiH_2Cl + HCl$, (III.1)
$SiH_2(NH_2)Cl \leftrightarrow Si(NH_2)H - HCl$, (I.7b)
$SiH_2(NH_2)Cl \leftrightarrow Si(NH_2)Cl + H_2$, (I.7a)
$SiH_2(NH_2)Cl \leftrightarrow SiHCl + NH_3$, (I.7c)
$SiH(NH_2)Cl_2 + NH_3 \leftrightarrow SiH(NH_2)_2Cl + HCl$, (II.8b)
$SiH(NH_2)Cl_2 + NH_3 \leftrightarrow Si(NH_2)_2Cl_2 + H_2$, (II.8b)
$SiH(NH_2)Cl_2 + SiH_2Cl_2 \leftrightarrow SiHCl_2NHSiH_2Cl + HCl$, (III.2)
$SiH(NH_2)Cl_2 \leftrightarrow Si(NH_2)Cl + HCl$, (I.8b)
$SiH(NH_2)Cl_2 \leftrightarrow SiCl_2 + NH_3$, (I.8c)
$SiH_2Cl_2 + SiH(NH_2) \leftrightarrow SiHCl_2\text{–}SiH_2(NH_2)$, (III.3a)
$SiH_2Cl_2 + SiH(NH_2) \leftrightarrow SiH_2Cl\text{–}SiH(NH_2)Cl$, (III.3b)
$SiH_2Cl_2 + SiHCl \leftrightarrow SiHCl_2\text{–}SiH_2Cl$, (III.4a)
$SiH_3Cl \leftrightarrow SiH_2 + HCl$, (I.2b)
$SiH_3Cl \leftrightarrow SiHCl + H_2$, (I.2a)
$SiHCl_3 \leftrightarrow SiCl_2 + HCl$, (I.4b)
$SiH_2Cl_2 + H_2 \leftrightarrow SiH_3Cl + HCl$, (III.8)
$SiH_2Cl_2 + HCl \leftrightarrow SiHCl_3 + H_2$, (III.7)
$SiHCl_3 + NH_3 \leftrightarrow SiH(NH_2)Cl_2 + HCl$, (II.4b)

$SiHCl_3 + NH_3 \leftrightarrow Si(NH_2)Cl_3 + H_2$, (II.4a)
$SiH_3Cl + NH_3 \leftrightarrow SiH_3(NH_2) + HCl$, (II.2b)
$SiH_3Cl + NH_3 \leftrightarrow SiH_2(NH_2)Cl + H_2$, (II.2a)

Reactions with $SiH_2(NH_2)Cl$ are listed first, because we may expect that $SiH(NH_2)Cl_2$ will be produced slower and in smaller amounts (through reaction (II.3a)) than $SiH_2(NH_2)Cl$ (through reaction (II.3b)). Next, reactions with NH_3 are kinetically more significant than similar reactions with DCS, because NH_3 is present in the initial gas mixture in considerably (to ten times) higher concentrations than DCS. For the same reasons as discussed above, we may also expect that a reaction with the elimination of HCl will be kinetically more significant than the similar reaction (another reaction channel of the given reaction) with the elimination of H_2. Consequently, we arranged all the reactions in the order of their expected kinetic significance. We suggest that the reactions listed above would compose a sufficiently full reaction scheme for this investigation. If kinetic calculations indicate that some product is accumulated in the reaction mixture in a sufficiently high concentration (comparable with the concentrations of the starting reagents), it will be necessary to expand the kinetic scheme by appending it with additional reactions with participation of this component. A similar expansion of the list of reactions to be included in the reaction scheme may be required if gas-surface reactions make the film formation kinetics sensitive to species present in low concentrations. Note that all the reverse reactions of the above sets were included simultaneously with their direct counterparts, because this does not require additional quantum-chemical calculations and does not increase the complexity of the kinetic scheme.

As was already mentioned, the activation barrier for the elimination of H_2 in reaction (II.3a) is considerably higher (by about 113 kJ/mol) than the activation barrier for the channel with the elimination of HCl (reaction (II.3b). Therefore, reaction (II.3a) can be excluded from the total kinetic scheme. For the same reason, we also excluded the chemically similar reactions between aminosilanes and ammonia and between aminosilanes and dichlorosilane that result in the elimination of H_2 (reactions (II.7a) and (II.8a). The exchange reaction (III.6) is also characterized by a rather high activation barrier; therefore, it could be discarded at the first level of consideration along with the reactions of its products, SiH_3Cl and $SiHCl_3$.

In order to enhance the model capabilities, the gas-phase chemical mechanism was extended by including the decomposition of silanes (dichlorosilane and aminochlorosilane, in our case) assisted by silylenes in the kinetic scheme [30, 32]. In the case of dichlorosilane, this decomposition scheme can be written as follows:

$SiH_2Cl_2 + SiHCl \rightarrow HCl_2SiSiH_2Cl$ (III.4)
$HCl_2SiSiH_2Cl \rightarrow SiH_3Cl + SiCl_2$ (III.9r)
$SiH_3Cl \rightarrow SiHCl + H_2$ (I.2a)

Energy diagrams for dichlorosilane decomposition initiated by SiHCl and $SiCl_2$ are presented in Fig 3. It can be seen in this figure that DCS decomposition assisted by $SiCl_2$ cannot compete with direct decomposition. This result is confirmed below by kinetics calculations within different mechanisms. Only the DCS decomposition assisted by SiHCl should be included into the resulting kinetics scheme. Similar decomposition scheme can be proposed for aminochlorosilane decomposition (the leading product at middle and high temperatures).
It was already found previously [27] that the high-pressure limit of the RRKM theory for some unimolecular reactions under consideration is attained only at a pressure of approximately 10^4 Torr. We verified this conclusion for a broader range of reactions included in our kinetic scheme. For all the dissociation and recombination reactions of the first and the second levels, we calculated rate constants by the RRKM theory at $T = 1000$ K (which falls within the temperature range typical for a single-wafer reactor) as functions of P within a broad range of pressures.

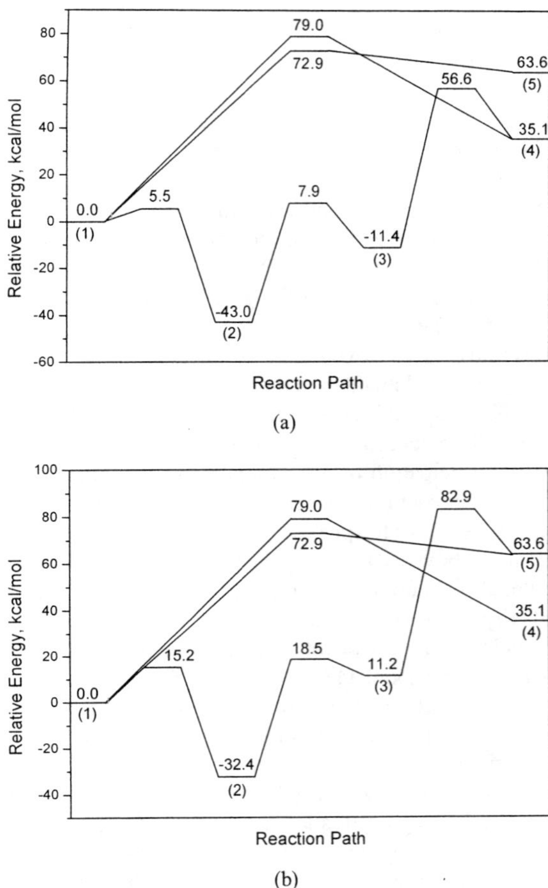

Fig. 3. Dichlorosilane decomposition initiated (a) by SiHCl (*1*) DCS + SiHCl, (*2*) $SiHCl_2$–SiH_2Cl, (*3*) $SiCl_2$ + SiH_3Cl, (*4*) $SiCl_2$ + SiHCl + H_2, and (*5*) 2SiHCl + HCl and (b) by $SiCl_2$ (*1*) DCS + $SiCl_2$, (*2*) SiH_2Cl–$SiCl_3$, (*3*) SiHCl + $SiHCl_3$, (*4*) $2SiCl_2$ + H_2, and (*5*) SiHCl + $SiCl_2$ + HCl.

The calculated curves are shown in Fig. 4 on a logarithmic scale. In consistence with [27], the high-pressure limit of the RRKM theory is attained for most reactions only at 10^3–10^4 Torr or even at higher pressures. Only for reactions (I.8c) and (I.8b) (decomposition of $SiH(NH_2)Cl_2$), the pressure dependence of log k is very flat in the entire range of P. All curves are quite close to straight lines and allow a power approximation $k \sim p^{\xi}$ within the range $10 < P < 100$ Torr (which covers the range of pressures typical for a single-wafer reactor [27]). Finally, we calculated all the rate constants for the bi- and unimolecular reactions under consideration within the ranges $T = 298$–1200 K and two fixed pressures 50 and 100 Torr.

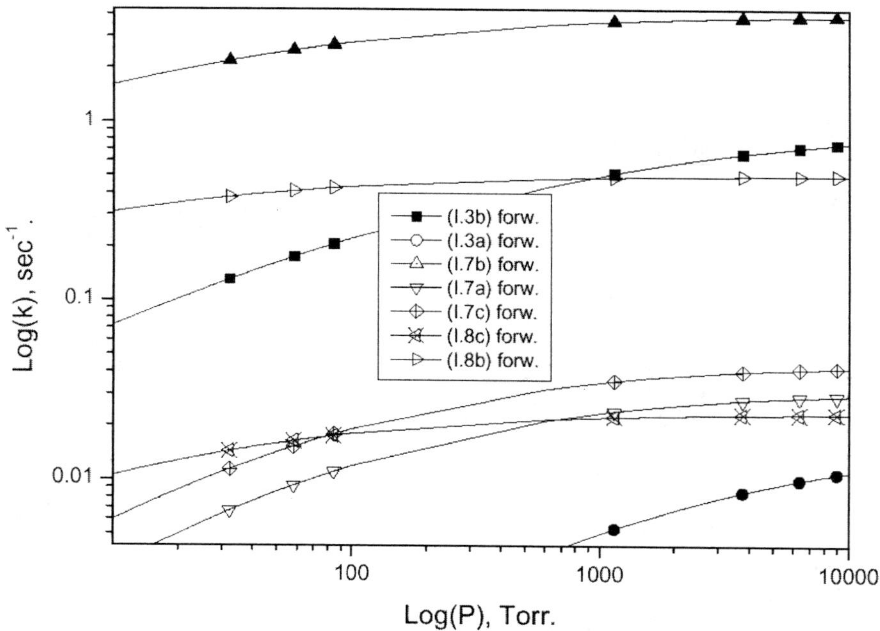

Fig. 4. Rate constants calculated for some unimolecular reactions by the RRKM theory at $T = 1000$ K as functions of P.

The results for reaction (i) were approximated by the equations $k(i) = A(i)T^{\xi(i)} \exp(-E_A(i)/RT)$, which give a better agreement of the approximated results with the calculated points for unimolecular reaction rates.

2.5. Gas-phase reactions: reactor modelling and sensitivity analysis

Using the calculated rate constants, we investigated the gas-phase kinetics based on the WSR model for typical CVD process conditions (see section 2.2). The effects of surface reactions on the gas-phase composition were neglected, because the relative volume concentration of surface Si and N atoms comprises only a small fraction of their gas-phase concentrations. Our estimations showed that less then 4% gas-phase silicon atoms are deposited onto the surface at a growth rate of 100 Å/sec.

The major advantage of this simplified reactor model lies in relatively small computational demands of the mathematical model, which allows one to consider and analyze extended chemical mechanisms with only moderate computational resources. In the framework of this model the flow through the reactor is characterized by the nominal residence time $\tau = V / I^*$, where V is the reactor volume and I^* is the flow rate.

Only scarce experimental or external theoretical data on gas-phase chemistry in a DCS : NH_3 mixture is available. The results of our simulation are compared with the experimental mass-spectrometric data obtained under typical CVD conditions [18] in Table 4. The first five columns contain the process conditions, the sixth column presents the major gas-phase component as reported in [18], and the last column presents the results of simulation based on the mechanism developed with the parameters calculated as described above.

Table 4. Comparison between experimental mass-spectrometric data [11] and calculated results

DCS:NH_3 input ratio	DCS:NH_3 flow rate (sccm)	T(K)	Pressure (Torr)	Residence time (sec)	Neutral species	Calculated
1 : 5	40 : 200	973	3.0	4.2	$SiH_2(NH_2)Cl$	$SiH_2(NH_2)Cl$
1 : 4	50 : 200	973	2.6	3.5	$SiH_2(NH_2)Cl$	$SiH_2(NH_2)Cl$
1 : 4	30 : 120	973	2.0	4.5	$SiH_2(NH_2)Cl$	$SiH_2(NH_2)Cl$
1 : 4	50 : 200	773	2.3	4.4	$SiH_2(NH_2)Cl$	$SiH_2(NH_2)Cl$
1 : 4	50 : 200	573	2.5	5.9	$SiH_2(NH_2)Cl$	SiH_2Cl_2
1 : 3	60 : 180	973	1.7	2.4	$SiH_2(NH_2)Cl$	$SiH_2(NH_2)Cl$
1 : 3	100 : 300	773	1.5	1.8	$SiH_2(NH_2)Cl$	$SiH_2(NH_2)Cl$
1 : 3	100 : 300	573	1.6	5.8	$SiH_2(NH_2)Cl$	SiH_2Cl_2
1 : 2	70 : 140	973	2.0	3.2	SiH_2Cl_2	$SiH_2(NH_2)Cl$
1 : 2	100 : 200	773	1.4	2.3	SiH_2Cl_2	SiH_2Cl_2
1 : 2	100 : 200	573	1.2	2.4	SiH_2Cl_2	SiH_2Cl_2

In order to check the influence of additional reactions on the kinetics in a DCS and ammonia mixture, kinetic calculations are performed for a sequence of extending chemical mechanisms. The simplest one:

$$SiH_2Cl_2 \leftrightarrow SiHCl + HCl$$
$$SiH_2Cl_2 \leftrightarrow SiCl_2 + H_2 \qquad \text{(A-1)}$$
$$SiH_2Cl_2 + NH_3 \leftrightarrow SiH_2(NH_2)Cl + HCl$$
$$SiH_2Cl_2 + NH_3 \leftrightarrow SiH(NH_2)Cl_2 + H_2$$

The next mechanism includes additionally reactions of the silylene-assisted decomposition scheme listed below:

$$SiH_2Cl_2 + SiHCl \leftrightarrow SiH_2Cl\text{–}SiHCl_2$$
$$SiH_2Cl_2 + SiCl_2 \leftrightarrow SiH_2Cl\text{-}SiCl_3$$
$$SiH_2Cl_2 + SiCl_2 \leftrightarrow SiHCl_2\text{-}SiHCl_2 \qquad \text{(A-2)}$$
$$SiH_2Cl\text{-}SiHCl_2 \leftrightarrow SiCl_2 + SiH_3Cl$$
$$SiH_2Cl\text{–}SiCl_3 \leftrightarrow SiHCl + SiHCl_3$$
$$SiHCl_2\text{–}SiHCl_2 \leftrightarrow SiHCl + SiHCl_3$$

Finally, the mechanism (A-1)-(A-2) is extended by the decomposition reactions of aminosilanes:

$$SiH_2(NH_2)Cl \leftrightarrow Si(NH_2)H + HCl$$
$$SiH_2(NH_2)Cl \leftrightarrow Si(NH_2)Cl + H_2 \qquad \text{(A-3)}$$
$$SiH_2(NH_2)Cl \leftrightarrow SiHCl + NH_3$$
$$SiH(NH_2)Cl_2 \leftrightarrow SiCl_2 + NH_3$$

The molar fractions of the major gas-phase components calculated for the chemical mechanism (A-1)-(A-3) is shown in Fig. 5. The effect of the silylene-assisted decomposition scheme on the dichlorosilane concentration is illustrated by Fig. 6. It can be seen that the silylene-assisted mechanism of decomposition in excess of ammonia has no strong effect on the gas-phase composition. The gas-phase mechanism is further reduced using sensitivity analysis. Some reactions were excluded based on the normalized sensitivity coefficients of concentrations to reaction rates.

The reduced mechanism is presented below as follows:

$SiH_2Cl_2 + NH_3 \leftrightarrow SiH_2(NH_2)Cl + HCl$, (II.3b)

$SiH_2Cl_2 + NH_3 \leftrightarrow SiH(NH_2)Cl_2 + H_2$, (II.3a)

$SiH_2Cl_2 \leftrightarrow SiHCl + HCl$, (I.3b)

$SiH_2Cl_2 \leftrightarrow SiCl_2 + H_2$. (I.3a)

$SiH_2Cl_2 + SiHCl \leftrightarrow SiH_2Cl\text{-}SiHCl_2$, (III.4a)

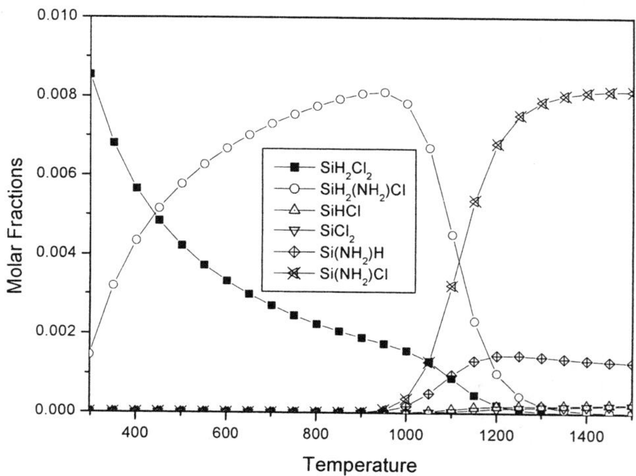

Fig. 5. Molar fractions of the major gas-phase components (P = 100 Torr, residence time 5 s, DCS : NH_3 ratio is 1 : 10).

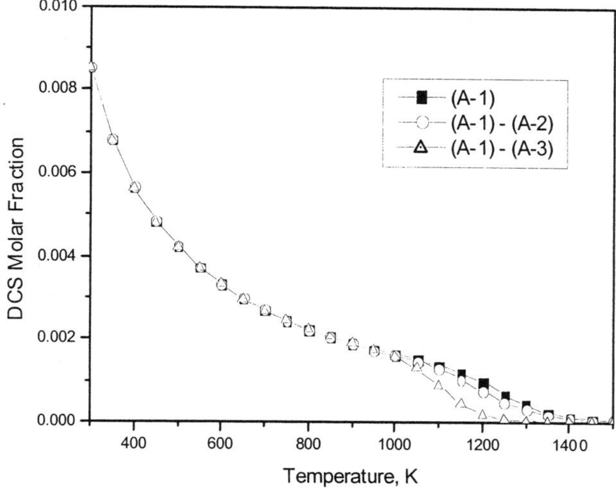

Fig. 6. Molar fraction of dichlorosilane calculated for different mechanisms (P = 100 Torr, residence time is 5 s, DCS : NH_3 ratio is 1 : 10).

$SiH_2Cl–SiHCl_2 \leftrightarrow SiCl_2 + SiH_3Cl$, (III.7)
$SiH_3Cl \leftrightarrow SiHCl + H_2$, (I.2a)
$SiH_2(NH_2)Cl + NH_3 \leftrightarrow SiH_2(NH_2)_2 + HCl$, (II.7b)
$SiH_2(NH_2)Cl \leftrightarrow Si(NH_2)H + HCl$, (I.7b)
$SiH_2(NH_2)Cl \leftrightarrow Si(NH_2)Cl + H_2$, (I.7a)
$SiH_2(NH_2)Cl \leftrightarrow SiHCl + NH_3$, (I.7c)
$SiH(NH_2)Cl_2 + NH_3 \leftrightarrow SiH(NH_2)_2Cl + HCl$, (II.8b)
$SiH(NH_2)Cl_2 \leftrightarrow SiCl_2 + NH_3$, (I.8c)

The molar fractions of silylenes calculated within this mechanism as functions of the residence time at $T = 1200$ K and $P = 100$ Torr are shown in Fig. 7.

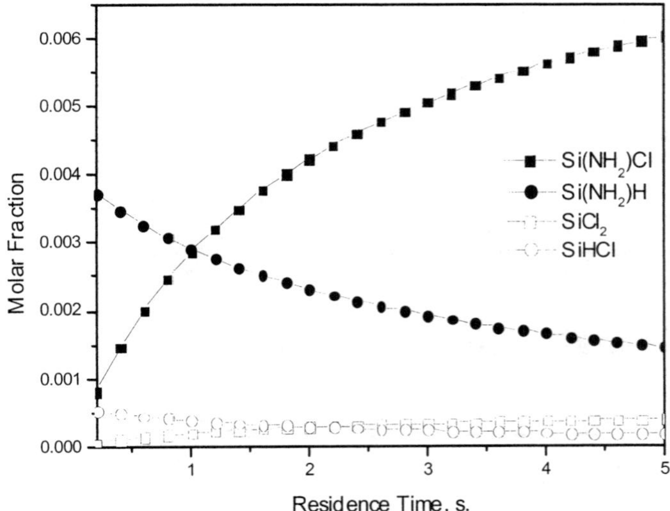

Fig. 7. Molar fractions of silylenes calculated as functions of the residence time ($P = 100$ Torr, $T = 1200$ K, DCS : NH_3 ratio = 1 : 10).

Gas-phase kinetics in a mixture of DCS and ammonia is a complicated process controlled by the temperature in reactor chamber and the residence time. The influence of the total pressure inside the reactor is not so critical. At higher pressures, all the decomposition processes are accelerated, but the overall gas-phase composition is modified only slightly.

The total concentration of silylenes is negligibly small at 800 K and remains much lower than the total concentration of aminosilanes even at 1000 K. $SiH_2(NH_2)Cl$ and/or SiH_2Cl_2 are the main reaction products at $T \leq 1100$ K. Thus, aminochlorosilane is the leading reaction product under common conditions in a single-wafer reactor at low temperatures and with an excess of ammonia in the reaction mixture. At high temperatures, some other products are also accumulated in the gas phase (mainly, aminochlorosilane or dichlorosilane decomposition products). At a short residence time, aminosilylene or silylene is dominating at high temperatures (see experimental work [20]). However, at longer residence times (more then 2–3 s), channels of the decomposition of (amino)silanes, which are kinetically less favourable, become dominating. In this case, the major products at high temperatures are $SiCl_2$ and $Si(NH_2)Cl$, which are more stable thermodynamically. In the case of an excess of ammonia (1 : 10), the products of dichlorosilane decomposition do not accumulate in significant amounts. A further step in understanding the mechanism of Si_3N_4 CVD is the inclusion of surface reactions in the model.

3. Theoretical Study of the Silicon Nitride Surface Structures and Mechanisms of Some Essential Surface Reactions

3.1. Introduction

Many experimental and theoretical works have been devoted to studying silicon nitride, its bulk and surface properties, deposition conditions, etc. The crystal structure of silicon nitride is well documented [51–56]. The nitridation of Si(100) and Si(111) surfaces and atomic layer deposition of Si_3N_4 on Si(100) was studied experimentally [57–68] and theoretically by DFT B3LYP calculations [69]. In particular, it was shown [65, 66] that the nitridation of a Si(111) surface with NH_3 gives rise to the formation of β-Si_3N_4 with the Si_3N_4(0001) surface. Empirical force fields were developed for silicon nitride and used for predicting its bulk properties such as equilibrium lattice parameters, phonon dynamics, thermodynamic properties, etc. [70–75]. Large-scale molecular dynamics simulations were performed to study the structure, dynamics, and mechanical behaviour of cluster-assembled Si_3N_4 and the silicon/silicon nitride interface [76–79]. An application of molecular dynamics to modelling silicon nitride film growth from gas-phase Si and N atoms on an amorphous silicon surface was reported in [80]. Several *ab initio* calculations have been performed to determine the properties and band structure of bulk silicon nitride [81–86].

However, in spite of the great technological importance of Si_3N_4 CVD from DCS and NH_3, no detailed theoretical study of Si_3N_4 surfaces and surface reactions had been published before we started our investigations in this area [27, 35, 87]. In this chapter, we present the results of a detailed *ab initio* cluster simulation of the silicon nitride surface structures and surface reactions related to silicon nitride CVD from DCS and ammonia.

3.2. Computational Details

Quantum chemical calculations with geometry optimization are carried out using the GAMESS [37] and Gaussian 98 [88] program packages with the 6-31G** basis set at the MP2 level [38]. Atoms in silicon nitride clusters were divided into two groups: “active” (positions optimized) and “inactive” (fixed coordinates). The active group included selected surface atoms ("active sites") and any chemisorbed groups or gas-phase molecules reacting with the surface. The inactive group included boundary crystal atoms chemically bound to the active-site atoms and hydrogen atoms saturating the broken bonds of the boundary atoms. The hydrogen atoms were placed along the deleted crystal Si–N bonds and fixed at standard Si–H or N–H distances (1.46 and 1.01 Å, respectively).

In each case, we have confirmed by vibrational analysis that all the stationary points found on the potential energy surfaces correspond to true minima (equilibrium structures) of true saddle points (transition states) with only one imaginary frequency whose vibrational mode actually described the motion from the reagent(s) to the product(s).

3.3. Surface structure and relaxation

Depending on the process conditions, Si_3N_4 CVD on a silicon surface may produce either amorphous or crystalline films with different degrees of crystallinity [52, 54, 55, 65, 66]. The crystal structures of both α- and β-modifications of Si_3N_4 consist of nearly planar (0001) layers (four and two layers per unit cell for α- and β-Si_3N_4, respectively) [51–56]. Surface clusters used in this study correspond to a β-Si_3N_4(0001) surface (see Fig. 8).

A bare β-Si_3N_4(0001) layer consists of seven-atom fragments (islands). Note that similar seven-atom islands with the same topology are the basic elements of crystal layers in the α-Si_3N_4 structure as well, though their geometrical structures in the α- and β-modifications of Si_3N_4 are slightly different (see, for example, ref. [55]).

An island has the same stoichiometry as the film and contains the central nitrogen atom (N_c), three silicon atoms bonded to N_c, and three terminal nitrogen atoms (N_t). Three SiN_t surface bonded pairs are linked through N_c, and three Si–N_c bonds lie in the same plane. Each Si or N_t atom has also bonds connecting it with the upper and underlying layers. The neighbouring islands in the same layer are linked to each other by bridges extending through the adjacent layers. The shortest of these bridging chains are diatomic Si–N_t bridges. Each SiN_t pair on a bare β-Si_3N_4(0001) surface has two dangling bonds, which form a π bond or a surface dimer, similar to that on the Si(100) surface (see, for example, [89]). We call such a dimer diatomic active site (DAS).

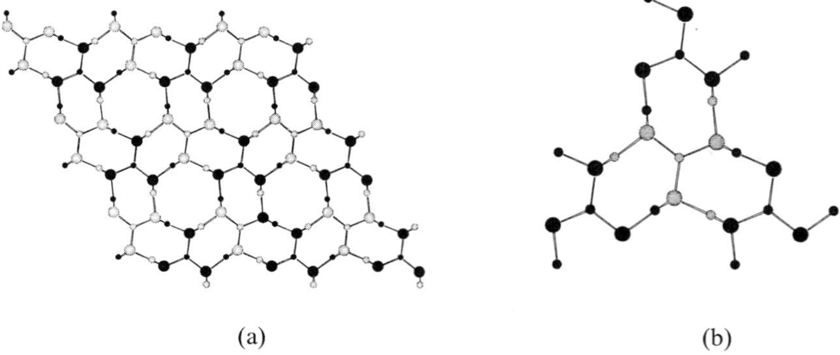

(a) (b)

Fig. 8. (a) Schematic view of two upper crystal layers of β-Si_3N_4(0001) and (b) a seven-atom Si_3N_4 island resting on three neighbouring islands of the preceding layer. Atoms of the upper and lower layers are displayed as light grey and dark grey circles, respectively; large circles designate silicon atoms, and small circles designate nitrogen atoms.

The simplest cluster describing a DAS (see Fig. 9a) has a single SiN_t surface pair of atoms as an active part of the cluster (single-DAS cluster model, 1-DAS). For comparison, we consider a larger cluster (Fig. 9b) that includes three DAS fragments bound to the central N_c atom into a seven-atom surface fragment as described above (seven-atom surface model, 3-DAS).[1] The nearest crystal environment (inactive part) of both clusters is constructed as described above.

The relaxed geometries of the bare and H-terminated Si_3N_4(0001) surfaces and the hydrogenation energies of surface Si=N bonds have been calculated using 1-DAS and (successively hydrogenated) 3-DAS models. Generally, the results obtained with both models are in good agreement with each other. The calculated lengths of the hydrogenated H–[Si–N_t]–H and formally double Si=N_t surface bonds are in the range 1.73–1.74 Å and 1.61 Å, respectively, for all the clusters under consideration. That is, the π-component of the double Si=N_t surface bond reduces the bond length by about 0.12–0.13 Å. Note for comparison that the lengths of the Si–N bonds lying in the (0001) plane in the bulk α- and β-Si_3N_4 crystals fall in the range of 1.72–1.74 Å [51–56]. According to X-ray structural data for $(t\text{-Bu})_2$Si=N–Si(t-Bu$)_3$ [90, 91] and $(t\text{-Bu})_2$Si=N–Si$(t\text{-Bu})_2$Ph [92], the lengths of the single and double SiN bonds are (to the second decimal place) 1.57 and 1.70 Å, respectively, with the same difference of 0.13 Å as was found in our calculations.

[1] The comparison here is made at the SCF level.

The energies of successive hydrogenation calculated for 3-DAS (−71, −70, and -69 kcal/mol) only slightly decrease in absolute value as the cluster becomes more saturated. That is, the hydrogenation energy of each Si=N_t bond is nearly independent of the state of the adjacent SiN bonds. The hydrogenation energy of the Si=N_t bond calculated for the small 1-DAS cluster is lower (67 kcal/mol) than the values calculated for the 3-DAS clusters, though it is rather close to the hydrogenation energy of the last Si=N_t bond in the (3-DAS)H_4 cluster. Finally we may infer that the 1-DAS cluster describes relatively well both the bond length changes and the reaction energy of an isolated Si=N_t double bond on the silicon nitride surface. Therefore, the 1-DAS cluster is used in further calculations as a basis for constructing various surface structures.

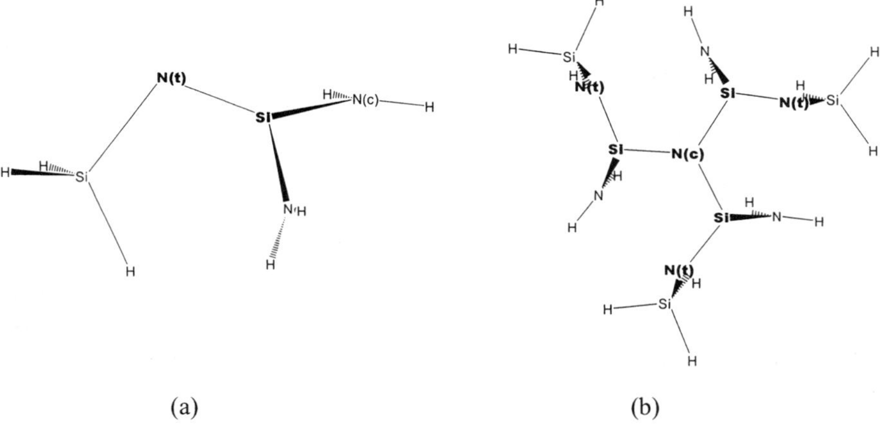

Fig. 9. (a) 1-DAS and (b) 3-DAS clusters used in calculations. Atoms that belong to the active part of the clusters are indicated in bold letters.

3.4. *Gas–surface reactions on the bare surface: dissociative chemisorption of H_2, HCl, NH_3, and SiH_2Cl_2*

The calculated reaction energies E_r, energies of formation of the intermediate reaction complexes E_c (non-dissociative adsorption), and activation energies (transition state energies) E_a for chemisorption of H_2, HCl, NH_3, and SiH_2Cl_2 on the 1-DAS cluster are presented in Table 5. In the chemical formulas below, only the active atoms of surface clusters are indicated explicitly. The "surface" part of active cluster atoms is enclosed in square brackets, and the surface groups chemisorbed on a cluster are indicated outside the brackets. Thus, 1-DAS (see Fig. 9a) is designated as [N=Si], whereas a 1-DAS-based cluster with R_1 and R_2 chemical groups chemisorbed on it is designated as R_1–[N–Si]–R_2 or simply R_1[N–Si]R_2.

Table 5. Energies (kcal/mol) of intermediate complexes E_c, activation energies E_a, and reaction energies E_r, for addition reactions of gas-phase molecules to the bare surface

No.	Surface reaction	E_c	E_a	E_r
SI.1	H_2 + [Si=N] = H[Si-N]H	−0.2	19.2	−61.0
SI.2*	HCl + [Si=N] = Cl[Si-N]H	−13.4	−10.2	−90.0
SI.3	NH_3 + [Si=N] = H_2N[Si-N]H	−34.9	−27.1	−77.5
SI.4a	SiH_2Cl_2 + [Si=N] = Cl[Si-N]SiH_2Cl	−8.1	−4.2	−86.7
SI.4b	SiH_2Cl_2 + [Si=N] = H[Si-N]$SiHCl_2$		−4.1	−84.7

(*) N–Cl bond formation is unfavourable and is not considered here.

Hydrogen chemisorption (reaction SI.1) is the least exothermic reaction. Two paths of SiH_2Cl_2 adsorption and dissociation on 1-DAS yielding the Si–Cl or Si–H bonds are energetically nearly equivalent. Relatively stable intermediate complexes are found for reactions SI.2–SI.4. The hydrogen molecule does not form a stable complex. Its reaction with the surface Si=N bond is the only one whose transition-state energy is above the energy of the initial state, that is, the separated H_2 and 1-DAS. The transition state energies for reactions SI.2–SI.4 are lower than the energies of the separated reagents; that is, there are barrierless reaction paths for the formation of the chemisorbed addition products in these reactions. The reaction energies E_r for the dissociative chemisorption reactions of aminochlorosilane $SiH_2(NH_2)Cl$, a product of the bimolecular gas phase reaction $SiH_2Cl_2 + NH_3 \rightarrow SiH_2(NH_2)Cl + HCl$ [27, 35], are also given below:

		E_r
(SI.5a)	$SiH_2(NH_2)Cl$ + [Si=N] = $ClH_2Si(HN)$–[Si–N]–H	–84.4
(SI.5b)	$SiH_2(NH_2)Cl$ + [Si=N] = Cl–[Si–N]–$SiH_2(NH_2)$	–83.5
(SI.5c)	$SiH_2(NH_2)Cl$ + [Si=N] = H–[Si–N]–$SiH(NH_2)Cl$	–83.6
(SI.5d)	$SiH_2(NH_2)Cl$ + [Si=N] = H_2N–[Si–N]–SiH_2Cl	–83.7

Chemically, reactions SI.5a, SI.5b, and SI.5c are analogues of reactions SI.3, SI.4a, and SI.4b, respectively, whereas reaction SI.5d combines features of reactions SI.3 and SI.4. The energies of reactions SI.5a–SI.5d are similar within 1 kcal/mol and close to the average value of reaction energies for reactions SI.3, SI.4a, and SI.4b.

We used the results of quantum-chemical calculations for reactions SI.1–SI.4b in order to estimate the equilibrium composition of the Si_3N_4 surface exposed to a mixture of DCS and NH_3. The equilibrium constants K_i for reactions SI.1–SI.4b were calculated from the partition functions Q^{ads} for the groups on the surface and the rotational and vibrational partition functions for the free gas molecules $Q^g_{r,v}$ and for the diatomic active sites on the surface Q^{DAS} using the relationship:

$$K_i = \left(\frac{2\pi\hbar^2}{m_i k_B T}\right)^{3/2} \frac{Q^{ads}}{Q^g_{r,v} Q^{DAS}} \exp\left(-\frac{\Delta E}{k_B T}\right) \tag{13}$$

Next, we calculated the surface coverage with the corresponding chemisorbed groups using the equations:

$$\theta_i = \frac{K_i[X_i]}{1+\sum_{i\geq 1} K_i[X_i]}, \quad \theta_{[Si=N]} = \frac{1}{1+\sum_{i\geq 1} K_i[X_i]} \tag{14}$$

where $[X_i]$ is the concentration of the ith gas-phase component, θ_i is the coverage for the corresponding surface group, and $\theta_{[Si=N]}$ is the fraction of vacant surface sites. For simplicity, we considered only equilibrium described by reactions SI.3, SI.4a, and SI.4b.

The results for a mixture of DCS and NH_3 are presented in Fig. 10. The surface is almost completely covered by chemisorbed groups at temperatures below 1400 K. The equilibrium concentrations of $-NH_2$ and $-SiH_2Cl$ groups chemisorbed at the surface become equal at T = 1000 K, whereas the equilibrium concentration of $-SiHCl_2$ groups remains substantially lower over the entire temperature range. Note that, according to our calculations, the chemisorbed SiH_2Cl groups dominate over the entire range of temperatures for a 1 : 1 mixture [87]. This result explains why a large excess of NH_3 is normally used in the CVD process (see, for example, [27, 14]). We may also infer that the fraction of vacant surface sites remains negligibly small at typical CVD conditions, implying that film growth is limited not by chemisorption but by reactions between chemisorbed species and other gas-phase or surface species.

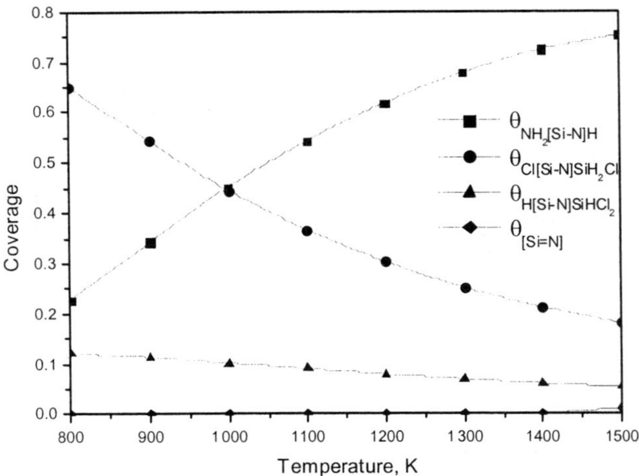

Fig. 10. Equilibrium concentrations of various surface groups as functions of temperature for DCS and NH_3 gas mixtures with partial pressures 1.0 and 10.0 Torr, respectively.

3.5. *Gas–surface reactions and surface reactions between surface groups*

According to the results presented above, the growing surface is almost completely covered by chemisorbed surface groups (chemically terminated surface). Each DAS may generate the following surface structures: H–[Si–N]–H, Cl–[Si–N]–H, NH_2–[Si–N]–H, Cl–[Si–N]–SiH_2Cl, H–[Si–N]–$SiHCl_2$, and NH_2–[Si–N]–SiH_2Cl. These surface groups may participate in further reactions with gas-phase molecules and with neighbouring surface groups to form a new seven-atom island of the next surface layer. The formation of bridges linking two of three islands serving as a base for the new one (see Fig. 8b) is a necessary step in this process. Therefore, along with reactions of surface groups with gas-phase molecules, we consider bridge closure reactions and reactions of such bridged structures.

Two chemically different types of reactions may result in the formation of a new Si–N bond at the surface: (1) reactions between saturated Si and N centres with the elimination of either HCl or H_2 and (2) addition reactions that involve a silylene-type centre. Among reactions of the first type, those leading to the desorption of HCl are energetically more favourable [27, 35].

The calculated results for various types of surface reactions are presented in Table 6. The main trends here are similar to those found above (section 2.3) for analogous gas-phase reactions [27, 35]. The activation barriers for the substitution reactions with the displacement of HCl are relatively low and lower than for similar reaction with the displacement of H_2. The displacement of H_2 is exothermic, whereas the displacement of HCl is endothermic. The activation barrier for the insertion of silylenes into the N–H bond is higher than for their insertion into the Si–H bond, but the reaction is more exothermic in the first case. The insertion reactions of $SiH(NH_2)$ have higher barriers and are less exothermic than the corresponding reactions of SiHCl.

Table 6. Activation E_a and reaction E_r energies for various surface reactions, kcal/mol

No.	Surface reaction*	E_a	E_r
	Reactions with surface atoms		
SII.1	$H[N–Si]–Cl + NH_3 = H[N–Si]–NH_2 + HCl$	25.1	10.3
SII.2	$H[N–Si]–H + NH_3 = H[N–Si]–NH_2 + H_2$	49.4	–7.5
SII.3	$H[Si–N]–H + SiH_2Cl_2 = H[Si–N]–SiH_2Cl + HCl$	11.6	0.0
SII.4	$H[Si–N]–H + SiHCl = H[Si–N]–SiH_2Cl$	9.4	–64.9
SII.5	$H[Si–N]–H + SiH(NH_2) = H[Si–N]–SiH_2NH_2$	19.3	–54.1
SII.6	$H[N–Si]–H + SiHCl = H[N–Si]–SiH_2Cl$	0.9	–43.9
SII.7	$H[N–Si]–H + SiH(NH_2) = H[N–Si]–SiH_2NH_2$	11.9	–29.0
	Reactions of chemisorbed surface groups		
SIII.1	$Cl[Si–N]–SiH_2Cl + NH_3 = Cl[Si–N]–SiH_2–NH_2 + HCl$	13.2	8.5
SIII.2	$H[Si–N]–SiHCl_2 + NH_3 = H[Si–N]–SiHCl–NH_2 + HCl$	21.5	6.4
SIII.3	$H[N–Si]–NH_2 – SiH_2Cl_2 = H[Si–N]–NH–SiH_2Cl + HCl$	7.6	0.3
SIII.4	$Cl[Si–N]–SiH_2Cl = Cl[Si–N]–(HSi:) + HCl$	72.2	63.1
SIII.5	$Cl[Si–N]–SiH_2Cl = Cl[Si–N]–(ClSi:) + H_2$	76.9	30.6
SIII.6	$H[Si–N]–SiHCl_2 = H–[Si–N]–(ClSi:) + HCl$	72.1	55.7
	Reactions of bridge closure**		
SIV.1	$\{Si\}–NH_2 + \{N\}–SiH_2Cl = \{Si\}–NH–SiH_2–\{N\} + HCl$	18.4	5.1
SIV.2	$\{Si\}–NH_2 + \{N\}–SiH_2Cl = \{Si\}–NH–SiHCl–\{N\} + H_2$	44.7	–22.6
SIV.3	$\{Si\}–NH_2 + \{N\}–SiH_2(NH_2) = \{Si\}–NH–SiH(NH_2)–\{N\} + H_2$	41.5	–18.5
	Reactions with bridged Si atom		
SV.1	$H[N–Si](H)(H) + NH_3 = H[N–Si](H)(NH_2) + H_2$	38.2	–17.4
SV.2	$H[N–Si](H)(H) = H[N–Si:] + H_2$	76.9	28.9
SV.3	$H[N–Si:] = [N{=}Si]H$	77.0	13.5
SV.4	$H[N–Si](H)(Cl) = H[N–Si:] + HCl$	70.8	56.6
SV.5	$H[N–Si](H)(NH_2) = H[N–Si:] + NH_3$	78.6	46.3

(*) Reactions are classified according to sequential film growth steps.
(**) Bridge closure reactions SIV.1–SIV.3 are calculated with the use of reduced models in which only the reacting surface groups are included in the active part of the clusters.

The insertion of silylenes into the Si–H bond will produce nonstoichiometric defects with Si–Si bonds. Hence, the presence of silylenes in the gas phase (especially, SiHCl) possibly results in a nonstoichiometric silicon-rich film. The formation of silylene type structures at the surface (reactions SIII.4–SIII.6 and SV.2–SV.5) is associated with high activation barriers and hardly contributes to the entire deposition process. The results for reaction SV.3 indicate that the :Si–N silylene and Si=N double-bond structures are separated by a high activation barrier.

3.6. *Implications for the mechanism of Si_3N_4 film growth from DCS and NH_3*

Within the model of film growth along the crystal *c*-axis (perpendicular to the $Si_3N_4(0001)$ surface) assumed in this work, each seven-atom surface island on the perfect $Si_3N_4(0001)$ surface grows independently of the other surface islands, bridging three neighbouring seven-atom islands in the previous layer (see fig 8b.). Let us now consider the construction of the seven-atom island resting on three neighbouring islands of the underlying layer as a sequential process of forming Si–N bonds. Twelve Si–N bonds should be formed in order to complete such an island: 6 bonds linking the new island with the underlying islands and 6 bonds within the seven-atom island itself (see Fig.8). According to the stoichiometry of the overall process ($3SiH_2Cl_2 + 4NH_3 = Si_3N_4 + 6HCl + 6H_2$), three SiH_2Cl_2 molecules and four NH_3 molecules must be consumed from the gas phase and six HCl and six H_2 molecules must be desorbed for each constructed seven-atom island.

Since seven gas phase molecules must participate in the full cycle of construction of a new seven-atom island, seven new Si-N bonds will be formed through gas-surface reactions and only five Si–N bonds will be formed via reactions between surface groups. However, only six HCl molecules are removed from the surface in this process. This simple count indicates that either one of the gas-phase molecules must react with the surface with the displacement of H_2 or one H_2 molecule must be removed from the surface through direct dissociation from a $-SiH_2-$ group or from a =SiH–NH– diatomic site. The abstraction (desorption) of an H_2 molecule from a surface $-SiH_2Cl$ group (reaction SIII.5) or from a bridged SiH_2 group (reaction SV.2) to form a silylene surface group proceeds with a very high activation barrier of ~77 kcal/mol. Of the gas–surface reactions involving the removal of H_2, the lowest activation energy is encountered in the reaction of a gas-phase NH_3 molecule with a bridged Si centre (reaction SV.1), which has an activation barrier of ~38 kcal/mol. Thus, we may infer that the removal of one H_2 molecule from the surface, either through a gas-phase reaction with H_2 elimination or through the desorption of an H_2 molecule from the surface, must be the rate-determining step of the overall Si_3N_4 film growth. Our results also demonstrate that the deviation from stoichiometry observed in the Si_3N_4 films deposited from a mixture of NH_3 and SiH_2Cl_2 are expected to be primarily associated with SiH_x and NH_x (x = 1, 2) groups buried in the film because of the high activation energy of H_2 desorption from the surface of the growing Si_3N_4 film. On the other hand, the deviations from stoichiometry associated with the formation of Si–Si bonds via reactions with either gas-phase or surface silylene groups are less probable.

4. Kinetic Monte Carlo atomic scale simulation of chemical vapour deposition of silicon nitride film

4.1. *Introduction*

This section presents the results of a 3D simulation of silicon nitride film growth based on the kinetic Monte Carlo approach. This method is widely used for the modelling of crystal growth processes. A theoretical investigation of silicon crystal growth based on this approach was reported in [93]. The kinetic Monte Carlo approach was used for modelling the chemical deposition of a diamond film [94, 95].
The time step for kinetic Monte Carlo simulations depends on the surface structure at the atomic level. Thus, this model allows us to predict the process of film growth during time intervals large enough for determining the main characteristics of growth. Besides, this model simulates the real behaviour of a system in hand on the atomic scale; hence, we can investigate all the possible atomic scale processes such as the formation of substitution defects, impurity migration, and so on. Here, we use kinetic Monte Carlo simulation on a fixed 3D network of silicon nitride crystal lattice sites. Atoms newly generated in the process of film growth are placed in positions that correspond to some virtual (empty) sites of the crystal lattice. Though this approach can describe the formation of defects and irregularities in the growing film, it cannot predict the growth of truly amorphous film. An extension of the kinetic Monte Carlo method suitable for describing amorphous film growth has been proposed recently in our work [96].

4.2. *Method of Kinetic Monte Carlo Simulation of Silicon Nitride Film Growth*

The model of the surface of a growing film used in our studies was discussed in detail in the previous section. Periodic boundary conditions in the crystallographic directions [100] and [010], respectively, are used in the simulations. It is assumed that the crystalline silicon and nitrogen atoms are fixed at certain crystal lattice sites and can be characterized by a certain set of numbers $\vec{X}$. These numbers include data on the occupations of crystal sites, the number of hydrogen or chlorine atoms possibly bonded to the atoms of the growing crystal, and data on the bonds formed between these atoms. This set of numbers determines the state of the crystal. All changes that occur in the system (interaction of surface groups with gas-phase molecules and between each other) are reduced to a finite list of reactions or events each described by a certain set of changes that occur in the crystal lattice. These changes are the addition of new Si and N atoms to the lattice, the removal of old Si and N atoms, the formation or breaking of bonds between crystal atoms, and the resulting decrease or increase in the number of H or Cl atoms bonded to these atoms, etc.

The choice of the next event is carried out in the following way: First, a set of possible reactions (events) for the current state $\vec{X}$ of the system is constructed. The rate of each reaction from this set $W_{\vec{X}\to\vec{X}'}$ equals the rate of the elementary reaction corresponding to the transition $\vec{X} \to \vec{X}'$ in the system. Thus, $\vec{X}'$ runs through all the possible states that can be obtained from $\vec{X}$ via one elementary reaction (event). Next, the total reactivity of the system in the state $\vec{X}$ is calculated by the formula

$$W_{\vec{X}} = \sum_{\vec{X}'} W_{\vec{X}\to\vec{X}'} \,. \tag{15}$$

At the next step, a random number R_1 uniformly distributed between 0 and 1 is generated. Running through the set of possible elementary reactions, we add up the corresponding rates until the sum becomes larger than the product $R_1 W_X$ according to the following inequality:

$$\sum_{i=0}^{m-1} W_{\vec{X}\to\vec{X}'} < R_1 W_{\vec{X}} < \sum_{i=0}^{m} W_{\vec{X}\to\vec{X}'} \,. \tag{16}$$

Thus, the reaction with the index m is selected as occurring at the current time step. This algorithm selects a reaction with a probability proportional to its rate $W_{\vec{X}\to\vec{X}'}$. After selecting a reaction, we should change the state of the system $\vec{X} \to \vec{X}_m$ and increase the current time by

$$\Delta t = -\frac{\ln R_2}{W_{\vec{X}}} \,, \tag{17}$$

where R_2 is a random number uniformly distributed between 0 and 1.

The Monte Carlo approach described above is statistically equivalent to approaches with a varying time increment [94]. If many fast reactions can proceed in the given state of the system, the time step will be small. In contrast, if slow reactions prevail, the time step will be larger.

4.3. *Calculations of Reaction Rates for Elementary Surface Reactions*

The reaction rates of elementary gas-surface reactions were calculated based on the transition state theory (TST) within the cluster approach. Various applications of the TST to calculations of surface reaction rates can be found in [97–99]. The calculation of the reaction rate constant k of a chemical reaction within the transition state theory is reduced to the calculation of the ratio of the partition functions of the transition state and the reagents

$$k = \frac{k_B}{2\pi\hbar}\frac{Q^{\neq}}{Q_R}\exp\left(-\frac{\Delta E}{k_B T}\right). \tag{18}$$

Here, k_B is Boltzmann's constant, T is the temperature, and ΔE is the activation barrier of the reaction. The method of the calculation of this ratio depends on the reaction type. The description of the possible types of surface reaction is given below.

In the course of adsorption and substitution reactions, gas-phase molecules interact with the surface. We may distinguish two possible types of such reactions: barrierless reactions and reactions with an activation barrier.

In the first case, the motion of gas-phase molecules to the surface may be considered the reaction coordinate. Therefore, the partition function of the transition state will differ from the partition function of the reagents by the factor corresponding to the one-dimensional translation partition function. Thus, we obtain the following equation for the reaction rate of a gas-phase molecule X with the surface:

$$k = \sqrt{\frac{k_B T}{2\pi\mu}} S[X], \tag{19}$$

where S is the area of the reacting surface, and μ is the mass of the gas molecule. Equation (19) has the same form as the equation for the frequency of collisions of the gas-phase molecules with the surface. Hence, the rate constant of a barrierless reaction between a gas-phase molecule and an active surface site is

$$K = N_A s \sqrt{\frac{k_B T}{2\pi\mu}}, \tag{20}$$

where N_A is the Avogadro number and s is the area of the active surface site. In the case of silicon nitride, the area of the diatomic active site at the bare (0001) surface is $s = 1.67\cdot 10^{-15}$ cm^2.

The partition functions for reactions with a barrier were calculated based on the hypothesis that interactions between different degrees of freedom could be neglected. To evaluate the crystal influence on the partition functions, we assumed that distant crystal atoms make the same contribution to the partition functions of the transition state and the reagents and, hence, may be neglected in Eq. (18). In our case when active surface sites were modelled using surface clusters, we eliminated the vibrations of these inactive atoms (which model the crystal environment of active surface atoms) by setting their masses at 10^5 amu. Under the above assumptions, Eq. (18) takes the following form:

$$K = N_A r \frac{\hbar^2}{\mu} \sqrt{\frac{2\pi}{\mu k_B T}} \frac{\prod_i q_i^{\neq}}{\prod_i q_i^R} \exp\left(-\frac{\Delta E}{k_B T}\right), \tag{21}$$

where $q_i^{\neq}$ and q_i^R are the vibrational and rotational partition functions for the transition state and reagents, respectively, and r is a kinematic factor taking into account the ratio of the vibrational frequencies of the transition state and reagents around the inactive atoms of the molecular cluster modelling the active site. In the case of two inactive atoms,

$$r = \sqrt{I^{\neq}/I^C}, \tag{22}$$

where $I^{\neq}$ and I^C are the moments of inertia of the transition state and the molecular cluster in respect to the axis connecting the inactive atoms. In the case of one inactive atom, we obtain

$$r = \sqrt{I_1^{\neq} I_2^{\neq} I_3^{\neq} / I_1^C I_2^C I_3^C}, \tag{23}$$

where $I_1^{\neq}$, $I_2^{\neq}$, $I_3^{\neq}$, and I_1^C, I_2^C, I_3^C are the principal moments of inertia of the transition state and the molecular cluster in reference to the position of the inactive atom. The kinematic factor r is equal to 1 for clusters with more then two inactive atoms.

It is assumed that the desorption rate constants are related to the adsorption rate constants by the equilibrium constants as follows:

$$k_{des} = k_{ads} k_{eq} = k_{ads} \frac{Q_{ads}}{Q_{des}} \exp\left(-\frac{\Delta H}{k_B T}\right), \tag{24}$$

where Q_{ads} is the partition function for the surface with the adsorbed molecule, Q_{des} is that for the surface with free gas-phase molecule, ΔH is the change of enthalpy during adsorption. The ratio Q_{ads}/Q_{des} was calculated under the above assumptions, and the following equation was found for the rate constant:

$$K_{des} = \frac{K_{ads}}{N_A}\left(\frac{\mu k_B T}{2\pi\hbar^2}\right)^{3/2} \frac{\prod_i q_i^{ads}}{\prod_i q_i^{des}} \exp\left(-\frac{\Delta H}{k_B T}\right), \tag{25}$$

where q_i^{ads} and q_i^{des} are the vibrational and rotational partition functions for the cluster with the adsorbed molecule and for the cluster with the free gas-phase molecules, respectively.

4.4. Results of Calculations

The set of possible elementary surface processes has been refined based on available experimental data [27]. Experimental data are available only for one set of process conditions $T = 750°C$, $P = 100$ Torr. The inlet flow rate for N_2 is equal to 7 slm, while the inlet flow rates for NH_3 and DCS are varied. The rates of elementary reactions have been fitted to reproduce film growth rates and the stoichiometric composition in agreement with the experimental data. The results of our simulations are given in Fig. 11. The new parameters reproduce the decrease in the film growth rate with increasing partial pressure of NH_3.

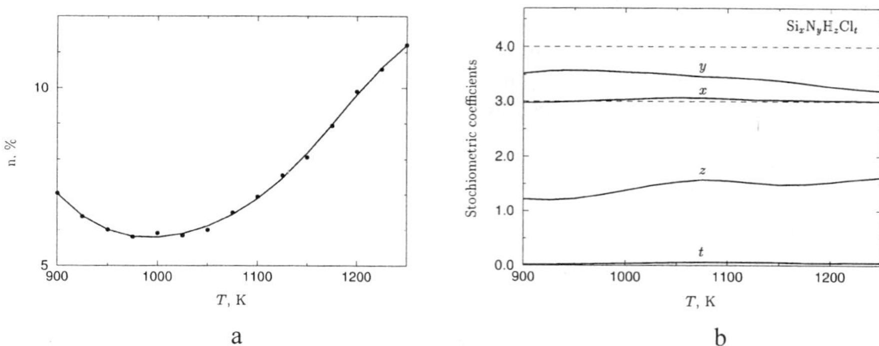

a b

Fig. 11. Kinetic Monte Carlo simulations of Si_3N_4 film growth: (a) vacancy concentration and (b) composition as functions of temperature; $P = 50$ Torr, DCS : NH_3 = 1 : 10.

Table 7. Comparison of experimental data and the data obtained by Monte Carlo simulation for film growth rates and the stoichiometric composition; deposition conditions: $T = 750°C$, $P = 100$ Torr, N_2 flow rate = 7 slm

No.	DCS flow	NH_3 flow	Dep. Rate		N/Si	
			Exp.	Simulated	Exp.	Simulated
1	30	300	2	2.1	---	1.43
2	30	600	1.8	1.4	---	1.43
3	30	900	1.7	1.3	1.62	1.48
4	150	300	9.6	12.2	1.39	1.30
5	150	600	5.5	9.8	---	1.34
6	150	900	5.5	8.3	---	1.37
7	270	300	22.4	20.6	1.24	1.25
8	270	600	10.6	17.3	---	1.28
9	270	900	6.7	15.0	1.54	1.32

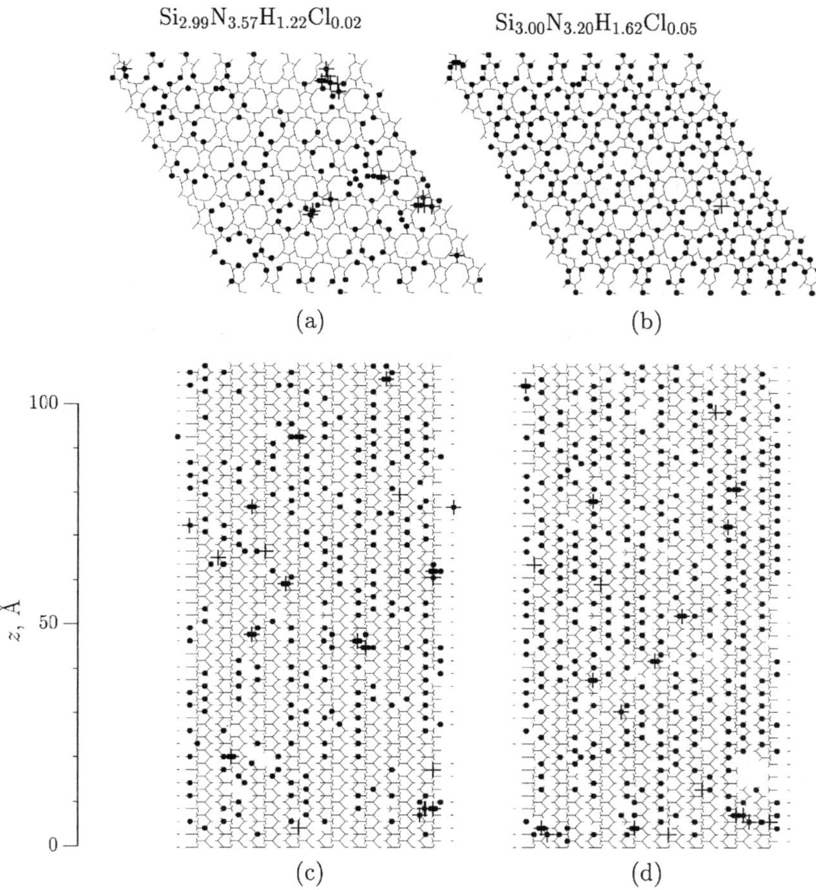

Fig. 12. Schematic representation of film sections at (a), (b) *xy* plane and (c), (d) *xz* plane for (a), (c) $T = 950$ K and (b), (d) $T = 1250$ K; the layer contains 8 x 8 seven-atom islands, H atoms are shown by circles, Cl atoms are shown by squares, sites with wrong atoms (Si instead of N and N instead of Si) are labelled by crosses.

In addition, simulations of silicon nitride film growth have been performed for the following process conditions: the pressure in the reactor chamber 50 Torr; inlet flow rates 7 slm, 30 sccm, and 300 sccm for N_2, DCS, and NH_3, respectively; the temperature of the gas mixture has been varied in the range from 900^0K to 1300^0K. These conditions correspond to the experimental data cited in [27]. For these process conditions, the concentrations of the gas-phase component have been determined based on the WSR model without deposition using the gas-phase mechanism from described in the previous sections [35]. The dependence of the film quality and composition on the process conditions is demonstrated in Fig. 11, where the concentration of vacancies (the ratio of unoccupied sites in the lattice to the total number of sites inside the lattice) and the composition of the growing silicon nitride film represented as $Si_xN_yH_zCl_t$ are shown as functions of temperature. The sections of the film for different temperatures are shown in Fig. 12.

5. General Conclusions and Outlook

The results described above demonstrate that a film growth process can be modelled in rather fine details including gas-phase reactions, surface structure, gas–surface reactions, the kinetics of elementary processes, and the kinetics of film growth. This integrated modelling of the entire process allows one to predict, at least qualitatively, the growth rate , the film composition, concentrations of various defects, the film structure at the atomic level, etc. as functions of deposition parameters, such as temperature, pressure, gas-phase component ratio, and others. This integrated approach includes theoretical *ab initio* calculations of the structure of the reagents, surfaces, surface groups, transition states, etc.; *ab initio* reaction rate calculations; formal kinetic analysis; and kMC simulations of thin film growth, which eventually give the predictions of the structure and the composition of the resulting film. It is clear that *ab initio* calculations may not be restricted to only quantum-chemical molecular or cluster calculations but may involve, for example, periodic-type DFT approaches, which are more common in the case of solid-state structures. It is also clear that the kMC technique based on the rigid lattice approximation is limited to describing only regular structures or structures that can be described, at least approximately, as defects in a regular periodic lattice. However, this approach fails when it is necessary to describe amorphous film growth.

This deficiency becomes especially important in the case an ionic system, like ZrO_2, when irregular or amorphous film structures should be included into consideration. Such systems have recently aroused great interest as promising alternative high-*K* dielectric materials for microelectronic applications (see, for example, [100]). An integrated approach that takes into account arbitrary irregular structures in kMC simulations as applied to the atomic layer deposition (ALD) of zirconium oxide is presented in [96, 101]. In this approach, the structure of the growing film at any kMC time step is determined using some energy minimization technique, which should be preferably as fast as possible. The simplest technique of this kind is the force-field approach using classical potential functions for the prediction of the growing film structure. The approach of this kind is developed in our recent work [96]. It may be believed that detailed atomistic description of various complex chemical and physical processes can be obtained based on this versatile integrated approach. The further development may include more elaborate techniques of structure optimizations based on plane-wave DFT methods (see, for example, [102, 103].

References

[1] *The International Technology Roadmap for Semiconductors*, 1999 (Semiconductor Industry Association, San Jose, CA, 1999), http://public.itrs.net/files/1999_SIA_Roadmap/Home.htm
[2] S.T. Rodgers and K.F. Jensen, *J. Appl. Phys.*, 1998, vol. 83, p. 524.
[3] K.F. Jensen, S.T. Rodgers, and R. Venkataramani, *Curr. Opin. Solid State Mat. Sci.* 3 (1998) 562.
[4] T.P. Merchant, M.K. Gobbert, T.S. Cale, and L.J. Borucki, *Thin Solid Films* 365 (2000) 368.
[5] H.N.G. Wadley, X. Zhou, R.A. Johnson, and M. Neurock, *Prog. Mater. Sci.*, 46 (2001) 329.
[6] G. Valente, C. Cavallotti, M. Masi, and S. Carra, *J. Cryst. Growth,* 2001, vol. 230, p. 247.
[7] A. Nakano, M.E. Bachlechner, R.K. Kalia, E. Lidorikis, P. Vashishta, G.Z. Voyiadjis, T.J. Campbell, S. Ogata, and F. Shimojo, *Comput. Sci. Eng.* 3 (2001) 56.
[8] P.D. Davides and L.I. Maissel, *J. Appl. Phys. 37* (1966) 574.
[9] W.A. Pliskin, *Thin Solid Films 2* (1968) 1.
[10] A. Lekhollm, *J. Electrochem. Soc. 119* (1972) 1122.
[11] K.H. Jack and W.I. Wilson, *Natural Phys. Sci. 238*(1972) 28.
[12] R.S. Rosler, *Solid State Technol. 20* (1977) 63.
[13] R.N. Katz, *Science 208* (1980) 841.
[14] C.-E. Morosanu, *Thin Solid Films,* 1980, vol. 65, p. 171.
[15] C.-E. Morosanu and E. Segal, *Mater. Chem. 7* (1982) 79.
[16] K.F. Roenigk and K. F. Jensen, *J. Electrochem. Soc.*, 1987, vol. 134, p. 1777.
[17] G. Peev, L. Zambov and Y. Yanakiev, *Thin Solid Films,* 1990, vol. 189, p. 275.

[18] T. Sorita, T. Satake, H. Adachi, T. Ogata, and K. Kobayashi, *J. Electrochem. Soc.*, 1994, vol. 141, no. 12, p. 3505.

[19] T. Ogata, T. Sorita, K. Kobayashi, Y. Matsui, K. Horie, and M. Hirayama, *Jpn. J. Appl. Phys.*, 1996, vol. 35, part 1, no. 3, p. 1690.

[20] Y. Kusakabe, K. Hanaoka, H. Komori, H. Ohnishi, and K. Yamanishi, *Jpn. J. Appl. Phys.*, 1997, vol. 36, part 1, no. 1A, p. 6.

[21] S. Ishidzuka, Y. Igari, T. Takaoka, and I. Kusunoki, *Appl. Surf. Sci.*, 1998, vol. 130–132, p. 107.

[22] S. Yokoyama, N. Ikeda, K. Kajikawa, and Y. Nakashima, *Appl. Surf. Sci.*, 1998, vol. 130–132, p. 352.

[23] J.W. Klaus, A.W. Ott, A.C. Dillon, and S.M. George, *Surf. Sci.*, 1998, vol. 418, p. L14.

[24] S. Koseki and A. Ishitani, *J. Appl. Phys.*, 1992, vol. 72, no. 12, p. 5808.

[25] S. Koseki and A. Ishitani, *Bull. Chem. Soc. Jpn.*, 1992, vol. 65, no. 11, p. 3174.

[26] S. Koseki, A. Ishitani, and Y. Fujimura, *Jpn. J. Appl. Phys.*, 1997, vol. 36, part. 1, no. 10, p. 6518.

[27] A.A. Korkin, J.V. Cole, D. Sengupta, and J.B. Adams, *J. Electrochem. Soc.*, 1999, vol. 146, p. 4203.

[28] M.-D. Su and H.B. Schlegel, *J. Phys. Chem.*, 1993, vol. 97, p. 9981.

[29] J.M. Wittbrodt and H.B. Schlegel, *Chem. Phys. Lett.*, 1997, vol. 265, p. 527.

[30] M.T. Swihart and R.W. Carr, *J. Phys. Chem. A*, 1997, vol. 101, no. 40, p. 7434.

[31] M.T. Swihart and R.W. Carr, *J. Phys. Chem. A*, 1998, vol. 102, no. 4, p. 785.

[32] M.T. Swihart and R.W. Carr, *J. Phys. Chem. A*, 1998, vol. 102, no. 9, p. 1542.

[33] K.F. Jensen, H. Simka, T.G. Mihopoulos, P. Futerko, and M. Hierlemann, in *Proceedings of the NATO Advanced Study Institute on Advances in Rapid Thermal and Integrated Processing, Acquafredda di Maratea, Italy, July 3–14, 1995,* F. Roozeboom, Ed., Dordrecht (The Netherland): Kluwer, 1996.

[34] M.D. Allendorf and C.F. Melius, Surface and Coatings Technology, 1998, vol. 108–109, p. 191.

[35] A.A. Bagatur'yants, K.P. Novoselov, A.A. Safonov, L.L. Savchenko, J.V. Cole, and A.A. Korkin, *Mater. Sci. Semicond. Process.* 3 (2000) 23.

[36] P.J. Robinson and K.A. Holbrook, *Unimolecular Reactions*, London: Wiley-Interscience, 1972.

[37] M.W. Schmidt, K.K. Baldridge, J.A. Boatz, S.T. Elbert, M.S. Gordon, J.H. Jensen, S. Koseki, N. Matsunaga, K.A. Nguyen, S.J. Su, T.L. Windus, M. Dupuis, and J.A. Montgomery, *J. Comput. Chem.*, 1993, vol. 14, p. 1347.

[38] C. Møller, M.S. Plesset, *Phys. Rev.*, 1934, vol. 46, p. 618.

[39] A.D. McLean and G.S. Chandler, *J. Chem. Phys.*, 1980, vol. 72, p. 5639.

[40] H. Simka, B.G. Willis, I. Lengyel, and K.F. Jensen, in *Progress in Crystal Growth and Characterization of Materials,* 1997, vol. 35, nos. 2–4, p. 117.

[41] B.G. Willis and K.F. Jensen, *J. Phys. Chem.*, 1998, vol. 102, p. 2613.

[42] H. Eyring, S.H. Lin, and S.M. Lin, Basics Chemical Kinetics. Wiley-Interscience, New York, 1980.

[43] R.E. Weston, Jr. and H.A. Schwarz, *Chemical Kinetics* (Prentice-Hall, Englewood Cliffs, 1972), p. 101.

[44] E.W. Schlag and G.L. Haller, J. Chem. Phys., 1965, vol. 42, p. 584.

[45] R.G. Gilbert and S.C. Smith, Theory of unimolecular and recombination reactions. Blackwell, Oxford, 1990.

[46] C. Kleijn *"Transport Phenomena in Chemical Vapor Deposition Reactors"*, TU Delft, the Netherlands, 260 p., 1991.

[47] Press W.H., Vetterling W.T., Teukolsky S.A., Flannery B.P. *Numerical Recipes in C* // Cambridge University Press, 994 p., 1992.

[48] Rabitz, H., Kramer, M., and Dacol, D. *Sensitivity Analysis in Chemical Kinetics.* // Ann. Rev. Phys. Chem., v. 34, pp. 419-461, 1983.

[49] Dickinson, R.P., Gelinas, R.J. *Sensitivity Analysis of Ordinary Differential Equation System - A Direct Method.* // J.Comp.Phys., v. 21, pp. 123-143, 1976.

[50] L.A. Curtiss, K. Raghavachari, G.W. Trucks, and J.A. Pople, *J. Chem. Phys.*, 1991, vol. 94, p. 7221.

[51] K. Kato, Z. Inoue, K. Kijima, I. Kawada, and H. Tanaka, *J. Am. Ceram. Soc. 58* (1975) 90.

[52] R. Grun, *Acta Cryst. B 35* (1979) 800.

[53] W.Y. Lee, J.R. Strife, and R.D. Veltri, *J. Am. Ceram. Soc. 75* (1992) 2803.

[54] P. Yang, H.-K. Fun, I. A. Rahman, and M.I. Saleh, *Ceram. Int. 21* (1995) 137.

[55] C.M. Wang, X.Q. Pan, M. Ruhle, F.L. Riley, and M. Mitomo, *J. Mater. Sci. 31* (1996) 5281.

[56] H. Toraya, *J. Appl. Cryst. 33* (2000) 95.

[57] F. Bozso and Ph. Avouris, *Phys. Rev. Lett. 57* (1986) 1185.

[58] R. Wolkow and Ph. Avouris, *Phys. Rev. Lett. 60* (1988) 1049.

[59] Ph. Avouris and R. Wolkow, *Phys. Rev. B 39* (1989) 5091.

[60] L. Kubler, J.L. Bischoff, and D. Bolmont, *Phys. Rev. B 38* (1988) 13113.

[61] G. Rangelov, J. Stober, B. Eisenhurt, and Th. Fauster, *Phys. Rev. B 44* (1991) 1954.
[62] S. Ishidzuka, Y. Igari, T. Takaoka, and I. Kusunoki, *Appl. Surf. Sci. 130–132* (1998)107.
[63] T. Watanabe, A. Ichikawa, M. Sakuraba, T. Matsuura, and J. Murota, *J. Electrochem. Soc. 145* (1998) 4252.
[64] M. Bjorkqvist, M. Gothelid, T.M. Grehk, and U.O. Karlsson, *Phys. Rev. B* 57 (1998) 2327.
[65] X.-s. Wang, G. Zhai, J. Yang, and N. Cue, *Phys. Rev. B 60* (1999) R2146.
[66] G. Zhai, J. Yang, N. Cue, and X.-s. Wang, *Thin Solid Films 366* (2000) 121.
[67] S. Yokoyama, N. Ikeda, K. Kajikawa, and Y. Nakashima, *Appl. Surf. Sci. 130–132* (1998) 352.
[68] J.W. Klaus, A.W. Ott, A.C. Dillon, and S.M. George, *Surf. Sci. 418* (1998) L14.
[69] Y. Widjaja, M.M. Mysinger, and C.B. Musgrave, *J. Phys. Chem. B 104* (2000) 2527.
[70] J.A. Wendal and W.A. Goddard III, *J. Chem. Phys. 97* (1992) 5048.
[71] P. Vashishta, R.K. Kalia, and I. Ebbsjö, *Phys. Rev. Lett. 75* (1995) 858.
[72] C.-K. Loong, P. Vashishta, R.K. Kalia, and I. Ebbsjö, *Europhys. Lett. 31* (1995) 201.
[73] W.-Y. Ching, Y.-N. Xu, J.D. Gale, and M. Rühle, *J. Am. Ceram. Soc. 81* (1998) 3189.
[74] F. de Brito Mota, J.F. Justo, and A. Fazzio, *Int. J. Quant. Chem. 70* (1998) 973.
[75] F. de Brito Mota, J.F. Justo, and A. Fazzio, *Phys. Rev. B 58* (1998) 8323.
[76] R.K. Kalia, A. Nakano, K. Tsuruta, and P. Vashishta, *Phys. Rev. Lett. 78* (1997) 689.
[77] M.E. Bachlechner, R.K. Kalia, A. Nakano, A. Omeltchenko, P. Vashishta, I. Ebbsjö, A. Madhukar, and G.-L. Zhao, *J. Eur. Ceram. Soc. 19* (1999) 2265.
[78] A. Omeltchenko, M.E. Bachlechner, A. Nakano, R.K. Kalia, P. Vashishta, I. Ebbsjö, A. Madhukar, and P. Messina, *Phys. Rev. Lett. 84* (2000) 318.
[79] M.E. Bachlechner, A. Omeltchenko, A. Nakano, R.K. Kalia, P. Vashishta, I. Ebbsjö, and A. Madhukar, *Phys. Rev. Lett. 84* (2000) 322.
[80] X.Y. Guo and P. Brault, *Surf. Sci.* 488 (2001) 133.
[81] A.Y. Liu and M.L. Cohen, *Phys. Rev. B 41* (1990) 10727.
[82] Y.-N. Xu and W.Y. Ching, *Phys. Rev. B 51* (1995) 17379.
[83] A. Reyes-Serrato, D.H. Galvan, and I.L. Garzon, *Phys. Rev. B 52* (1995) 6293.
[84] G.L. Zhao and M.E. Bachlechner, *Phys. Rev. B 58* (1998) 1887.
[85] G. Pacchioni and D. Erbetta, *Phys. Rev. B 60* (1999) 12617.
[86] W.Y. Ching, L. Ouyang, and J.D. Gale, *Phys. Rev. B 61* (2000) 8696.
[87] A.A. Bagatur'yants, K.P. Novoselov, A.A. Safonov, J.V. Cole, M. Stoker, and A.A. Korkin, *Surf. Sci.* 486 (2001) 213.
[88] M.J. Frisch et al., *Gaussian 98, Revision A.7*, (Gaussian, Inc., Pittsburgh PA, 1998).
[89] D.J. Doren, in: Advances in Chemical Physics, Eds. I. Progogine and S.A. Rice (Wiley, NY, 1996) p. 1.
[90] N. Wiberg, K. Schurz, G. Reber, and G. Muller, *Chem. Commun.* (1986) 591.
[91] G. Reber, J. Riede, N. Wiberg, K. Schurz, and G. Muller, *Z. Naturforsch., B: Chem. Sci. 44* (1989) 786.
[92] J. Niesmann, U. Klingebiel, M. Schafer, and R. Boese, *Organometallics 17* (1998) 947.
[93] Levine S.W., Engstrom J.R., Clancy P., *Surf. Sci.*, vol. 401, p. 112, 1998
[94] C.C. Battaile, D.J. Srolovitz, and J.E. Butler, *J. Appl. Phys.*, vol. 82, p. 6293, 1997.
[95] Battaile C.C., Srolovitz D.J., Butler J.E., *J. Cryst. Growth*, v. 194, p. 353, 1998.
[96] A.A. Knizhnik, A.A. Bagaturyants, I.V. Belov, B.V. Potapkin, and A.A. Korkin, *Comp. Mater. Sci.* 24 (2002) 120.
[97] Gonzalez-Lafont A., Truong T.N., and Truhlar D.G., *J. Chem. Phys.*, v. 95(12), p. 8875, 1991.
[98] Pitt I.G., Gilbert R.G., Ryan K.R., *J. Chem. Phys.*, v. 102(8), p. 3461, 1995.
[99] Mills G., Jonsson H., Shenter G.K., *Surf. Sci.*, v. 324, p. 305, 1995.
[100] G.D. Wilk, R. M. Wallace, and J. M. Anthony, *J. Appl. Phys.* 87 (2000) 484.
[101] V.V. Brodskii, E.A. Rykova, A.A. Bagatur'yants, and A.A. Korkin, *Comp. Mater. Sci.* 24 (2002) 210.
[102] Bockstedte M., Kley A., Neugerbauer J. and Scheffler M., *Comp. Phys. Comm.* 107 (1997) 187.
[103] CAMP Open Software project (URL http://www.fysik.dtu.dk/CAMPOS).

Author Index